SECOND EDITION

Engineering Mechanics

Dr. M.P. Poonia
Vice-Chairman
All India Council for Technical Education (AICTE)

Dr. D.S. Bedi
Formerly
Professor Emeritus
Department of Mechanical Engg.
Institute of Engineering and Technology
Bhaddal, Dist. Ropar Punjab

KHANNA BOOK PUBLISHING CO. (P) LTD.

Publisher of Engineering and Computer Books

4C/4344, Ansari Road, Darya Ganj, New Delhi-110002
Phone : 011-23244447-48 **Mobile:** +91-9910909320
E-mail : contact@khannabooks.com
Website : www.khannabooks.com

Dear Readers,

To prevent the piracy, this book is secured with HIGH SECURITY HOLOGRAM on the front title cover. In case you don't find the hologram on the front cover title, please write us to at contact@khannabooks.com or whatsapp us at +91-99109 09320 and avail special gift voucher for yourself.

Specimen of Hologram on front Cover title:

Moreover, there is a SPECIAL DISCOUNT COUPON for you with EVERY HOLOGRAM.

How to avail this SPECIAL DISCOUNT:

Step 1: Scratch the hologram
Step 2: Under the scratch area, your "coupon code" is available
Step 3: Logon to www.khannabooks.com
Step 4: Use your "coupon code" in the shopping cart and get your copy at a special discount
Step 5: Enjoy your reading!

Price: ₹ 525.00

ISBN: 978-93-86173-26-3

Engineering Mechanics
by M.P. Poonia, D.S. Bedi

First Edition: 2013
Reprint Edition: 2025

To view complete list of books, Please scan the QR Code:

Published by:
Khanna Book Publishing Co. (P) Ltd.
CIN: U22110DL1998PTC095547

Visit us at: **www.khannabooks.com**
Write us at: **contact@khannabooks.com**

Printed in Bharat.

PREFACE

This book "Engineering Mechanics" has been written on the express demand of our old students who are occupying senior faculty positions in various universities. Based on the expertise, which the authors have gained through his long teaching careers, the subject matter of this book has been dealt keeping in view the needs of not only the bright students but also for the average as well as the below average students. Particularly the language used is very simple so that it can be understood by one and all.

The attention of the students is drawn to the topics of bending moments and twisting moments, which are not properly explained in most of the other books. These bending moments and twisting moments have been explained with the help of vectors, which are used to present these quantities in such a way that one can easily distinguish between these two, as what is bending moments and what is twisting moments.

At the end, authorswish to invite healthy criticisms about the text so that in the further editions of this book, the text can be improved and mistakes, if any, can be removed.

Authors

ACKNOWLEDGEMENT

We hereby express our sincere gratitude to Mr. Deepam Goyal (Gold Medalist from Panjab University Chandigarh), a Research Scholar in Department of Mechanical Engineering, National Institute of Technical Teachers Training & Research (NITTTR), Sector - 26, Chandigarh - 160 019, for having exhaustively reviewed this manuscript resulting in our successful publishing.

BRIEF TABLE OF CONTENTS

CONTENTS

2. FRICTION 64

3. SHEARING FORCE AND BENDING MOMENT IN BEAMS 123

6. KINEMATICS OF RIGID BODY 282

7. KINETICS OF RIGID BODIES 334

CHAPTER 1

TWO DIMENSIONAL FORCE SYSTEM

1.1 INTRODUCTION TO ENGINEERING MECHANICS AND BASIC DEFINITIONS

The branch of physical science that deals with the state of rest or state of motion of a body under the action of certain forces is known as MECHANICS. In the branch when the body is at rest or when the forces acting on it are such that the body is in static equilibrium, then this branch is popularly known as statics. On the contrary, if the analysis deals with the study of a body when it is in motion, it is known as dynamics. Dynamics is further subdivided into "kinematics" and "kinetics". When a body is in motion, and the forces which cause the motion are not considered is called kinematics and when the forces are also taken into consideration, then this branch of science is called the kinetics. Fig 1.1, is depicts the classification of Engineering Mechanics.

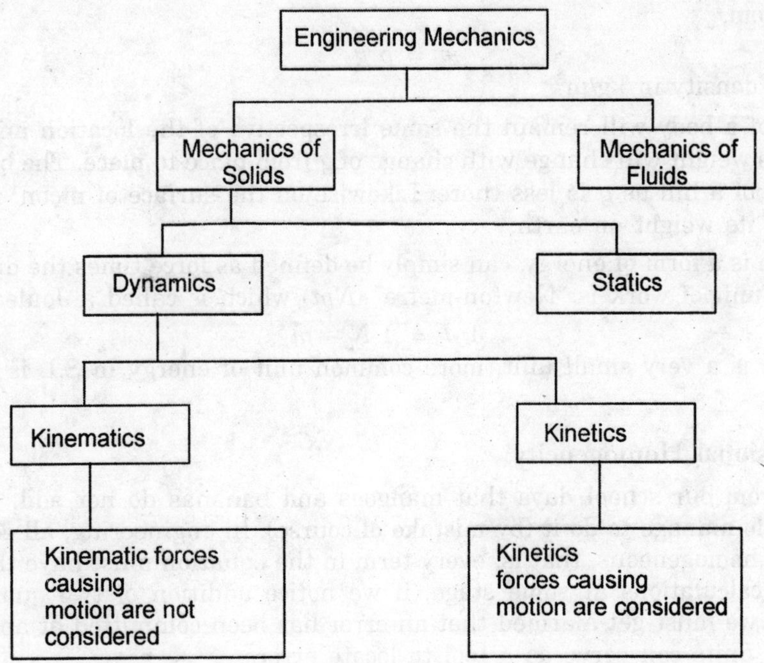

Fig. 1.1

1.2 BASIC DEFINITIONS, DIMENSIONS AND UNITS

Any physical quantity can be characterised by dimensions. Arbitray magnitudes assigned to dimensions length L, time t, mass m, and temperature T are known as primary dimensions, while others such as velocity v, volume V and energy U are expressed in terms of primary dimensions and are called secondary or derived dimensions.

Despite great efforts in the scientific community to unify the world with a single system of units, two sets of units are still in common use today. But the most popular and the one being followed in India and most of the countries, including England, is the metric S.I. (from *Le* System International *d'* units) which is also known as System International (in short S.I.). The other system known as English system which is also known as United States Customary System is prevalent only in United States. Even in U.S.A., the S.I. system is slowly being adopted in order to capture the export market of other countries.

The S.I. system is based on decimal relationship between units. The prefixes like Tera, T (10^{12}), Giga, G (10^9), Mega, M (10^6), Kilo, k (10^3), Centi, c (10^{-2}), Mili, m (10^{-3}), micro, μ (10^{-6}), nano, n (10^{-9}) and pico, p (10^{-12}) are used to express multiples of units.

In S.I., the units of mass, length and time are kilogram (kg), metre (m) and second (s) respectively. The unit of force is Newton (N) and is defined as the force required to accelerate a mass of 1 kg at a rate of 1 m/s^2. The term weight is usually incorrectly used to express mass. Weight is actually a force. It is the gravitational force acting on the body and is estimated from Newton's second law and is directed towards the centre of earth. Mathematically,

$$W = mg$$

Where m is the mass of the body and g is the local gravitational acceleration $g = 9.81$ m/s^2 at sea level and 45° latitude.

Similarly, the weight of a unit volume of a substance is called the specific weight w and is calculated from,

$$w = \rho g$$

where ρ is the density in kg/m^3

The mass of a body will remain the same irrespective of the location anywhere in the universe but its weight will change with change of g from place to place. The body will weigh less on the top of a hill as g is less there. Likewise on the surface of moon, its weight will be one-sixth of its weight on earth.

Work which is a form of energy, can simply be defined as force times the distance moved. Therefore, the unit of work is "Newton-metre" (Nm) which is called a Joule (J).

$$1\ J = 1\ N - m$$

Since Joule is a very small unit, more common unit of energy in S.I. is the kilo Joule (1 kJ = 10^3 J)

1.2.1 Dimensional Homogeneity

We all know from our school days that mangoes and bananas do not add. But somehow, sometimes we do manage to do it (by mistake of course). In engineering, all equations must be dimensional homogeneous, that is, every term in the equation must have the same units. While making calculations at some stage, if we notice addition of two quantities having different units, we must get alarmed that an error has been committed at an earlier stage. Thus, checking units can serve as a tool to locate errors.

It is very important to keep in mind that a formula which is not dimensionally homogeneous, is definitely wrong but a dimensionally homogeneous formula may not necessarily be correct.

1.2.2 Basic Defintions

1.2.2.1 Particle

A particle is a body of infinity small size or volume and the mass of the particle is considered to be concentrated at a point.

1.2.2.2 Standard Metre

Standard metre is the length of a standard bar of platinum-iridium kept at International Bureau of Weights and Measures. In order to overcome the difficulty of reproduction of standard metre, it is now defined in terms of the wavelength of light emitted by Krypton 86 atom. It is defined as 1690763.73 wavelength of light emitted by Krypton – 86 atom.

1.2.2.3 Scalar Quantity

A scalar quantity is one which is completely specified by magnitude. For example, length, mass and temperature.

1.2.2.4 Vector Quantity

A quantity which is completely specified by magnitude and direction is known as a vector quantity. For example force, velocity, acceleration. Vector quantity can be represented by arrowhead, the length of which is proportional to its magnitude and the arrowhead points towards its direction of action. In Fig. 1.2, the arrowhead *AB* represents a force vector, the length of arrowhead represents on some scale the magnitude of the force and the direction of *AB* gives the direction of the line of action of the force with respect to the reference axis. Here, its direction is at an angle θ with *x*-axis.

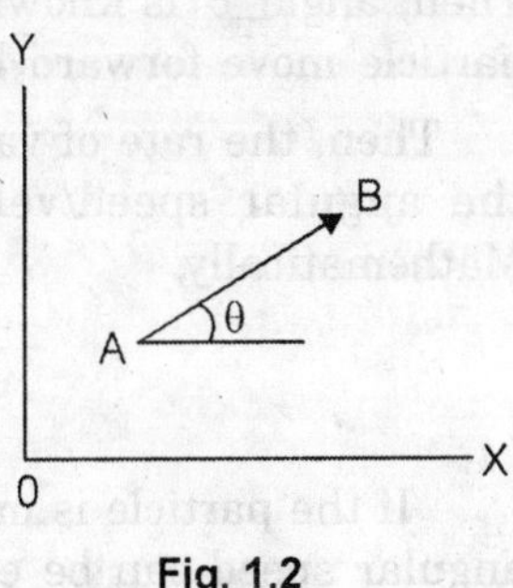

Fig. 1.2

1.2.2.5 Displacement

Displacement is defined as the distance moved by a body when it changes its position from an initial location '*A*' to the final location '*B*'. Fig. 1.3 shown the initial and final locations of the body as '*A*' and '*B*' respectively. The movement from *A* to *B* may not necessarily be along the straight line *AB*, it could be along the curve. The total distance moved along the curve will be the displacement. It is a scalar quantity.

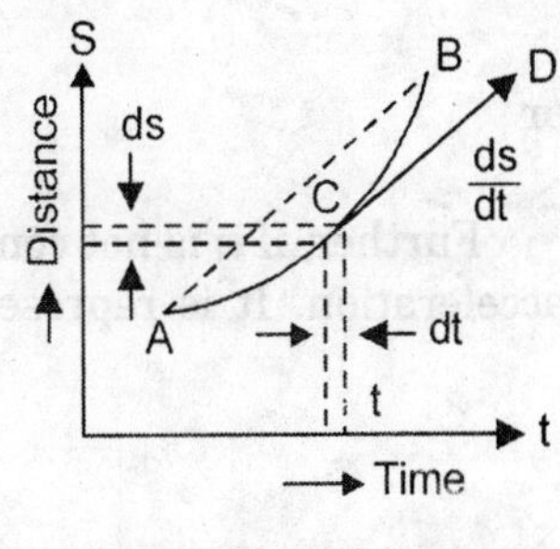

Fig. 1.3

1.2.2.6 Velocity (*V*)

Rate of change of displacement with respect to time at any distance is called the velocity. It is directed towards the tangent to the displacement path, Fig. 1.3 shows the displacement verses time curve. At the *T*, the velocity $\frac{ds}{dt}$ is directed from *C* to *D* and its magnitude is $\frac{ds}{dt}$. Evidently, it is a vector quantity, as its magnitude and direction are specified.

1.2.2.7 (*a*) Linear Acceleration

Acceleration is defined as the ratio of change of linear velocity. Fig. 1.4 shows the variation of velocity with time. At the instant C, the acceleration is given by $\frac{dv}{dt}$ and it is directed along the tangent C at that instant. Mathematically, if a is the acceleration, then

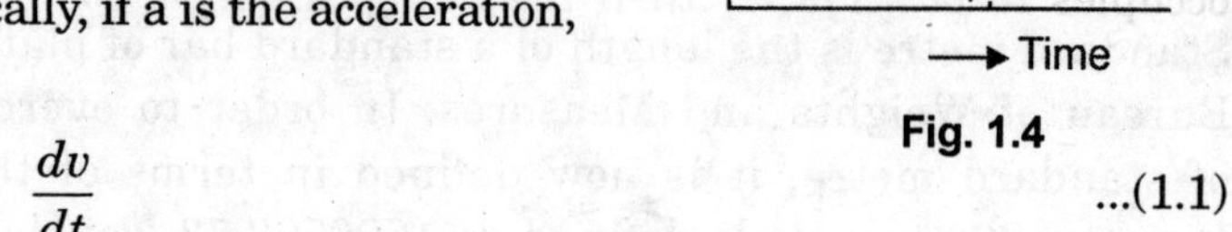

Fig. 1.4

$$a = \frac{dv}{dt} \qquad \text{...(1.1)}$$

where v is the velocity.

(*b*) Angular Displacement, Velocity and Acceleration

Let a particle P be moving in a circular path with O as circle. At any instant designated by time 't', it is at position P such that the angle it has traversed from the reference line OX beθ. Then, angle 'θ' is known as the angular displacement. Let the particle move forward by an angle dθ in time dt secs.

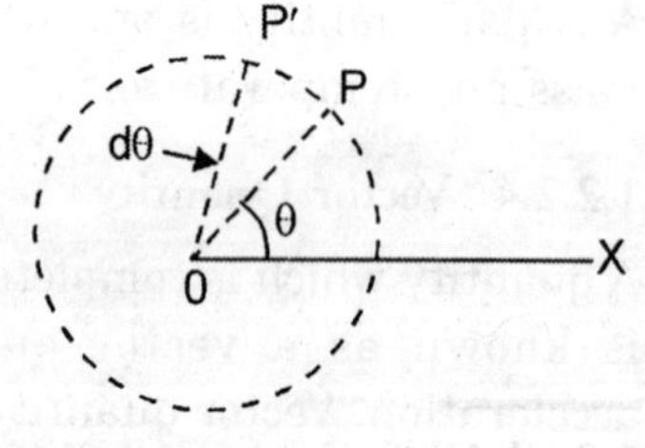

Fig. 1.5

Then, the rate of variation of 'θ' with respect to 't' is called the angular speed/velocity of rotation represented by ω. Mathematically,

$$\omega = \frac{d\theta}{dt} \text{ rad/sec} \qquad \text{...(1.2)}$$

If the particle is moving with uniform angular speed w rad/sec, then the corresponding angular speed can be expressed in revolutions per minute (r.p.m.) as N.

$$\omega = \frac{2\pi N}{60} \qquad \text{...(1.3}a\text{)}$$

or

$$N = \frac{60\,\omega}{2\pi} \text{ r.p.m.} \qquad \text{...(1.3}b\text{)}$$

Further, if ω is not constant, then the rate of variation of ω with time is called the angular acceleration. It is represented by α, given by,

$$\alpha = \frac{d\omega}{dt} = \frac{d}{dt}\left(\frac{d\theta}{dt}\right)$$

or

$$\alpha = \frac{d^2\theta}{dt^2} \qquad \text{...(1.4)}$$

1.2.2.8 Momentum

The product of the mass of the body and its velocity is ealled its momentum.

$$\text{Momentum} = \text{mass} \times \text{velocity}$$
$$= \text{m.v}$$

1.2.2.9 Rigid Body

A rigid body is one in which the relative positions of any two of its particles do not change under the action of forces. Even through the body as a whole may change its position due to the action of forces. For example, consider a body under the action of a system of forces moves from the position P_1 to position P_2 in space. The line joining the points A and B when the body occupied initial position is represented by AB. When under the action of forces, it occupies to position $P_{2,}$ then the new position of AB will be A_1B_1. Then,

$$A_1B_1 = AB$$

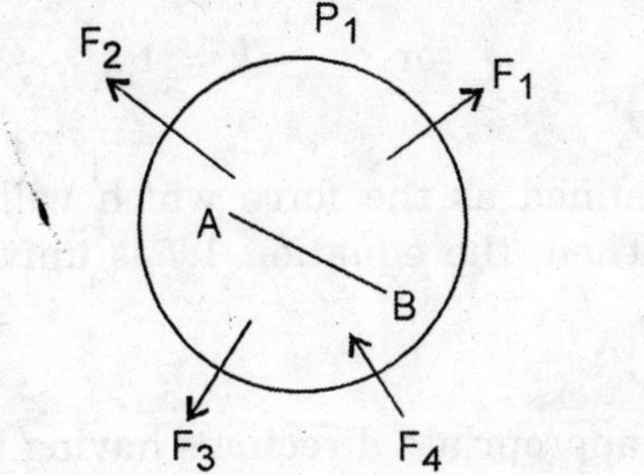

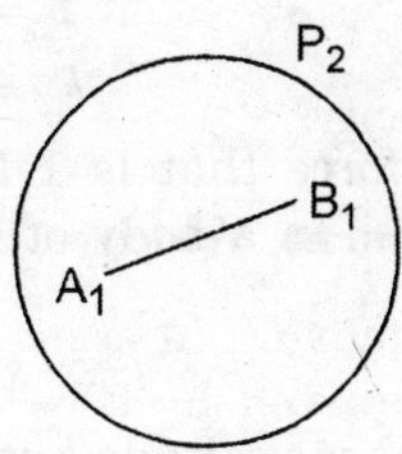

Fig. 1.6

1.3 LAWS OF MECHANICS

There are six basic laws, popularly known as Laws of Mechanics on which the subject of mechanics has been developed. These are explained below:

1.3.1 Newton's First Law of Motion

Statement: It states that a body will continue to be in the state of rest or of uniform motion in a straight line unless it is compelled to change that state by some external agency acting on it. This leads to the existence and definition of force which changes or tends to change the condition of rest or uniform motion in straight line of the body.

1.3.2 Newton's Second Law of Motion

It states that the rate of change of momentum of a body is directly proportional to the impressed force and it takes place in the direction of the force applied. Mathematically, it is expressed as :

Force ∝ Rate of change of momentum

∝ Rate of change of (mass × velocity)

or $$F \propto \frac{d}{dt}\text{ (m.v.)} \qquad \text{...(1.5)}$$

where F = Force, m = mass of body, v = velocity

Since mass does not change, we get from equation 1.5,

$$F \propto m\frac{dv}{dt}$$

or $F = km \cdot a$ ...(1.6)

where a = acceleration of the body

and k = constant

Now if m = 1 kg and a = 1 m/s^2

and F = 1 N

Then $1 = k \times 1 \times 1$ or $k = 1$

That is $F = ma$ ...(1.7)

Thus, the unit force that is 1 Newton is defined as the force which will generate an acceleration of 1 m/s^2 in a body of mass 1 kg. Then, the equation 1.7 is universally true.

1.3.2.1 Corollary

If two forces of equal magnitude but directed in appropriate directions having the same line of action, act at the same point on the body, the combined effect of these equal and opposite forces will be zero and the body will continue to be in the state of rest or uniform time motion. In other words, these forces will nullfy the effect of each other or they will have no effect on the equilibrium state of the body.

1.3.3 Newton's Third Law of Motion

It states that for every action, there is an equal and opposite reaction. Let us suppose a book is lying on a table. The gravitational force of the earth, which will be equal to the weight of the book W, will tend to move the book downwards but the presence of the table surface does not allow it do so and the book continues to rest on the table. In other words, the table is applying an upward force equal to W in magnitude. This upward force on the block by the table top is called the Reactive force or simply the Reaction R.

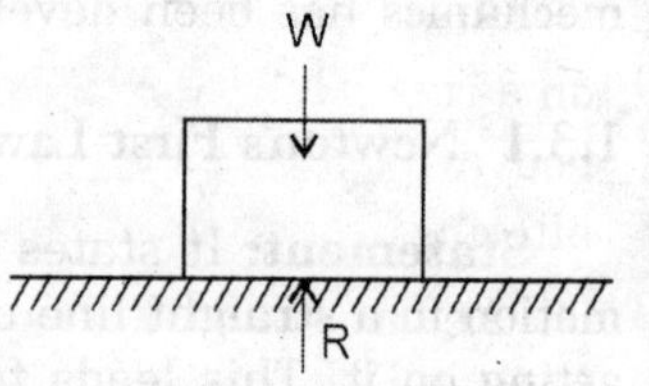

Fig. 1.7

1.3.4 Newton's Law of Gravitation

This law states that the force of attraction between two bodies is proportional to their masses and inversely proportional to the square of the distance between them. Thus, if F is the force of attraction and m_1 and m_2 are the respective masses separated by distance 'd', then,

$$F = G \cdot \frac{m_1 m_2}{d^2} \quad \text{...(1.7)}$$

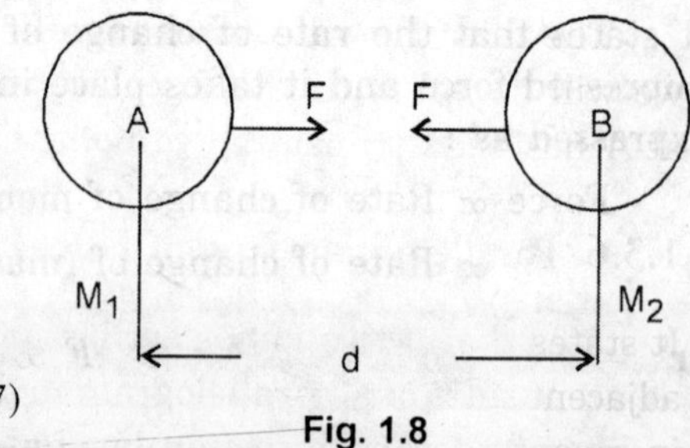

Fig. 1.8

Where G is the constant of proportionality and is known as Gravitational constant.

The value of the constant G is 6.67×10^{-11} N m^2/kg^2 or m^3/kg s^2.

Corrollary

Equation 1.7 leads us to the definition of weight of a body. Weight is defined as the force of attraction of the earth on a body.

Let the mass of the body be M

$$\text{Mass of earth} = 5.9761 \times 10^{24} \text{ kg} = M_e$$

$$\text{Radius of earth} = 6.371 \times 10^6 \text{ m}$$

$$= \text{distance of the body on the surface of each from its centre}$$

$$= d$$

$\therefore$ The force of attraction of earth on the body

or
$$W = G\frac{M M_e}{d^2}$$

$$= 6.67\times10^{11}\left(\frac{Nm^2}{kg^2}\right)\frac{M\times5.9761\times10^{24}\ kg^2}{(6.371\times10^6)^2 m^2}$$

$$= (9.81\ M)\ N$$

$\therefore$
$$W = Mg$$

where,
$$g = 9.81 \text{ m/sec}^2$$

1.3.5 Principle of Transmissibility of Forces

Statement: It states that a force acting at point A on a rigid body can be shifted to any other point of the body which lies on the line of action of the force without altering the effect of the force on the body.

Let us suppose a force F is acting on a body at a point A and the line of action is AB. Let at a point C, which is on the line AB, we apply two equal and opposite forces F_1. and F_2 both of magnitude F, as shown in Fig. 1.9.

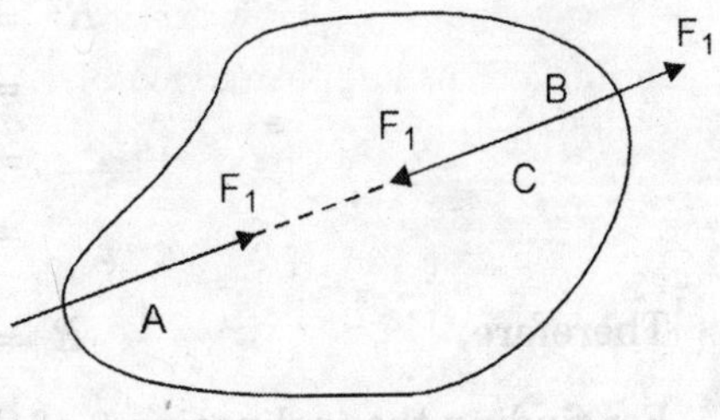

Fig. 1.9

The force F_2 will cancel the effect of force F as F_2 and F are equal and opposite. Thus, we will be left with the force F_1 which is equal to F, too. Thus, the original force F acting at A has been transfered to point C.

1.3.6 Parallelogram Law of Forces

It states that if two forces acting simultaneously on a body at a point are represented by the adjacent sides of a parallelogram in direction and magnitude, then the resultant is represented in magnitude and direction by the diagonal of the parallelogram which passes through the point of intersection of the two sides which represent the two forces.

Suppose that two forces P and Q are simultaneously acting on a body at a point 'O' and are inclined to each other at an angle α. Draw to scale the force vector P as $\overline{DA}$ along X-axis on some convenient scale (Fig. 1.10). Then take the force vector Q along the line $\vec{OB}$ at an angle α. Drawing BC parallel to OA and AC parallel to OB, complete the parallelogram $OACB$. Then as per this law the line $\overline{OC}$ represents the resultant force vector in direction and magnitude which is inclined at an angle θ with force vector $\vec{P}$ or X-axis. Its magnitude and direction *i.e.*, line of action can be determined as follows:

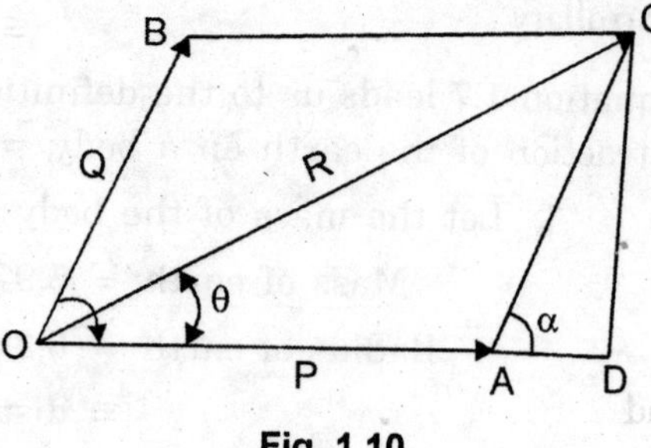

Fig. 1.10

From the right angle triangle OCD, where D is the foot of the perpendicular drawn from point C on X-axis, we have,

$$OC^2 = OD^2 + CD^2 \quad ...(1.8)$$

Now $OD = OA + AD$

$$= P + AC \cos \alpha \quad (\because \angle CAD = \alpha, \text{ is } \Delta\, CAD)$$

or $$OD = P + Q \cos \alpha \quad ...(1.9)$$

and $$CD = AC \sin \alpha$$

$$= Q \sin \alpha \quad ...(1.10)$$

Therefore, from equation (1.8), (1.9) and (1.10) we get

$$R^2 = (P + Q \cos \alpha)^2 + (Q \sin \alpha)^2$$
$$= P^2 + Q^2 \cos^2 \alpha + 2PQ \cos \alpha + Q^2 \sin^2\alpha$$
$$= P^2 + Q^2 (\cos^2 \alpha + \sin^2 \alpha) + 2PQ \cos \alpha$$
$$= P^2 + Q^2 + 2PQ \cos \alpha$$

Therefore, $$R = \sqrt{P^2 + Q^2 + 2PQ\cos\alpha} \quad ...(1.11)$$

For finding the inclination θ of the resultant force R with the X-axis or direction of force $\vec{P}$, we have from the triangle $C\ O\ D$,

$$\tan \theta = \frac{CD}{OD} = \frac{Q\sin\alpha}{P + Q\cos\alpha} \quad ...(1.12a)$$

$$\therefore \quad \theta = \tan^{-1}\left[\frac{Q\sin\alpha}{P + Q\cos\alpha}\right] \quad ...(1.12b)$$

1.3.6.1 Case 1

When the given forces P and Q are equal in magnitude, then equations (1.11) and (1.12) yield

$$R = \sqrt{P^2 + P^2 + (2P)\,Q\cos\alpha}$$

$$= \sqrt{2P^2 + 2P^2\cos\alpha}$$

$$= \sqrt{2}\, P\sqrt{1+\cos\alpha}$$

$$= \sqrt{2}\, P\sqrt{2\cos^2\left(\frac{\alpha}{2}\right)} \qquad \left(\because 1+\cos\alpha = 2\cos^2\frac{\alpha}{2}\right)$$

$$R = 2P\cos\frac{\alpha}{2} \qquad \text{...(1.13)}$$

and $$\tan\theta = \frac{Q\sin\alpha}{P+Q\cos\alpha} = \frac{P\sin\alpha}{P(1+\cos\alpha)}$$

$$= \frac{2\sin\frac{\alpha}{2}\cos\frac{\alpha}{2}}{2\cos^2\frac{\alpha}{2}} = \tan\frac{\alpha}{2}$$

i.e., $$Q = \frac{\alpha}{2} \qquad \text{...(1.14)}$$

From equation (1.14), we conclude that in case the given forces are equal, then their resultant bisects the angle between them.

Case 2 : When the angle between the given forces P and Q is 90°, then we get from equation (1.11) and (1.12),

$$R = \sqrt{P^2+Q^2+2PQ\cos 90^\circ}$$

or $$R = \sqrt{P^2+Q^2} \qquad \text{...(1.15)}$$

and $$\tan\theta = \frac{Q\sin\alpha}{P+Q\cos\alpha}$$

Here, $\alpha = 90^\circ$ $\therefore \sin\alpha = 1$ and $\cos\alpha = 0$

That is $$\tan\theta = \frac{Q}{P}$$

or $$\theta = \tan^{-1}\left(\frac{Q}{P}\right) \qquad \text{...(1.16)}$$

1.3.6.1 Corollary I Triangle Law of Forces

Statement : If two forces acting simultaneously on a body are represented by the sides of a triangle taken in order, their resultant is represented by the closing side of the triangle taken in opposite order.

Case 2 : When the given forces P and Q are acting along the same direction and are along the same line *i.e.*, when $\alpha = 0$, then,

$$R = \sqrt{P^2+Q^2+2PQ\cos(0)}$$

$$= \sqrt{P^2+Q^2+2PQ}$$

$$= \sqrt{(P+Q)^2} = P+Q$$

Thus, when the given forces are collinear and are along the same direction, then the resultant is the sum of the two forces,

$$R = P + Q \quad \text{...(1.17a)}$$

As $$\tan\theta = \frac{Q\sin(0)}{P + Q\cos 0} = 0 \quad \text{...(1.17b)}$$

i.e., $$Q = 0$$

The resultant is along the direction of P or Q

Case 3 : When given forces P and Q are acting in opposite direction but along the same line. Let be acted at 180° to the direction of P

Then, $$R^2 = P^2 + Q^2 + 2PQ\ (\cos 180°)$$

$$= P^2 + Q^2 - 2PQ$$

or $$R^2 = (P^2 + Q^2) - 2PQ$$

$$= (P - Q)^2$$

or $$R = P - Q \quad \text{...(1.18a)}$$

and $$\tan\theta = \frac{Q\sin(180°)}{P - Q\cos 180°} = \frac{0}{P + Q}$$

$$= 0$$

$$Q = 0 \quad \text{...(1.18b)}$$

Thus, the resultant force is $(P - Q)$ and this resultant vector is directed along P.

1.3.6.1 Corollary II Triangle Law of Forces

Statement : If two forces acting simultaneously on a body are represented by the sides of a triangle taken in order, their resultant is represented by the closing side the triangle taken in opposite order.

Let two forces P_1 and P_2 be acting on the body. Let, $\overrightarrow{AB}$ represent P_1 at some scale in magnitude and taking parallel to P_1. Then, from B draw another line BC parallel to P_2 and take its length to represent P_2 in magnitude on same scale. Then, $\overrightarrow{BC}$ vector represents the force P_2. Then, as per the triangle law of forces. The closing side AC in the direction A to C represents the resultant of P_1 and P_2

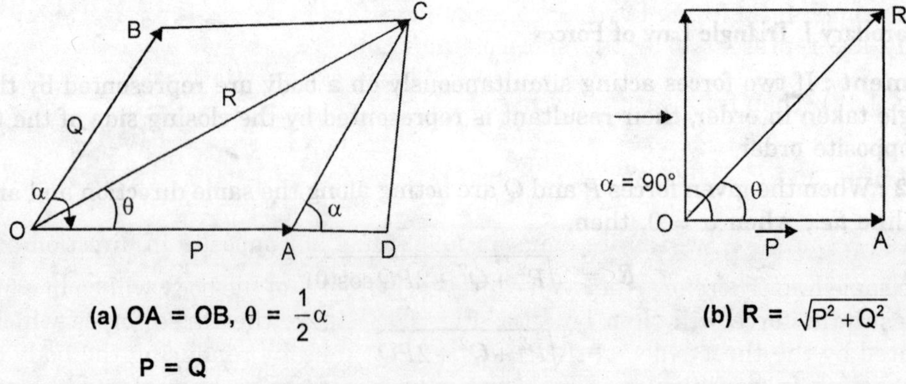

(a) $OA = OB$, $\theta = \frac{1}{2}\alpha$

$P = Q$

(b) $R = \sqrt{P^2 + Q^2}$

Fig. 1.11

Proof : Draw a line AD parallel to BC and make $AD = BC$. This way the given forces are represented by vectors $\overrightarrow{AB}$ and $\overrightarrow{AD}$, (which are the two adjaceant sides of a parallelogram $ABCD$. Therefore, from parallelogram law of forces, the resultant R will be given by the $\overline{AC}$.

1.3.6.2 Corollary II

In case there are more than two forces acting Simultaneously on the body, the resultant can be found by applying what is known as polygon law of forces, which states that if the forces can be represented by the sides of a polygon taken in order, the closing side of the polygon in the opposite order represents the resultant. Let there be added one more force P_3. Then from the point C, draw a vector $\overline{CE}$ parallel to force P_3 and its length equal to the magnitude of P_3 on the same scale. Then the resultant of P_1, P_2 and P_3 will be $\overline{AE} = R_2$ which is actually the resultant of R_1 and P_3 and so on for any number of forces, the method can be extended to any number of forces.

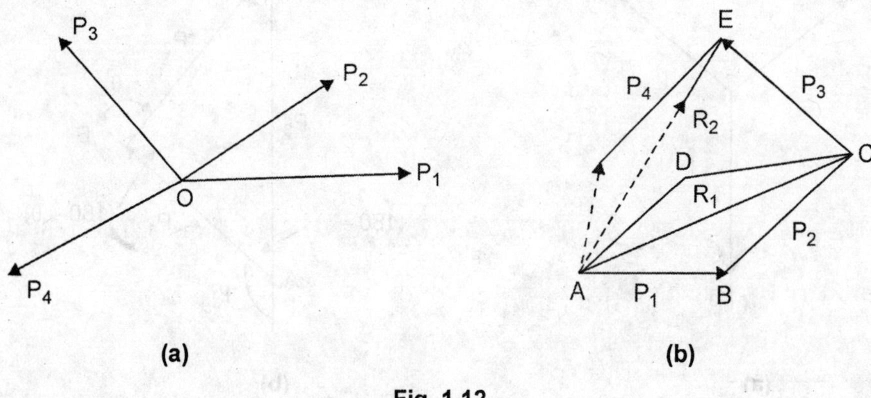

Fig. 1.12

1.3.6.3 Corollary III

If three forces acting simultaneously on a body can be represented by the sides of a triangle taken in order in direction and magnitude, the body will be in equilibrium.

Let us suppose the three forces P_1, P_2 and P_3 acting on a body are represented by the sides of the trangle ABC (Fig. 1.12). Then from earlier discussion, $\vec{AC}$ is the resultant of P_1 and P_2. That is P_1 and P_2 can be replaced by $(= \vec{AC})$. Then, we are left with two forces $\vec{AC}$ and $\vec{CA}$ which are equal in magnitude but opposite in direction and acting along the same line. Therefore, they will cancel each other and the body will be in equilibrium, as the resultant force will then be zero. Fig. 1.13 shows the three forces which can be represented by the three sides of the triangle ABC, thereby rendering the body on which they are acting in equilibrium. Then sine rule is applied to triangle as shown in Fig. 1.13.

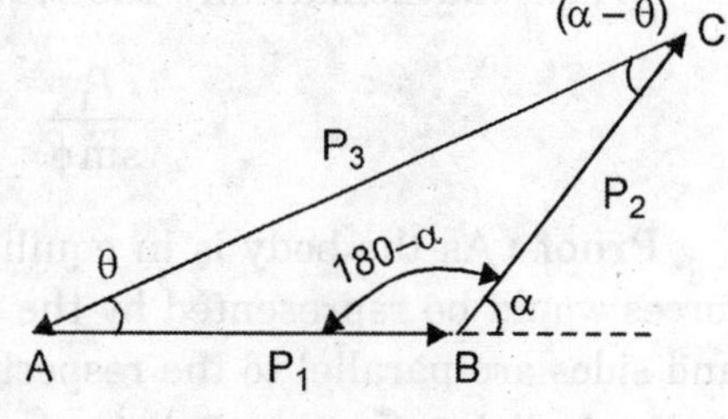

Fig. 1.13

$$\frac{AB}{\sin(\alpha - \theta)} = \frac{BC}{\sin\theta} = \frac{CA}{\sin(180° - \alpha)}$$

or $$\frac{P_1}{\sin(\alpha - \theta)} = \frac{P_2}{\sin\theta} = \frac{P_3}{\sin(180° - \alpha)} \quad ...(1.17)$$

1.3.7 Lami's Theorem

Statement : If a body is in equilibrium under the action of three forces, each force is proportional to the sine of the angle between the other two forces. Let the three forces acting on the body be represented by P_1, P_2 and P_3 as shown in fig. 1.14 *k*.

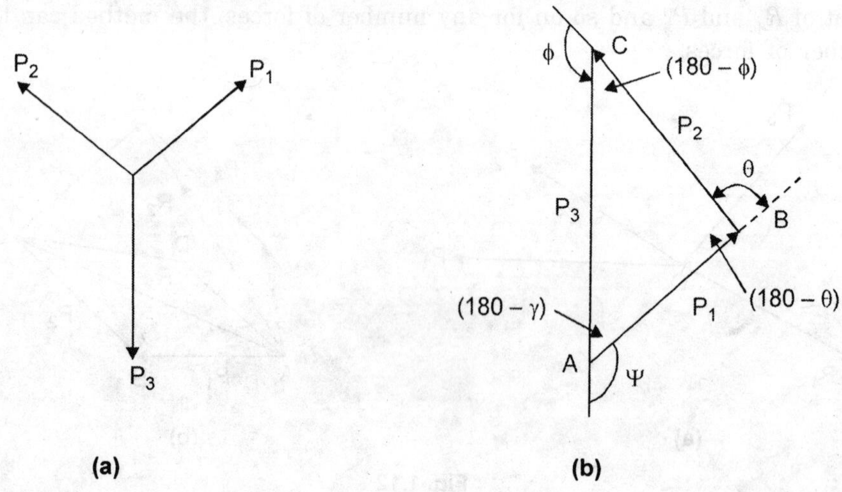

Fig.1.14

Then, mathematically the statement of the theorem is,

$$\frac{P_1}{\sin\phi} = \frac{P_2}{\sin\psi} = \frac{P_3}{\sin\theta} \quad ...(1.18)$$

Proof : As the body is in equilibrium under the action of three forces, then these three forces would be represented by the sides of a triangle in which the forces are taken in order and sides are parallel to the respective directions of the forces. Therefore, draw $\overline{AB}$ in same scale parallel to P_1, from B draw the vector $\overline{BC}$ parallel to P_2 and then side CA of the triangle ABC must represent the force P_3. In other words, the vector $\overline{CA}(= P_3)$ must close at point.

Then, from triangle law of forces we have,

$$\frac{P_1}{\sin(180 - \phi)} = \frac{P_2}{\sin(180 - \psi)} = \frac{P_3}{\sin(180 - \theta)}$$

or $$\frac{P_1}{\sin\phi} = \frac{P_2}{\sin\psi} = \frac{P_3}{\sin\theta} \quad ...(1.19)$$

Note of Caution

While applying the Lami's equation, an important point to keep in mind is measure the angles θ, φ and ψ between the respective forces only when all the three forces are either directed towards or all of them directed away from the point of concurrence, as shown in Fig. (1.15*a* or *b*).

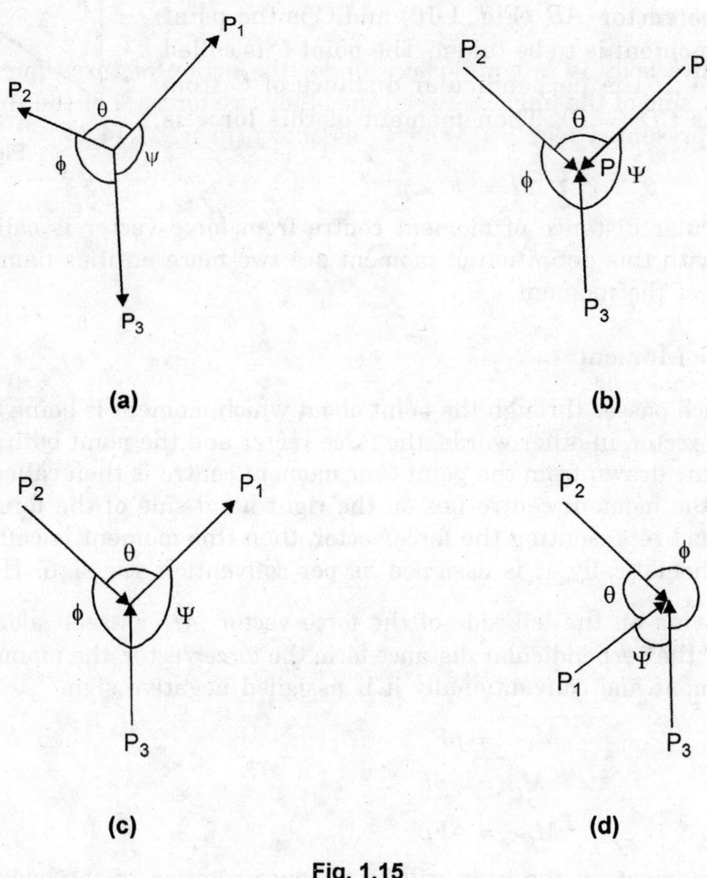

Fig. 1.15

To the system shown in fig.1.15(*c*), the Lami's equation can not be applied. However, the correct situation of Fig. (*c*), is shown in Fig. (*d*)

1.4 FREE BODY DIAGRAM

In order to analyse the effect of a system of forces being encountered by different components of a machine or a structure, it is essential to isolate a part or a group of parts from rest of structure for this purpose, the part or the body is taken and its diagram is drawn on which all the forces including the self weight and reactions at the points of constant with the adjacent components or parts are shown. Then this diagram is called the Free Body Diagram of that part component.

1.5 MOMENT OF A FORCE

The product of the force and the perpendicular distance of the line of action of the force from a point is called the movement of the force about that point. Let the force F be represented by the vector $\vec{AB}$ (Fig. 1.16) and C is the point about which the moment is to be taken. The point C is called the moment entre. If the perpendicular distance of C from force vector AB is CD (= P). Then moment of this force is given by,

Fig. 1.16

$$M = F \times p \qquad ...(1.5)$$

The perpendicular distance of moment centre from force vector is called the 'moment arm'. Associated with this definition of moment are two more entities namely, plane of the moment and axis of the moment.

1.5.1 Plane of the Moment

It is the plane which passes through the point about which moment is being taken and which contains the force vector, in other words, the force vector and the point both lie in this plane. Normal to this plane drawn from the point C or moment centre is then called the axis of this moment. In case the moment centre lies on the right hand side of the force vector as seen along the arrowhead representing the force vector, then this moment is called the clockwise moment and mathematically, it is assigned as per convention *+ve* sign. However, moment centre had been lying on the left side of the force vector $\vec{AB}$ as seen along AB (*e.g.*, the point C') having p' the perpendicular distance form the force vector, the moment will be called anticlockwise moment and conventionally it is assigned negative sign.

$$M_{C'} = pF \qquad ...(1.6a)$$

$$M_C = +pF$$

$$M_{C'} = -Fp' \qquad ...(1.6b)$$

Effect of the moment on the body will be to cause rotation of the body in the plane of the moment provided the body is free to do so. The rotation will be clockwise or anticlockwise depending upon the relative location of the axis with respect to the direction of the force vector or whether the moment of the force is clockwise or anticlockwise.

Sign Convention of Moment

Clockwise moment will be taken as positive moment and the anticlockwise as negative.

1.5.2 Vectorial Representation of a Moment

As explained above, moment is the product of force and perpendicular distance of the point from the force vector. Since the distance p is a scalar quantity and force is a vector, it is concluded that moment is also a vector quantity. Therefore, it is to be represented by an arrowhead, the length of which will be proportional to the magnitude of the moment and the direction of the arrowhead is marked in the direction of the axial movement of right handed screw under the action of the moment with respect to a fixed nut placed in the moment. Evidently, the arrowhead which will represent the moment will be normal to the plane of the moment. Thus, if the plane of the moment is horizontal plane, the moment vector being normal to the plane will be along the axis of the moment and will be along vertical line, thus, it will be visible in the elevation. In the case of anticlockwise moment, the arrowhead will be directed upwards and directed downwards in the case of clockwise moment.

1.5.3 Physical Significance of Moment of Force

In Fig. 1.16 Force Vector F is represented by $\overline{AB}$ to some scale and the moment is to be taken about the point C. Join the point C with the ends A and B of the force vector $\overline{AB}$. Then, we have from equation (1.5),

$$M = F \cdot p$$

$$= \overline{AB} \cdot p$$

$$= 2\left(\frac{1}{2}\overline{AB} \cdot p\right)$$

$$= 2[\text{Area of } \Delta\, ABC] \qquad \text{...(1.7}a\text{)}$$

or $$\text{Area of } \Delta\, ABC = \left(\frac{M}{2}\right) \qquad \text{...(1.7}b\text{)}$$

Fig. 1.16

From equation (1.7), we conclude that area of the triangle formed by joining the point C to the ends of the arrowhead representing the force vector is 'half' the value of the moment of the force about the point under consideration, called the moment centre. Alternatively, the moment is twice the area of the $\Delta\, ABC$.

Further in case the moment centre lies on line of action of the force, the area of the $\Delta\, ABC$ will be zero. Thus, in that case the moment of the force will also be zero. Alternatively, the perpendicular distance from moment centre to the force vector, which is also known as moment arm, will be zero thereby rendering the moment value as zero.

1.6 COUPLE

Two parallel forces equal in magnitude but opposite in direction and separated by a finite distance is called a couple. Fig. 1.17 shows a couple since the forces are of same magnitude and are directed in opposite direction the resultant force in any direction will be zero. Thus, there will be no translations effect of the couple on the body .

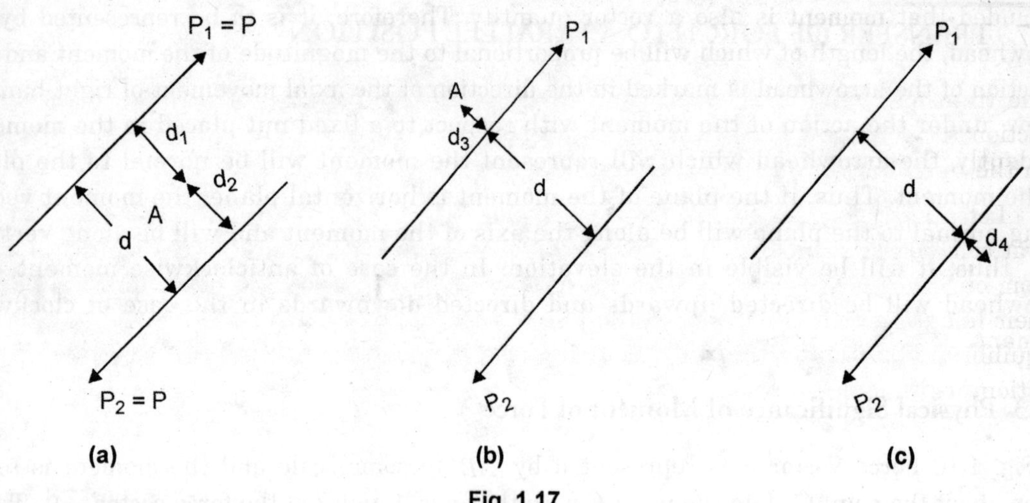

Fig. 1.17

However, the couple will have certainly rotational effect on the body. Let the two forces which constitute the couple be both equal to P in magnitude. They are equal and opposite and separated by a distance d (Fig. 1.17a) just to differentiate these forces while analysing by designating them as P_1 and P_2 such that ($P_1 = P_2 = P$). In order to find the resultant moment at any point in the plane of the moment, we have three options that the point about which moments are to be taken lies within the two given forces or outside. Firstly, let it lie within the forces say at A (Fig. 1.17a).

$$M_A = P_1 d_1 + P_2 d_2$$

where, d_1 and d_2 are distance of point A from P_1 and P_2 respectively.

$$P = P(d_1 + d_2) = Pd \quad \text{...(1.8)}$$

In second case, let the point B moment centre lie at B which is on the left of force P_1. Let the distance of B from force P_1 be d_3.

Then
$$M_B = -P_3 d_3 + P_2(d_3 + d)$$
$$= P(-d_3 + d_3 + d) \quad (\because P = P_1 = P_2)$$
$$= Pd \quad \text{...(1.9)}$$

The third option is that the moment centre lies at C on the left hand side of the force vector $\overrightarrow{P_2}$ so that $\overrightarrow{P_2}$ has a tendency to give anti clockwise moment about B (Fig. 1.17c).

$$M_B = P_1(d + d_4) - P_2 d_4$$
$$= P_1(d + d_4 - d_4) = P \cdot d \quad \text{...(1.10)}$$

From equation 1.8 and (1.10), we see that the effect of couple is same everywhere and the moment of the couple (Pd) is constant everywhere irrespective of the location of the centre of moment and its value is equal to $P \cdot d$. If P is in Newtons and d in metres, known as moment the of couple, the moment of the couple will be Pd, Thus, the couple can be shifted anywhere in the plane of the couple, which is plane containing the two forces P_1 and P_2.

1.7 TRANSFER OF FORCE TO A PARALLEL POSITION

The transmissibility of a force along the line of action of the force has been discussed in section 1.3.5. We now proceed to discuss the transfer of the force to a point which is not lying on the line of action in the following paragraphs:

Let a force P be acting on a body at point A (Fig. 1.18*a*). We wish to transfer the force P at some other point B. Let $P_1 = P_2 = P$ be two equal and opposite forces which we apply from our own side at point B. The magnitudes of both P_1 and P_2 are equal to that of P and their lines of action is parallel to the line of action of given force P. This will not alter the equilibrium position of the body. Let 'd' be the perpendicular distance between the lines of action of the forces P and P_2 or P_1.

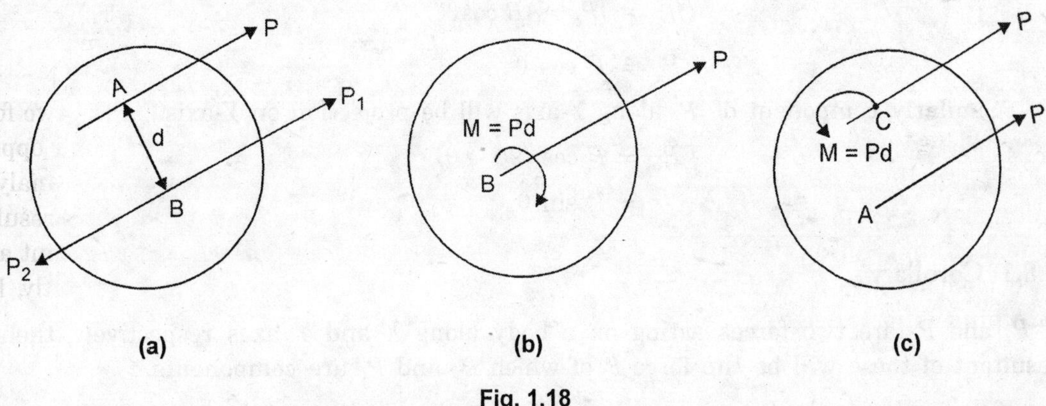

Fig. 1.18

Now consider the forces P and P_2. These constitute a couple of clockwise moment M *given by,*

$$M = P \cdot d \qquad ...(1.11)$$

The effect of this couple will be to induce rotation (clockwise). The force P_1 will induce linear motion in the direction of P_1 which is parallel to P.

Thus, we conclude that the original force P at A is replaced by an equal force P_1(= P) at B in a direction parallel to the direction of original force P alongwith a clockwise couple of moment = $P \times d$, which tends to introduce rotation in the plane of the couple (plane passing through the Force Vector P and the point B). Since couple can be transferred to any point in the moment plane, we conclude that while transferring P from point A to point B, a couple of moment Pd is superimposed at point B, as shown in Fig. (1.18*a*).

☛ It may be noted that if the force P has to be transferred to point C, which is on the left side of force vector P, then line superimposing couple would be anticlockwise as shown in Fig. (1.18*c*).

1.8 RESOLUTION OF A FORCE

Resolution of a force along two arbitrarily chosen mutually perpendicular directions means "finding the components of the given force along two chosen perpendicular directions."

Let the given force be represented by the Vector $\vec{P}$ (Fig. 1.19) which is inclined at an angle θ with arbitarily chosen X-axis. The component of $\vec{P}$ along the X-axis P_x is the projection of vector $\overrightarrow{OA}(=\vec{P})$ on X-axis and is given by,

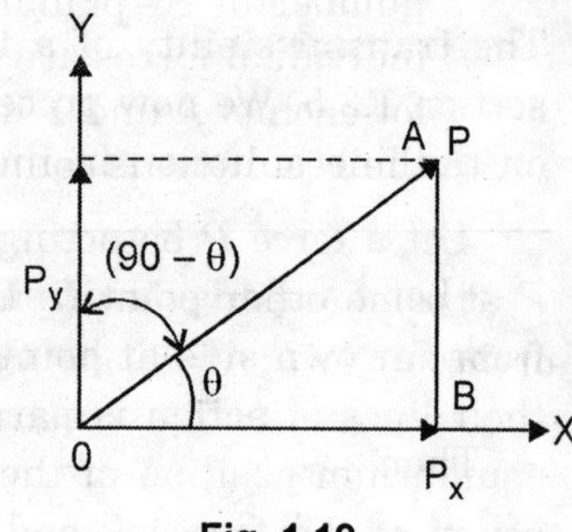

Fig. 1.19

$$\overrightarrow{OB} = P_x = \overrightarrow{AB}\cos\theta$$

i.e., $$P_x = P\cos\theta \qquad ...(1.12)$$

Similarly, component of $\vec{P}$ along Y-axis will be projection on Y-axis.

or $$P_y = P\cos(90 - \theta)$$

$$= P\sin\theta \qquad ...(1.13)$$

1.8.1 Corollary

If P_x and P_y are two forces acting on a body along X and Y axes respectively, then the resultant of these will be the force P of which P_x and P_y are components.

If P_x and P_y are the forces along X and Y axes, the resultant of these is given by equation (1.15).

$$R = \sqrt{P_x^2 + P_y^2}$$

$$= \sqrt{(P\cos\theta)^2 + (P\sin\theta)^2}$$

$$= \sqrt{P^2(\cos^2\theta + \sin^2\theta)} = \sqrt{P^2}$$

or $$R = P$$

Fig. 1.20

The inclination of R with X-axis (say ϕ)

$$\phi = \tan^{-1}\left(\frac{P_y}{P_x}\right)$$

$$= \tan^{-1}\left(\frac{P\sin\theta}{P\cos\theta}\right) = \tan^{-1}(\tan\theta) = \theta$$

i.e. the resultant is the same force P of which P_x and P_y were the components.

☛ Corollary of article 1.8.1 is extensively used in the cases where resultant of large number of co–planar concurrent forces acting on the body is to be found. Then the individual forces are resolved along two arbitrarily chosen perpendicular directions and for each of X and Y-directions, the algebraic sum of the components are calculated and the resultant is found with the help of equation 1.12.

Let ΣX = Algebraic sum of component of individual forces along X-axis

or ΣY = Algebraic sum of components of individual forces along Y-axis

Then $$R = \sqrt{(\Sigma X)^2 + (\Sigma Y)^2} \qquad ...(1.12)$$

and its Inclination will be X-axis in,

$$\theta = \tan^{-1}\left(\frac{\Sigma Y}{\Sigma X}\right) \qquad ...(1.13)$$

1.9 RESULTANT OF CO–PLANAR CONCURRENT FORCES

Let there be n number of force P_1, P_2, P_3 ... P_n acting along direction α_1, α_2, α_3 ...α_n with the X-axis (Fig. 1.21). Let $P_{X_1}, P_{X_2} ... P_{X_n}$ be the respective components of these forces along X-axis and $P_{Y_1}, P_{Y_2} ... P_{Y_n}$ respective components along Y-axis.

Then resolving each of these forces along two perpendicular direction X and Y axis and taking the algebraic sum of these along the respective X and Y axis we get,

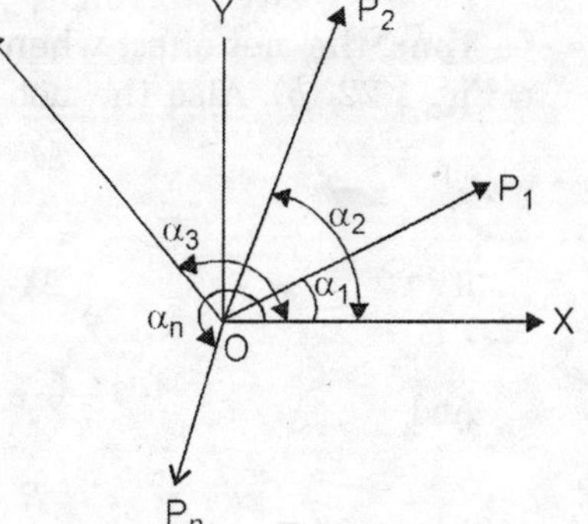

Fig. 1.21

$$\Sigma X = P_{x_1} + P_{x_2} + P_{x_3} + ... + P_{x_n}$$

and $$\Sigma Y = P_{y_1} + P_{y_2} + P_{y_3} + ... + P_{y_n}$$

Then from article 1.8.1 we have,

$$R = \sqrt{(\Sigma X)^2 + (\Sigma Y)^2}$$

and the inclination of resultant vector R with X-axis,

$$\tan\theta = \frac{\Sigma Y}{\Sigma X}$$

1.10 RESULTANT OF CO–PLANAR NON-CONCURRENT FORCES

Let the system of forces P_1, P_2 and P_3 be acting at points A, B and C as shown in Fig. (1.22). Let the axis of reference be marked as OX and OY (Fig. 1.22) Suppose d_1, d_2 and d_3 are the perpendicular distances from origin O to the forces P_1, P_2 and P_3 respectively.

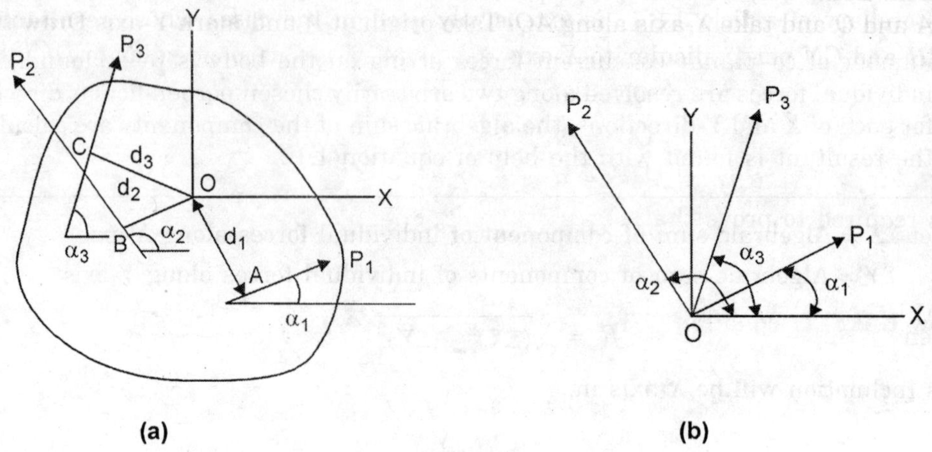

Fig. 1.22

Then transfer the forces P_1, P_2 and P_3 to the origin O:

(*i*) Transfer of P_1 from A to $O \equiv$ Force P_1 at O and inclined at α_1 with X-axis and anticlockwise couple of magnitude $P_1 d_1$

(*ii*) Transfer of P_2 from B to $O \equiv P_2$ at O and making an angle α_2 with X-axis + clockwise couple $P_2 d_2$

(*iii*) Similarly P_3 at C transferred to point $O \equiv P_3$ at angle α_3 + clockwise couple at O of magnitude P_3

Thus, the net effect when all the three forces P_1, P_2 and P_3 are transferred to O as shown in Fig. 1.22 (*b*). Also the net couple at O will be,

$$M_r = -M_1 + M_2 + M_3$$

$$= -P_1 d_1 + P_2 d_2 + P_3 d_3$$

If ΣX = Algebric sum of the components of P_1, P_2 and P_3 along OX

$$= P_1 \cos \alpha_1 + P_2 \cos \alpha_2 + P_3 \cos \alpha_3$$

And $$\Sigma Y = P_1 \sin \alpha_1 + P_2 \sin \alpha_2 + P_3 \sin \alpha_3$$

$$R = \sqrt{(\Sigma X)^2 + (\Sigma Y)^2}$$

And the inclination of resultant with X-axis is,

$$\tan \theta = \frac{\Sigma Y}{\Sigma X}$$

1.11 VARIGNON'S THEOREM (PRINCIPLE OF MOMENTS)

Statement : Algebraic sum of the moments of a system of co–planar forces about a point (called the moment centre) in the plane is equal to the moment of their resultant about the same point.

Proof : In Fig. (1.23) P_1 and P_2 are two forces represented by the vectors $\overline{AB}$ and $\overline{AC}$ and their resultant is $\vec{R}$ represented by $\overline{AD}$. Q is any point about which the moments are to be taken. Draw *QE*, *QG* and *QF* perpendicular on the force vectors P_1, P_2 and Resultant *R*. Join *A* and *Q* and take *X*-axis along *AQ*. Take origin at *A* and mark *Y*-axis Draw the lines *DMK*, *BH* and *CN* perpendicular to *X*-axis. *BL* and *CM* are perpendicular to *DK*.

Let
$$QE = d_1$$
$$QG = d_2$$
$$QF = d$$

It is required to prove that,

$$\vec{R}d = \vec{P_1}d_1 + \vec{P_2}d_2 \quad \text{...(1.14)}$$

Now L.H.S. of equation (1.14),

$$= \vec{AD}\,QF$$
$$= \overrightarrow{AD}(AQ\sin\theta) \quad \text{(From } \Delta\, AQF)$$

or
$$\vec{R}D = (\overrightarrow{AD}\sin\theta)\,AQ \quad \text{..(1.15)}$$

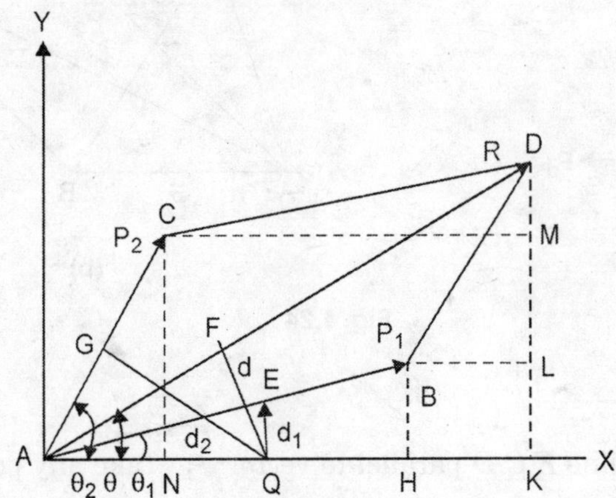

Fig. 1.23

But from Δ *ADK*,

$$AD\sin\theta = DK$$
$$= LK + LD \quad \text{...(1.16)}$$

But
$$LD = \text{Projection of } BD \text{ on vertical line}$$
$$= \text{Projection of } AC \text{ on vertical line}$$
$$= CN = \overrightarrow{AC}\sin\theta_2 \quad \text{...(1.17)}$$

Also,
$$KL = BH = \overrightarrow{AB}\sin\theta_1 \quad \text{...(1.18)}$$

From (1.15), (1.16), (1.17) and (1.18),

$$\vec{R}d = ((\vec{AD}\sin\theta)AQ$$

Now
$$AD\sin\theta = DK = (DL + LK)$$

$$\therefore \quad \overrightarrow{R}d = (\overrightarrow{AB}\sin\theta_1 + \overrightarrow{AC}\sin\theta_2)AQ$$

$$\text{But} \quad = \overrightarrow{P_1}AQ\sin\theta_1 + \overrightarrow{P_2}AQ\sin\theta_2 \quad \text{...(1.19)}$$

$$\text{In} \quad \underline{\Delta AQE} = AQ \sin \theta_1 = d_1 \quad \text{...(1.20}a\text{)}$$

and from $\Delta\ AQG$ $\quad AQ \sin \theta_2 = d_2$ ...(1.20*b*)

i.e., equation (1.19), (1.20) reduces to,

$$\overrightarrow{R}d = \overrightarrow{P_1}\,d_1 + \overrightarrow{P_2}\,d_2 \text{ Hence, proved} \quad \text{...(1.21)}$$

1.11.1 Alternate Graphical Proof of Varignon's Theorem

Proof : In Fig. 1.24(*a*) are given the forces represented by Vectors $\overrightarrow{AB}$ and $\overrightarrow{AC}$. Let E be the point about which moments are to be taken.

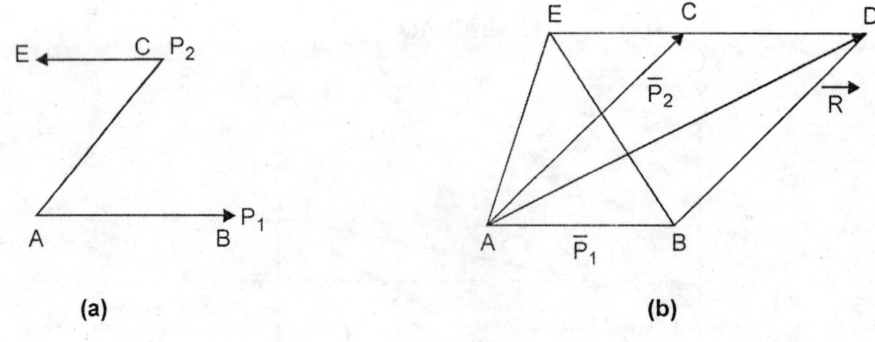

Fig. 1.24

Construction

From the point E, draw a line $E\ C\ D$ parallel to vector $\overline{P_1}$. Take any point C on this line and from C draw a line parallel to CA and make at some scale $\overrightarrow{AC} = \overrightarrow{P_2}$. (Fig. 1.24)

From A, draw the line AB parallel EC and make $\overrightarrow{AB} = \overrightarrow{P_1}$ on the same scale. Then complete the parallelogram $ABCD$ and join the points A and D. Such that $\overline{AD} = \overrightarrow{R}$ where $\overrightarrow{R}$ is the resultant of forces P_1 and P_2. Join the points E to A and E to B.

Then from equation (1.7) we have,

$$M_1 = \text{Moment of } P_1 \text{ about } E$$
$$= 2 \text{ (area of } \Delta\ ABE)$$
$$M_2 = \text{Moment of } P_2 \text{ about } E$$
$$= 2(\text{area of } \Delta\ ACE)$$
$$\therefore \quad M_1 + M_2 = 2(\text{Area of } \Delta\ ABE + \Delta ACE)$$
$$\text{But} \quad \text{Area of } \Delta\ ABE = \text{Area of } \Delta\ ABD$$

($\because$ Both triangles lie between the same parellel lines with the same base)

$$\therefore \quad M_1 + M_2 = 2(\text{Area of } \Delta\, ABD + \Delta\, ACE)$$
$$= 2(\text{Area } \Delta\, ACD + \Delta\, ACE)$$
$$= 2\, \Delta\, AED$$
$$= M_R = \text{Moment of } \vec{R} \text{ about } E, \text{ Hence, the proof}$$

1.12 RESULTANT OF PARALLEL FORCES

There are three different cases involving parallel forces. These are dealt with below.

1.12.1 Two Like Parallel Forces

In Fig. 1.25 shows two like parallel forces P_1 and P_2 acting on a body at points A and B. It is required to find the resultant force R and its point of application.

Let the origin O be arbitrarily chosen on line AB and use X-axis be taken along OB. Resolving the forces along X-and Y-axis and taking their algebraic sum we have,

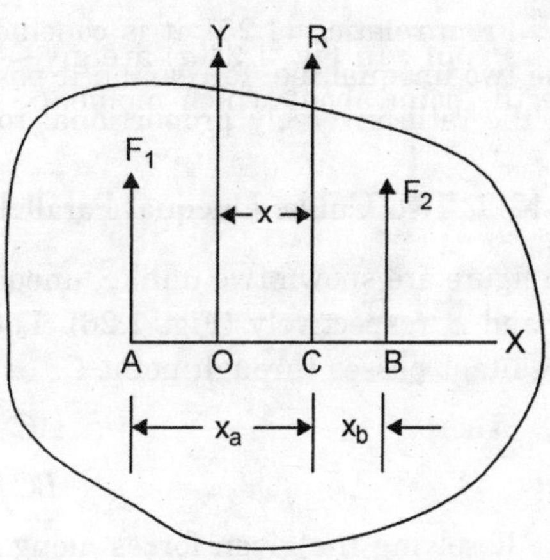

Fig. 1.25

$$\Sigma X = F_1 \cos 90° + F_2 \cos 90°$$
$$= 0$$

And
$$\Sigma Y = F_1 \cos 0° + F_2 \cos 0°$$
$$= F_1 + F_2$$

$$\therefore \quad R = \sqrt{(\Sigma X)^2 + (\Sigma X)^2}$$
$$= \sqrt{0 + (F_1 + F_2)^2}$$

or
$$R = F_1 + F_2 \qquad \text{...(1.22)}$$

Equation (1.22) only gives the magnitude of the resultant force. Still its location or point of application of force is to be determined.

Let the result be acting at the point C which is at a distance x_a from the point A and x_b from B and at a distance x from origin.

Then we have, $\quad AC = x_a,\ BC = x_b$

are $\quad OC = x$

Now from Varignon's theorem of moments, we know that the moment of the resultant R about any point is equal to the algebraic sum of the moments of the composing forces. Therefore, taking moments about 'O' we have (Ref. Fig. 1.25).

$$-R\,(OC) = -F_2\ (OB) + F_1(AO)$$
$$-xR = -F_2\,(OC + CB) + F_1(AC - AO)$$
$$= -F_2\,(x_b + x) + F_1\,(x_a - x)$$
$$= -x\,(F_2 + F_1) - F_2x_b + F_1x_a$$

or
$$-x\,R + x\,(F_1 + F_2) = F_1x_a - F_2x_b \qquad \text{...(1.23)}$$

Now substituting the value of R from equation (1.22).

$$-x(F_1+F_2)+x(F_1+F_2) = F_1x_a - F_2x_b$$

$$0 = F_1x_a - F_2x_b$$

or $$F_1x_a = F_2x_b \qquad ...(1.24)$$

or clockwise moment of F_1 about C

$$= \text{Anticlockwise moment of } F_2 \text{ about } C$$

Also from (1.24) $$\frac{x_a}{x_b} = \frac{F_2}{F_1} \qquad ...(1.25)$$

From relation (1.25), it is concluded that the resultant R, which is equal to the sum of the two unequal like forces and it posses through the point C which divides the distance AB in the ratio inversely proportional to the magnitude of the forces F_1 and F_2

1.12.2 Two Unlike Unequal Parallel Forces

In figure are shown two unlike, unequal forces F_1 and F_2 ($F_1 > F_2$) acting on a body at points A and B respectively (Fig. 1.26). Let the origin O be chosen on the line AB. Assume the resultant passes through point C.

Where,
$$AC = x_a$$
$$BC = x_b$$

Resolving the given forces along X and Y directions (since the forces act parallel to Y-axis).

$$\Sigma X = 0$$

$$\Sigma Y = F_1 - F_2$$

or $$R = \sqrt{(\Sigma X)^2 + (\Sigma X)^2}$$

$$= \sqrt{(0 + F_1 - F_2)^2}$$

Fig. 1.26

or $$R = F_1 - F_2 \qquad ...(1.26)$$

For finding the point of application of the resultant force R, taking moment of R about the point C and equating it to the algebric sum of moments of F_1 and F_2 about C.

i.e., $$F_2x_b - F_1x_a = R \times 0$$

i.e., $$F_2x_b = F_1x_a$$

or $$\frac{F_1}{F_2} = \left(\frac{x_b}{x_a}\right) \qquad ...(1.27)$$

From equation (1.27), we conclude that the point of application of the resultant lies out side the line joining the points A and B on the side of bigger force ($= F_1$) and the distance of C from the points of application of the given forces is inversely proportional to magnitudes of the forces.

1.12.3 Two Unlike Equal Parallel Forces

Two unlike equation parallel forces separated by a finite distance constitute a couple which has earlier been discussed in section 1.6.

1.13 RESULTANT OF A FORCE AND A COUPLE

Let a couple consisting of two equal and opposite forces ($= F_1$) having moment arm d and single force P acting at a point A be acting on the body Fig. 1.27.

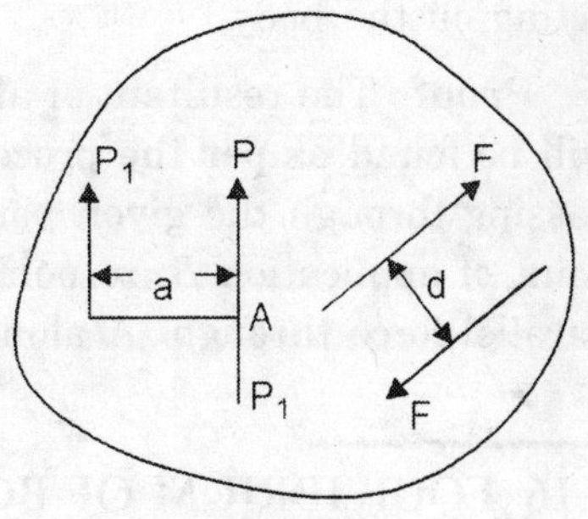

Fig. 1.27

Moment of couple Fd. Let this couple Fd be replaced by another couple (in the same plane) $P \times a$ such that,

$$P_1 a = Fd$$

or $$P_1 = \frac{Fd}{a} \quad \text{(where } P_1 = P\text{)} \qquad ...(1.28a)$$

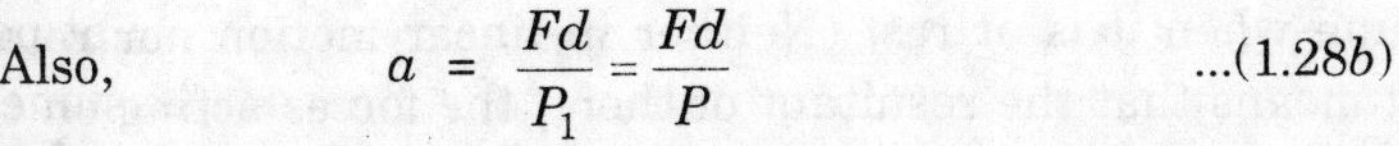

Also, $$a = \frac{Fd}{P_1} = \frac{Fd}{P} \qquad ...(1.28b)$$

If one force P, of this couple is made to act through the point A and in the opposite direction to the original P (as shown in Fig. 1.27), then these two equal and opposite forces will cancel each other. Then, the second force $= P$ which was part of the couple will be left. Thus, the resultant of the couple and the force P at A is a single force $= P$ but acting at a distance equal to 'a' as given in equation (1.28b) from the point A.

i.e., $$a = \frac{Fd}{P} \qquad ...(1.28b)$$

1.14 RESULTANT OF MORE THAN TWO COPLANAR PARALLEL FORCES

Fig. 1.28 shows six co–planar parallel forces. Out of six forces F_1, F_2 and F_5 are acting on the body in one direction (downward here) and other three forces are acting in the opposite (upward here) direction.

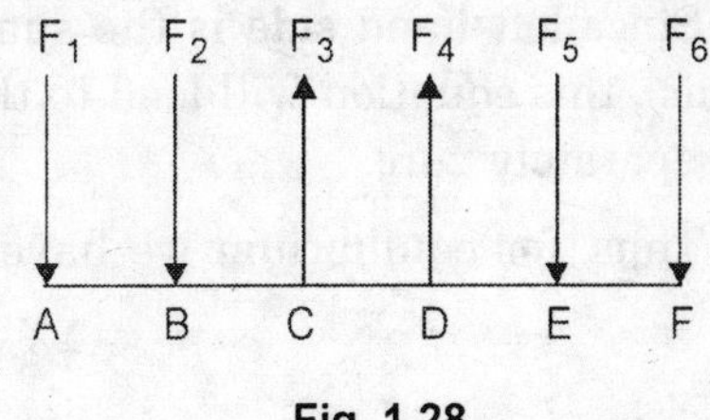

Fig. 1.28

Let R_1 be the resultant of first set of parallel forces which are directed downwards and R_2 is the resultant of other set. R_1 and R_2 can be found by the method outlined in article 1.12.1. Then following possibilities can occur:

(*i*) R_1 and R_2 are equal and opposite and acting along the same line. Then the net resultant of all the six forces will be zero.

(*ii*) R_1 and R_2 are equal opposite but are acting along lines which are at finite distance apart. Then, these will constitute a couple and will be dealt in the manner outlined in article 1.6.

1.15 EQUIVALENT SYSTEM

Statement : Let a body be under the action of a large number of co–planar forces acting at different points and various directions. Then, the quivalent system for the given system of co–planar forces with respect to a given point is a "combination of a force passing through the given point and a moment about that point. The force is the resultant of all the forces acting on the body.

Proof : The resultant of all the given co–planar non concurrent forces acting on the body will be found as per the procedure outlined in section 1.10. The resultant force will not be passing through the given point (say A). Then this resultant R can be transferred from its point of application B, to point A as per the procedure given in section 1.7 so as to give a parallel force through (A along) with a moment at A.

1.16 EQUILIBRIUM OF BODIES

A body is said to be in equilibrium when it is at rest (Neither in linear motion nor rotary motion). Then mathematically, it means that the resultant of there, the forces acting on the body and the albebraic sum of their moments is zero.

That $$R = 0$$

and $$M = 0$$

Also, we know $$R^2 = (\Sigma X)^2 + (\Sigma Y)^2 \quad ...(1.29)$$

$$\Sigma M = 0 \quad ...(1.30)$$

Where R is the resultant force, ΣX, ΣY are the algebraic sum of the components along any chosen X and Y axis respectively ΣM is the algebraic sum of the moments about a point in the plane of forces.

In equilibrium, $R = 0$, therefore, from (1.29), we have,

$$(\Sigma X)^2 + (\Sigma Y)^2 = 0$$

Since left hand side is the sum of the positive quantities, (being summation of square terms), this equation will lead to the conclusion that individually both $(\Sigma X)^2$ and $(\Sigma Y)^2$ must be separately zero.

Thus, for equilibrium we have the following three conditions to be fulfilled,

$$\Sigma X = 0, \quad \Sigma Y = 0$$

and $$\Sigma M = 0 \quad ...(1.31)$$

1.16.1 System of Forces Having Three Concurrent Forces

In case, the body is in equilibrium under the action of three concurrent forces, then the problem can also be analysed with the help of Lami's theorem or triangle of forces law.

EXAMPLE

1.1 : *(a) Two forces P and Q act on a body. Find the greatest possible value of their resultant R, what will the be the inclination of P and Q?*
(b) Also, determine when R will be least possible.

Solution : (*a*) The resultant R will be greatest when P and Q both will have the same line of action and along same direction.

That is $$R = P + Q$$

(*b*) The resultant R will be least possible when P and Q will act along the same line with their directions opposite to each other assuming $P > Q$.

That is $$R = P - Q$$ **Ans.**

1.2 : *(a) Two concurrent 100 N forces and 50 N act on body along directions at 0° and 60° to X-axis respectively. Find the magnitude and direction of the resultant.*
(b) Find α when r = 0.5, (c) whe r = 1.

Solution : Given $P = 100$ N, $Q = 50$ N

$$\alpha = 60°$$

Then
$$R = \sqrt{P^2 + Q^2 + 2PQ\cos\alpha}$$
$$= \sqrt{(100)^2 + (50)^2 + 2(100)\,(50)\cos 60}$$
$$= \sqrt{1000 + 2500 + 5000}$$
$$= \sqrt{17500} = 132.29 \text{ N}$$

If inclination of R with the direction of P is θ, then,

$$\tan\theta = \frac{Q\sin\alpha}{P + Q\cos\alpha}$$
$$= \frac{50\sin 60}{100 + 50\cos 60} = \frac{50 \times \sqrt{3}}{2\left(100 + \frac{50}{2}\right)}$$
$$= \frac{50\sqrt{3}}{2 \times 125} = \frac{\sqrt{3}}{5}$$
$$= 0.3464$$
$$\therefore \quad \theta = \tan^{-1}(0.3464) = 19.106°$$

1.3 : *Two concurrent forces F_1 and F_2 act on a body. The magnitude of their resultant is half their sum. Find this inclination between them in terms of the ratio $r = F_1/F_2$.*

Solution: Given $\frac{F_1}{F_2} = r$

$$R^2 = F_1^{\,2} + F_2^{\,2} + 2F_1 F_2 \cos\alpha$$

and
$$R = \frac{1}{2}(F_1 + F_2)$$

It is required to calculate the value of α.

$$\left(\frac{F_1 + F_2}{2}\right)^2 = F_1^{\,2} + F^2 + 2F_1F_2 \cos\alpha$$

Also,
$$F_1 = r\,F_2$$

$$\left[\frac{(1+r)F_2}{2}\right]^2 = r^2F_2^{\,2} + F_2^{\,2} + 2rF_2^{\,2} \cos\alpha$$

Therefore,
$$2rF_2^{\,2} \cos\alpha = \left[\left(\frac{1+r}{2}\right)^2 - (r^2+1)\right]F_2^2$$

$$\cos\alpha = \frac{\left(1+r^2+2r\right)-4(r^2+1)}{4(2r)}$$

$$= \frac{-3r^2+2r-3}{8r}$$

$$= \frac{1}{4} - \frac{3}{8}\frac{\left(r^2+1\right)}{r} \quad \text{...(A)}$$

(1) when $r = 0.5$

$$\cos\alpha = \frac{1}{4} - \frac{3}{8}\frac{(1.25)}{0.5} = \frac{1}{4} - \frac{3}{4}(1.25)$$

$$= -\frac{2.75}{4} = -\,0.6875$$

$\therefore$ $\alpha = \text{Cos}^{-1}\,(-.6875) = 133.43°$

i.e., When $r = 1.0$ we have from *(A)*

$$\cos\alpha = \frac{1}{4} - \frac{3}{8}\frac{(1+1)}{1} = 0.25 - .75 = -\,0.5$$

$\therefore$ $\alpha = 120°$

EXAMPLE

EXAMPLE

1.4 : *The resultant of two forces F_1 and F_2 which are inclined to each other at an angle α, is at right angles to F_1. The resultant of F_1 and F_3 which are also inclined attain angle α to each other, is at right angle to F_3. Find the relationship between F_1, F_2 and F_3.*

Solution : In this first case,

Given F_1 and F_2 inclined at a, this resultants is al right angles is F_1

$$\therefore \quad \theta = 90°$$

But $$\tan\theta = \frac{F_2 \sin\alpha}{F_1 + F_2 cos\alpha} = \tan 90°$$

$$F_1 + F_2 \cos\alpha = 0 \quad \text{...(A)}$$

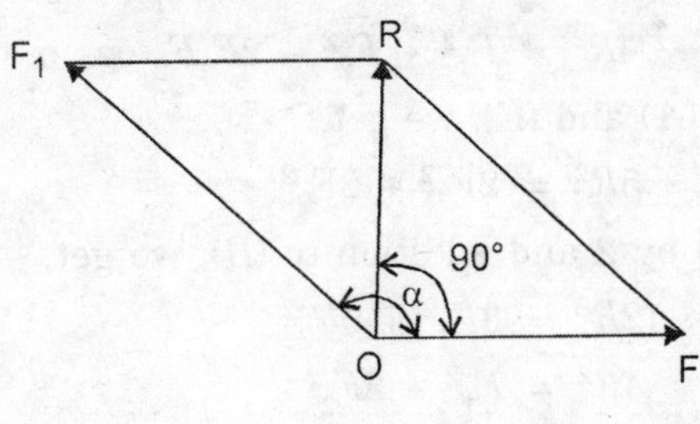

(a)

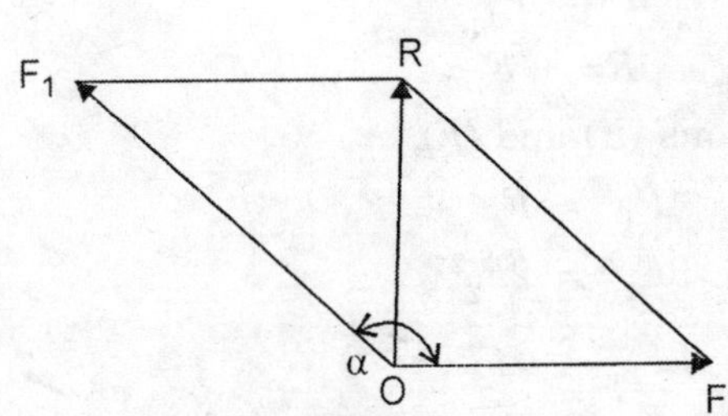

(b)

Fig. 1.29

In this second case,

$$\tan 90° = \frac{F_1 \sin\alpha}{F_3 + F_1 \cos\alpha} = -\alpha$$

$$\therefore \quad F_3 + F_1 \cos\alpha = 0$$

$$F_1 = -F_2 \cos\alpha$$

Dividing these, $$\frac{F_1}{F_3} = \frac{F_2}{F_1}$$

or $$F_1^2 = F_2 F_3$$

That is F_1 is geometric mean between F_2 and F_3

1.5 : *Two forces F_1 and F_2 acting at a point in a body have the resultant R.If this force F_2 is doubled, the resultant R also gets doubled. Again if this directing of F_2 is reversed, then also R is doubled. Find the ratio F_1: F_2:*

Solution : We know that,

$$R^2 = F_1^{\,2} + F_2^{\,2} + 2F_1 F_2 \cos\alpha \qquad \text{...(A)}$$

When F_2 is doubled, R gets doubled

or $$(2R)^2 = F_1^{\,2} + (2F_2)^2 + 2F_1(2F_2)) \cos\alpha$$

or $$4R^2 = F_1^{\,2} + 4F_2^{\,2} + 4F_1F_2 \cos\alpha \qquad \text{...(B)}$$

Again when F_2 is reversed, R is doubled

$$(2R)^2 = F_1^{\,2} + (-F_2)^2 + 2F_1 (-F_2)\cos\alpha$$

or $$4R^2 = F_1^{\,2} + F_2^{\,2} - 2F_1F_2 \cos\alpha \qquad \text{...(C)}$$

Adding equation (A) and (C), we get,

$$5R^2 = 2F_1^{\,2} + 2F_2^{\,2} \qquad \text{...(D)}$$

Now multiply (C) by 2 and addition to (B), we get,

$$12R^2 = 3F_1^{\,2} + 6F_2^{\,2}$$

or $$4R^2 = F_1^{\,2} + 2F_2^{\,2} \qquad \text{...(E)}$$

Subtract (E) from (D)

$$R^2 = F_1^{\,2}$$

Or $$R = \pm F_1 \qquad \text{...(F)}$$

Then from equations (E) and (F),

$$4F_1^{\,2} = F_1^{\,2} + 2F_2^{\,2}$$

or $$3F_1^{\,2} = 2F_2^{\,2}$$

or $$\frac{F_1}{F_2} = \sqrt{\frac{2}{3}}$$

or $$F_2 = \sqrt{\frac{3}{2}}F_1$$

$\therefore$ $$F_1: F_2; R :: F_1 : \sqrt{\frac{3}{2}}F_1 : F_1$$

or $$F_1 : F_2 : R :: 1 : \sqrt{\frac{3}{2}} : 1$$ **Ans.**

1.6 : *Two forces of value 60N and P acting at a point, have resultant of 80N which is inclined at 60° to the 60 N force. Find the value of P and its inclination with 60N force.*

Solution : The vector diagram for the given forces and the resultant R is given in Fig. (1.30) from this Δ of forces, we have,

EXAMPLE

$$\frac{60}{\sin(\theta - 60)} = \frac{80}{\sin(180 - \theta)} = \frac{P}{\sin 60}$$

$$\therefore \qquad \frac{60}{80} = \frac{\sin(\theta - 60)}{\sin(180 - \theta)} \qquad \text{...(A)}$$

$$\text{or} \qquad \frac{3}{4} = \frac{\sin\theta\cos 60 - \cos\theta\sin 60}{\sin\theta}$$

$$\text{or} \qquad 3\sin\theta = 4(\sin\theta)\times\frac{1}{2} - (\cos\theta)\frac{\sqrt{3}}{2}\times 4$$

$$= 2\sin\theta - 2\sqrt{3}\cos\theta$$

$$\text{or} \qquad \sin\theta = -2\sqrt{3}\cos\theta$$

$$\text{or} \qquad \tan(180 - \theta) = 2\sqrt{3} = 3.4641$$

$$\text{Therefore,} \qquad (180 - \theta) = 37.889°$$

$$\text{or} \qquad \theta = 180 - 37.889$$

$$\theta = 106.111°$$

C, R = 80N, (θ – 60°), 60, (180 – θ), θ, A, 60 N, B

Fig. 1.30

Again from equation (*A*),

$$P\sin(180 - \theta) = 80\sin 60°$$

$$P\sin(37.889°) = 80 \times \frac{\sqrt{3}}{2}$$

$$\therefore \qquad P = \frac{40\sqrt{3}}{\sin(37.889°)} = \frac{40\sqrt{3}}{0.6141}$$

$$= 112.82 \quad \textbf{Ans.}$$

1.7 : *Find the magnitude of horizontal force P to be applied on a body of mass 700 kg so that this body is held in a smooth 20° inclined plane AB (Fig. 1.31)*

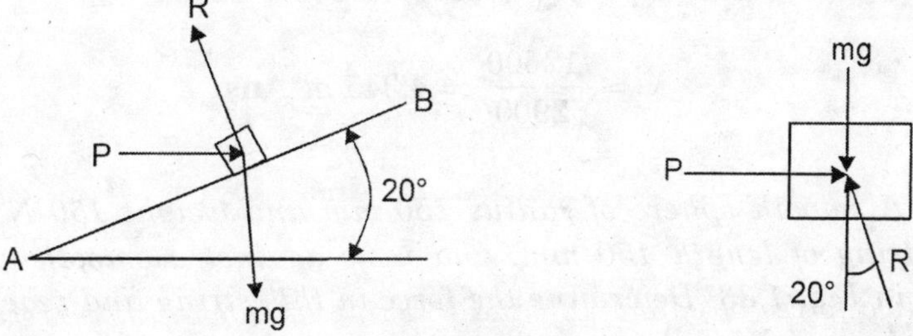

Fig. 1.31

Solution : Let R be the normal reaction of this plane of this. Then drawing the free body diagram Fig. 1.31 kg. we see that therefore, there concurent fores P. mg and R, then applying this Lam's equation we have,

$$\frac{mg}{\sin(90° + 20°)} = \frac{P}{\sin(180° - 20°)} = \frac{R}{\sin 90°}$$

Now $\dfrac{700 \times 9.81}{\sin 110°} = \dfrac{P}{\sin(160°)} = \dfrac{R}{1}$

That $P = \dfrac{(7w \times 9.81)\sin 160°}{\sin 110°}$

$= \dfrac{700 \times 9.81 \times 0.3420}{0.9396}$

$= 2499.5\ N$ **Ans.**

1.8 : *Four forces P_1, P_2, P_3 and P_4, are acting on a rod 9 m long as shown in Fig. 1.32. Find the resultants and its point of action.*

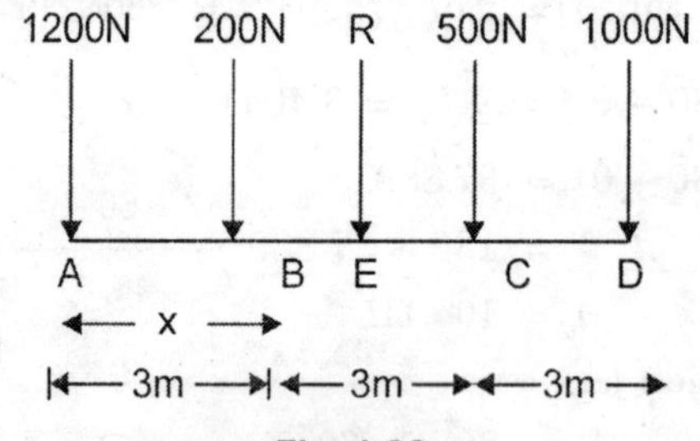

Fig. 1.32

Solution : Here, $\Sigma X = 0$

$\Sigma Y = 1200 + 200 + 500 + 1000$

$= 2900\ N.$

Now $R^2 = (\Sigma x)^2 \times (\Sigma y)^2 = 0 + (2900)^2$

$\therefore$ $R = 2900\ N$

Let the resultant R pass through point E which is at distance x from the point A. Then from Varignon's theorem, we take moments about point A,

$R \times x = 1200 \times 0 + 200 \times 3 - 500 \times 6 + 1000 \times 9$

or $2900\ x = 600 + 3000 + 9000 = 12600$

$\therefore$ $x = \dfrac{12600}{2900} = 4.345\ m$ **Ans.**

1.9 : *A smooth sphere of radius 150 mm and weight 150 N hangs by a string of length 150 mm. and rests against a smooth wall as shown in Fig. 1.33. Determine the force in this string and reaction of the wall.*

Solution : Given $AB = 150$ mm

$DC =$ radius of sphere

$= 150$ mm

$W = 150\ N$

For the equilibrium of ball, the three forces viz. tension in string T, reaction R of the wall on ball and its weight W, are concurrent. Therefore, from Lami's theorem we have,

$$\frac{W}{\sin(180^\circ - \theta)} = \frac{R}{\sin(90^\circ + \theta)} = \frac{T}{\sin 90} \quad ...(A)$$

Now from $\Delta\, ADC$ in which,

$$AC = AB + BC$$
$$= 150 + 150$$
$$= 300 \text{ mm}$$
$$DC = 150 \text{ mm}$$

Therefore, $$\cos\theta = \frac{CD}{AC} = \frac{150}{300} = \frac{1}{2}$$

or $$\theta = 60^\circ$$

Fig. 1.33

Therefore, from equation A,

$$\frac{W}{\sin(180 - 60)} = \frac{R}{\sin 150} = T$$

$$\therefore \quad T = \frac{W}{\sin 60^\circ} = \frac{2W}{\sqrt{3}} = \frac{2 \times 150}{\sqrt{3}}$$
$$= 173.2 \text{ N}$$

and $$R = \frac{W}{\sin 60^\circ} \times \sin 150^\circ = \frac{150 \times 0.5 \times 2}{\sqrt{3}}$$
$$= 86.60\ N \text{ Ans.}$$

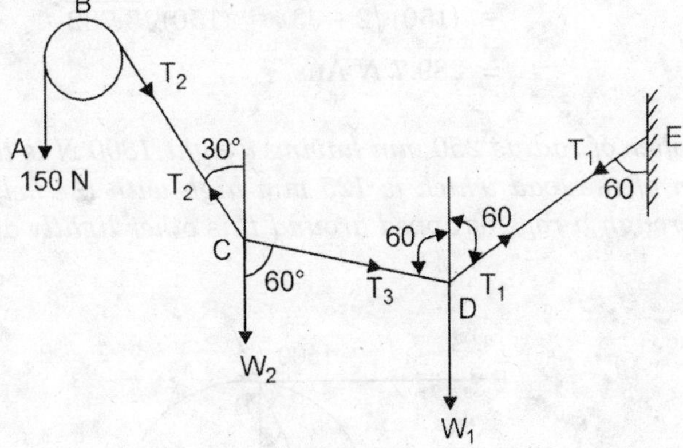

Fig. 1.34

1.10 : *The end E of a smooth string ABCDE is fixed in wall. It carries two weights W_1, and W_2 attached to it at counts D and C as shown in Fig. 1.34. To the end A of this string passing over a smooth pulley B, is attached a weight. 150N. If the inclination of the string portion DE is 60° with the Vertical and that of BC 30° with the vertical, find the tension in* *portions CD and DE and also, the weights W_1 and W_2. Also, find the load borne by the pulley.*

Solution: As the pulley is smooth, the tension in the portion CB is T_2 = 150N. Considering the equilibrium at point C and applying Lami's theorem,

$$\frac{T_2}{\sin 60°} = \frac{T_3}{\sin 150°} = \frac{W_2}{\sin 150°}$$

i.e.,
$$T_3 = T_2 \frac{\sin 150°}{\sin 60°} = 150\left(\frac{0.5 \times 2}{\sqrt{3}}\right)$$
$$= 86.60\ N$$

And
$$W_2 = T_3 = 86.6\ N$$

Again for the equilibrium of point *D*,

$$\frac{T_3}{\sin 120°} = \frac{W_1}{\sin 120°} = \frac{T_1}{\sin 120°}$$

i.e.,
$$T_3 = W_1 = T_1$$

or
$$W_1 = T_1 = 80.6\ N$$

Load on the pulley will be the resultant of the load of 150 *N* attached to end *A* and the tension T_2 and these are at 30° to each other,

$$R = \sqrt{(150)^2 + (150)^2 + 2(150)(150)\cos 30°}$$

or
$$R = (150)\sqrt{1 + 1 + 2 \times \frac{\sqrt{3}}{2}}$$
$$= (150)\sqrt{2 + \sqrt{3}} = (150)\sqrt{3.732}$$
$$= 289.7\ N \text{ Ans.}$$

1.11 : *A roller of radius 250 mm having weight 1800 N is to be pulled over the brin of the load which is 125 mm high with the help of a pull P applied through a rope wrapped around this other lightly as shown in Fig. 1.35*

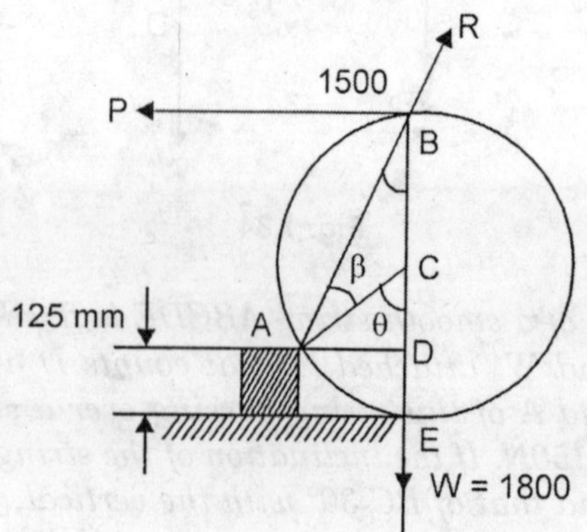

Fig. 1.35

E X A M P L E

Solution: Given $AC = BC = CE = r$

$$= 250 \text{ mm}$$

$$DE = 125 \text{ mm}$$

or $CD = 125$ mm

Thus, in Δ *A* and *D*,

$$\cos\theta = \frac{CD}{AC}, \frac{125}{250} = \frac{1}{2}$$

or $\theta = 360°$

$$\theta = 60°$$

Therefore, in triangle *ABC* $\beta = \frac{60}{2} > 30°$ Since the roller is just on the point of being lifted up from the ground, it is in equilibrium under the action of three forces W_1 *R* and *P*· The reaction R, of road brim must pass through the point of intersection of *W* and *P*, that is the point *B*.

Therefore, applying Lami's theorem,

$$\frac{1800}{\sin(90+\beta)} = \frac{R}{\sin 90°}$$

$$= \frac{P}{\sin(180-\beta)}$$

But $\beta = 36°$

Therefore, we get,

$$\frac{1800}{\sin 120} = \frac{R}{\sin 90}$$

$$= \frac{P}{\sin 150}$$

or $\frac{1800}{0.5} = R = \frac{2P}{\sqrt{3}}$

Therefore, $R = 3600\ N$

and $P = \frac{3600\times\sqrt{3}}{2} = 3117.65\ N$

1.12 : *A rectangular plate ABCD, 300mm × 150 mm is acted upon by three forces $F_1 = 1000$ N, $F_2 = 1500$ N and $F_3 = 800N$ as shown in Fig. 1.36. Find the resultant of the the given co–planar forces and x and y intercepts of the resultant. Take X-axis along AB and Y-axis along AD.*

Solution 1.12. The given forces are co-planar, therefore, resolving the forces along *X* and *Y*, we have,

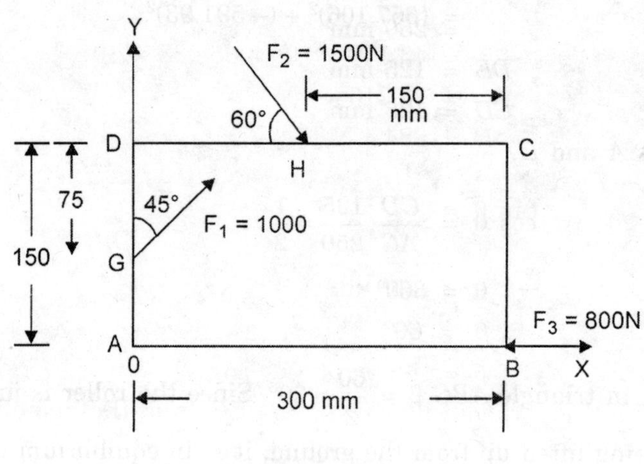

(a)

(b)

Fig. 1.36

$$\Sigma X = F_1 \sin 45° + F_2 \cos 60° - F_3$$

$$= 100 \times \frac{1}{\sqrt{2}} + 1500 \times \frac{1}{2} - 800$$

$$= 707.106 + 750 - 800$$

$$= .657.106\ N$$

$$\Sigma Y = F_1 \cos 45° - F_2 \sin 60° - F_3 \cos 90°$$

$$= 1000 \times \frac{1}{\sqrt{2}} - 1500 \times \frac{\sqrt{3}}{2} - 0.0$$

$$= 707.106 - 1299.038$$

$$= -591.3\ N$$

$$R^2 = (\Sigma x)^2 + (\Sigma y)^2$$
$$= (657.106)^2 + (-591.93)^2$$
$$= 43.178 \times 10^4 + 35.038 \times 10^4$$
$$= 78.216 \times 10^4$$

$\therefore$ $$R = 884.4\ N$$

Now $$\tan\theta = \frac{\Sigma Y}{\Sigma X}$$

$$= \frac{-591.93}{657.183} = -0.9008$$

$$\theta = 42.01^\circ \text{ (as shown)}$$

Let x be the the perpendicular distance of O from the line of action of R, then taking moments of F_1, F_2 and F_3 about O

$$R(x) = F_{1x}(OG) + F_{2x}(OD) + F_{2y}(DH)$$

Where, $(F_1)_x$ = Component of F_1 along x-axis

$(F_2)_x$ = Component of F_2 along x-axis

$(F_2)_y$ = Component of F_2 along Y-axis

$\therefore$ $$R(x) = (F_1 \sin 45^\circ)(AG) + (F_2 \cos 60^\circ)(OD) + (F_2 \sin 60^\circ)\,DH$$

$$= 1000 \times \frac{1}{\sqrt{2}} \times 75 + 1500 \times \frac{1}{2}(150) + \left(1500 \times \frac{\sqrt{3}}{2}\right)(150)$$

$$= 53032.5 + 112500 + 194855.72$$

or $$R(x) = 360388.6\ N\text{-mm}$$

Inserting the value of R in it, we get,

$$x = \frac{360388.6}{884.4} = 407.5 \text{ mm}$$

In Fig. (1.36b), vector R has been drawn. Let its inclination with X-axis be θ,

$$\tan\theta = \frac{\Sigma y}{\Sigma x} = \frac{-591.93}{657.103} = -0.9008$$

$\therefore$ $$\theta = 42.01^\circ \text{ (as shown in Fig. (1.36 b)}$$

Let the intercept of R on x-axis be 'a' as shown in Fig. (1.36 b), then

$$a \sin\theta = x$$

or $$a = \frac{x}{\sin\theta} = \frac{407.5}{\sin(42.01)^\circ}$$

It "b" is the intercept on y-axis, then from Fig. (1.36 b)

$$\frac{b}{a} = \tan\theta = 0.9008$$

$$\therefore \quad b = \alpha \times .9008 = 582.4 \times .9008$$

$$= 527.2 \text{ mm}$$

1.13 : *A weightless ring A can slip over a smooth wire which is bent in a circular shape and is fixed in a vertical plane from the ring hangs W. To this ring is tied a string which passes through a hook fixed at the vertex of the wire circle. And a pull P is applied such that the inclination of the radial line joining C to A makes an angle θ with the vertical and the string portion AB can be treated as chord of the circle (Fig. 1.37). If r is the radius of the circle, find the value of angle θ and the reaction of the wire on the ring A .*

Solution 1.13: In the triangle ABC,

$$BC = CA = r$$

$$\therefore \quad \angle CBA = \angle CAB = \beta$$

$$\therefore \quad \theta = 180 - 2\beta$$

or
$$\frac{\theta}{2} = (90 - \beta)$$

or
$$\beta = \left(90 - \frac{\theta}{2}\right) \qquad \text{...}(A)$$

Fig. 1.37

The ring A is in equilibrium and on it are acting three concurrent forces W, P, and R. Therefore, applying Lami's theorem we have,

$$\frac{R}{\sin(\theta+\beta)} = \frac{P}{\sin(180-\theta)} = \frac{W}{\sin(180-\beta)}$$

or
$$\frac{R}{\sin(\theta+\beta)} = \frac{P}{\sin\theta} = \frac{W}{\sin\left[180 - \left(90 - \frac{\theta}{2}\right)\right]}$$

$$= \frac{W}{\sin\left(90 + \frac{\theta}{2}\right)} \qquad \text{...}(B)$$

or
$$\frac{P}{\sin\theta} = \frac{W}{\cos\frac{\theta}{2}}$$

or
$$P = \frac{2\sin\frac{\theta}{2}\cos\frac{\theta}{2}}{\cos\frac{\theta}{2}} W$$

$$P = 2W\sin\frac{\theta}{2}$$

EXAMPLE

$$\therefore \quad \sin\frac{\theta}{2} = \frac{P}{2W} \quad \text{...(C)}$$

or $$\sin\frac{\theta}{2} = \frac{P}{2W}$$

or $$\frac{\theta}{2} = \sin^{-1}\left(\frac{P}{2W}\right) \quad \text{...(D)}$$

Now equation B $$R = \frac{P}{\sin\theta} - \sin(\theta+\beta)$$

$$= \frac{P}{\sin\theta}\sin\left(\theta + 90 - \frac{\theta}{2}\right)$$

$$= \frac{P}{\sin\theta}\sin\left(90 + \frac{\theta}{2}\right)$$

$$= \frac{P\cos\frac{\theta}{2}}{2\sin\frac{\theta}{2}\cos\frac{\theta}{2}} = \frac{P}{2\sin\frac{\theta}{2}}$$

$$= \frac{P}{2\left(\frac{P}{2W}\right)} = W$$

1.14 : *A ladder AB 4m long of weight 500 N rests against a smooth wall and on a smooth floor. Its lower end A is attached with a rope to the junction of wall and floor. A man of weight 750 N climbs on the ladder upto a point C which is 3 m from the point A. if the inclination of the ladder with the horizontal is 60°, Find the tension in the rope. Also, find the reactions R_A and R_B.*

Solution: Let R_A and R_B be the reactions of floor and wall on the ladder (Fig. 1.38). Let the tension in rope be T. The ladder is in equilibrium under the action of R_A, T, W_1, W_2 and R_B. Resolving the forces along the horizontal direction,

$$T = R_B \quad \text{...(A)}$$

Taking moments about the point A and equating the sum to zero, we get,

$W_1(AE) + W_2(AF) - R_B\,(AK) = 0$

$500 \times 2\cos 60° + 750 \times 3\cos 60° - R_B\, 4\sin 60 = 0$

$$\therefore \quad R_B \times 4 \times \frac{\sqrt{3}}{2} = 500 \times \frac{2}{2} + 750 \times 3 \times \frac{1}{2}$$

$$3.464\, R_B = 500 + 1125$$

$$\therefore \quad R_B = \frac{1625}{3.464} = 469.1\ N$$

$$\therefore \quad T = R_B = 469.1\ N \quad \textbf{Ans.}$$

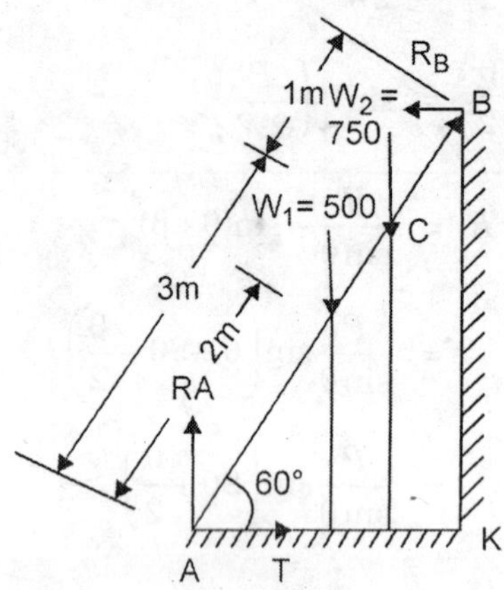

Fig. 1.38

Again taking the algebraic sum of vertical forces and equating it to zero, we get,

$$R_A = 750 + 500 = 1250\ N \textbf{ Ans.}$$

1.15 : *A wheel of radius 400 mm and weight 1 kN rests against a rigid block 200 mm thick. A pull P is applied to the wheel through its centre C to just turn the wheel over the corner of the block. Assuming all the surfaces are smooth, Find the least value of the pull P and the reaction of the block on the wheel.*

Solution: Given data,

Radius of wheel = 400 mm = CB

Thickness of block = AD = 200 mm

The reaction of the block on the wheel will pass through centre C of the wheel Let the pull P be inclined to the radial line AC at an angle θ. The wheel is in equilibrium under the action of three forces P, R and the weight W where it is just on the point of turning about the point A

Taking moments about the point A,

$$W(AF) - P(AK) = 0 \qquad ...(A)$$

Now from Δ ACF,

$$AF^2 = AC^2 - CF^2$$

or $$AF^2 = (400)^2 - (BC - AD)^2$$

$$= (400)^2 - (400 - 200)^2$$

$$= 16000 - 40000 = 120000$$

or $$AF = 3.464 \times 10^2 = 346.4 \text{ mm}$$

E X A M P L E

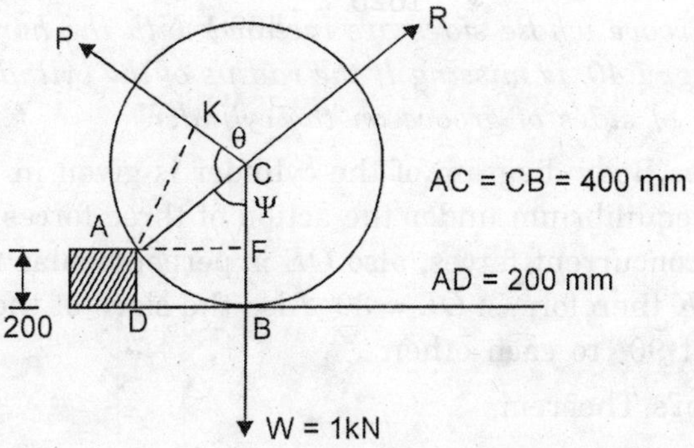

E X A M P L E

Fig. 1.39

And from ΔACK,

$$AK = AC (\sin \theta)$$
$$= 400 \sin \theta$$

$\therefore$ From equation (A),

$$P(400 \sin \theta) = (1\ kN)\ (346.4)\ \text{mm}$$

or

$$P = \frac{346.4}{400 \sin \theta}\ kN$$

$$P = \frac{0.866}{\sin \theta} \quad \text{...(B)}$$

From equation (B), we conclude that P will be minimum, when $\sin \theta$ is maximum

i.e, when $\theta = 90°$

Then, $P_{min} = \frac{0.8666}{\sin 90°} = 0.866\ kN$

Let the angle ACF be Ψ. Then from the triangle ACF,

$$\cos \Psi = \frac{CF}{AC} = \frac{BC - FB}{AC} = \frac{400 - 200}{400}$$

$$= \frac{1}{2}$$

i.e. $\Psi = 60°$

Now resolving the forces, along the direction of R

$R - P \cos \theta - W \cos \Psi = 0$

Substituting values, of P, θ and Ψ we get,

$R - (0.866) \cos 90° - 1 \cos 60 = 0$

or $R = 1\cos 60° = 0.5\ kN.$

1.16 : *A circular cylinder having smooth surfaces, weighing 1000 N rests on a V-groove whose sides are inclined with the horizontal at 60° and 30° an Fig. 1.40. is missing If the radius of the cylinder is 400 mm, find reactions of sides of groove on the cylinder.*

Solution: Free Body diagram of the cylinder is given in Fig. 1.40. The cylinder is in equilibrium under the action of three forces R_1, R_2 and W all these are concurrent forces, also OE is perpendicular to AC and OL is at 90° to AK therefore, $< OL = 30°$ Also the sides of the V-groove AC and AB are at 90° to each other.

Applying Lami's Theorem,

$$\frac{W}{\sin 90°} = \frac{R_1}{\sin(180-30)} = \frac{R_2}{\sin 120°}$$

It is given that $W = 1000\ N$

$$\therefore \qquad R_1 = \frac{1000 \sin 150°}{\sin 90°} = 500N$$

and

$$R_2 = \frac{1000}{\sin 90°} \times \sin 120° = 866.6 \text{ N}$$

1.17 : *Fig 1.41 shows an A-Type ladder, the two legs BA and CA are hinged at the top point A. In order to Aviod shipping of the feet of the legs a rope is tied to the legs as shown in Fig. (1.40) such that when in use the legs, are inclined with the horizontal at 60° If a man weighing 750 N is standing at the a distance of 3 m from B and each leg of the ladder weighs 220 N, find the reaction at B, C and at A. Also, determine the tension in the rope.*

Solution: The free body diagram of the leg AB in shown in Fig. 1.41 (*b*).

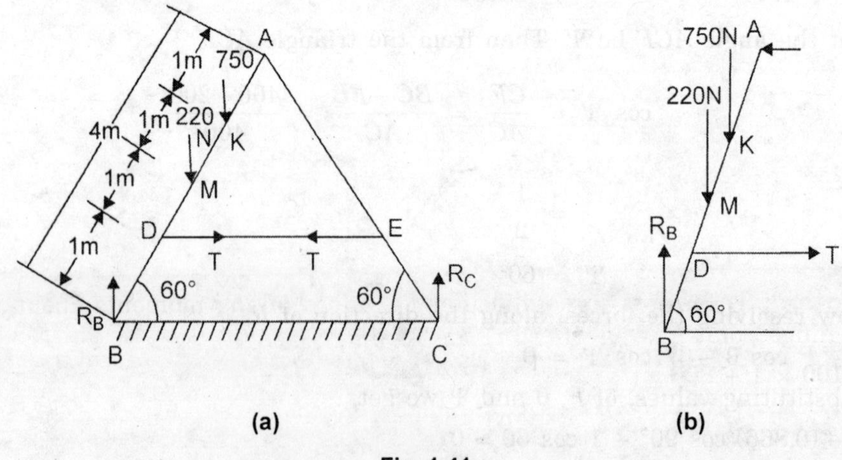

Fig. 1.41

EXAMPLE

Given data $BD = 1m$

$BA = 4m$

$BK = 3m$

$BM = 2m$

For the equilibrium of left leg, resolving forces vertical direction

$$R_B = 220 + 750 = 970\ N$$

Again resolving along Horizontal direction,

$$R_A = T$$

Now taking moments about B,

$$T(BD \sin 60°) + 220\ (BM \cos 60°) + 750(BK \cos 60°) - R_A\ (AB \sin 60°) = 0$$

or $$T\left(1 \times \frac{\sqrt{3}}{2}\right) + 220\left(2 \times \frac{1}{2}\right) + 750\left(3 \times \frac{1}{2}\right) - T\left(4 \times \frac{\sqrt{3}}{2}\right) = 0$$

$$0.866\ T + 220 + 1125.0 = 4 \times 0.866\ T$$

or $$1345 = 3 \times 0.866\ T$$

$\therefore$ $$T = 5177.7\ N = R_A$$

Again considering the equilibrium of whole ladder and taking the components along vertical direction,

$$R_B + R_C = 220 + 750 + 220$$

$$R_C = 1190 - R_B$$

$$= 1190 - 970 = 220\ N$$

1.18 : *A rigid rod AB weighing 100N is 2 m long and carries a load of 800 N at C. It is hinged at point A. It is kept horizontal by applying an upward force P at B. Find the value of P.*

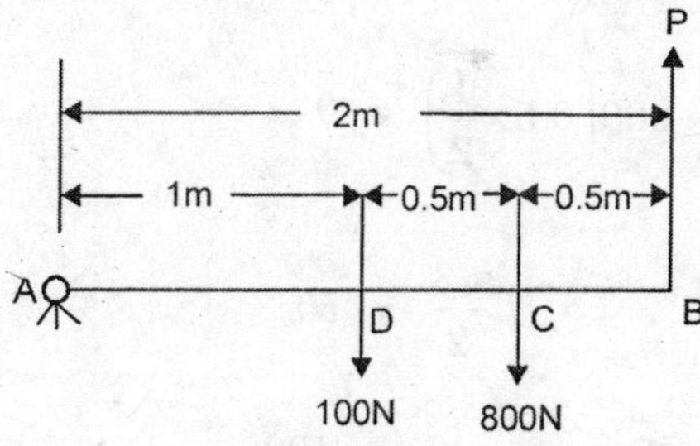

Fig. 1.42

Solution: The rod is in equilibrium, therefore taking moments about the hinge point A we get,

$$100 \times 1 + 800 \times 1.5 = P \times 2$$

or $$2P = 100 + 1200 = 1300$$

$\therefore$ $$P = \frac{1300}{2} = 650\ N$$

1.19 : *A frame picture AB is hung on a wall with the help of a string BC tied to a hook at C and resting on a batton at A. If the weight of the picture is 200N and BC is horizontal, find the tension in the string and reaction at A. The inclination of picture AB with the vertical is 30°. Take AB = 1.0 m (Refer Fig. 1.43)*

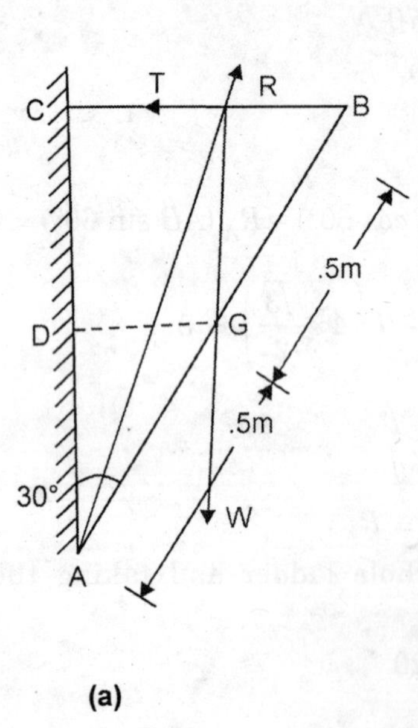

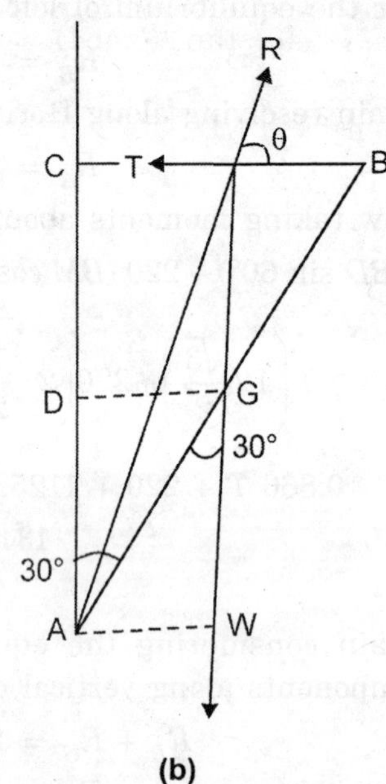

Fig. 1.43

Solution : Fig. 1.43 shows the free-body diagram of the picture. Taking moments of forces about the point A,

$$-T \times (AC) + W\,(GD) = 0$$

$$-T \times AB \cos 30° + W\,(AG) \sin 30° = 0$$

or
$$-T \times \frac{1 \times \sqrt{3}}{2} + 200\left(0.5 \times \frac{1}{2}\right) = 0$$

$$T\frac{\sqrt{3}}{2} = 50$$

$$T = \frac{100}{\sqrt{3}}$$

$$= 57.735\ N$$

Resolving the forces along the horizontal direction

$$T = R \cos \theta \qquad ...(A)$$

and RE solving the forces along vertical direction

$$W = R \sin \theta \qquad ...(B)$$

EXAMPLE

EXAMPLE

Dividing (*B*) and (*A*),

$$\tan\theta = \frac{W}{T} = \frac{200}{57.735}$$

$$= 3.464$$

$$\therefore \qquad \theta = \tan^{-1}(3.464) = 73.89°$$

Then from (*B*),

$$R = \frac{W}{\sin\theta} = \frac{200}{\sin 73.89}$$

$$= \frac{200}{0.9607} = 208.18 \text{ N}$$

1.20 : *A cord ABC 4.5 m long is tied to two hooks at A and C on the vertical walls as shown in Fig. 1.44. The walls are 3.5 m apart and point C is 1.5 m lower in level than point A. A ring carrying a load of 500 N can slide freely on the cord. Determine the location of the ring in terms of distances AD and DB, when it will be in equilibrium. Also, find the tension in the cord.*

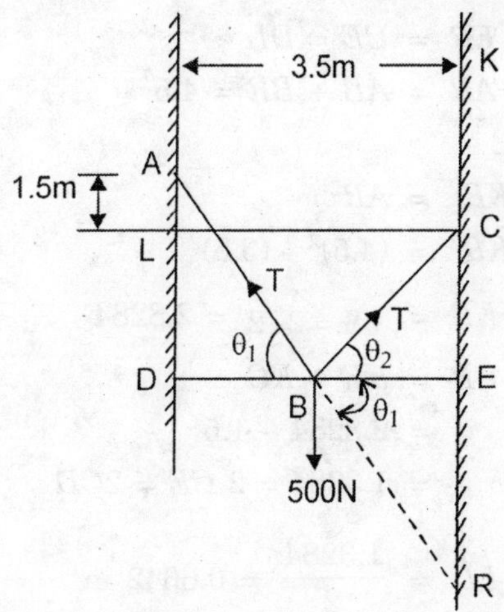

Fig. 1.44

Solution : At the equilibrium position of the ring, let the inclination of the two portions of the cord *AB* and *BC* be θ_1 and θ_2 respectively with the horizontal.

Then for the equilibrium of the ring, resolving the forces along the horizontal direction,

$$T\cos\theta_2 - T\cos\theta_1 = 0$$

i.e, $\cos\theta_2 = \cos\theta_1$

or $\theta_1 = \theta_2$ ...(*A*)

Again resolving the forces along the vertical direction we get,

$$T\sin\theta_1 + T\sin\theta_2 = 500$$

i.e., $2T \sin \theta_1 = 500$

or $$T = \frac{250}{\sin \theta_1} \quad ...(B)$$

From Δ *ABD* and Δ *BCE,* ...(*C*)

$$AB \cos \theta_1 = BD$$

and $$BE = AC \cos \theta_2 \quad ...(D)$$

Adding equations (*C*) and (*D*),

$$BD + DE = (AB + AC) \cos \theta \qquad (\because \theta_1 = \theta_2 = \theta)$$

$$3.5 = 4.5 \cos \theta$$

$\therefore$ $$\cos \theta = \frac{3.5}{4.5} = 0.7777$$

or $$\theta = \cos^{-1}(0.777)$$

$$\theta = 38.94°$$

Again Δ *BCE* and Δ *BRE* are congruent.

Therefore, $BC = BR$

and $ER = CE = DL$

or $AR = AB + BR = 4.5$

Now from Δ *AKR,*

$$AK^2 + KR^2 = AR^2$$

or $$KR^2 = (4.5)^2 - (3.5)^2$$

$$KR = \sqrt{8} = 2\sqrt{2} = 2.8284$$

$\therefore$ $$CR = KR - KC$$

$$= 2.8284 - 1.5$$

$$= 1.3284 = 2\ CE = 2CD$$

$\therefore$ $$LD = \frac{1.3284}{2} = 0.6642 \text{ m}$$

$\therefore$ $$AD = AL + LD = 1.5 + .6642$$

$$= 2.1642 \text{ m}$$

From Δ BAD, $$\tan \theta = \frac{AD}{BD}$$

or $$BD = \frac{AD}{\tan(38.94°)} = \frac{2.1642}{0.808}$$

$$= 2.6785 \text{ m}$$

From equation *B*, $$T = \frac{250}{\sin 38.94°} = \frac{250}{0.6285} = 397.77\ N$$

EXAMPLE

EXAMPLE

1.21 : *Two smooth cylinders, each of 150 mm diameter and 200 N weight are stacked in a channel shaped space with vertical walls 200 mm away as shown in Fig. 1.45. Find the reactions at A, B, C and D.*

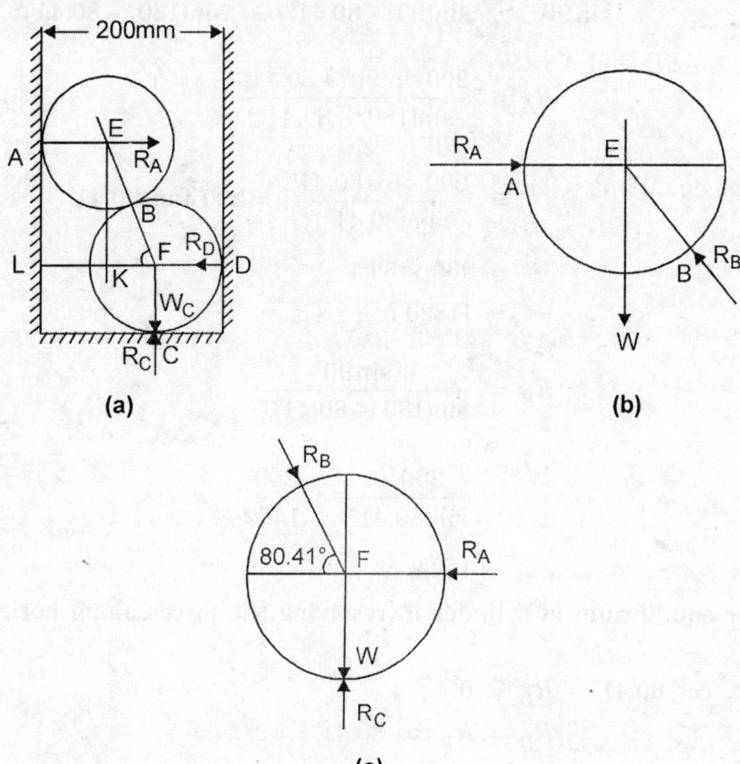

Fig. 1.45

Solution 1.21: Let the inclination of center line EF with the horizontal be θ.

Geometrically, $EF = 300$ mm

Then $KF = EF \cos\theta$

$KL = AE$ = radius of cylinder

$$= \frac{150}{2} = 75 \text{ mm}$$

Similarly, $FD = \dfrac{150}{2} = 75$

Therefore, $LD = LK + KF + FD$

or $200 = 75 + EF\cos\theta + 75$

or $\cos\theta = \dfrac{50}{300} = \dfrac{1}{6} = 0.1666$

or $\theta = \cos^{-1}(0.1666)$

$= 80.41°$

Considering the equilibrium of cylinder E and applying Lami's equation,

$$\frac{R_B}{\sin 90°} = \frac{R_A}{\sin(90° + 80.41°)} = \frac{W}{\sin(180° - 80.41°)}$$

$$\therefore \quad R_A = \frac{200 \sin(90° + 80.41°)}{\sin(180° - 80.41°)}$$

$$= \frac{200 \times \cos 80.41°}{\sin(80.41°)} = 200 \tan 80.41°$$

$$= 200 \times 59.1$$

$$= 11820\ N$$

$$\text{and} \quad R_B = \frac{W \sin 90°}{\sin(180° - 80.41°)}$$

$$= \frac{200 \times 1}{\sin(80.41°)} = \frac{200}{.1462}$$

$$= 136.8\ N$$

(*b*) For equilibrium of cylinder F, resolving the forces along horizontal line,

$$R_B \cos 80.41° - R_D = 0$$

$$\therefore \quad R_D = R_B \cos 80.41°$$

$$= 136.8 \times 0.1666$$

$$= 22.79\ N$$

Resolving forces along vertical direction,

$$R_B \sin 80.41° + W - R_C = 0$$

$$R_C = 130.8 \sin 80.41° + 200$$

$$= 134.09 + 200 = 334.09\ N$$

1.22 : *Three cylinders, each weighing 500 N and of 200 mm diameter are arranged in such a manner that A and C are placed on a smooth horizontal table and touch each other at D and the third cylinder B is placed over then as shown in Fig. 1.46. To check the sideward movement of the cylinders A and C, a string is lied to the axes of A and C. Find the reactions R_1 and R_2 of the table, the reaction among the cylinders and the tension in the string.*

Solution: Let the tension in the string be T. Since the purpose of the string is not to allow the cylinders A and C to move sidewards, There will be no reaction behind these cylinders, then the five body diagrams of the rollers will be as given in Fig. (1.46b).

EXAMPLE

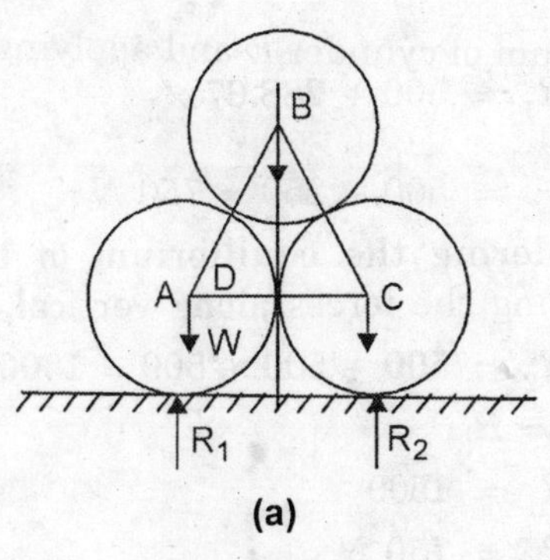

(a)

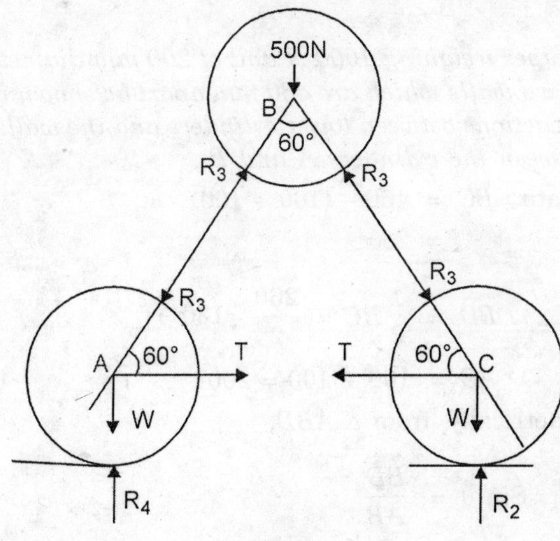

(b)

Fig. 1.46

Considering the equilibrium of cylinder A and applying the Lami's equation we get,

$$\frac{W}{\sin 60^\circ} = \frac{R_3}{\sin 150^\circ}$$

$$R_3 = \frac{500 \times 2}{\sqrt{3}} \times \frac{1}{2} = 2.88.67\ N$$

(*b*) For the equilibrium of cylinder A and resolving the forces along horizontal direction,

$$T = R_3 \cos 60^\circ$$

$$= \frac{500}{\sqrt{3}} \times \frac{1}{2} = 144.33N$$

Again resolving the forces along vertical,

$$R_1 - W - R_3 \sin 60 = 0$$

$$R_1 = 500 + 288.67 \times \frac{\sqrt{3}}{2}$$

$$= 500 + 250 = 750\ N$$

Alternatively considering the equilibrium of the three cylinders collectively and resolving the forces along vertical,

$$R_1 + R_2 = 500 + 500 + 500 = 1500$$

and R_m symmetry $R_1 = R_2$

$\therefore \quad 2R_1 = 1500$

or $\quad R_1 = 750\ N$

1.23 : *Three pipes weighing 1000 N and of 200 mm diameter, each are stacked within two walls which are 460 mm apart as shown in Fig. 1.47. Determine the reactions between lower cylinders and the walls. Also, find the reaction between the cylinders A and B.*

Solving given data $BC = 460 - (100 + 100)$

$$= 260$$

$$\therefore \quad BD = \frac{1}{2} BC = \frac{260}{2} = 130$$

$$AB = 100 + 100 = 200$$

Solution: Geometrically, from $\Delta\ ABD$,

$$\text{Sin } \theta = \frac{BD}{AB}$$

$$= \frac{130}{200} = 0.65$$

$$\therefore \quad \theta = \sin (0.65)$$

$$= 40.54°$$

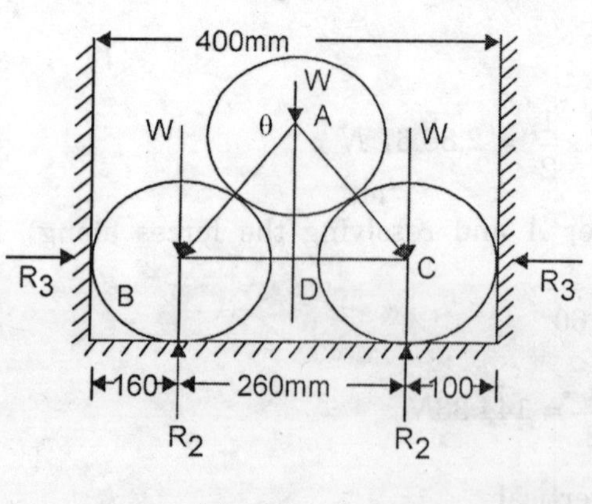

(a)

(b)

Fig. 1.47

For the equilibrium of cylinder A and resolving the forces vertically,

$$W = 2R_1 \cos\theta$$

$$\therefore \quad R_1 = \frac{W}{2\cos\theta} = \frac{1000}{2\times\cos(40.54°)}$$

$$= 657.94\ N$$

Considering the free-body diagram of cylinder B and resolving the forces in vertical direction,

$$W - R_2 + R_1 \cos\theta = 0$$

$$1000 + 657.94\cos(40.54) = R_2$$

$$1000 + 500 = R_2$$

$$R_2 = 1500$$

Again resolving the forces along horizontal direction,

$$R_3 - R_1 \sin\theta = 0$$

$$R_3 = R_1 \sin(40.54)$$

$$= 657.94 \times .65$$

$$= 427.66\ N$$

1.24 : *A uniform rod 300 mm long weighing 50 N is placed in a smooth hemispherical bowl physical level of radius 100 mm. Find (a) the inclination of the rod axis with the horizontal when the rides in equilibrium with the reaction R_B and R_D.*

Solution : In Fig. 1.48 $ADBC$ is the hemispherical bowl of radius = AC = CB = 100 mm

The rod DE of length 300 mm is placed in the bowl and is in equilibrium under the action of 3 forces viz. weight W, Reactions R_B and R_D

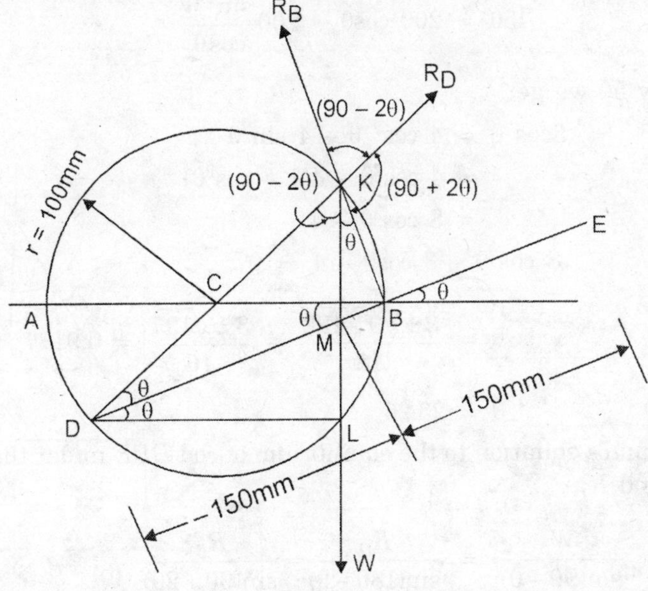

Fig. 1.47

The reaction R_D will be along the radial line DC and the reaction R_B will be perpendicular to the rod DBE. Completing the circle of which the bowl is part we notice that the point of intersection of R_B and R_D, is point K, will be such that DCK is a diameter because $\angle DBK = 90°$ and B is lying on the circle. Evidently the third force, W, will be vertical and pass through K. Thus, kM will be along the vertical direction, where M is the mid-point of rod DE.

Let the inclination of the rod with horizontal be θ.

Thus, $\angle EBX = \angle BDL = \theta$

Again in $\Delta\, CDB$ $CD = CB$

$\therefore$ $\angle CDB = \angle CBD = \theta$

In ΔDBK, DCK is diameter,

$\therefore$ $\angle DBK = 90°$

$$KB = DK \sin\theta = 200 \sin\theta \quad ...(A)$$

and

$$DB = DK \cos\theta = 200 \cos\theta \quad ...(B)$$

Since BK is perpendicular to OB and MK is perpendicular to AB,

$$\angle MKB = 0$$

Therefore, from traingle MKB,

$$BM = KB \tan\theta \quad ...(B)$$

$$= 200 \sin\theta \tan\theta$$

$$= 200\frac{\sin^2\theta}{\cos\theta} \quad ...(C)$$

Now

$$DM = DB - MB$$

$$150 = 200\cos\theta - 200\frac{\sin^2\theta}{\cos\theta}$$

Dividing by 50 we get,

$$3\cos\theta = 4\cos^2\theta - 4\sin^2\theta$$

$$= 4\cos^2\theta - 4(1 - \cos^2\theta)$$

$$= 8\cos^2\theta - 4$$

i.e.,

$$8\cos^2\theta - 3\cos\theta - 4 = 0$$

$$\cos\theta = \frac{3 \pm \sqrt{9+128}}{16} = \frac{3 \pm \sqrt{137}}{16} = 0.9187$$

$\therefore$

$$\theta = 23.26°$$

Applying Lami's equation to the equilibrium of rod DBE under the action of W, R_B and R_D,

$$\frac{W}{\sin(90-\theta)} = \frac{R_D}{\sin(180-\theta)} = \frac{R_B}{\sin(90+2\theta)}$$

EXAMPLE

$$\frac{W}{\sin(90-\theta)} = \frac{R_D}{\sin\theta} = \frac{R_B}{\cos 2\theta}$$

$$\therefore \quad R_B = \frac{50\cos 2\theta}{\cos\theta} = \frac{50\times\cos(46.52°)}{\cos(23.26°)}$$

$$= \frac{50\times 0.6881}{0.9187} = 37.45\ N$$

and
$$R_D = \frac{W}{\cos\theta}\sin\theta = W\tan\theta$$

$$= 50 \times 0.4298 = 21.49\ N$$

1.25 : *A uniform bar AB of negligible weight, 1m long carries two weights W_1 = 100N and W_2 = 200 N as shown in Fig. 1.49. The ends A and B of the rod rest on two smooth inclined planes at 45° and 30° respeetinely. The weight W_1 is at a distance 300 mm from,*

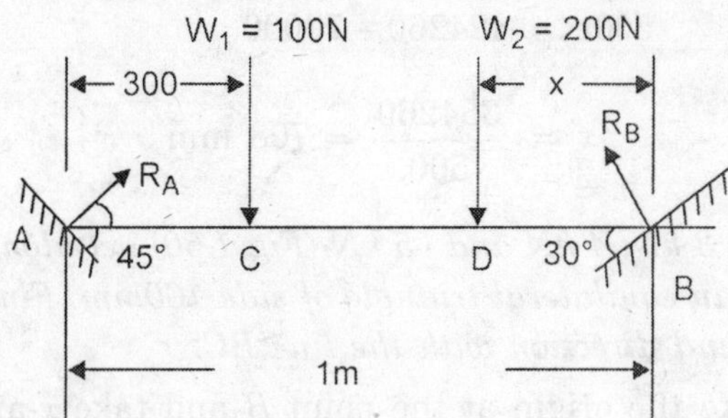

Fig. 1.49

the end A while the weight W_2 is at a distance x from B. Find the value of x if the rod AB is horizontal.

Solution : The bar AB is in equilibrium under the action of four forces namely R_A, R_B, W_1 and W_2 and the reactions R_A and R_B will be normal to the respective surfaces. Then, resolving the forces along horizontal and vertical directions.

Along Horizontal direction, $R_A \cos 45° = R_B \sin 30°$

or
$$\frac{R_A}{\sqrt{2}} = \frac{R_B}{2}$$

or
$$R_B = R_A\left(\frac{2}{\sqrt{2}}\right) = \sqrt{2}R_A \qquad ...(A)$$

Along vertical direction,

$$W_1 + W_2 = R_A \sin 45° + R_B \cos 30°$$

$$100 + 500 = \frac{R_A}{\sqrt{2}} + \frac{R_B}{2}$$

$$= \frac{R_A}{\sqrt{2}} + \frac{\sqrt{2}R_A}{2} \quad \text{(From (A))}$$

$$600 = \sqrt{2}R_A$$

$$\therefore \quad R_A = \frac{600}{\sqrt{2}} = 424.26$$

From (A), $$R_B = \sqrt{2}R_A = \sqrt{2} \times \frac{600}{\sqrt{2}} = 600\text{N}$$

(*b*) Taking moment about point *B*,

$$R_A(\sin 45°)AB - W_1\,(CB) - W_2(x) = 0$$

$$424.26 \times 1000 - 100 \times 700 - W_2 x = 0$$

i.e, $$500x = 424260 - 70000$$

$$x = \frac{354260}{500} = 708 \text{ mm}$$

1.26 : *Forces 3 kN, 4 kN and –5 kN (Fig 1.50) act along the sides AB, BC and CA of an equilateral triangle of side 100mm. Find the resultant in magnitude and direction with the line BC.*

Solution : Take the origin at the point *B* and take *x*-axis along *BC*,

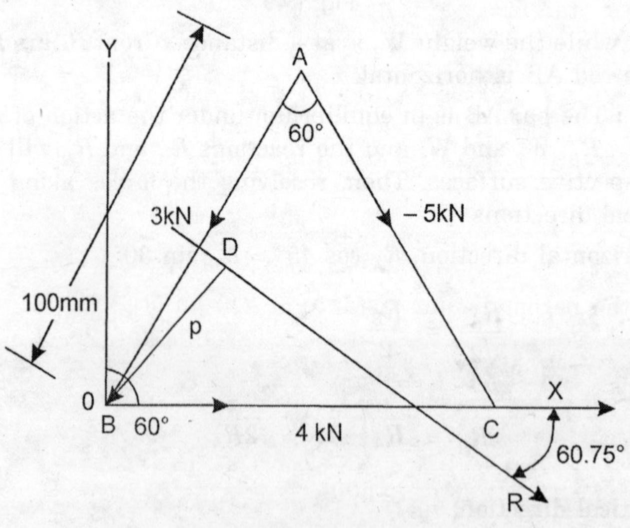

(a)

E X A M P L E

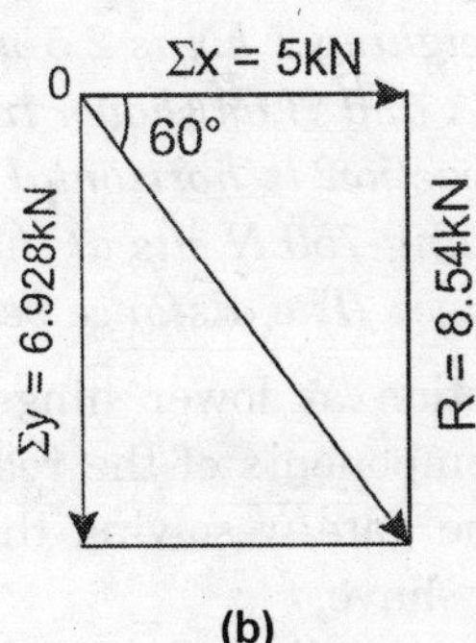

(b)

Fig. 1.50

Resolving the forces along x-axis and Y-atis,

$$\Sigma X = 4.0 + 5 \times \cos 60° - 3\cos 60$$

$$= 4 + 5 \times \frac{1}{2} - \frac{3}{2}$$

$$= 5 \; kN$$

$$\Sigma Y = -3 \cos 30° - 5\cos 30°$$

$$= -8 \times \frac{\sqrt{3}}{2} = -4\sqrt{3} = -6.928 \; kN$$

$$R^2 = \Sigma X^2 + \Sigma Y^2$$

$$= (5.2 + (-6.925)^2$$

$$25 + 48 = 73$$

$$\therefore \quad R = \sqrt{73} = 8.544 \; kN$$

$$\tan \theta = \frac{\Sigma Y}{\Sigma X} = \frac{-6.928}{5} = -1.7856$$

$$\therefore \quad \theta = 60.75°$$

(*b*) Taking moments about B,

$$\Sigma M = 0 \times 3 \; kN + 0 \times 4 \; kN + 5 \; (100 \sin 60°)$$

$$= 500 \times \frac{\sqrt{3}}{2}$$

If p is the perpendicular distance of R from origin B.

$$p \times R = 500 \times \frac{\sqrt{3}}{2}$$

$$p = 500 \times \frac{\sqrt{3}}{2} \times \frac{1}{8.544}$$

$$= 50.68 \text{ mm}$$

EXAMPLE

1.27 : *A factory gate weighing 1 kN is 2.5 m high and 3 m wide. It is supported on two hinges A and B as shown in Fig. 1.51. Assuming that on the lower hinge the reaction is horizontal, find the reactions at the hinges when a man weighing 750 N sits at the gate 700 mm away from the vertical line of the hinges. The distance between the hinges is 1.8 m.*

Solution : Let the reaction at lower hinge be R_B *At* hinge *A*, the horizontal and vertical components of the reaction R_A are R_{Ax} and R_{Ay} For the equilibrium of the gate, resolving the forces in the horizontal and vertical directions we have,

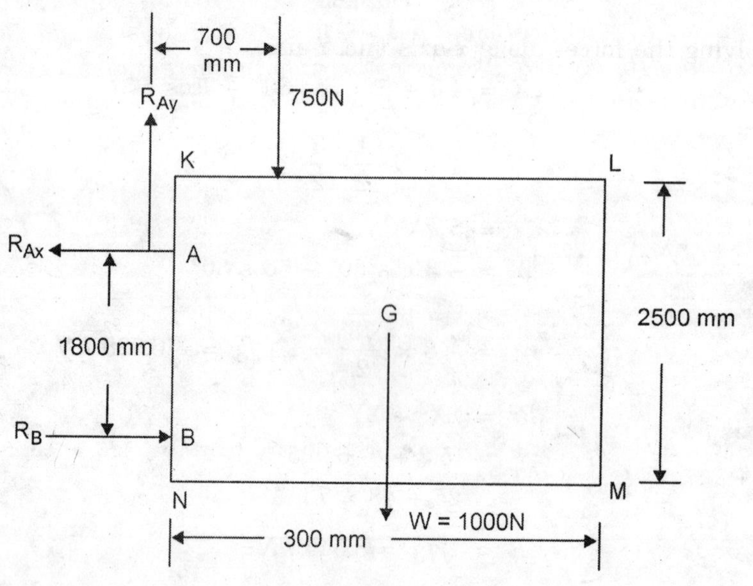

Fig. 1.51

$$R_{Ax} = R_B \quad ...(A)$$

$$R_{Ay} = W_1 + W_2$$

$$= 1000 + 750 = 1750\ N \quad ...(B)$$

Now taking moments about the point *A*,

$$R_B(1800) = 1000 \times 150 + 750 \times 700$$

$$= 15 \times 10^4 + 52.5 \times 10^4$$

$$= 67.5 \times 10^4 \text{ N}$$

$$\therefore \quad R_B = \frac{67.5 \times 10^4}{1800} = 375 \text{ N}$$

Then from equation *A*,

$$R_{Ax} = R_B = 375\ N$$

EXAMPLE

E X A M P L E

Therefore, resultant reaction at hinge A,

$$R_A = \sqrt{R_{Ax}^2 + R_{Ay}^2} = \sqrt{(375)^2 + (1750)^2}$$

$$= \sqrt{140625 + 3062500}$$

$$= \sqrt{3203125}$$

$$R_A = 1789.73\ N$$

1.28 : *A 6m beam AB is hinged at the left hand support A and at the end B it is supported on rollers so that the reaction at end B is vertical It is loaded by a vertical point load of 10 kN at 2m from A and inclined point load of 15 kN at a distance of 3m from A. There is also a uniformly distributed load over 2 m length of beam (EB) as shown in fig 1.52. Find the reaction at A and B.*

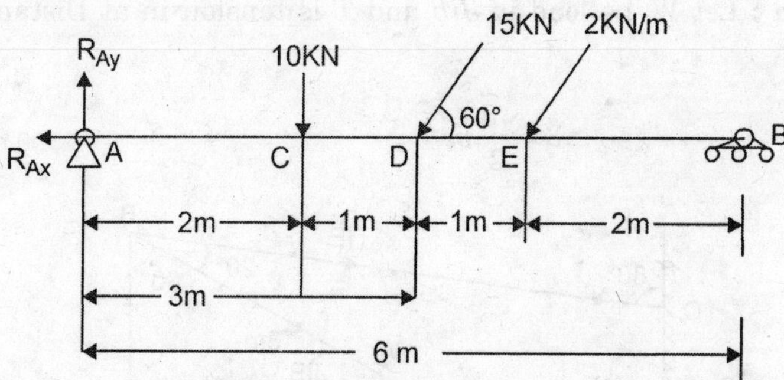

Fig. 1.52

Solution : Let R_{Ax} and R_{xy} be the horizontal and vertical components of reaction R_A. and R_B is the reaction at point B.

For the equilibrium of the beam, resolve the forces along horizontal and vertical directions,

$$R_{Ax} = 150 \cos 60° = 7.5\ kN \quad ...(A)$$

$$R_{Ay} + R_B = 10 + 15 \sin 60° + (2 \times 1)$$

$$= 10 + 12.99 + 2 = 24.99\ kN \quad ...(B)$$

Taking moments about point A,

$10 \times 2 + 15 \sin 60° \times 3 + (2 \times 1)(5 + 0.5) = R_B \times 6$

$$\therefore \quad R_B = \frac{1}{2}[20 + 38.97 + 11]$$

$$\text{or} \quad R_B = \frac{69.97}{6} = 11.66\ kN \quad ...(C)$$

From equation (B) and (C),

$$R_{Ay} = 24.99 - 11.66$$
$$= 13.33\ kN$$

Thus,
$$R_A = \sqrt{(R_{Ax})^2 \times (R_{Ay})^2}$$
$$= \sqrt{(7.5)^2 + (13.33)^2}$$
$$= \sqrt{56.25 + 177.6889}$$
$$= \sqrt{233.94} = 15.29\ kN$$

1.29 : *The Jib of the Jib crane while lifting a load of 15 kN. is inclined to the vertical post of the crane at 60°. The length of the Jib is 3m . The tie is inclined at 80° with the vertical (Fig. 1.53). Find the load carried by the Jib and the tension in the Jib.*

Solution : Let W_j be load in *Jib* and T is tension in at Distance AD = 3 cos 30°

$$= 3 \times \frac{\sqrt{3}}{2} m$$

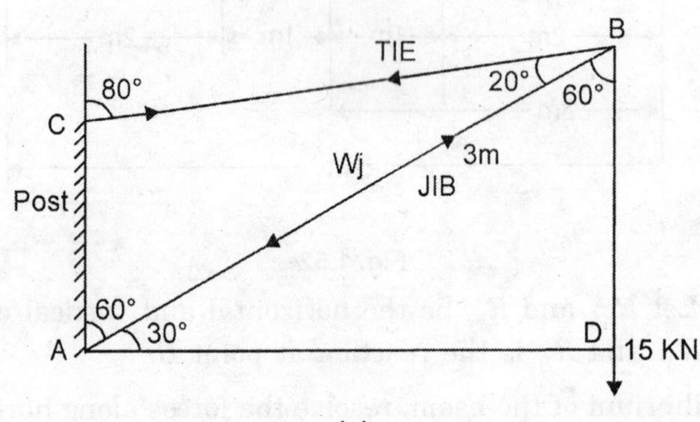

(a)

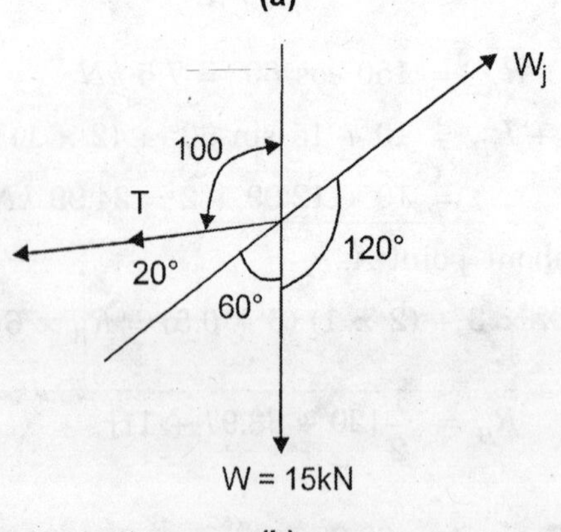

(b)

Fig. 1.53

E X A M P L E

E X A M P L E

For the equilibrium of forces at B. Applying Lami's theorem,

$$\frac{W}{\sin 160°} = \frac{W_j}{\sin 80°} = \frac{T}{\sin 120°}$$

$$\therefore \quad W_j = \frac{15 \times \sin 80°}{\sin 160°}$$

$$= \frac{15 \times 0.9848}{0.0349} = 42.33\ kN$$

and
$$T = \frac{15 \times \sin 120}{\sin 160}$$

$$= 15 \times \frac{\sqrt{3}}{2} \times \frac{1}{.0349} = 372.2\ kN$$

1.30 : *The beam AB of a mobile crane (Fig. 1.54) is 10m long and* weighs 1.5 kN with its center of gravity G at a distance of 5 m . A. *At certain point of time it lifts a load 5 kN. Find out the position shown (a) the tension in the rope BC, (b) the reaction at the hinge point A of the beam.*

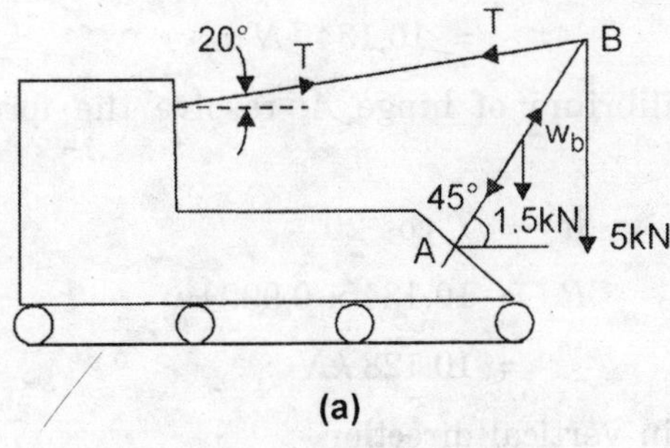

(a)

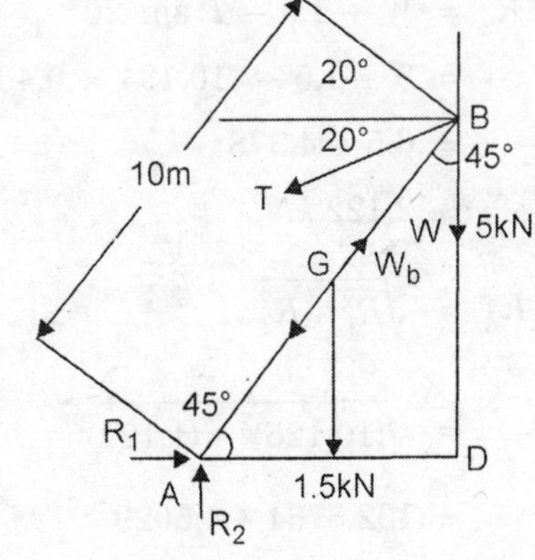

(b)

E X A M P L E

Solution : fig 1.54 (b) shows the free body diagram of the beam Let R_1 and R_2 be the horizontal and vertical components of the reaction R_A at the point *A*. and the tension the tie rope be *T*.

Then for the equilibrium of the beam, taking moments about the point *A* we get,

$$(T \sin 20)(AD) - (T \cos 20)(BD) + (1.5)(5 \cos 45°) + 5(10 \cos 45°) = 0$$

$$T(0.4320)(10 \cos 45°) - T(.9994)(10 \sin 45°) + \left(7.5 \times \frac{1}{\sqrt{2}}\right) + 50 + \frac{1}{\sqrt{2}} = 0$$

$$T\left[\left(4.320 \times \frac{1}{\sqrt{2}}\right) - 9.994 \times \frac{1}{\sqrt{2}}\right] + \frac{57.5}{\sqrt{2}} = 0$$

or $$\frac{-5.674}{\sqrt{2}}T + \frac{57.5}{\sqrt{2}} = 0$$

or $$T = \frac{57.5}{5.674}$$

$$= 10.134 \; kN$$

(b) For the equilibrium of hinge *A*, resolve the forces in horizontal direction,

$$R_x = T \cos 20°$$

or $$R_1 = 10.134 \times 0.9994$$

$$= 10.128 \; kN$$

Again resolving in vertical direction,

$$R_2 = W + 1.5 - T \sin 20°$$

$$= 5 + 1.5 - 10.134 \times 0.432$$

$$= 6.5 - 4.378$$

$$= 2.122 \; kN$$

Therefore, $$R_A = \sqrt{R_1^2 + R_2^2}$$

$$= \sqrt{(10.125)^2 + (1.122)^2}$$

$$= 102.5764 + 4.5029$$

$$= \sqrt{107.08}$$

$$R_A = 10.348 \; kN$$

REVIEW QUESTIONS

1.1. State Newton's Law of motion.

1.2. State Newton's Law of gravitation and hence, derive the value of 'g' the acceleration due to gravity.

1.3. Explan the principle of transmissibility of forces.

1.4. State the triangle law of forces and show how the Lami's theorem can be derived from this.

1.5. Show that the moment of a forces about a point is twice the area of triangle formed by joining the force vector to the point about which moment is taken.

Here prove the varignons theorem.

PROBLEM FOR PRACTICE

1. A ladder *AB* 4m long, weighing 500 *N* rests, at a smooth Floor and against a smooth wall. *A* man weighing 750 *N* climbs on the ladder upto point *C* which is 3m from the end *A* which is on the floor. The inclination of the ladder is 60° with the horizontal. Find the least magnitude of the horizontal force to be applied at the point *A* so that the ladder is prevented from slipping Also, find the reactions at the points *A* and *B*. (**Ans.** 469.1 *N*; R_A = 1250*N*, R_B = 469.1 *N*)
2. Three cylinders, each weighing 100*N* and each of 150mm diameter are resting in a channel of width 375 mm as shown in Fig. 1.55 Determine the reaction at the points *D*, *E K* and *L*. (**Ans.** $R_D = R_E$ = 56.7 *N*; $R_K = R_L$ = 150 *N*)

Fig. 1.55

3. A wheel weighing 1500 *N* and of 500 mm diameter rests against a rectangular block 200 mm high lying on the ground. It is to be pulled over the block by applying a force F at the end of a rope wound around the circumference of the wheel (Fig. 1.56). Find the value of the force F which will be sufficient just to roll the wheel over the block. Assume there is no friction. (**Ans.** 1225 *N*)

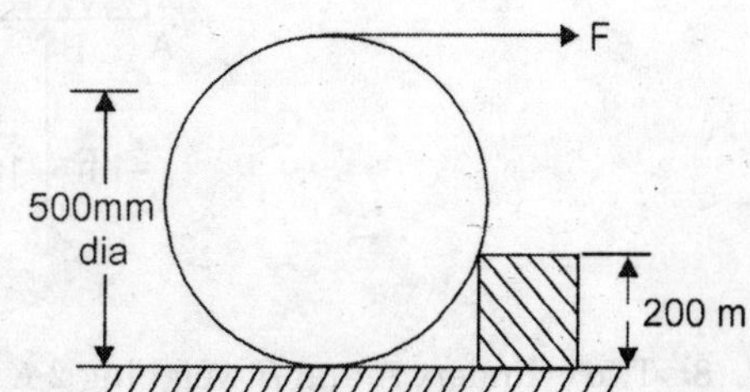

Fig. 1.56

4. In problem no. 3, if the force *F* is applied in the horizontal direction at the centre of the wheel,then what will be the value of *F*? (**Ans.** 7348.47 *N*)

5. Two equal cylinders each of 200 mm diameter and weight 100 N are stored in a channel shaped space 250 mm in width (Fig. 1.57). Find the reaction. R_B, R_A and R_D

(**Ans.** R_B = 200N, R_A = 150 N , R_D = 112 N)

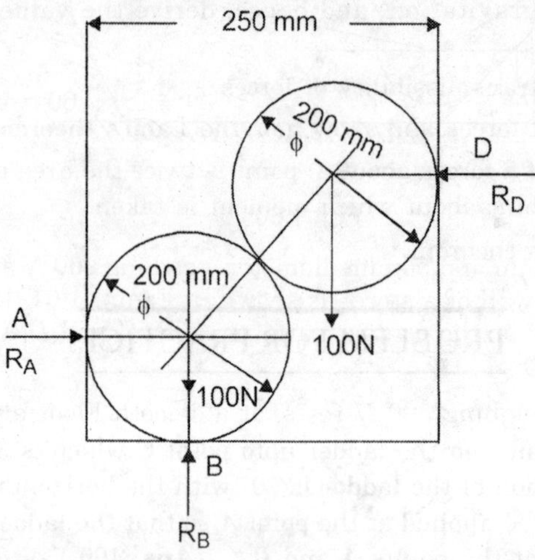

Fig. 1.57

6. A beam simply supported at its end is loaded as shown in Fig. 1.58. Find the reactions at the support. **(Ans. R_A = 6.5 kN, R_B = 9.5 kN)**

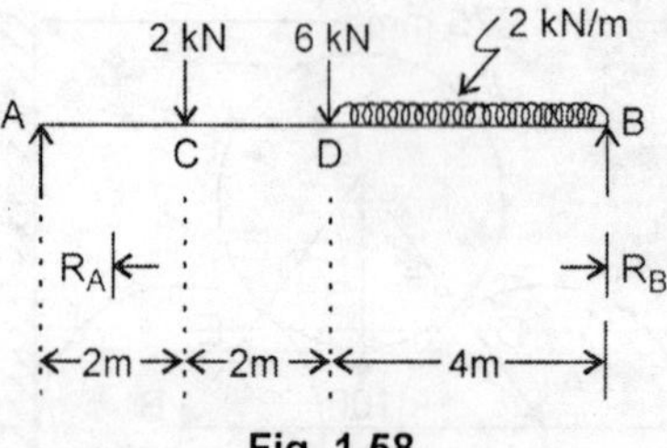

Fig. 1.58

7. An overhanging beam ABCDE is supported at points B and D. The loading is shown in figure 1.59. Find the reaction at the two supports. (**Ans.** R_B = 80 N, R_D = 80 N)

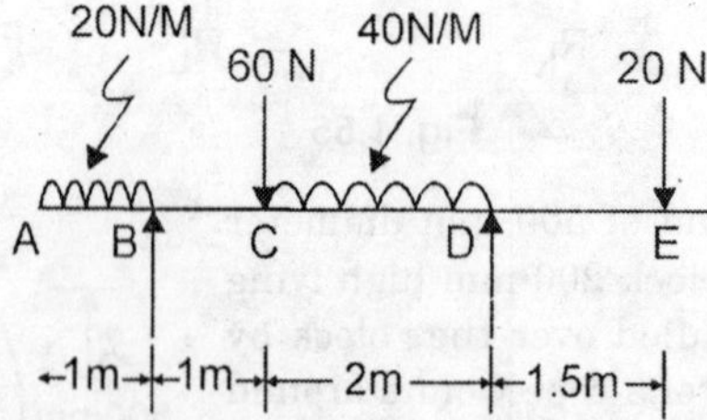

Fig. 1.59

8. Two cylinders A and B weighing 2 KN and 4 kN respectively rest an two smooth inclined planes as shown in Fig. (1.60). These cylinders are connected to each other by a rigid bar of negligible weight hinged to the cylinders at their respective ends by smooth pins. Find the magnitude of a force F to be applied at the axis of cylinder A as shown in figure 1.60 so as to hold the system of the cylinders in the position shown. (**Ans.** 1071.7)

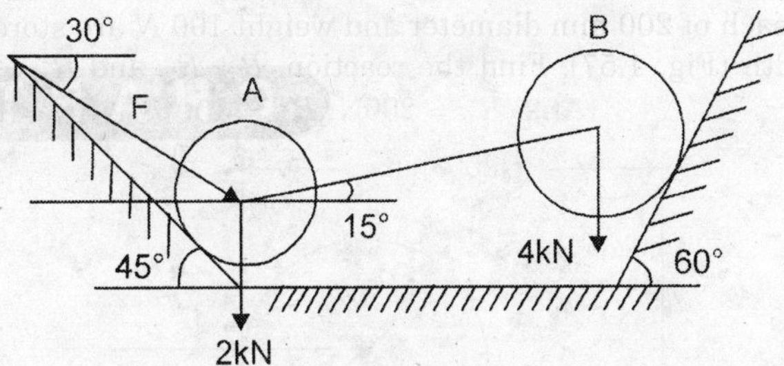

Fig. 1.60

9. Two cylindrical pipes 100 mm and 50 mm diameter weighing 200 *N* and 50 *N* are stacked within the brackets fitted on the wall of a store, as shown in Figure 1.61. If the surfaces are assumed smooth, find the reactions at the contact surfaces *A*, *B*, *C* and *D*[Ans. R_A = 37.5 *N* R_B = 62.5 *N*, R_C = 287.5, R_D = 353.5 *N*.

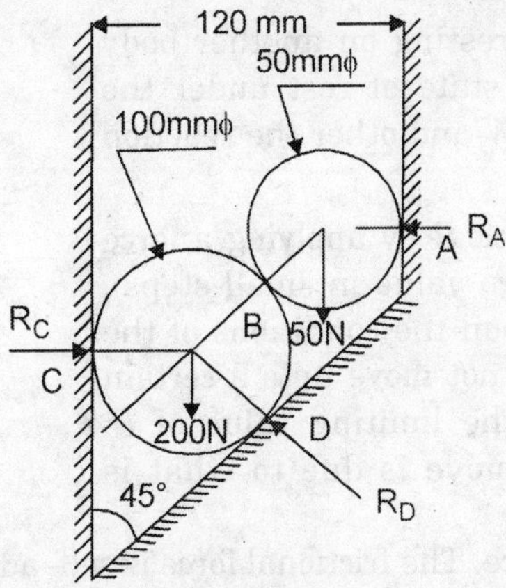

Fig. 1.61

10. A plane lamina in the shape of an aquilateral triangle of side 200 mm is acted upon by four forces of magnitude 30 *kN*, 50 *kN*, 50 *kN* and 60 *kN* as shown in Fig. 1.62. Find the magnitude and direction of the resultant of these forces and give the position of the resultant force vector.

(**Ans.** *R* = 57.3541 *kN* θ = 6.70° and at 11.51mm from *A* 11.51 mm)

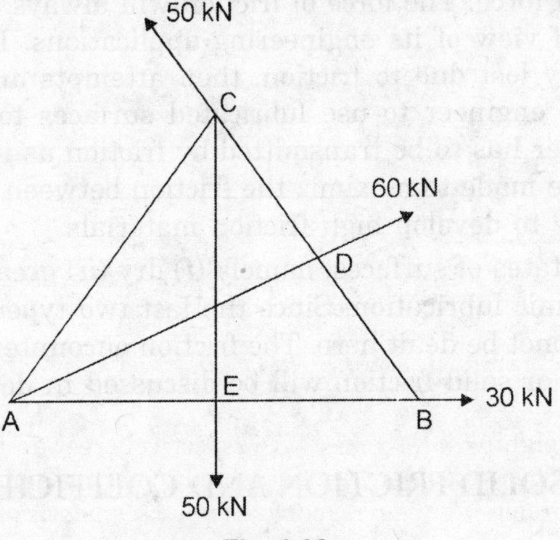

Fig. 1.62

❑❑❑

CHAPTER 2

FRICTION

2.1 INTRODUCTION

Let a block B of weight W be resting on another body A. It will be in equilibrium or state of rest under the action of two forces, one weight W and other the reaction R of the body A on block B.

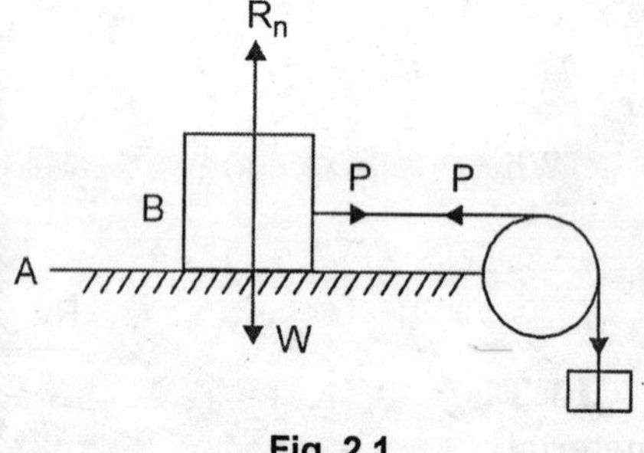

Fig. 2.1

If we now try to pull the block B by applying a force P (gradually increasing from zero value in small steps), it is observed that depending upon the conditions of the mating surfaces the block B will not move upto a certain value of the magnitude of P. The limiting value of P, when the block just starts to move is due to what is called the limiting value of frictional force. The frictional force is self–adjusting upto the limiting value.

This phenomenon exists primarily due to the presence of hills and valleys called the asperities present on the mating surfaces. These may not be visible with naked eye but can be seen under microscope. When the mating surfaces of two bodies slide with respect to each other, the asperities of one surface get interlocked with those of the other surface and thus offer resistance to the free movement of one with respect to the other surface. This resisting force is called the frictional force. The force of friction will always oppose the motion. It has two facts from the point of view of its engineering applications. For the first aspect when power transmitted is partly lost due to friction, then attempts are made to minimise this powerloss which leads the engineer to use lubricated surfaces to reduce friction. For the other aspect when the power has to be transmitted by friction as in belt and rope drives, or friction drives, attempts are made to maximiz the friction between the mating surfaces, this in term has paved the way to develop high friction materials.

There are three main states of surfaces, namely (*i*) dry (*ii*) greasy or partially lubricated (*iii*) full film or hydrodynamic lubrication. Since the last two types are beyond the scope of present syllabus, these will not be dealt here. The friction encountered in relative sliding dry surfaces called dry friction or solid friction will be discussed in detail.

2.2 LAWS OF DRY OR SOLID FRICTION AND COEFFICIENT OF FRICTION

The laws of dry friction are the outcome of the extensive experimental investigations. These are given below:

(i) The frictional force always acts in the direction opposite to that in which the body tends to move.

(ii) The force of friction is independent of the area of contact between the mating surfaces.

(iii) The frictional force is directly proportional to the normal reaction between the surfaces in contact.

(iv) The functional force depends upon the nature of the materials of the surfaces in contact.

(v) The force of friction is independent of the relative velocity of sliding of one surface with respect to the other surface.

Let R_n be the normal reaction between the contact surfaces of two bodies (Fig. 2.1) and F is the force at which the body B just starts sliding over the body A. Then, as per the third law as stated above, the force F will be proportional to R_n.

Mathematically,

$$F \propto R_n$$

or

$$F = \mu R_n \qquad \text{...(2.1a)}$$

Where μ is a constant and is called the coefficient of friction.

or

$$\mu = \frac{F}{R_n} \qquad \text{...(2.1b)}$$

In Table (2.1) are given the average values of co–efficient of friction μ for different materials. These values are based on experimental observations.

Table 2.1. Average values of μ for various commonly used materials

S.No.	Material	μ (coefficient of friction)	
		Dry	**Greased**
1.	Mild Steel on Mild Steel	0.57	0.1 to 0.18
2.	Hardened Steel on Hardened Steel	0.41	0.9 to 0.13
3.	Hardened Steel on Babit Metal	0.34 to 0.35	0.06 – 0.18
4.	Mild Steel on Lead	0.95	0.29
5.	Mild Steel on C.I.	0.25	0.13
6.	CI on CI	0.15	0.06 – 0.07
7.	Bronze on CI	0.22	0.08
8.	Wood on Wood	0.45	—
9.	Leather on Wood	0.3 to 0.5	—
10.	Tyre rubber on cement concrete Road	0.64 to 0.85	—

2.3 LIMITING ANGLE OF FRICTION

Fig. 2.2 shows which is a body B in contact with another body A. The body B is free to slide over the body A and the surfaces are dry without any lubricant. Due to the weight of the body B (say W), there will be normal reaction of the body A on body B equal to W.

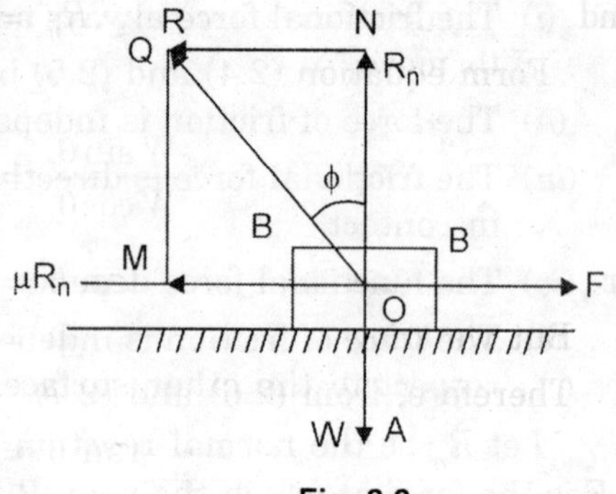

Fig. 2.2

Let a horizontal force, F, which is gradually increasing from zero value, be applied on the body B. In the beginning there will be no relative motion between them. However, when the magnitude of the force F is equal to the frictional force, which is acting in the opposite direction to F, the body B will be in static equilibrium position at this stage which is called the limiting value of frictional force. It may be noted that due to the presence of body A, there are two forces acting on the body B *viz.* normal reaction R_n, and frictional force $F_r = \mu R_n$. The resultant of R_n and $\mu\, R_n$ is called the resultant reaction of body A on Body B and is represented by R (Fig. 2.2). Let the inclination of R with the normal reaction be ϕ.

Then
$$\tan\phi = \frac{RN}{ON} = \frac{F_r}{R_n} = \frac{\mu R_n}{R_n} = \mu = \text{constant} \qquad ...(2.2a)$$

As the body B was just on the point of moving with respect to body A, $\mu\, R_n$ is therefore, the limiting frictional force. Accordingly ϕ is known as the limiting angle of friction.

If value of μ is known, the limting angle of friction ϕ is calculated from,

$$\phi = \tan^{-1}(\mu) \qquad ...(2.2b)$$

2.3.1 Corollary-I

In case $\tan\phi < \mu$, then the body B will move over the body A irrespective of the magnitude of the force F.

2.3.2 Corollary-II

In case $\quad \tan\phi < \mu$

or $\quad \phi < \tan^{-1}(\mu) \qquad ...(2.3a)$

Then no motion of Body B on Body A is possible, how-so-ever large the value of F may be.

2.3.3 Corollary-III

Tan ϕ can not be greater than μ or ϕ can not be greater than $(\tan^{-1}\mu) \qquad ...(2.3b)$

2.4 ANGLE OF REPOSE AND LIMITING ANGLE OF FRICTION

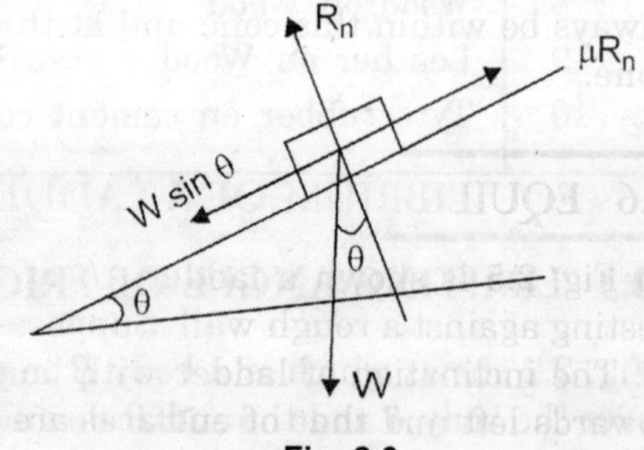

Fig. 2.3

In Fig. 2.3 in shown the body B is resting on a plane surface of body A. B is free to slide on the plane surface of A. At any instant, the inclination of the with the horizontal plane is say θ. If the angle θ is increased, gradually a stage will come when the body B will just start moving down the plane and at that instant the frictional force $F = R_n\,\mu$, where μ, = co–efficient of friction. Resolving the forces acting on the body A along and normal to the plane we get for equilibrium,

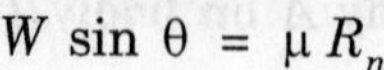

$$W\sin\theta = \mu\, R_n \qquad ...(2.4)$$

and $$R_n = W \cos \theta \quad ...(2.5)$$

Form equation (2.4) and (2.5) by dividing,

$$\frac{W \sin\theta}{W \cos\theta} = \frac{\mu R_n}{R_n} = \mu$$

or $$\tan \theta = \mu \quad ...(2.6)$$

But we know $$\mu = \tan \phi \quad ...(2.7)$$

Therefore, from (2.6) and (2.7),

$$\tan \theta = \tan \phi$$

or $$\theta = \phi \quad ...(2.8)$$

Thus, when the plane is inclined at angle equal to friction angle the body B will start down At this situation,

$$\theta = \tan^{-1} (\mu) = \phi \quad ...(2.9)$$

At the limiting value of θ at which the plane is inclined to the horizontal is called the angle of repose. From equation (2.9), it is evident that angle of repose = ϕ = Friction angle.

2.5 FRICTION CONE

In Fig. 2.4 is shown a block B of weight W lying on a table A. If F is the limiting force which, while gradually increasing from zero value, just causes the body B move along X-axis lying on the $X - Y$ plane. Then R_1 is the resultant frictional force which is inclined at friction angle $\phi = (\tan^{-1} \mu)$ on the left side of the Z-axis. However, if the applied force is directed towards OX', then the resultant frictional force R_2 will be directed at angle ϕ to the right of Z-axis. Similarly, if the direction of applied force vector F is changed from zero to 360°, still F vector lying in the horizontal plane X-Y, the resultant frictional force will go on rotating about Z-axis still making an angle ϕ with it. Thus, the surface generated by the resultant frictional force vector will be the surface of a cone with its apix lying at origin O and OZ as the axis of cone. This cone with its apex angle as 2ϕ is known as FRICTION CONE. The resultant frictional force vector will always be within this cone and at the limiting friction value, it will lie on the surface of the cone.

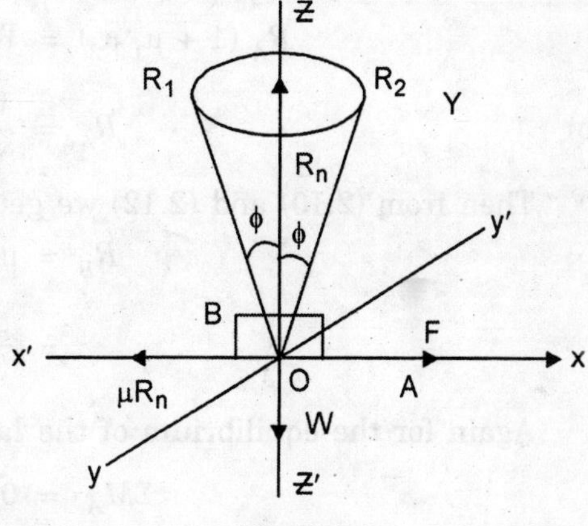

Fig. 2.4

2.6 EQUILIBRIUM OF A LADDER

In Fig. 2.5 is shown a ladder AB of weight W_1 and length l lying on a rough floor AC and resting against a rough wall. Suppose a man of weight W_2 moves up the ladder and is at point D. The inclination of ladder with horizontal is θ. The tendency of foot of ladder will be slip towards left and that of end B downwards. Accordingly, the frictional forces $\mu_1 R_A$ *at* A and $\mu_2 R_B$ at. B will be directed towards right and upwards respectively. In order to avoid slipping of the foot A of the ladder force P is applied at A.

Considering the equilibrium of the ladder and resolving the forces along horizontal and vertical directions we have,

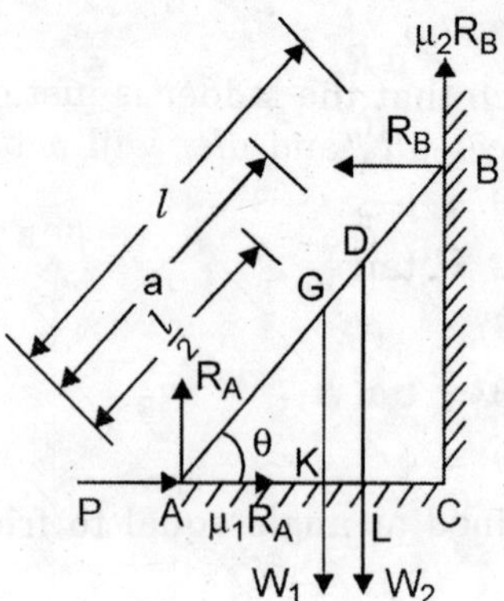

Fig. 2.5

$$\mu_1 R_A + P - R_B = 0 \quad \text{...(2.10)}$$

$$R_A + \mu_2 R_B = W_1 + W_2 \quad \text{...(2.11)}$$

Substituting value of R_B from (2.10) into (2.11),

$$R_A + \mu_2(\mu_1 R_A + P) = W_1 + W_2$$

$$R_A (1 + \mu_1 \mu_2) = W_1 + W_2 - \mu_2 P$$

or $$R_A = \frac{W_1 + W_2 - \mu_2 P}{1 + \mu_1 \mu_2} \quad \text{...(2.12)}$$

Then from (2.10) and (2.12) we get

$$R_B = \mu_1 R_A + P \quad \text{...(2.13a)}$$

$$= \frac{\mu_1[W_1 + W_2 - \mu_2 P]}{(1 + \mu_1 \mu_2)} + P \quad \text{...(2.13b)}$$

Again for the equilibrium of the ladder, taking moments about A.

$$\Sigma M_A = 0$$

or $$W_1 (AK) + W_2(AL) - R_B (BC) - \mu R_B(AC) = 0$$

or $$W_1\left(\frac{l}{2}\cos\theta\right) + W_2(a\cos\theta) - R_B(l\sin\theta) - \mu_2 R_B(l\cos\theta) = 0$$

$$W_1 \frac{l\cos\theta}{2} + W_2(a\cos\theta) = l\, R_B\, [\sin\theta + \mu_2 \cos\theta]$$

or $$R_B = \frac{\left(\frac{l}{2} W_1 \cos\theta + a W_2 \cos\theta\right)}{l[\sin\theta + \mu_2 \cos\theta]} \quad \text{...(2.14)}$$

Equating (2.13) and (2.14), we get after simplification the value of the force P required to avoid slipping of ladder,

$$P = \frac{(1 + \mu_1 \mu_2)}{(\tan\theta + \mu_2)}\left[W_1 + 2\frac{a}{l} W_2\right] - \mu_1 (W_1 + W_2) \quad \text{...(2.15)}$$

2.1 : *A ladder of weight W and length 'l' is placed in the store such that it makes angle θ with the floor. If the co-efficient of friction between wall and ladder and between floor and ladder is 0.2, find the minimum value of v so that it does not slip.*

Solution : Let θ be such that the ladder is just on the point of slipping. Then the frictional forces μR_A and μR_B will act as shown in Fig. 2.6.

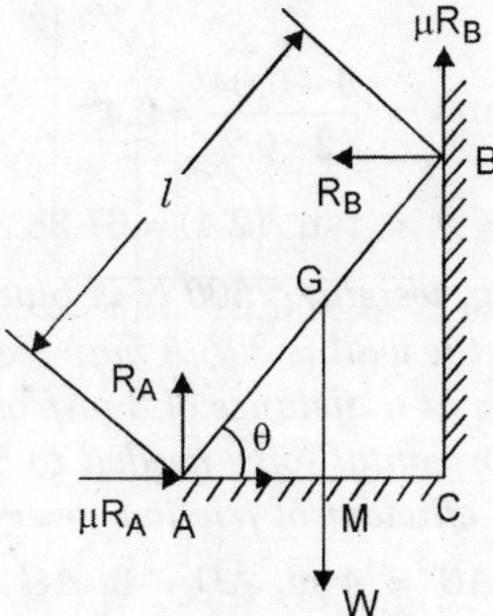

Fig. 2.6

For the equilibrium of ladder, resolving forces in vertical and horizontal directions we get,

$$R_A + \mu R_B = W \qquad ...(A)$$

and

$$\mu R_A = R_B \qquad ...(B)$$

Taking moments about A gives,

$$W(AM) - \mu R_B\,(AC) - R_B(BC) = 0$$

$$W\left(\frac{l}{2}\cos\theta\right) - \mu R_B\, l\cos\theta - R_B\, l\sin\theta = 0$$

$$\left(\frac{W}{2} - \mu R_B\right) l \cos\theta = R_B\, l \sin\theta$$

$$\left(\frac{W}{2} - \mu R_B\right) = R_B \tan\theta$$

or

$$\tan\theta = \frac{\frac{W}{2} - \mu R_B}{R_B} \qquad ...(C)$$

Then equations (A) and (B),

$$R_A + \mu^2 R_A = W$$

or

$$R_A = \frac{W}{1+\mu^2} \qquad ...(D)$$

and

$$R_B = \frac{\mu W}{1+\mu^2} \qquad ...(E)$$

E X A M P L E

From equation (*C*) and (*E*),

$$\tan\theta = \frac{\dfrac{W}{2} - \dfrac{\mu^2 \cdot W}{1+\mu^2}}{\dfrac{\mu W}{1+\mu^2}} = \frac{(1+\mu^2)-2\mu^2}{2\mu} = \frac{1-\mu^2}{2\mu}$$

$\therefore$ $$\tan\theta = \frac{1-0.04}{2\times 0.2} = 2.4$$

Thus, $$\theta = \tan^{-1}(2.4) = 67.38°$$

2.2 : *A ladder 4m long, weighing 500 N is placed against a wall such that its inclination with the wall is 30°. A man weighing 700 N climbs up the ladder such that it is at a distance of 3 m from the foot of the ladder. Find the value of the horizontal force needed to be applied at the foot to avoid slipping. Take co–efficient of friction everywhere as 0.2.*

Solution : Given $AB = 4$ m, $AD = 3$, $AG = 2$

$$\theta = 60°$$

$$\mu = 0.2$$

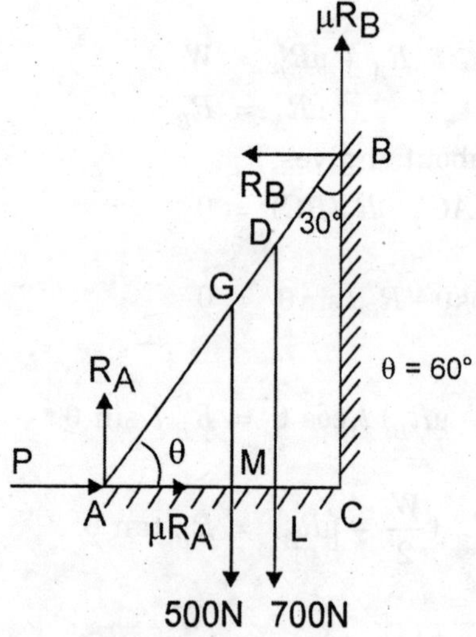

Fig. 2.7

Let the horizontal force needed at *A* be *P* to avoid slipping. For equilibrium of ladder ΣX, ΣY and ΣM are zero.

$\therefore$ For $$\Sigma X = 0$$

$$P + \mu R_A - R_B = 0$$

or $$R_B = (P + \mu R_A) \quad \text{...(A)}$$

And for $$\Sigma Y = 0$$

$$R_A + \mu R_B = 500 + 700 = 1200 \quad \text{...(B)}$$

For $\Sigma M_A = 0$

$$500 \times (AM) + 700(AL) - R_B(BC) - \mu R_A(AC) = 0$$

or $500 \times 2 \cos\theta + 700 \times 3 \cos\theta = R_B[4 \sin\theta + \mu \times 4 \cos\theta]$

or $$(1000 + 2100) \cos\theta = 4R_B\left[\frac{\sqrt{3}}{2} + 0.2 \times \frac{1}{2}\right]$$

or $$3100 \times \frac{1}{2} = 4R_B\ (0.866 + 0.1)$$

or $$1550 = 3.864\ R_B$$

or $$R_B \frac{1550}{3.864} = 401.14\ N \quad \text{...(C)}$$

∴ From (B) and (C),

$$R_A = 1200 - 0.2 \times 401.14$$
$$= 1119.77\ N$$

Then from equation (A),

$$P = R_B - \mu R_A$$
$$= 401.14 - 0.2 \times 1119.77$$
$$= 401.14 - 223.96$$
$$= 177.18\ N \ \textbf{Ans.}$$

2.3 : *A ladder AB of 5 m lenght and weighing 400 N is resting against a wall making an angle of 45° with the ground. The co–efficient of friction between wall and ladder and floor and ladder is 0.5. Find how far a man weighing 600 N can climb without slipping.*

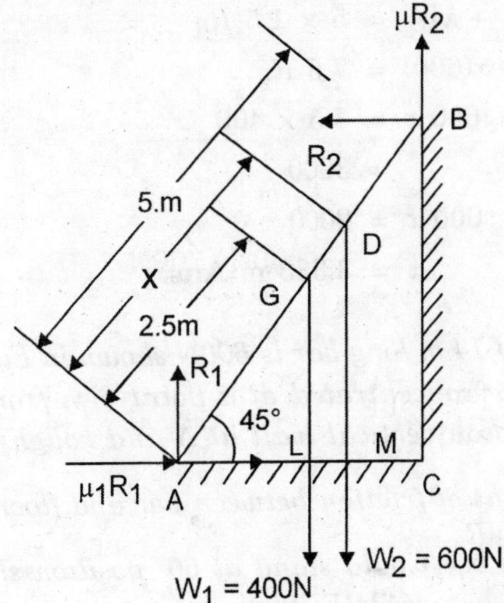

Fig. 2.8

EXAMPLE

Solution : For equilibrium of ladder,

$$\Sigma X = 0, \quad \Sigma Y = 0$$

and $$\Sigma M_A = 0$$

For $\Sigma X = 0$

$$\mu_1 R_1 - R_2 = 0$$

$$0.5R_1 = R_2$$

or $$R_1 = 2R_2 \quad ...(A)$$

For $\Sigma Y = 0$

$$R_1 + \mu R_2 = W_1 + W_2$$

$$= 400 + 600 = 1000$$

or $$R_1 + 0.5\,R_2 = 1000$$

$$2R_2 + 0.5\,R_2 = 10000 \quad \text{(From } A\text{)}$$

$$R_2 = \frac{1000}{2.5} = 400 \text{ N}$$

and $$R_1 = 2 \times 400 = 800 \text{ N}$$

For $\Sigma M_A = 0$

$$W_1(AL) - W_2(AM) = R_2(BC) + \mu R_2(AC)$$

$$W_1(2.5 \times \cos 45) + W_2(x \cos 45) = R_2 5(\cos 45°) + 0.5\,R_2\,5(\sin 45°)$$

or $$W_1(2.5)\frac{1}{\sqrt{2}} + W_2\frac{x}{\sqrt{2}} = R_2 \times \frac{5}{\sqrt{2}} + .5 \times R_2 \frac{5}{\sqrt{2}}$$

$$= \frac{5}{\sqrt{2}} R_2(1.5)$$

$$2.5 \times W_1 + xW_2 = 5 \times 1.5\,R_2$$

or $$2.5 \times 400 + x(600) = 7.5\,R_2$$

or $$1000 + 600\,x = 7.5 \times 400$$

$$= 3000$$

$$600\,x = 2000$$

or $$x = 3.333 \text{ m}$$ **Ans.**

EXAMPLE **2.4 :** *The weight of 14 m long bar is 600N shown in Fig. 2.9 and it may be considered to the concentrated at a point 6 m from the bottom. It rests against a smooth vertical wall at A and rough horizontal floor at B. The co–efficient of friction between bar and floor is $\frac{1}{3}$. Establish by calculation if the bar would stand at 60° position shown.*

(AMIE 1989)

EXAMPLE

Solution : Given data $AB = 14$ m

$AC = 8$ m, $CB = 6$ m

$W = 600$ N

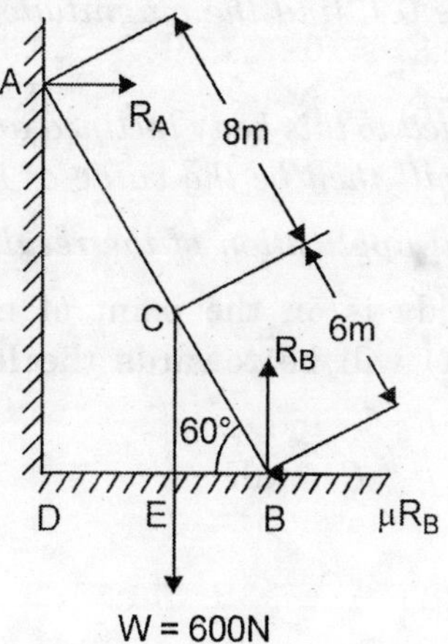

Fig. 2.9

Since the vertical wall is smooth, there will be no frictional force at the point A and the reaction R_A will be normal to wall.

Calculation of actual frictional force available at B (= F_B say)

$$F_B = \mu R_B$$

Since $\Sigma X = 0$

$\therefore$ $F_B = R_A$

and $\Sigma V = 0$

$\therefore$ $R_B = W = 600$

$\therefore$ $F_B = 600 \times \frac{1}{3} = 200$ N

Individual force required for equilibrium

Let it be F'_B then $F'_B = R'_A$

Also, for equilibrium $\Sigma M = 0$ Taking moments about B

$$600 \times (EB) = R'_A AD$$

$$600 \times 6 \cos 60 = R'_A \times 14 \sin 60$$

$$R'_A = \left(600 \times 6 \times \frac{1}{2}\right)\left(\frac{2}{\sqrt{3}}\right) = 148.46 = F'_B$$

Since F'_B which is the frictional force required for equilibrium is being only 148.46 N is less than the available frictional force F_B which is 200 N, the bar will not slip but stay in position inclined at 60° to horizonal.

2.6.1 Rough Planes

E X A M P L E

2.5 : *(a) A body weighing 100 N is lying on a rough horizontal plane. A pull P is applied to the body at an angle of 30° with the horizontal. If the co–efficient of friction is 0.1, find the magnitude of P which will just set the body in motion.*

(b) If the push P is applied to this body inclined at 30° with the horizontal, instead of pull, what will then be the value of P?

(c) Discuss the practical application of the result.

Solution : (*a*) The body is on the point of motion to the right the frictional force $F(= \mu R)$ will be towards the left. If R is the normal reaction, Then,

$$F = \mu R \qquad \text{...(A)}$$

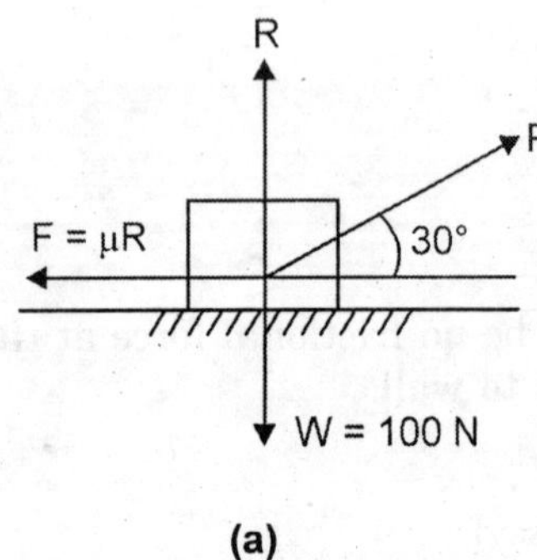

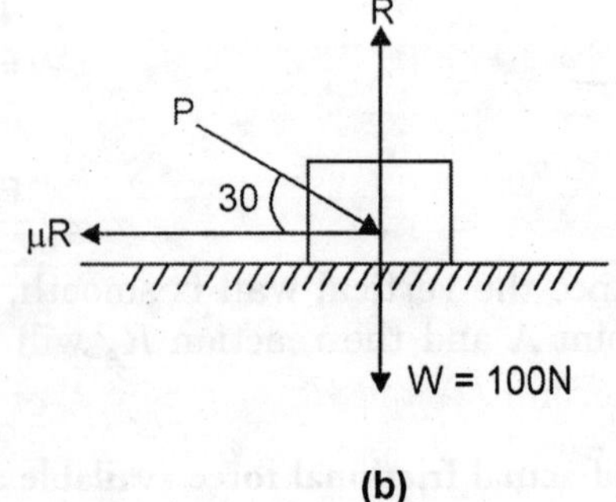

Fig. 2.10

For the equlibrium of the body,

$$\Sigma X = 0 \quad \text{and} \quad \Sigma Y = 0$$

For $\Sigma X = 0$ we get by resolving the forces along horizontal direction,

$$F = P \cos 30°$$

or $$\mu R = P \cos 30° = P\frac{\sqrt{3}}{2} \qquad \text{...(B)}$$

For $\Sigma V = 0 P \sin 30 + R = W$

or $$R = W - \frac{P}{2} = 100 - \frac{P}{2} \qquad \text{...(C)}$$

From equation (*B*) and (*C*),

$$0.1 \times R = P\frac{\sqrt{3}}{2}$$

or $$R = P\frac{\sqrt{3}}{0.2} = 5P\sqrt{3}$$

Substituting the value of R in equation (C),

$$\therefore \quad 5\sqrt{3}P = 100 - \frac{P}{2}$$

$$8.66\,P = 100 - 0.5\,P$$

or
$$9.16\,P = 100$$

$$P = \frac{100}{9.16}$$

$$= \mathbf{10.92}\text{ N}$$

(b) From Fig. (2.10b), for equlibrium of body

For $\Sigma Y = 0$
$$R = W + P \sin 30$$

$$= 100 + \frac{P}{2} \qquad \text{...}(D)$$

For $\Sigma X = 0$
$$\mu R = P \cos 30$$

or
$$0.1\,R = P\frac{\sqrt{3}}{2} = 0.866\,P \qquad \text{...}(E)$$

i.e.,
$$R = 8.66\,P$$

From (D) and (E),
$$8.66\,P = 100 + \frac{P}{2}$$

$$8.16\,P = 100$$

$$P = \frac{100}{8.16}$$

$$= \mathbf{12.25}\text{ N}$$

☛ Important Result and its Application

The important inference from these results is that we need more force while pushing than pulling it. The process of cutting wood with a handsaw and of steel by hand hacksaw is based on the above result. Steel being much stronger than wood requires more force than wood to cut it Therefore, the cutting of steel piece with hand hacksaw forward stroke is made the cutting stroke and to check buckling of the blade due to compressive reactive force in the blade, a frame is used on which the blade is fitted and is tightened to provide initial tensile force in blade so as to overcome compressive reactive force developed in blade while cutting steel.

On the contrary wood is very soft material, we need very little force to cut it. Therefore, in the case of wood, the reverse stroke of the saw is made the cutting stroke the cutting teeth are made accordingly. In addition the saw material will be in tension, therefore there is no need to provide frame.

EXAMPLE

2.6 : *(a) A block weighing 200 N rests on a rough inclind plane at 30° with the horizontal. A horizontal force P acts on the body. If the co–efficient of friction is 0.2 find (b) the value of P when the block is just on the point of moving up (c) when the block is just on the point of moving down.*

Solution: *In Fig. 2.11 (a) is shown the free body diagram of the block. As the body tends to move up, the direction of frictional force will be downwards.*

For the equilibrium of the block resolving the forces normal to and along the plane,

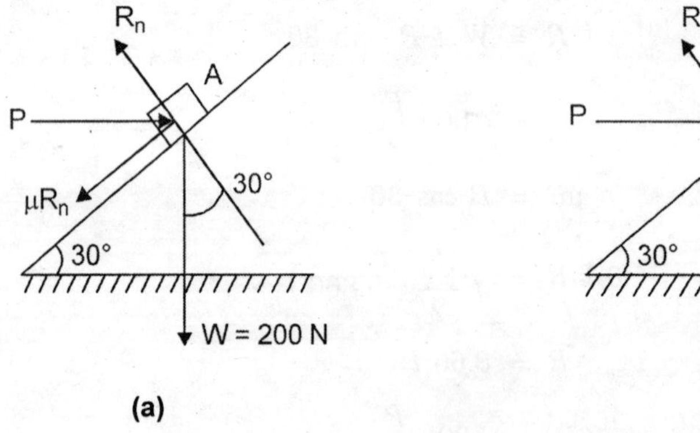

Fig. 2.11

For $\Sigma Y = 0$ $\quad R_n = W \cos 30 + P \cos 60$

or $\quad R_n = 200 \times \frac{\sqrt{3}}{2} + P \cos 60$

$$= 173.2 + \frac{P}{2}$$

For $\Sigma X = 0$

$$P \cos 30 - \mu R_n - W \cos 60 = 0$$

$$P\frac{\sqrt{3}}{2} - 0.2\left(173.2 + \frac{P}{2}\right) - 200 \times \frac{1}{2} = 0$$

$$0.866\ P - 134.64 - 0.1\ P - 100 = 0$$

or $\quad 0.766\ P = 134.64$

$\therefore \quad P = \frac{134.64}{0.766} = 175.77\text{N}$

(*b*) As the body is just on the point of moving down, the frictional force will be acting upwards. In Fig. 2.11 (*b*) is drawn the free body diagram. For the equilibrium of the body, resolving the forces along and normal to the block

For $\Sigma X = 0$

$$\mu R_n + P\cos 30 - W\cos 60 = 0$$

$$0.2R_n + P\frac{\sqrt{3}}{2} - \frac{W}{2} = 0$$

or $$0.2R_n + P\frac{\sqrt{3}}{2} - 100 = 0 \qquad ...(A)$$

For $\Sigma Y = 0$

$$R_n - W\cos 30 - P\cos 60 = 0$$

$$R_n = 200\times\frac{\sqrt{3}}{2} + \frac{P}{2}$$

$$= 173.2 + \frac{P}{2} \qquad ...(B)$$

From (A) and (B),

$$0.2\,(173.2 + 0.5\,P) + P(0.866) - 100 = 0$$

$$34.66 + 0.1P + 0.866\,P = 100$$

or $$.966\,P = 100 - 34.64$$

$\therefore$ $$P = \frac{64.36}{.966} = \mathbf{97.515\ N\ Ans.}$$

2.7 : *Two blocks A and B weighing 200 N and 500 N respectively are resting on a rough floor as shown in Fig. 2.12. The block A is tied to wall with a string. The co–efficient of friction between the blocks A and B is 0.25 and between block B and floor is 0.3. Find the value of P required just to move the block B: (a) if P is horizontal (b) if P is inclined upwards at 30° to the horizontal. What will be the tension in the string?*

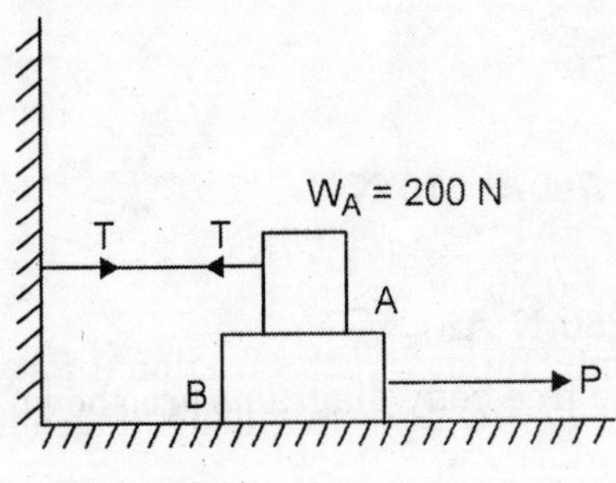

(a)

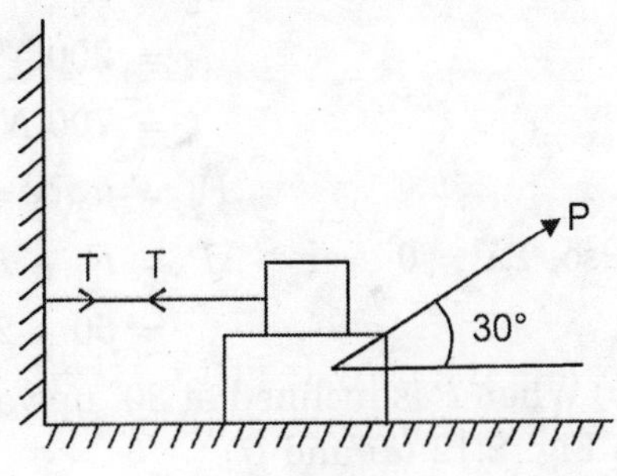

(d)

EXAMPLE

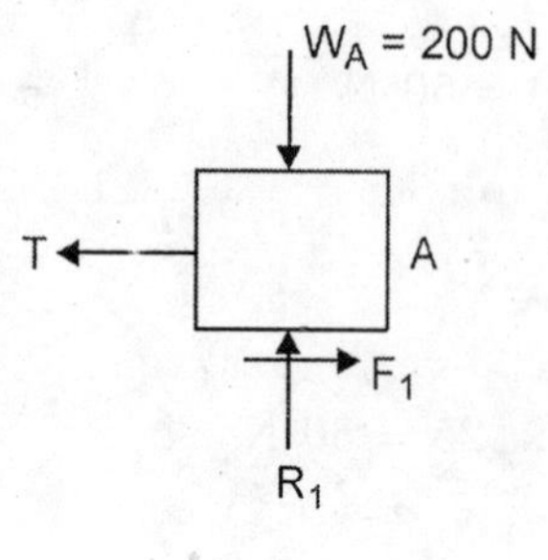

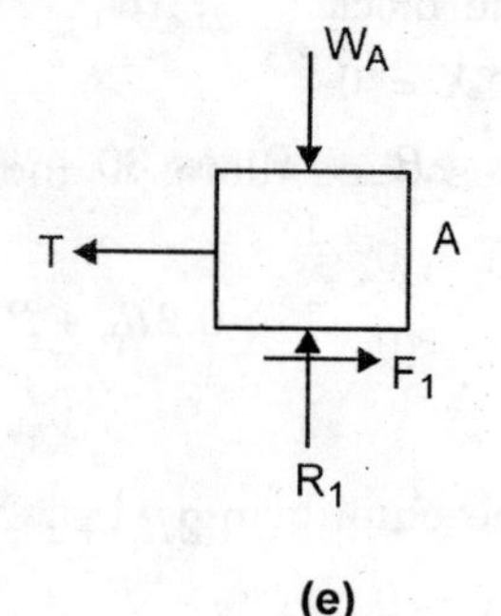

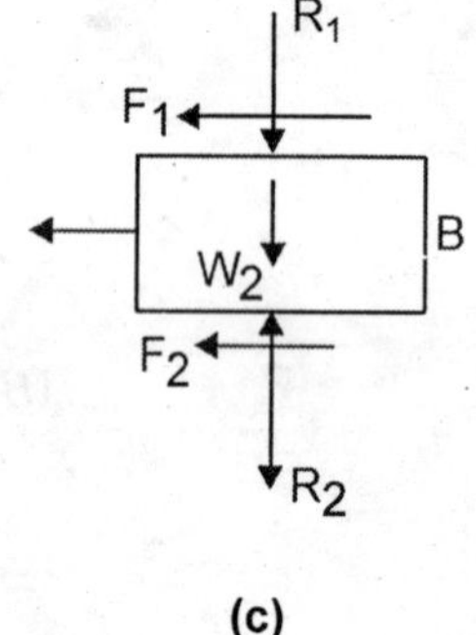

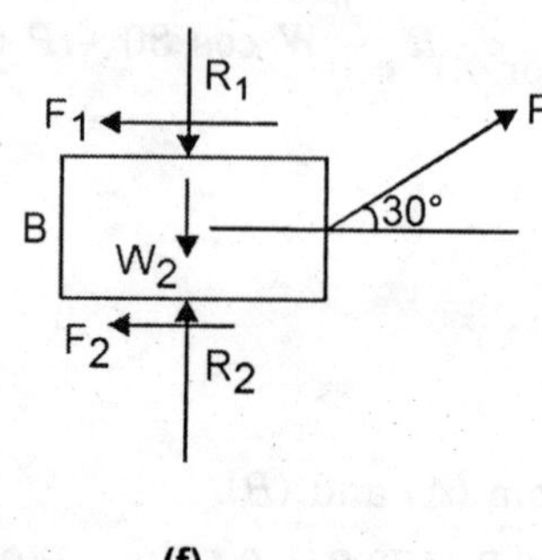

Fig. 2.12

Solution : (*a*) The free body diagrams of the blocks *A* and *B* are shown in Fig. .212 (*b*) and (*c*) for the first case.

For the equilibrium of block *A*,

$\Sigma Y = 0$ and $\Sigma X = 0$

$$W_A = R_1$$

$$\therefore \quad R_1 = W_A = 200 \text{ N}$$

$$\therefore \quad F_1 = \mu_1 R_1 = .25 \times 200 = 50 \text{ N}$$

Also, $\Sigma X = 0$

or $T = F_1 = 50$ N **Ans.**

For the equilibrium of block *B*,

$\Sigma Y = 0$ or $R_2 = R_1 + W_2$

$$= 200 + 500$$

$$= 700 \; N$$

$$\therefore \quad F_2 = \mu_2 R_2 = 0.3 \times 700 = 210 \text{ N}$$

Also, $\Sigma X = 0$ or $P = F_1 + F_2$

$= 50 + 210 = 260$ N **Ans.**

(*b*) When *P* is inclined at 30° upwards, the free body diagrams are shown is Fig. 2.12 (*e*) and (*f*).

For the equilibrium of block *A*,

$$R_1 = W_A = 200 \text{ N}$$

$$\therefore \quad F_1 = \mu_1 R_1 = 0.25 \times 200 = 50 \text{ N}$$

Now resolving along X and Y directions,

For $\Sigma X = 0$ $\quad T = F_1 = \mu_1 R_1$

For $\Sigma Y = 0$ $\quad R_1 = W_A = 200$ N

$$\therefore \quad T = F_1 = \mu_1 R_1 = 200 \times .25 = 50 \text{ N}$$

For the equilibrium of block B,

For $\Sigma X = 0$ $\quad F_2 + F_1 = P \cos 30°$ ...(*A*)

$$\mu_2 R_2 + 50 = P \cos 30$$

as For $\Sigma Y = 0$

$$+ R_1 + W_2 = R_2 + P \sin 30$$

$$-\frac{P}{2} + R_2 = 200 + 500 = 700 \text{ N}$$

or $\quad R_2 = 700 + \dfrac{P}{2}$

Then equation A reduces to,

$$\left(\frac{\mu_2 P}{2} + \mu_2 R_2\right) + F_1 = P \cos 30$$

or $\quad 0.3 \times 700 + 50 = P \cos 30° - 0.3\dfrac{P}{2}$

$$210 + 50 = P\left(\frac{\sqrt{3}}{2} - 0.15\right)$$

or $\quad 260 = P(0.716)$

or $\quad P = \dfrac{260}{0.716} = 363.13$ N **Ans.**

2.8 : *Two blocks A and B weighing 100 N and 200 N respectively are lying on inclined plane CD as shown in Fig. 2.13. The plane CD is inclined at 45° from horizontal. The block A is tied with the help of a string to the support DE which is fixed to the plane CD. A force P is applied on the block B so as just check it from slipping down the plane CD. The fore P is inclined at 15 degree to the plane CD as shown. Find the value of P and the tension in the string Assme co-efficient of friction every where is 0.2.*

EXAMPLE

EXAMPLE

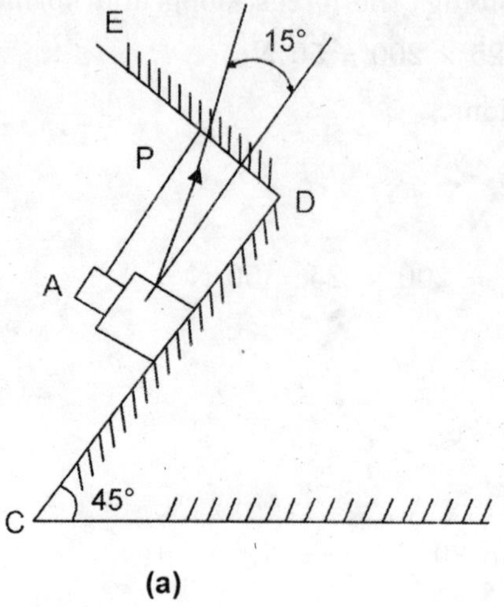

(a)

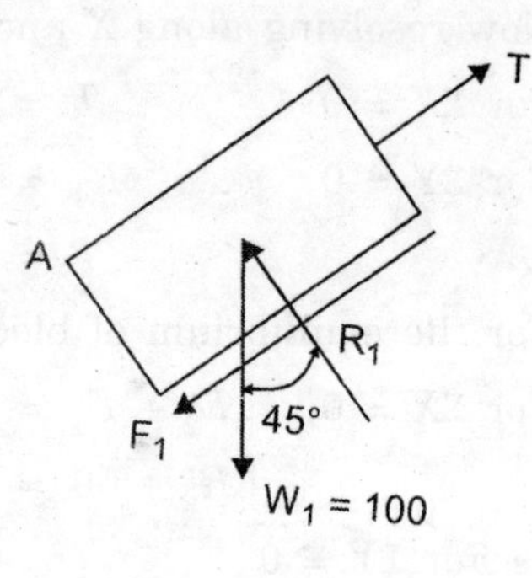

(b)

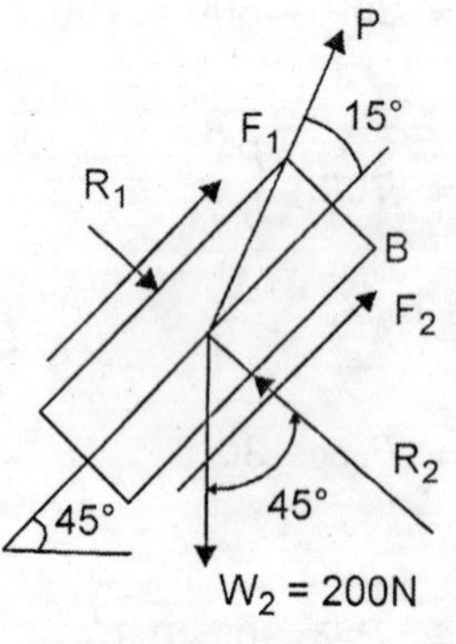

(c)

Fig. 2.13

Solution 2.8 : The free-body diagrams of blocks A and B are given in Fig. 2.13 (b) and (c) respectively.

Considering the equilibrium of block A and resolving the forces along the plane CD and normal to CD

Along CD $\quad T - F_1 - W_1 \cos 45° = 0$

or $\quad T - \mu R_1 - W_1 \cos 45° = 0$

or $\quad T - 0.2R_1 - 100 \times 0.707 = 0$

or $\quad T - 0.2\,R_1 - 70.7 = 0 \qquad ...(A)$

Along normal to CD $\quad R_1 = W_1 \cos 45°$

$= 100 \times 0.707$

$R_1 = 70.7 \qquad ...(B)$

From (A) and (B), $\quad T = 0.2 \times 70.7 + 100 \times .707$

$= 84.84$ Ans.

For the equilibrium of block B. Resolving the forces along and normal to the plane CD

Along plane CD,

$$P \cos 15° + F_1 + F_2 = W_2 \sin 45°$$

$$P(.9659) + \mu R_1 + \mu R_2 = 200 \times 0.707$$

$$0.9619\, P + 0.2 \times 70.7 + 0.2\, R_2 = 141.4$$

$$0.9619\, P + 0.2R_2 + 14.14 = 126.26 \quad ...(C)$$

Also, resolving forces normal to CD,

$$P \sin 15 - R_1 + R_2 - W_2 \cos 45 = 0$$

$$P(0.2555) - 70.7 + R_2 - 200 \times 0.707 = 0$$

$$.2588\, P + R_2 = 141.4 + 70.7$$

$$= 212.1$$

or

$$R_2 = 212.1 - 0.2588\, P$$

Multiply (D) by 0.2,

or

$$0.2\, R_2 = 42.42 - 0.05176\, P \quad ...(D)$$

Therefore (C) and (D)

$$0.9619\, P + 42.42 - 0.05176\, P = 126.26$$

$$.91014P = 83.84$$

$$P = \frac{83.84}{0.91014}$$

$$= 92.118\, N \textbf{ Ans.}$$

From equation (A) and (B),

$$T = 84.84 \text{ N } \textbf{Ans.}$$

2.9 : *Two blocks A and B weighing 200 N and 600 N respectively are placed on a rough plane as shown is Fig. 2.14 The inclination of the plane CD with the horizontal can be changed. The block A is tied with the support DE, with a string, which is fixed to the plane CD If the co–efficient of friction every where is* $\frac{1}{3}$. *Find at what value of* θ, *the inclination of plane CD with the horizontal will the block B be on the point of slipping drown?*

E X A M P L E

E X A M P L E

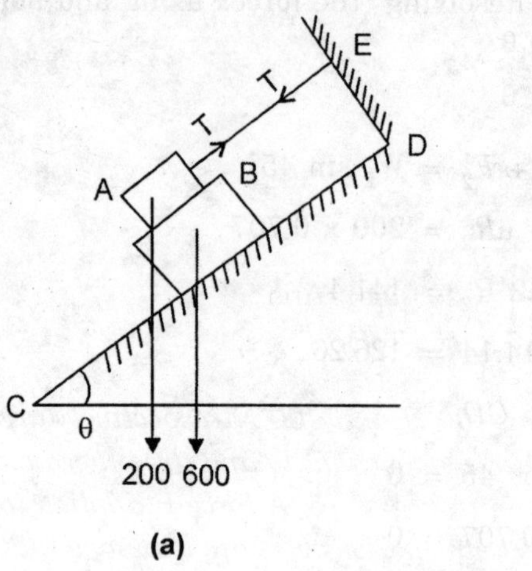

(a)

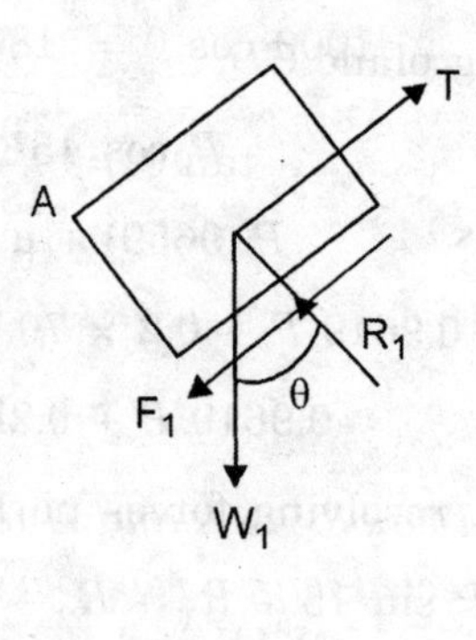

(b)

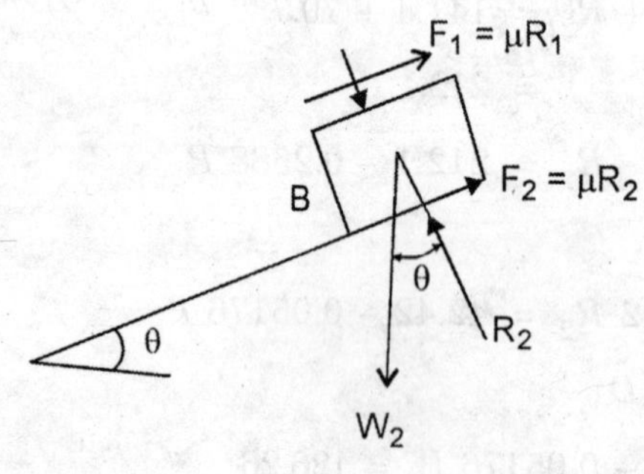

(c)

Fig. 2.14

Solution : Let the inclination of the place CD be θ with the horizontal. The free body diagrams of the bodies A and B are shown in Fig. (2.14*b*) and (2.14*c*) for the equilibrium of block A, resolving forces along the direction CD and normal to it we get,

$$T = F_1 + W_1 \sin\theta \qquad ...(A)$$

$$= \mu_1 R_1 + W_1 \sin\theta$$

and

$$R_1 = W_1 \cos\theta \qquad ...(B)$$

$$= 200 \cos\theta \qquad ...(B)$$

Similarly, for the equilibrium of block B we get,

$$\mu R_2 + \mu R_1 = W_2 \sin\theta$$

$$\frac{1}{3}(R_2 + R_1) = 600 \sin\theta$$

$$R_1 + R_2 = 1800 \sin\theta \qquad ...(C)$$

and

$$R_2 = R_1 + W_2 \cos\theta$$

$$= 200 \cos\theta + 600 \cos\theta \qquad \text{(From } B\text{)}$$

$$= 800 \cos\theta \qquad ...(D)$$

From equation (*C*) and (*D*),

$$200 \cos \theta + 800 \cos \theta = 1800 \sin \theta$$

or $$1000 \cos \theta = 1800 \sin \theta$$

or $$\tan \theta = \frac{10}{18} = \frac{5}{9}$$

$\therefore$ $$\theta = \tan^{-1}\left(\frac{5}{9}\right) = 29.0546° \text{ Ans.}$$

2.10 : *Two blocks A and B weighing 700 N and 400 N respectively are interconnected with a rope passing over a smooth pully and are placed on the rough planes AB and BC as shown in Fig 2.15. A force P applied to the block B inclined at an angle of 15° to the plane BC is applied such that the block B is just on the point of moving up the plane CB. Find the value of P Take co–efficient of friction every where is 0.2.*

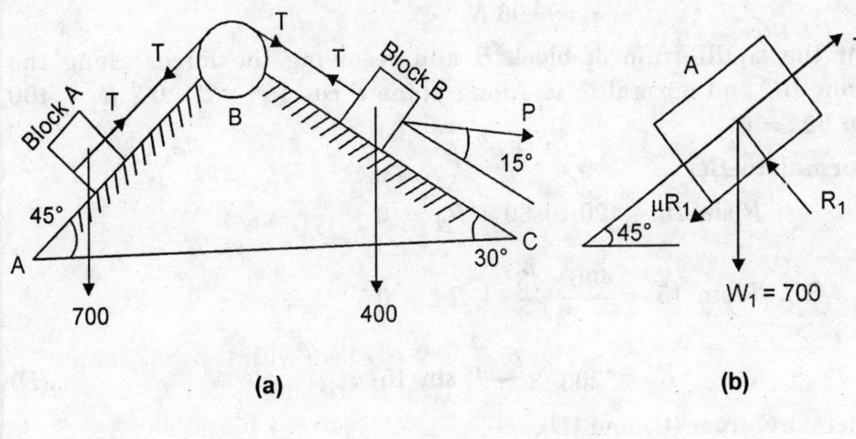

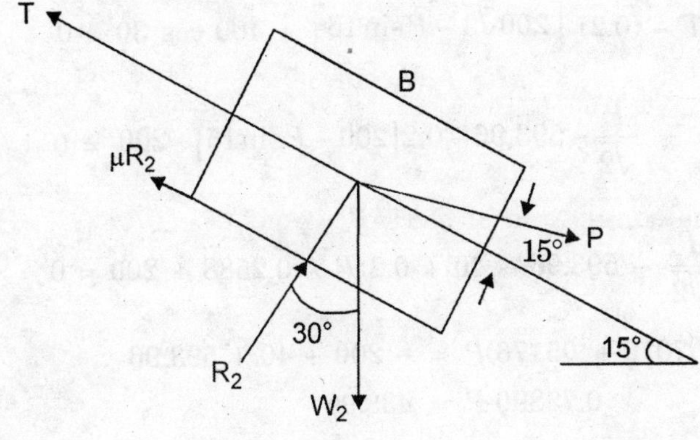

Fig. 2.15

E X A M P L E

Solution : The free-body diagrams of the block A and B are given in Fig. 2.15 (b) and Fig. 2.15 (c). For the equilibrium of block A, resolving the forces along the plane AB and normal to it we have,

$$T - F_1 - 700 \cos 45° = 0$$

$$T - 0.2\, R_1 - \frac{700}{\sqrt{2}} = 0$$

and $$R_1 = 700 \cos 45° = \frac{700}{\sqrt{2}} \qquad ...(B)$$

$$T - 0.2 \times \frac{700}{\sqrt{2}} - \frac{700}{\sqrt{2}} = 0$$

or $$T = 1.2 \times \frac{700}{\sqrt{2}}$$

$$= 593.96\ N$$

For the equilibrium of block B and resolving the forces along the plane BC and normal to it. Along plane $P \cos 15° - T - 0.2\, R_2 + 400 \sin 30° = 0$ ...(C)

Normal to *BC*

$$P \sin 15 - 400 \cos 30 + R_2 = 0$$

$$P \sin 15 - \frac{400 \times \sqrt{3}}{2} + R_2 = 0$$

$\therefore$ $$R_2 = 200\sqrt{3} - P \sin 15° \qquad ...(D)$$

Therefore, from (C) and (D),

$$P \cos 45 - T - (0.2) \left[200\sqrt{3} - P \sin 15\right] + 400 \cos 30 = 0$$

or $$\frac{P}{\sqrt{2}} - 593.96 - 0.2[200 - P \sin 15] + 200 = 0$$

$$\frac{P}{\sqrt{2}} - 593.96 - 40 + 0.2\, P \times 0.2588 + 200 = 0$$

$$(.7072 + .05176)P = -\ 200 + 40 + 593.96$$

$$0.75890\ P = 433.96$$

or $$P = \frac{433.96}{0.75890}$$

$$= 571.82\ N \textbf{ Ans.}$$

2.11 : *A block A weighing 100 N rests on a rough inclined place whose inclination with the horizontal is 45°. This block is connected to another block B weighing 300 N resting on a rough horizontal plane by a weightless rigid bar inclined at 30° the horizontal (Fig. 2.16). Find the horizontal force required to be applied to the block B to just move the block A in the upward direction. Assume limiting angle of friction as 15° at all surfaces where there is sliding. (AMIE 1984)*

Solution:

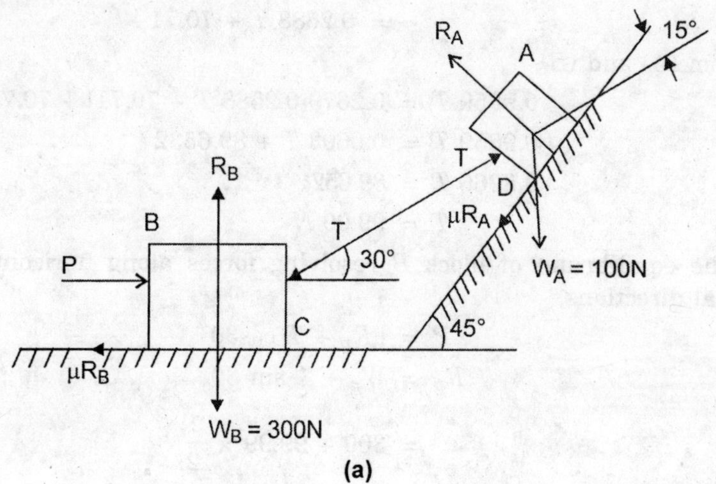

(a)

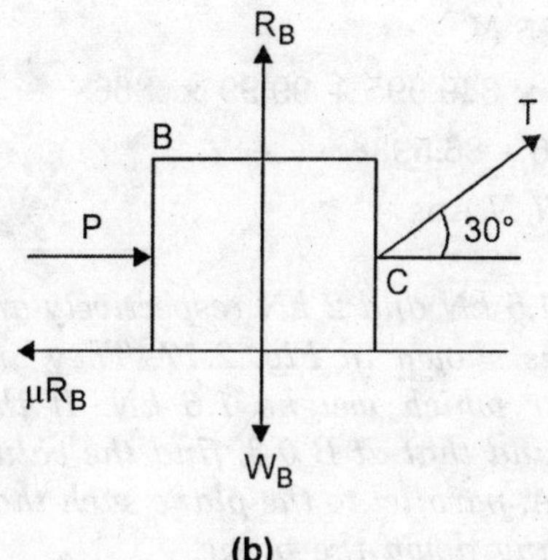

(b)

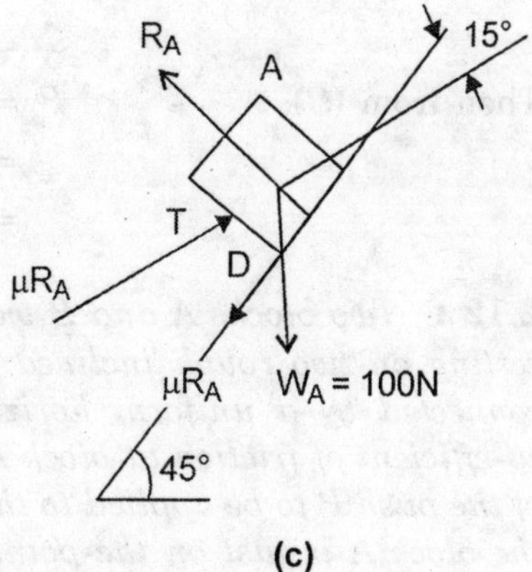

(c)

Fig. 2.16

Given data,

$$W_A = 100\ N$$
$$W_B = 300\ N$$
$$\phi = 15°$$
$$\mu = \tan\phi$$
$$= \tan 15 = 0.2679$$

Let P be the required force to be applied and T the thrust in the connecting bar CD.

For the equilibrium of the block A and resolving forces along the plane EF and normal to EF,

EXAMPLE

E X A M P L E

Along the plane EF, $T \cos 15° = \mu R_A + W_A \cos 45°$

$$0.9659\,T = 0.2679\,R_A + 100 \times \left(\frac{1}{\sqrt{2}}\right)$$

$$= 0.2679\,R_A + 70.71 \qquad ...(A)$$

And normal to plane EF, $R_A = T \sin 15 + W_A \sin 45$

$$= T\,0.2588 + 100 \sin 45$$

$$= 0.2588\,T + 70.71 \qquad ...(B)$$

$\therefore$ From (A) and (B),

$$0.9659\,T = 0.2679[0.2588\,T + 70.71] + 70.71$$

or $$0.9659\,T = 0.0693\,T + 89.6532$$

or $$0.8966\,T = 89.652$$

or $$T = 99.99\,N$$

For the equilibrium of block B resolving forces along horizontal and vertical directions,

$$P = \mu R_B + T \cos 30 \qquad ...(C)$$

and $$R_B = W_B + T \sin 30 \qquad ...(D)$$

$$= 300 + 99.99 \times \frac{1}{2}$$

$$= 300 + 49.995$$

$$= 349\,.995\,N$$

Then from (C), $$P = 0.2679 \times 349.995 + 99.99 \times .866$$

$$= 93.7636 + 86.5938$$

$$= 180.357\,N \text{ Ans.}$$

2.12 : *Two blocks A and B weighing 1.5 kN and 2 kN respectively are resting on two rough inclined planes as shown in Fig. 2.17. They are connected by a uniform horizontal bar which weighs 1.6 kN. If the co–efficient of friction at block A is 0.2 and that at B 0.3, find the value of the push P to be applied to the block A, parallel to the plane such that the block A is just on the point of moving down the plane.*

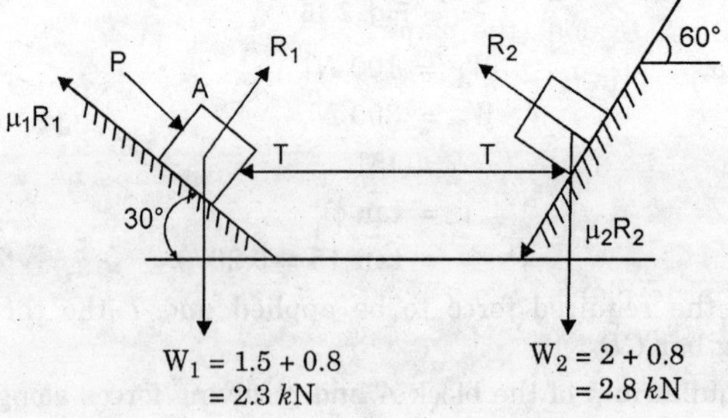

(a)

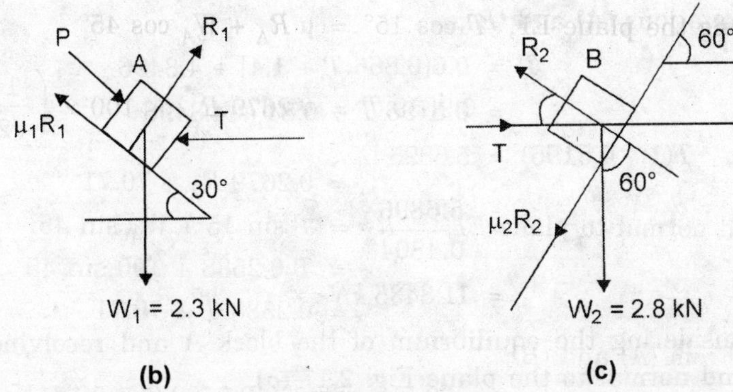

Fig. 2.17

Solution: Given data,

$$\mu_1 = 0.2$$

and $$\mu_2 = 0.3$$

Weight of block A, $$= W_A = 1.5 \, kN$$

$$W_B = 2.0 \, k \, N$$

Weight of bar $$W_1 = 1.6 \, kN$$

Since the bar is uniform, half of its weight will be supported by block A and half by block B,

The total drownward load at block A will be,

$$1.5 + \frac{1.6}{2} = 1.5 + 1.8$$

$$= 2.3 \, kN$$

and Total downward at block,

$$B = 2.0 + 0.8$$

$$= 2.8 \, kN$$

Accordingly, the frce body diagrams of the blocks A and B are given in Fig. 2.17 (b) and (2.17 (c).

Considering the equilibrium of block B and resolving the forces along and normal to the plane, Assuming the thrust in the bar as T

For Block B along the plane,

$$T \cos 60° - m_2 R_2 - (2.8) \sin 60° = 0$$

$$\frac{T}{2} - 0.3 \text{ R}_2 - 2.8 \text{ 5 } 0.866 = 0$$

or $$T = 0.6 \, R_2 + 4.8496 \qquad ...(A)$$

Normal to plane,

$$R_2 - T \sin 60 - 2.8 \cos 60° = 0$$

or $$R_2 = 0.866 \, T + 1.4 \qquad ...(B)$$

EXAMPLE

EXAMPLE

From equations (*A*) and (*B*),

$$T = 0.6[0.866\ T + 1.4] + 4.8496$$

$$= 0.5196\ T + 0.84 + 4.8496$$

or $$T(1 - 0.5196) = 5.6896$$

$\therefore$ $$T = \frac{5.6896}{0.4804}$$

$$= 11.8435\ kN$$

Now considering the equilibrium of the block *A* and resolving forces along and normal to the plane Fig. 2.17 (*b*),

$$P - \mu_1 R_1 + 2.3 \cos 60 - R_1 \cos 30 - T \cos 30 = 0$$

or $$P - 0.2\ R_1 + \frac{2.3}{2} - 11.8435\ (\ 0.866)\ 0.866\ R_1 = 0$$

$$P - 1.066\ R_1 = 10.2565 - 1.150$$

or $$P - 1.066\ R_1 = 9.1065 \quad ...(C)$$

Resolving forces on *A* normal to plane,

$$R_1 - T \sin 30 - 2.3 \sin 60 = 0$$

or $$R_1 - \frac{11.8435}{2} - 2.3 \times .866 = 0$$

or $$R_1 = 5.9218 + 1.9918$$

$$= 7.9136 \quad ...(D)$$

Therefore from (*C*), and (*D*),

$$P = 9.1065 + 1.066(7.9136)$$

$$= 9.1065 + 8.4359$$

or $$P = 17.5419\ kN\ Ans.$$

2.7 MOTION ON ROUGH INCLINED PLANE

2.7.1 Motion UP The Plane

A body of weight *W* is resting on a rough plane inclined at angle with the horizontal. A force *P* is applied in the horizontal direction as shown in Fig. 2.18 so that the body is just an the point of moving *up*. The frictional force (μR) will be directed downwards. It is required to find the value of *P*.

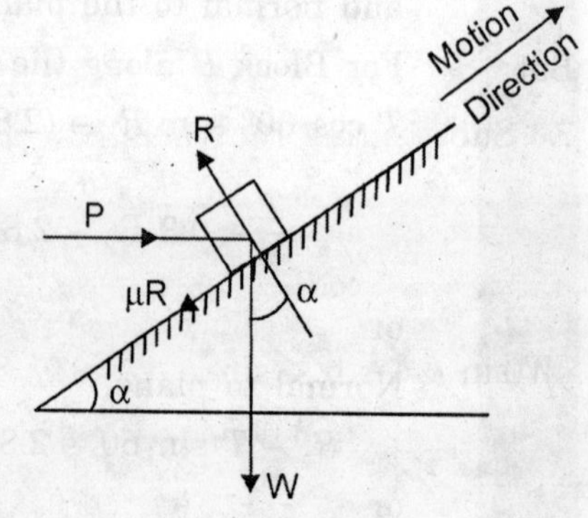

Fig. 2.18

For the equilibrium of body., resolving the forces along and normal to the plane. We get,

Along the plane,

$$P \cos \alpha - \mu R - W \sin \alpha = 0 \quad ...(2.15)$$

Normal to plane,

$$R - P \sin \alpha - W \cos \alpha = 0$$

or $$R = P \sin \alpha + W \cos \alpha \quad ...(2.16)$$

From equations (2.15) (*b*) and (2.16),

$$P \cos \alpha - \mu[P \sin \alpha + W \cos \alpha] - W \sin \alpha = 0$$

or $$P(\cos \alpha - \mu \sin \alpha) = W [\sin \alpha + \mu \cos \alpha] \quad ...(2.17)$$

Now $$\mu = \tan \phi \quad ...(2.18)$$

Therefore from equations (2.17) and (2.18),

$$P\left(\cos\alpha - \frac{\sin\phi}{\cos\phi}\sin\alpha\right) = W\left[\sin\alpha + \frac{\sin\phi}{\cos\phi}\cos\alpha\right]$$

or $$P [\cos \alpha \cos \phi - \sin \alpha \sin \phi] = W [\sin \alpha \cos \phi + \cos \alpha \sin \phi]$$

or $$P [\cos (\alpha + \phi)] = W [\sin (\alpha + \phi)]$$

or $$P = W \tan (\alpha + \phi) \quad ...(2.19)$$

2.7.2 Motion Down the Plane

In this case a force P is applied to the body in horizontal direction, as shown in Fig. 2.19. The body is on the point of moving down the plane, frictional force will be in the upward direction.

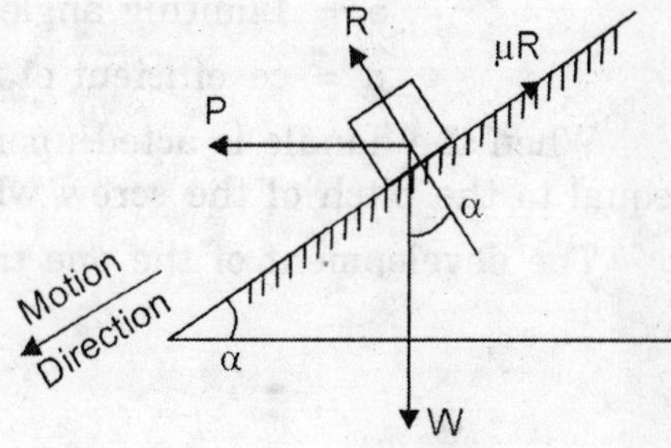

Fig. 2.19

For the equilibrium if the body resolving the force along and normal to the plane, we get,

$$\mu R - P \sin (90 - \alpha) - W \sin \alpha = 0$$

and $$R = W \cos \alpha - P \cos (90 - \alpha)$$

or $$\mu R = P\cos \alpha + W \sin \alpha \quad ...(2.20)$$

$$R = W \cos \alpha - P \sin \alpha \quad ...(2.21)$$

From equations (2.20) and (2.21),

$$\mu(W \cos \alpha - P \sin \alpha) = P \cos \alpha + W \sin \alpha$$

$$W(\mu \cos \alpha - \sin \alpha) = P(\text{sos}\ \alpha + \mu \sin \alpha)$$

But $$\mu = \tan \phi$$

∴ Substituting the value of μ, then equation reduces to,

$$W\left(\frac{\sin\phi}{\cos\phi}\cos\alpha - \sin\alpha\right) = P\left(\cos\alpha + \frac{\sin\phi}{\cos\phi}\sin\alpha\right)$$

or $$W(\sin \phi \cos \alpha - \sin \alpha \cos \phi) = P(\cos \alpha \cos \phi + \sin \alpha \sin \phi)$$

$$W \sin (\alpha - \phi) = P \cos (\alpha - \phi)$$

or $$P = W \tan (\alpha - \phi) \quad ...(2.22)$$

2.8 SCREW JACK

Screw jack is a device which is used to lift heavy load with a relatively small effort with the help of a screw having square threads.

Screw jack consists of a nut and a screw having square threads and C handle at the and of which the effort is applied. The nut is rigidly connected to the body. The load is placed on the head of the screw. The effort is applied at the end of a handle.

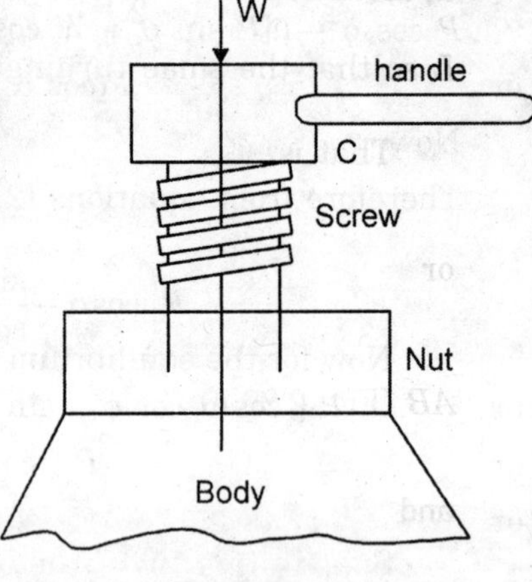

Fig. 2.20 (a)

Let, W = Weight to be lifted

P' = Effort applied at the mean residue of screw

P = Effort applied at the end of handle of length L

p = Pitch of the screw

d = Mean diameter of screw

α = Helix angle of the screw

ϕ = Limiting angle of friction

μ = co–efficient of friction

When the handle is acted upon by a force *P*, the load will be lifted through a distance. equal to the pitch of the screw when the effort *P* completes one rotation.

The development of the one thread of the screw is given is Fig. 2.20. (*b*)

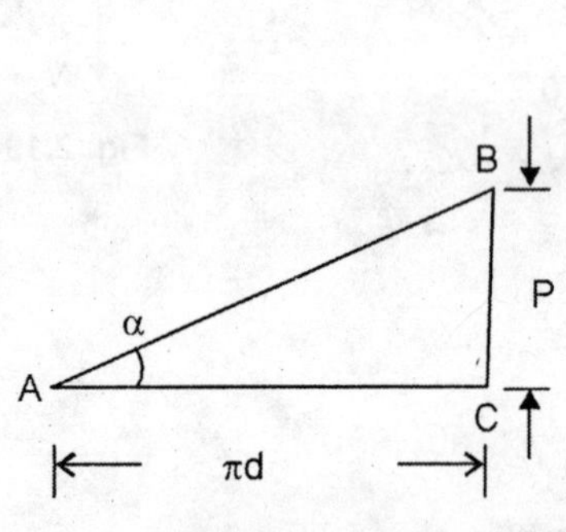

Fig. 2.20 (b)

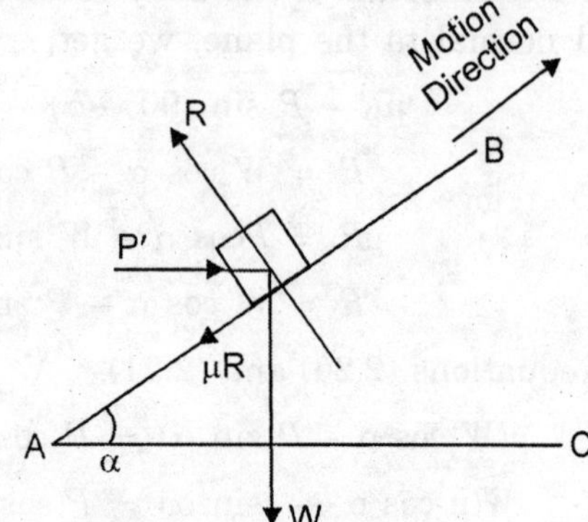

Fig. 2.20 (c)

The development is exactly similar to an inclined plane. AC represents the circumference of the thread and distance *CD* is the pitch (*p*) of the screw and angle *BAC,* the helix angle of threads.

Evidently, $\tan \alpha, = \dfrac{p}{\pi d}$ ...(2.23)

Where, π = pitch of the screw threads

d = mean clin of screw threads

Case I – LOAD BEING LIFTED UP

∴ As the load W is being lifted up, the frictional force will be acting in the downward direction. The free body diagram is shown in Fig. 2.20c. Let P' be the equivalent force applied at mean radius $\left(=\frac{d}{2}\right)$ which is equivalent to P being applied at the end of the lever of length L so that the smae turning moment is applied,

That is $$p'\frac{d}{2} = PL$$

or $$P' = \frac{2L}{d}P \qquad ...(2.24)$$

Now for the equilibrium of weight and resolving the forces along and normal to the plane AB (Fig. 2.20 c)

$$P' \cos\alpha - \mu R - W \sin\alpha = 0 \qquad ...(2.25)$$

and $$R - P' \sin\alpha - W\cos\alpha = 0 \qquad ...(2.26)$$

Substituting the value of R from (2.26) into (2.25),

$$P' \cos\alpha - \mu(P' \sin\alpha + W\cos\alpha) - W\sin\alpha = 0$$

$$P' (\cos\alpha - \mu\sin\alpha) - W(\mu\cos\alpha + \sin\alpha) = 0$$

Now since $\mu = \tan\phi$,

or $$P'\left(\cos\alpha - \frac{\sin\phi}{\cos\phi}\sin\alpha\right) = W\left(\frac{\sin\phi}{\cos\phi}\cos\alpha + \sin\alpha\right)$$

or $$P'(\cos\alpha\cos\phi - \sin\alpha\sin\phi) = W(\sin\phi\cos\alpha + \cos\phi\sin\alpha)$$

$$P'\cos(\alpha + \phi) = W\sin(\alpha + \phi)$$

$$P' = W\tan(\alpha + \phi) \qquad ...(2.27)$$

Therefore from equation (2.24) and (2.27),

$$\frac{2L}{d}P = W\tan(\alpha + \phi)$$

or $$P = \frac{dW}{2L}\tan(\alpha + \phi) \qquad ...(2.28)$$

$$P = \frac{dW}{2L}\,\frac{\tan\alpha + \tan\phi}{1 - \tan\alpha\tan\phi}$$

$$= \frac{dW}{2L}\left[\frac{\tan\alpha + \mu}{1 - \mu\tan\alpha}\right]$$

But $$\tan\alpha = \frac{p}{\pi d}$$

∴ $$P = \frac{Wd}{2L}\left[\frac{\frac{p}{\pi d} + \mu}{1 - \frac{p}{\pi d}\mu}\right]$$

$$P = \frac{dW}{2L}\left[\frac{p+\pi\mu d}{\pi d - p\mu}\right] \qquad ...(2.29)$$

Case II – LOAD BEING LOWERED

When the load in lowered the function a force μR will be acting upwards as known in Fig. 2.19(*d*) Therefore, the equilibrium of block and resolving forces along and normal to plane.

Fig. 2.20 (d)

$$\mu R - P' \cos\alpha - W \sin\alpha = 0 \qquad ...(2.30)$$

and
$$R - W\cos\alpha + P'\sin\alpha = 0 \qquad ...(2.31)$$

For equation (2.30) and (2.31),

$$\mu\,[\,W\cos\alpha - P'\sin\alpha] - P'\cos\alpha - W\sin\alpha = 0$$

or
$$W(\mu\cos\alpha - \sin\alpha) = P'(\cos\alpha + \mu\sin\alpha)$$

or
$$W\left(\frac{\sin\phi}{\cos\phi}\cos\alpha - \sin\alpha\right) = P'\left(\cos\alpha + \frac{\sin\phi}{\cos\phi}\sin\alpha\right)$$

or
$$W(\sin\phi\cos\alpha - \cos\phi\sin\alpha) = P'(\cos\alpha\cos\phi + \sin\alpha\sin\phi)$$

or
$$W(\sin\alpha - \phi) = P'\cos(\alpha - \phi)$$

i.e.,
$$P' = W\tan(\alpha - \phi) \qquad ...(2.32a)$$

Then
$$\frac{2LP}{d} = W\tan(\alpha - \phi)$$

or
$$P = \frac{dW}{2L}\tan(\alpha - \phi) \qquad ...(2.32b)$$

From equation 2.32(a) and 2.32(b),

$$\frac{2LP}{d} = W\left[\frac{\tan\alpha - \tan\phi}{1+\tan\alpha\tan\phi}\right]$$

$$= \left[\frac{\dfrac{p}{\pi d} - \mu}{1+\dfrac{p}{\pi d}\mu}\right]$$

$$= W\left[\frac{p-\pi\mu d}{\pi d + \mu P}\right]$$

or
$$P = \frac{Wd}{2L}\left[\frac{p-\pi\mu d}{\pi d + \mu p}\right] \qquad ...(2.33)$$

2.9 EFFICIENCY OF SCREW JACK

From equation (2.28), we see that the effort required at the end of normal of the screw jack is given by,

$$P = \frac{dW}{2L}\tan(\alpha + \phi) \qquad \text{...(2.28)}$$

Where ϕ is the limiting friction angle between screw and nut and $\phi = \tan^{-1}\mu$. However, an ideal case when the friction co–efficient is zero *i.e.* in the absence of friction, the effort P_{ideal} will be obtained by putting $\phi = 0$

i.e.,

$$P_{ideal} = \frac{dW}{2L}\tan\alpha \qquad \text{...(2.34)}$$

$\therefore$ Efficiency of screw jack $= \dfrac{P_{ideal}}{P_{original}}$

$$\eta = \frac{\frac{dW}{2L}\tan a}{\frac{dW}{2L}\tan(\alpha + \phi)}$$

$$\eta = \frac{\tan\alpha}{\tan(\alpha + \phi)} \qquad \text{...(2.35)}$$

$$= \frac{\sin\alpha}{\cos\alpha}\cdot\frac{\cos(\alpha + \phi)}{\sin(\alpha + \phi)}$$

Subtracting each side from unity,

$$1 - \eta = 1 - \frac{\sin\alpha\cos(\alpha + \phi)}{\cos\alpha\sin(\alpha + \phi)}$$

$$= \frac{\cos\alpha\sin(\alpha + \phi) - \sin\alpha\cos(\alpha + \phi)}{\cos\alpha\sin(\alpha + \phi)} \qquad \text{...(2.36)}$$

Now by using the trigonometric identity,

$$\sin A \cos B - \cos A \sin B = \sin (A - B)$$

Now put $A = \alpha + \phi$

and $B = \alpha$

$$\therefore \quad \sin(\alpha + \phi)\cos\alpha - \cos(\alpha + \phi)\sin\alpha = \sin[(\alpha + \phi) - \alpha] = \sin\phi$$

Therefore, (2.36) reduces to,

$$1 - \eta = \frac{\sin\phi}{\cos\alpha\sin(\alpha + \phi)}$$

$$= \frac{2\sin\phi}{2coa\alpha\ \sin(\alpha+\phi)}$$

or $$1 - \eta = \frac{2\sin\phi}{\sin[(\alpha+\phi)+\alpha]+\sin[(\alpha+\phi)-\alpha]}$$

$\therefore$ $$2\cos A\sin B = \sin(A+B) + \sin(A-B)$$

Put $$A = \alpha + \phi$$

and $$B = \alpha$$

or $$1 - \eta = \frac{2\sin\phi}{\sin(2\alpha+\phi)+\sin\phi} \qquad ...(2.36)$$

When η is maximum $(1 - \eta)$ will be minimum. Further *LHS* of equation (2.36) will be minimum when the denominator of *RHS* is maximum. Since ϕ is constant $\sin\phi$ is also constant, and it leads us to conclude that the denominator of *RHS* of (2.36) will be maximum when $\sin(2\alpha + \phi)$ is maximum.

i.e, when $$2\alpha + \phi = \frac{\pi}{2}$$

or where $$\alpha = \frac{\pi}{4} - \frac{\phi}{2} \qquad ...(2.37)$$

Now, substituting this value of α into equation (2.35),

$$\eta_{max} = \frac{\tan\left(\frac{\pi}{4}-\frac{\phi}{2}\right)}{\tan\left(\frac{\pi}{4}-\frac{\phi}{2}+\phi\right)}$$

$$= \frac{\tan\left(\frac{\pi}{4}-\frac{\phi}{2}\right)}{\tan\left(\frac{\pi}{4}+\frac{\phi}{2}\right)}$$

$$= \left(\frac{1-\tan\frac{\phi}{2}}{1+\tan\frac{\phi}{2}}\right)\left(\frac{1-\tan\frac{\phi}{2}}{1+\tan\frac{\phi}{2}}\right)$$

$$= \frac{\left(1-\tan\frac{\phi}{2}\right)^2}{\left(1+\tan\frac{\phi}{2}\right)^2}$$

or,

$$\eta_{max} = \frac{\left(1 - \dfrac{\sin\frac{\phi}{2}}{\cos\frac{\phi}{2}}\right)^2}{\left(1 + \dfrac{\sin\frac{\phi}{2}}{\cos\frac{\phi}{2}}\right)^2}$$

$$= \frac{\left(\cos\frac{\phi}{2} - \sin\frac{\phi}{2}\right)^2}{\left(\cos\frac{\phi}{2} + \sin\frac{\phi}{2}\right)^2}$$

$$= \frac{\left(\cos^2\frac{\phi}{2} + \sin^2\frac{\phi}{2} - 2\sin\frac{\phi}{2}\cos\frac{\phi}{2}\right)}{\left(\cos^2\frac{\phi}{2} + \sin^2\frac{\phi}{2} + 2\sin\frac{\phi}{2}\cos\frac{\phi}{2}\right)}$$

i.e.,

$$\eta_{max} = \left(\frac{1 - \sin\phi}{1 + \sin\phi}\right) \quad ...(2.38)$$

EXAMPLE

2.13 : *The pitch and mean diameter of the square therads of a screw jack are 24mm and 60mm respectively. If the co-efficient friction is 0.1, find the effort required to be applied at the end of the handle 600 mm long to lift load of 1 kN. Also, calculate the effort required if this load were to be lowered.*

Solution : $W = 100\ N$, $L = 600$ mm, $p = 24\ mm$

$$d = 60 \text{ mm}$$

and $$\mu = 0.1 = \tan\phi$$

$\therefore$ $$\tan\alpha = \frac{p}{\pi d} = \frac{24}{\pi \times 60} = \frac{4}{10\pi} = 0.12732$$

or $$\alpha = \tan^{-1}(0.12732)$$

$$= 7.2561°$$

and $$\phi = \tan^{-1}(0.1)$$

$$= 5.7106°$$

Wo have from equation (2.28),

$$P = \frac{dW}{2L}\tan(\alpha + \phi)$$

$$= \frac{60 \times 1000}{2 \times 600}\tan(7.2561 + 5.7106)$$

EXAMPLE

$$= 50 \tan (12.9667°)$$
$$= 11.5\ N\ Ans.$$

(*b*) In case of lowering the load, we have from equation (2.32 b),

$$P = \frac{dW}{2L}\tan(\alpha - \phi)$$

$$= \frac{60 \times 100}{2 \times 600}\tan(7.2567 - 5.7106°)$$

$$= 50 \tan (1.5455°)$$
$$= 1.27\ N\ \text{Ans.}$$

2.14 : *The pitch and mean diameter of square threads of a screw jack are 18 and 90 mm respectively. The co–efficient of friction at the screw threads is 0.095. Calculate the effort to be applied at the end of the handle 500 mm long to lift a load of 5 kN. Also, calculate the efficiency of the screw jacke.*

Solution : Given

$$p = 18 \text{ mm}$$
$$d = 90 \text{ mm}$$
$$W = 5\ kN$$
$$\mu = 0.095$$
$$L = 500 \text{ mm}$$
$$\mu = \tan \phi = 0.0\ 95$$
$$\therefore \quad \phi = \tan^{-1}(0.095) = 5.4268°$$
$$\tan \alpha = \frac{p}{\pi d} = \frac{18}{\pi \times 90} = 0.06366$$
$$\therefore \quad \alpha = \tan^{-1}(0.06366)$$
$$= 3.6425°$$

$\therefore$ From equation (2.28),

$$P = \frac{dW}{2L}\tan(\alpha + \phi)$$

$$= \frac{90 \times 5000}{2 \times 500}\tan(3.6425° + 5.4268°)$$

$$= 450 \tan (9.0693°)$$
$$= 450 \times (0.1596 = 71.88\ A)$$

(*b*) Efficiency

$$\eta = \frac{\tan\alpha}{\tan(\alpha + \phi)} = \frac{\tan(3.6425°)}{\tan(3.6425 + 5.4268°)}$$

$$= \frac{0.06366}{0.1596} = 0.3888 \text{ or } 38.88\%$$

2.10 EQUILIBRIUM OF WEDGE

Wedge is a metalic piece triangular in shape with two sides at right angles to each other. The diagram of the wedge *ABCD* is shown in Fig. 2.21. The plane face *DC* is guided to move leftward by applying effort on the face *AD*. The surface *AB* is inclined and has sliding contact with equally inclined face *EF* of the block *EFHK* of weight *W* to be lifted. The block *EFHK* is guided to side up in the guides. When the effort *P* is applied on the wedge, there will be normal reaction R_{TN} of the guide on the face *DC* alongwith the frictional force $\mu_1 R_{TN}$ to give total resultant reaction R_1 on the wedge inclined at an angle $\tan^{-1}(\mu_1)$ with the normal (Fig. 2.21 (c)). Similarly, on the face *AB* of the wedge, there will be total reaction R_2 inclined at $\tan^{-1}(\mu_2)$ with the normal to the surface *EF* as shown in Fig. 2.21. Then R_2 can be resolved into two components–one vertical along the axis of block *EFHK* which will be actively raise the block. The other to component of R_2 which will be horizontal will give rise to bending moments which in turn will result in giving reactions R_2 and R_m at the contact points between the block EFHK and its guides because of the running clearance between the sliding pairs.

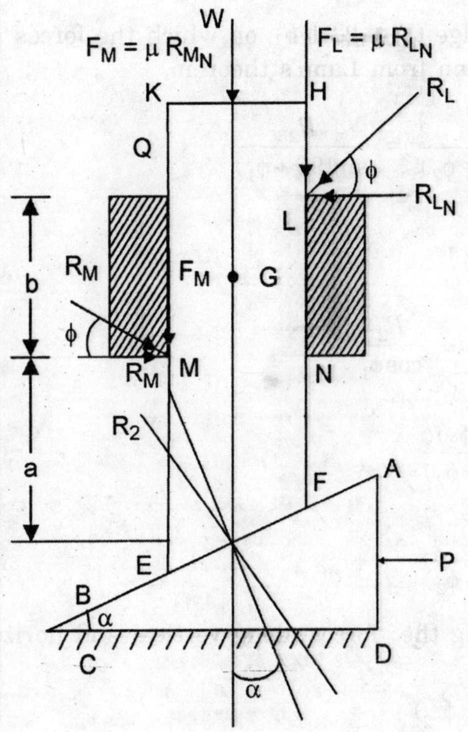

(a)

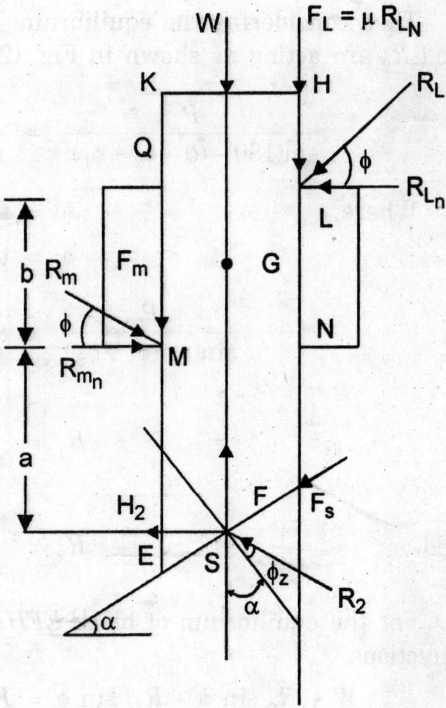

(b)

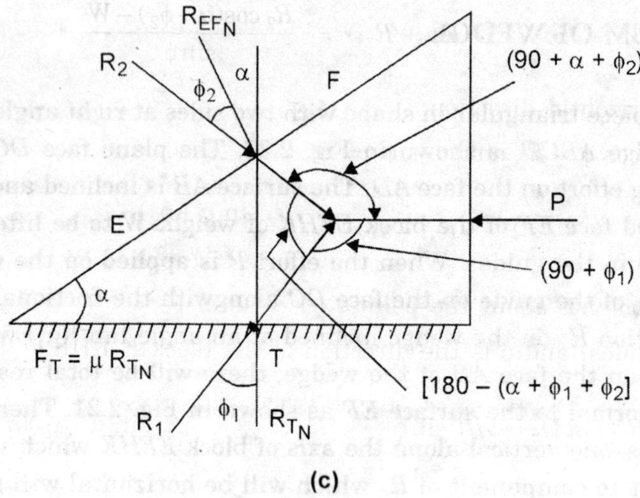

(c)

Fig. 2.21

Let the length of the bearing $QLNM$ be $= b$

Diameter of the bearing $= c$

Distance of point S from the edge of the bearing MN = overhang

$= a$

Then considering the equilibrium of the wedge (Fig. 2.21 *b*), on which the forces P, R_1 and R_2 are acting as shown in Fig. (220 (b). Then from Lami's theorem,

$$\frac{P}{\sin(180-(\alpha+\phi_2+\phi_1)]} = \frac{R_1}{\sin(90+\alpha+\phi_2)} = \frac{R_2}{\sin(90+\phi_1)}$$

Where, $\phi_1 = \tan^{-1}(\mu_1)$

$$\phi_2 = \tan^{-1}(\mu_2) \qquad ...(2.39)$$

or
$$\frac{P}{\sin(\alpha+\phi_1+\phi_2)} = \frac{R_1}{\cos(\alpha+\phi_2)} = \frac{R_2}{\cos\phi_1}$$

$\therefore$
$$R_1 = \frac{P\cos(\alpha+\phi_2)}{\sin(\alpha+\phi_1+\phi_2)} \qquad ...(2.40)$$

and
$$R_2 = \frac{P\cos\phi_1}{\sin(\alpha+\phi_1+\phi_2)} \qquad ...(2.41)$$

For the equilibrium of block $EFHK$, resolving the forces along vertical and horizontal directions,

$$W + R_2 \sin\phi + R_M \sin\phi = R_2 \cos(\alpha + \phi_2)$$

$$(R_L + R_M)\sin\phi = R_2\cos(\alpha + \phi_2) - W$$

Where, $\phi = \tan^{-1}(\mu)$

and μ is co-efficient of friction between bearing and the sliding block $EFHK$.

or $$(R_L + R_M) = \frac{R_2 \cos(\alpha + \phi_2) - W}{\sin\phi} \qquad ...(2.42)$$

Similarly, for horizontal direction,

$R_M \cos\phi - R_L \cos\phi - R_2 \sin(\alpha + \phi_2) = 0$

or $$(R_M - R_L) = \frac{R_2 \sin(\alpha + \phi_2)}{\cos\phi} \qquad ...(2.43)$$

Again taking moments about the point,

G (CG of the guides) and c is the direction of the block E RHK

$$-(R_M \cos\phi)\frac{b}{2} - (R_M \sin\phi)\frac{c}{2} + (R_L \sin\phi)\frac{c}{2} - (R_L \cos\phi)\frac{b}{2} + R_2 \sin(\alpha + \phi_2)(a + \frac{b}{2}) = 0$$

or $$-R_M\left[\frac{b}{2}\cos\phi + \frac{c}{2}\sin\phi\right] + R_L\left(\frac{c}{2}\sin\phi - \frac{b}{2}\cos\phi)\right) + R_2\left(a + \frac{b}{2}\right)\sin(\alpha + \phi_2) = 0$$

or $$-\left(\frac{b}{2}\cos\phi\right)(R_M + R_L) + \left(\frac{c}{2}\sin\phi\right)(R_L - R_m) = -R_2\left(a + \frac{b}{2}\right)\sin(\alpha + \phi_2) \qquad ...(2.44a)$$

Then from equation (2.41), (2.42), (2.43) and (2.44),

$$-\left(\frac{b}{2}\cos\phi\right)\frac{1}{\sin\phi}[R_2 \cos(\alpha + \phi_2) - W] + \left(\frac{c}{2}\sin\phi\right)\frac{[-R_2 \sin(\alpha + \phi_2)]}{\cos\phi}$$

$$= -R_2\left(a + \frac{b}{2}\right)\sin(\alpha + \phi_2)$$

or $$-\frac{b}{2\tan\phi}[R_2 \cos(\alpha + \phi_2) - W] - \frac{c}{2}R_2 \tan\phi \sin(\alpha + \phi_2) = -R_2\left(a + \frac{b}{2}\right)\sin(\alpha + \phi_2)$$

$$-\frac{b}{2}\frac{1}{\tan\phi}[R_2 \cos(\alpha + \phi_2)W] - \frac{c}{2}R_2 \tan\phi \sin(\alpha + \phi_2) = -R_2\left(a + \frac{b}{2}\right)\sin(\alpha + \phi_2)$$

or $$R_2\left[\frac{-b}{2}\frac{1}{\tan\phi}\cos(\alpha + \phi_2) - \frac{c}{2}\tan\phi \sin(\alpha + \phi_2) + \left(a + \frac{b}{2}\right)\sin(\alpha + \phi_2)\right] = -\frac{Wb}{2\tan\phi}$$

or $$-R_2\left[\frac{b}{2}\frac{1}{\tan\phi}(\cos(\alpha + \phi_2) + \frac{c}{2}\tan\phi \sin(\alpha + \phi_2) - a\sin(\alpha + \phi_2) - \frac{b}{2}\sin(\alpha + \phi_2)\right] = \frac{-Wb}{2\tan\phi}$$

But $\tan\phi = \mu$ and multiplying throughout with $2\tan\phi$

$$\therefore R_2\left[b\cos(\alpha + \phi_2) + c\mu^2 \sin(\alpha + \phi_2) - 2\alpha\mu \sin(\alpha + \phi_2) - \mu b \sin(\alpha + \phi_2)\right] = Wb$$

$$\therefore \quad W = R_2\left[\cos(\alpha+\phi_2)+\sin(\alpha+\phi_2)\left\{\frac{c\mu^2}{b}-\frac{2\alpha\mu}{b}-\mu\right\}\right]$$

$$= \frac{R_2}{b}\left[b\cos(\alpha+\phi_2)+(c\mu^2-2a\mu-b\mu)\sin(\alpha+\phi_2)\right] \quad ...(2.44b)$$

Substituting value of R_2 from equation (2.41),

$$P = \frac{W\sin(\alpha+\phi_1+\phi_2)}{\cos\phi_1}\left[\frac{b}{b\cos(\alpha+\phi_2)+\mu(c\mu-2a\ ?b)\sin(\alpha+\phi_2)}\right]$$

$$P = \frac{W\sin(\alpha+\phi_1+\phi_2)}{\cos\phi_1}\left[\frac{b}{b\cos(\alpha+\phi_2)+\mu(c\mu-2a-b)\sin(\alpha+\phi_2)}\right]$$

or
$$P = \frac{W\sin(\alpha+\phi_1+\phi_2)}{\cos\phi_1\cos(\alpha+\phi_2)}\left[\frac{b}{b+\mu(c\mu-2a-b)\tan(\alpha+\phi_2)}\right]$$

or
$$P = \frac{W\sin(\alpha+\phi_1+\phi_2)}{\cos\phi_1\cos(\alpha+\phi_2)}\left[\frac{1}{1+\mu(\mu\frac{c}{b}-\frac{2a}{b}-1)\tan(\alpha+\phi_2)}\right] \quad ...(2.45)$$

2.10.1 Limiting Value of α

We see from equation (2.45) that for P to have positive value, the denominator of the *R.H.S.* of equation (2.45) must be greater than zero or it must be positive,

i.e.,
$$1+\mu\left(\mu\frac{c}{b}-\frac{2a}{b}-1\right)\tan(\alpha+\phi_2)\geq 0$$

or
$$\mu\left(\mu\frac{c}{b}-\frac{2a}{b}-1\right)\tan(\alpha+\phi_2)\geq -1.0$$

or
$$\mu\left(\frac{2a}{b}+1-\frac{2a}{b}-1\right)\tan(\alpha+\phi_2)\leq 1.0$$

or
$$\boxed{\tan(\alpha+\phi_2)\leq\frac{1}{\mu\left(\frac{2a}{b}+1-\mu\frac{c}{b}\right)}} \quad ...(2.46)$$

or
$$\frac{\tan\alpha+\tan\phi_2}{1-\tan\alpha\tan\phi_2}\leq\frac{1}{\mu f} \quad ...(2.47)$$

Where,
$$f = \left(\frac{2a}{b}+1-\mu\frac{c}{a}\right)$$

and $$\tan\phi_2 = \mu_2$$

or $$\frac{\tan\alpha + \mu_2}{1 - \mu_2 \tan\alpha} \leq \frac{1}{\mu f}$$

or $$\mu f \tan\alpha + \mu \mu_2 f = 1 - \mu_2 \tan\alpha$$

or $$(\tan\alpha)[\mu f + \mu_2] \leq 1 - \mu M_2 f$$

or $$\tan\alpha \leq \frac{(1 - \mu\mu_2 f)}{\mu f + \mu_2} \quad \text{...(2.48)}$$

for + ve α $$1 \geq \mu\mu_2 f$$

or $$1 \geq \mu\mu_2\left[\frac{2a}{b} + 1 - \mu\frac{c}{b}\right]$$

$$\frac{1}{\mu\mu_2} > \frac{1}{b}(2a - \mu c) + 1$$

$$\frac{1}{\mu\mu_2} - 1 \geq \frac{1}{b}(2a - \mu c) \quad \text{...(2.49)}$$

Therefor, the wedge action to be successful the equation (2.49) must be satisfied and thus, the limting value of α is to be found from equation (2.49. If it is more than this limiting value, it will not work.

EXAMPLE

2.15 : *A block of weight W is to be lifted with the help of a wedge (Fig.2.21). It the ratio $a/b = 5.00$, $b = c$ and ,$\mu = \mu_2 = 0.2$, find the limiting value of the wedge angle α.*

Solution: From equation (2.48),

$$\tan a \leq \frac{1 - \mu\mu_2 f}{\mu_2 + \mu f}$$

Where, $$f = \left(\frac{2a}{b} + 1 - \mu\frac{c}{b}\right)$$

Given data $$\frac{a}{b} = 5$$

$$c = b$$

Then, $$f = (2 \times 5 + 1 - 0.2 \times 1)$$

$$= 10.8$$

$$\tan\alpha \leq \frac{1.0 - 0.2 \times 0.2 \times 10.8}{0.2 + 0.2 \times 10.8}$$

$$\leq 0.2407$$

or $$\alpha = \tan^{-1}(0.2407)$$

$$= 15.04^0 \quad \textbf{Ans.}$$

EXAMPLE

2.16 : *A block of weight 5 kN is be lifted with the help of a wedge of angel 10°. The block is guided to move in the guides (Fig. 2.21), having diameter c = 25 mm and length b = 25 mm. The overhang a = 100 mm. If the co–efficient of friction between the sliding faces is 0.2, find the value of the effort P to be applied on the wedge.*

Solution : Refer to Fig. 2.20 Given date

$$W = 2\ kN$$
$$\alpha = 10°$$
$$c = 25 \text{ mm}$$
$$b = 25 \text{ mm}$$
$$e = 100 \text{ mm}$$
$$\mu = \mu_1$$
$$= \mu_2 = 0.2$$
$$\phi_1 = \phi_2 = \mu$$
$$= \tan^{-1}(.2) = 11.31°$$

From equaiton 2.45,

$$P = \frac{W\sin(\alpha+\phi_1+\phi_2)}{\cos\phi, \cos(\alpha+\phi_2)}\left[\frac{1}{1+\mu\left(\mu\frac{c}{b}-\frac{2a}{b}-1\right)\tan(\alpha+\phi_2)}\right]$$

$$= \frac{5\sin(10+11.31+11.31)}{\cos(11.31)\cos(10+11.31)}\left[\frac{1}{1+0.2\left(.2\times\frac{25}{25}-\frac{2\times100}{25}-01\right)\tan(21.3)}\right]$$

$$= \frac{5\sin(32.62°)}{\cos(11.31)\cos(21.31)}\left[\frac{1}{1+.2(.2-8-1)\tan)(21.31)}\right]$$

$$= \frac{5\times0.5824}{\cos(11.31)\cos(21.31)}\left[\frac{1}{1-.2\times8.8(0.3900)}\right]$$

or $$P = \frac{5\times0.5824}{0.9806\times0.9316}\left[\frac{1}{1-0.6864}\right]$$

$$= \frac{5\times0.5824}{0.9806\times0.9316}\times\frac{1}{0.3136}$$

$$= 10.22\ kN \text{ Ans.}$$

2.17 : *Two symmetrically placed wedges are used to lift a load of 1 kN as shown in Fig. 2.22. The wedge angle of catch is 10° and the limiting friction angle of each shiding surface is 12°. Find The effort to be applied to each wedge*

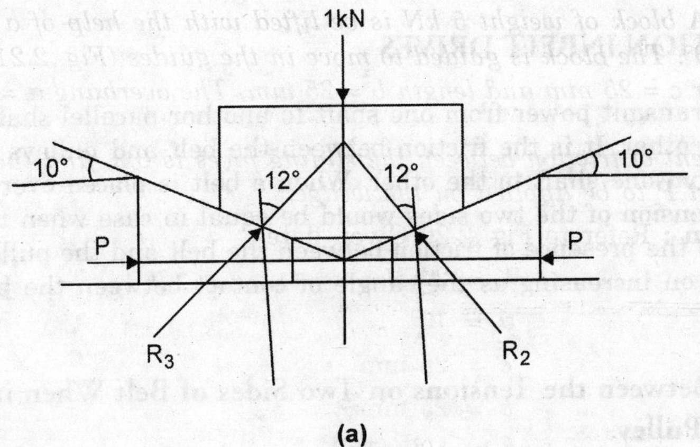

(a)

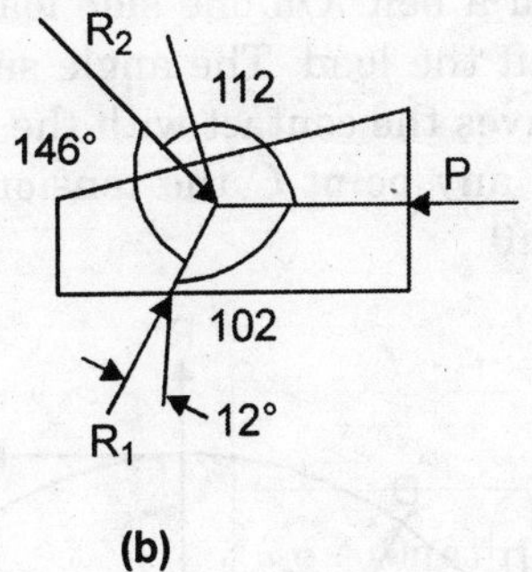

(b)

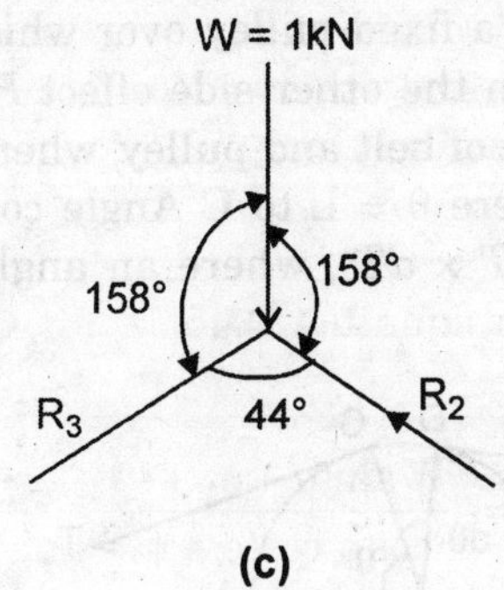

(c)

Fig. 2.22

From the force body diagram of the block,

$$\frac{R_2}{\sin 169°} = \frac{W}{\sin 22}$$

Here, $W = 1\ kN$

$$R_2 = 1 \times \frac{\sin 169°}{\sin 22°} = \frac{1(0.1908}{0.3746}$$

$$= 0.5093\ kN$$

Again from the force body diagram of wedges,

$$\frac{P}{\sin 146°} = \frac{R_2}{\sin 102} = \frac{R_1}{\sin 112}$$

$$\therefore \quad P = R_2 \frac{\sin 146°}{\sin 102°}$$

$$= \frac{0.5093 \times .5591}{0.9711}$$

$$= 0.2911\ kN \text{ Ans.}$$

EXAMPLE

2.11 BELT FRICTION INBELT DRIVES

Belts are used to transmit power from one shaft to another parallel shaft placed at a large distance from each other. It is the friction between the belt and pulleys that is responsible for transmitting from one shaft to the other. When a belt is placed over the circumference of the pulley, the tension of the two sides would be equal in case when there is no friction, But actually due to the presence of friction between the belt and the pulley face the tension in the belt will go on increasing as the angle of contact between the belt and the pulley increases.

2.11.1 Relation Between the Tensions on Two Sides of Belt When it is Wrapped Around a Pulley

Fig. 2.23 shows a fixed pulley over which is wrapped a belt. On one side load W is hanging from the belt, on the other side effect P applied is lift the load. The angle subtended by the points of contact of belt and pulley, where the belt leaves the contact with the pulley or angle of embrace. where θ = L to C Angle contact. Let at any point C the tension in the belt be T and at D is $(T \times dT)$, where an angle $C\ O\ D' = d\theta$

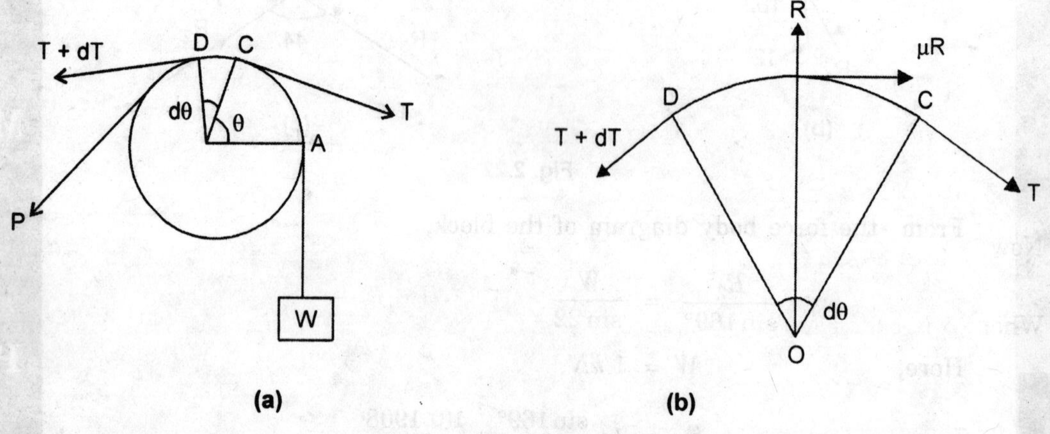

Fig. 2.23

For face body diagram of the element, CD is shown in Fig. (2.23)b. For the equilibrium of the element CD resolving the forces radially and along tangential direction we get,

$$R - 2T\sin\frac{d\theta}{2} - (T \times dT)\sin\frac{d\theta}{2} = 0 \quad ...(2.50)$$

and

$$\mu R + T\cos\frac{d\theta}{2} - (T + dT)\cos\frac{d\theta}{2} = 0 \quad ...(2.51)$$

When $d\theta$ is small, $\sin\frac{d\theta}{2} \simeq \frac{d\theta}{2}$

and $\cos\frac{d\theta}{2} \simeq 1$

∴ Then equations (2.50) and (2.51) reduce to,

$$R - T\frac{d\theta}{2} - T\frac{d\theta}{2} - dT\frac{d\theta}{2} = 0$$

or

$$R - 2T\frac{d\theta}{2} = 0$$

As product of dq and dT is very small to be neglected,

$$R = Td\theta \qquad ...(2.52)$$

Aso from (2.51),

$$\mu R + T - (T + dT) = 0$$

or

$$\mu R = dT$$

or From (2.52),

$$\mu(Td\theta) = dT$$

or

$$\frac{dT}{T} = \mu d\theta$$

Interacting from points A and B,

$$\int_{T_1}^{T}\frac{dT}{T} = \int_{0}^{\theta}\mu d\theta$$

$$ln\left(\frac{T_2}{T_1}\right) = \mu\theta$$

or

$$\frac{T_2}{T_1} = e^{\mu\theta}$$

or

$$T_2 = T_1 e^{\mu\theta} \qquad ...(2.53)$$

Now $T_1 = W \qquad T_2 = P$

i.e.,

$$P = We^{\mu\theta} \qquad ...(2.54)$$

Where θ is called angle of contact or angle of embrace

EXAMPLE

2.18 : *While lifting a weight with the help of a rope passing over a pulley a worker applies an effort of 500N. If the co–efficient of friction between the rope and fixed pulley is 0.25, and the angle of contact θ = 135° (Fig. 2.24), what load he can lift under the following conditions: (i) when the load is just on the point of moving up (ii) when the load is just on the point of moving down.*

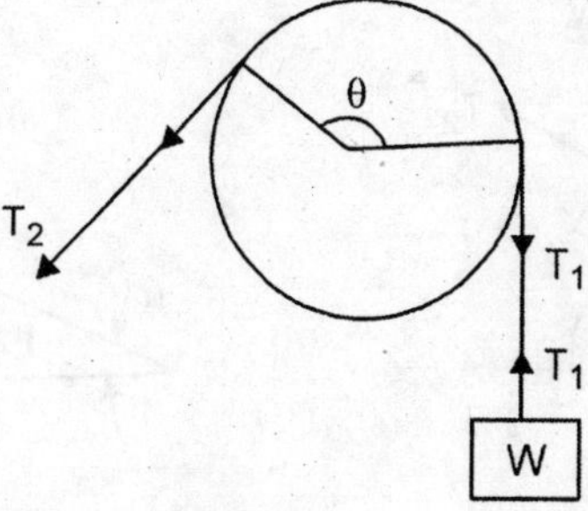

Fig. 2.24

E X A M P L E

Solution : (*a*) when the load is just on the point of moving up,

and $T_2 > T_1$

Then $T_2 = T_1 e^{\mu\theta}$

$$500 = We^{0.25\left(\frac{135\times\pi}{180}\right)}$$

$$500 = W\,e^{0.589} = W(1.80)$$

or $$W = \frac{500}{1.80} = 277.469\ N$$

(*b*) When the load is just on the point of moving down, then $T_1 > T_2$

$$T_1 = T_2 e^{\mu\theta}$$

$$\theta = \frac{135\times\pi}{180} = \frac{3\pi}{4}$$

$$W = P\ e^{\mu\theta}$$

$$= 500\times e^{\left(.25\times\frac{3\pi}{4}\right)} = 500\times 1.802$$

i.e., $W = 910\ N$

2.19 : *In Fig (2.25)(a) the co–efficent of friction is 0.2 between the rope and fixed pulley and between other surfaces of contact $\mu = 0.3$. Determine the maximum weight W to prevent the downward motion of the 100N body. (U.P tech the 2001)*

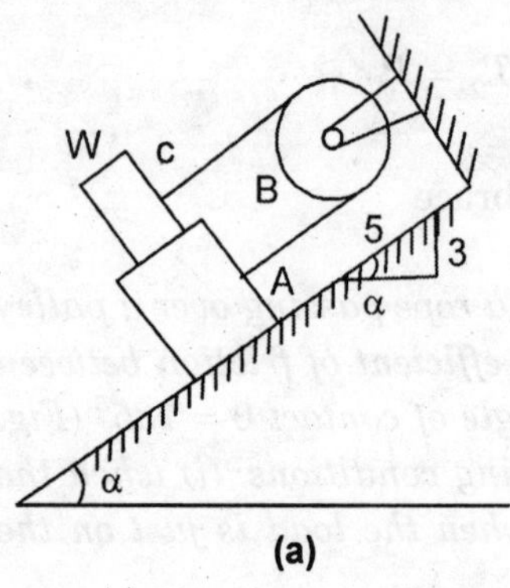

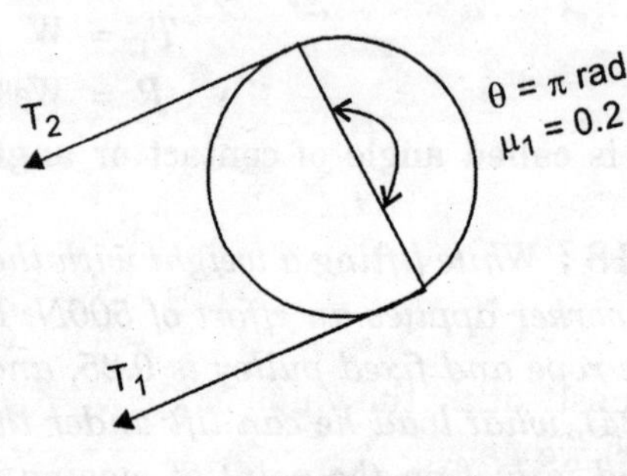

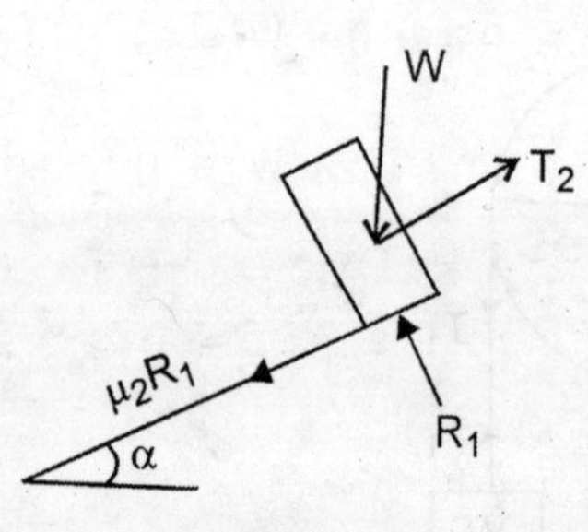

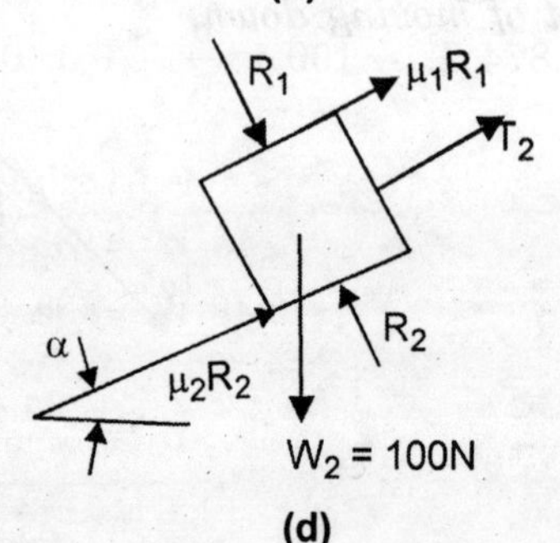

Fig. 2.25

Solution : Since the 100 N body is just on the point of moving down the frictional force on it will be acting in the upward direction . Accordingly the tension in the portion AB will be less than that in BC. Then the free body diagrams of various components are drawn in Fig.(2.25b), (2.25c) and (2.24 d) as shown.

For the equilibrium of weight W, we have after resolving the forces along and normal to the plane,

$$T_2 = \mu_2 R_1 + W \sin \alpha \qquad ...(A)$$

and $$R_1 = W \cos \alpha$$

$$= W \times \left(\frac{4}{5}\right) = 0.8\ W \qquad ...(B)$$

But $$T_1 = T_2\ e^{\mu\theta}$$

$$= T_2\ e^{0.2\pi}$$

$$= 1.874\ T_2 \qquad ...(C)$$

From equations (A), and (B) and (C),

$$T_2 = 0.3 \times 0.8W + W \times \frac{3}{5}$$

$$= .24\ W + 0.6\ W$$

$$= 0.84\ W \qquad ...(D)$$

Again considering the free body diagram of the 100 N body (Fig. 2.25 d) and resolving the forces along and normal to the plane, we get,

$$T_1 - 100 \sin \alpha + \mu_2 R_2 + R_1 \mu_1 = 0 \qquad ...(E)$$

and $$R_1 - R_2 + 100 \cos \alpha = 0$$

or $$R_2 = R_1 + 100 \times \frac{4}{5}$$

$$= 0.8\ W + 80 \qquad ...(F)$$

Substituting the value of T_1, R_1, R_2 for equations (B), (C), (D) and F note (E) we get,

$$1.874\ T_2 - 100 \times \frac{3}{5} + 0.3\ (0.8\ W + 80) + 0.3 \times 0.8\ W = 0$$

or $$1.874 \times .84 - 60 + 0.24\ W + 24 + 0.24\ W = 0$$

or $$48\ W = 35 - 1.874 \times .84$$

$$0.48\ W = 34.4258$$

or $$W = \frac{34.4258}{0.48}$$

$$= 71.7205\ N \text{ Ans.}$$

EXAMPLE

2.12 POWER TRANSMISSION BY BELT DRIVE

Belt drive consists of two pulleys, one on driver shaft and the other on driven shaft with belt passing over them .The arrangement shown in Fig. 2.26 is called the open belt arrangement. The direction of rotation of the two shafts is some but the speed ratio of the two shafts is inversely proportional to the radii of the pulleys because the linear velocty of the belt is equal to the peripheral speed of the pulley provided when there is no slip.

i.e., v = linear velocity of belt

$$= r_1\omega_1$$

$$= r_2\omega_2$$

$$= \frac{2\pi R_1 N_1}{60} = \frac{2\pi R_2 N_2}{60}$$

Where ω_1 = angular speed of driving pulley in rad/sec

ω_2 = Angular speed of driven pulley in rad/sec

N_1 = angular speed of driving pulley in r.p.m

N_2 = angular speed of driven pulley in r.p.m

θ_1 = angle of contact or lap on driving pulley

θ_2 = angle of contact on driven pulley

a = angle O_0O_1K when O_2K is perpendicular to O, A

Fig. 2.26

L = Length of the belt $= AB + BCD + DE + EFA$

$$= 2AB + BCD + EFA$$

$$= 2KO_2 + r_2\theta_2 + r_1\theta_1$$

Now $$KO_2 = \sqrt{(O_1O_2)^2 - (O_1K)^2}$$

$$= \sqrt{d^2 - (r_1 - r_2)^2}$$

When d = distance between shafts,

And $$\cos\alpha = \frac{r_1 - r_2}{d}$$

or $$\alpha = \cos^{-1}\left(\frac{r_1 - r_2}{d}\right)$$

and $$\theta_1 = 2\pi - 2\alpha = 2\left[\pi - \cos^{-1}\left(\frac{r_1 - r_2}{d}\right)\right]$$

$$\theta_2 = 2\alpha = 2\cos^{-1}\left(\frac{r_1 - r_2}{d}\right)$$

$$\therefore \quad L = 2\sqrt{d^2 - (r_1 - r_2)^2} + 2r_2\cos^{-1}\left(\frac{r_1 - r_2}{d}\right) + 2r_1\left[\pi - \cos^{-1}\left(\frac{r_1 - r_2}{d}\right)\right] \quad ...(2.55)$$

Power Transmission

Since the angle of lap on smaller pulley is less than that on larger pulley, the slip will occur on smaller pulley if it has to oceur. Therefore, for calculating the power transmission the angel of contact on smaller pulley θ_2 is taken

Since $\theta_2 < \theta_1$

Therefore, if T_1 is the tension on right side which will be AB and T_2 in the portion

$$T_1 = T_2^{e^{\mu\theta_2}} \quad ...(2.56)$$

Now toque in smaller $= (T_1 - T_2)\, r_2 - Nm$

Velocity pulley of belt $= r_2\omega_2 = r_1\omega_1$

$$= \frac{2\pi N_1}{60}\left(\frac{D_1}{2}\right)$$

$$= \frac{\pi D_1 N_1}{60} - \frac{2\pi r_2 N_1}{60} \quad m/\text{sec}$$

$\therefore$ Work done per sec $\cup = v_2(T_1 - T_2)$

$$= \frac{\pi D_2 N_2}{60}(T_1 - T_2)$$

$$P = \frac{\pi D_2 N_2}{60}\left(T_2 e^{\mu\theta_2} - T_2\right)$$

$$= \frac{2\pi N_2 D_2 T_2}{60}\left(e^{\mu\theta_2} - 1\right) \text{ watt} \quad ...(2.57)$$

2.13 CENTRIFUGAL TENSION IN BELT

Let the peripheral speed of belt over the pulley be 'v', then for the small length AB of belt

$$AB = rd\theta$$

If m = mass of belt per unit length, then the centrifugal force on the portion AB

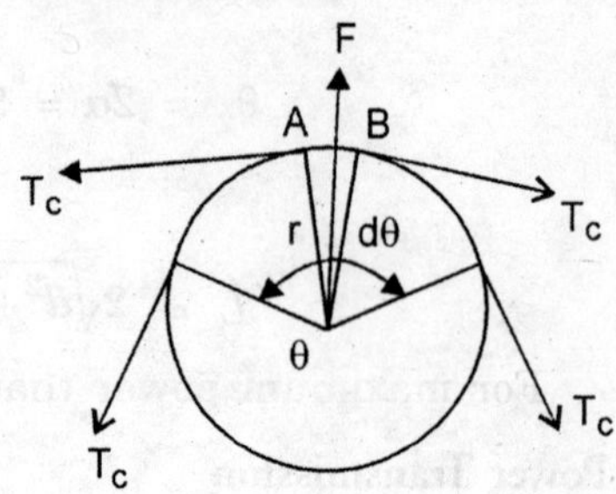

Fig. 2.27

is $$F = (mAB)\frac{v^2}{r}$$

Where, ω = angular speed in rad/sec

r = radus of rotation

$$\therefore \quad F = (mAB)\frac{\omega^2 r^2}{r}$$

$$F = m(rd\theta)\frac{v^2}{r} = mv^2\ d\theta \qquad ...(2.57)$$

For uniform speed of belt F is balanced by T_C (where T_C is centrifugal tension) on the two sides of portion AB

Therefore, resolving forces along the radial direction,

$$F = 2\ T_C\ \sin\frac{\delta\theta}{2}$$

$$\neq 2T_C\ \frac{\delta\theta}{2} = T_c\ d\theta \qquad ...(2.58)$$

$\therefore$ From equation (2.57) and (2.58),

$$mv^2 d\theta = T_C d\theta$$

or $$T_C = mv^2$$

$$= mr^2\omega^2 \qquad ...(2.59)$$

Therefore, max tension to be the belt is,

$$T_{max} = T_1 + T_C$$

Now $$P = (T_1 - T_2)v$$

$$= \left(T_1 - \frac{T}{e^{\mu\theta_2}}\right)v$$

or $$P = T_1\left(1 - \frac{1}{e^{\mu\theta}}\right)v \qquad ...(2.60)$$

Now $$T_{max} = T_1 + T_C$$

or $$T_1 = (T_{max} - T_C) \qquad ...(2.61)$$

From equation (2.60) and (2.61),

$$P = (T_{\max} - T_C)\left[1 - \frac{1}{e^{\mu\theta}}\right]v$$

$$= (T_{\max} - mv^2)\left[1 - \frac{1}{e^{\mu\theta}}\right]v$$

$$= (vT_{\max} - mv^3)\left[1 - \frac{1}{e^{\mu\theta}}\right]$$

For maximum power that can be transmittal by a belt,

$$\frac{\partial P}{\partial v} = 0$$

or
$$\frac{\partial}{\partial v}\left[\left(vT_{\max} - mv^3\right)\left(1 - \frac{1}{e^{\mu\theta}}\right)\right] = 0$$

or
$$T_{\max} - 3mv^2 = 0$$

$$T_{max} = 3\ mv^2$$

$$= 3T_C$$

i.e.,
$$T_C = \frac{1}{3}\ T_{max} \qquad \text{...(2.62a)}$$

Also
$$v = \sqrt{\frac{T_{\max}}{3m}} \qquad \text{...(2.62b)}$$

Now
$$T_{max} = T_1 + T_C$$

$$= T_1 + \frac{T_{\max}}{3}$$

i.e.,
$$T_1 = T_{\max} - \frac{T_{\max}}{3}$$

$$= \frac{2}{3}T_{\max} \qquad \text{...(2.63)}$$

Thus, the condition for maximum power transmission is,

$$T_C = \frac{1}{3}T_{\max}$$

or
$$T_1 = \frac{2}{3}T_{\max}$$

Corresponding expression for maximum power from equation 2.60 and 2.62,

$$P_{max} = T_1\left[1 - \frac{1}{e^{\mu\theta}}\right]v$$

$$= \frac{2}{3}T_{\max}\left[1 - \frac{1}{e^{\mu\theta}}\right]\sqrt{\frac{T_{\max}}{3m}} \qquad \text{...(2.64)}$$

E X A M P L E

2.20 : *An open belt drive connects. two pulleys 3000 mm diameter and 600 mm diameter on two parallel shafts 3m away from each other The maximum belt tension that the belt can sustain is 1900 N. The bigger pulley is the driver pulley and running at 250 r.p.m. The co–efficient of friction is 0.25. Calculate (a) the length of the belt (b) power transmitted and the torque on each pulley.*

Solution : Given data $r_1 = 1500\ mm$

$$r_2 = 300\ mm$$

$$d = 3000\ mm$$

$$L = 2\sqrt{d^2 - (r_1 - r_2)^2} + 2r_2 \cos^{-1}\left(\frac{r_1 - r_2}{d}\right) + 2r_1\left(\pi - \cos^{-1}\left(\frac{r_1 - r_2}{d}\right)\right)$$

$$L = 2\sqrt{(3000)^2 - (1500-300)^2} + 2\times 300\cos^{-1}\left(\frac{1500-300}{3000}\right)$$

$$+2\times 1500\left[\pi - \cos^{-1}\left(\frac{1500-300}{3000}\right)\right]$$

$$= 200\sqrt{(30)^2 - (12)^2} + 600\times\left(\frac{\pi}{180}\times 66.2°\right) + 3000\left[\pi - \frac{\pi}{180}\times 66.2\right]$$

$$= 5400 + 600 \times 1.155 + 3000 \times 1.9862$$

$$= 5400 + 693 + 5958.6$$

$$= 7191.6\ mm = 7.196\ m$$

(*b*) Lap angle on smaller pulley (Fig. 2.25),

$$\theta_2 = 2\alpha = 2\cos^{-1}\left[\frac{r_1 - r_2}{d}\right]$$

$$= 2\cos^{-1}\left[\frac{1500-300}{3000}\right]$$

$$= 2\times\frac{\pi}{180}(66.2°) = 2.31 rad$$

$$\therefore \quad \frac{T_1}{T_2} = e^{0.5 \times 2.31} = e^{0.5775} = 1.7463$$

or

$$\frac{T_1}{T_2} = 1.7463$$

$$v = \frac{\pi D_1 N_1}{60}$$

$$= \frac{\pi}{60}\times\frac{3000\times 250}{1000} m/s$$

$$= \frac{\pi}{20} \times 250 = 39.27\ m/s$$

$$\therefore \quad \text{Power} = (T_1 - T_2)v = \left(T_1 - \frac{T_1}{e^{\mu\theta}}\right)v$$

$$= T_1\left[1 - \frac{1}{1.7463}\right]39.27 = T_1\left(\frac{0.7463}{1.7463}\right)39.27$$

Now $T_1 = T_{max} = 1900\ N$

$$\therefore \quad \text{Power} = 1900 \times \frac{.7463}{1.7463} \times 39.27 N - m/\text{sec}$$

$$= 55902.6\ W$$

$$= 55.90.2\ kW$$

(*c*) Torque on Driver pulley,

$$T_2 = \frac{T_1}{1.7463}$$

$$= 1088$$

$$\text{Torque} = (T_1 - T_2)\ r_1$$

$$= (1900 - 1088)\frac{1500}{1000}\ N - m$$

$$= 2568\ Nm$$

Torque on smaller pulley

$$= (T_1 - T_2)r_2$$

$$= -(1900 - 1088)300$$

$$= 513.6\ Nm$$

2.21 : ***Determine the maximum power that can be transmitted using a belt of 100 mm × 10mm with an angle of lap of 160°. The density of belt is 10^{-3} gm/mm^3 and co–efficient of friction many be taken as 0.25. The tension in the belt should not exceed 1.5 N/mm^2 (AMIE 1984).***

Solution : Given data,

$$\theta = 160° = \frac{160 \times \pi}{180} = 2.79 \text{ rad}$$

$$\rho = \text{density of belt}$$

$$= 10^{-3} gm/mm^3 = 10^{-6}\ kg/mm^3$$

$$\therefore \quad m = \text{mass per unit length}$$

$$= \pi(\text{volume of } 1m \text{ length})$$

$$= \rho \times 1000 \times (100 \times 10)$$

$$= 10^{-6} \times 10^6 = 1 kg/mm^3$$

E X A M P L E

EXAMPLE

Allowable stress sir in belt

$$= 1.5\ N/mm^2$$

$$\therefore \quad T_{max} = 1.5 \times \text{area}$$

$$= 1.5 \times 100 \times 10 = 1500\ N$$

$$\therefore \quad T_C = \text{Centrifugal tension}$$

$$= \frac{T_{max}}{3} = \frac{1500}{3} = 500\ N$$

Now $\quad T_{max} = T_1 + T_C$

$$\therefore \quad T_1 = \text{Tension on tight side}$$

$$= T_{max} - T_C$$

$$= 1500 - 500 = 1000N$$

Now $\quad \dfrac{T_1}{T_2} = e^{\mu\theta} = e^{0.25 \times 2.79}$

$$= e^{.6975} = 2.00$$

$$\therefore \quad T_1 = 2T_2$$

$$T_1 = \frac{1000}{2} = 500\ N$$

Now $\quad T_C = mv^2$

$$\therefore \quad v^2 = \frac{T_C}{m} = \frac{500}{1} = 500$$

or $\quad v = \sqrt{500} = -22.36\ m/s$

Therefore, $\quad P = (T_1 - T_2)v$

$$= (1000 - 500)22.36\ W$$

$$= \frac{11180}{1000}\ kW = 11.18\ kW \text{ Ans.}$$

2.14 INITIAL TENSION IN BELT

When a belt is wrapped around two pulleys for transmitting power, it is wrapped around pulleys tightly so that there is ample grip between the belt and pulleys. The tension in the belt when it is stationary is called the initial tension and is represented by To.

While running let $\quad T_1$ = Tension on tight sids of belt

T_2 = Tension on slack side of belt.

Assuming the belt to be elastic, Let α be the co-efficient of linear stretch her unit tension. Therefore, while running the the belt length will change on both sides as on tight side the tension increases from T_0 to T_1. Howerver, it the belt material is assumed elastic, the total length of the belt will remain constant.

i.e., $$\alpha(T_1 - T_0) = \alpha(T_0 - T_2)$$

or $$2T_0 = T_1 + T_2$$

or $$T_0 = \frac{T_1 + T_2}{2} \quad \text{...(2.65)}$$

In case centrifugal tension is also taken into account,

$$T_0 = \frac{T_1 + T_2 + 2T_e}{2} \quad \text{...(2.66)}$$

EXAMPLE

2.22 : *A belt is stretched over two identical pulleys of diameter D metres the initial tension in the belt through out is 2.4 kN when the pulleys are at rest. In using these pulleys and the belt to transmit torque, it is found that the increase in tension on one side in equal to the decrease on the other side. Find the maximum torgue that can be transmitted by the belt given, the co–efficient of friction between the belt and pulleys is 0.30.*

(U.P.T.U. 2002, 2003)

Solution : Given data

Diameter of each pulley = D

Initial tension = T_0 = 2.4 kN

μ = 0.30

Let T_1 = Tension on tight side

T_2 = Tension *on* slack side

DT = Increase tension on light side when sunning

DT = decrease in tension or slack side when running

Then while running $T_1 = T_0 + DT$

and $T_2 = T_0 - dT$

$$\therefore \quad T_1 + T_2 = (T_0 + \Delta T) + (T_0 - \Delta T)$$

$$= 2T_0$$

or $$T_0 = \frac{T_1 + T_2}{2}$$

But it is given that T_0 = 2.4 kN

= 2400 N

$$\therefore \quad T_1 + T_2 = 2T_0 = 2 \times 2100$$

$$= 4800.0\ kN \quad \text{...(A)}$$

But we know $$\frac{T_1}{T_2} = e^{m\theta} = e^{0.3} \times \theta$$

Here, θ = 180° = π rad

EXAMPLE

$\therefore \quad \dfrac{T_1}{T_2} = e^{0.3\pi} = 2.566$

or $\quad T_1 = 2.566\, T_2$ ...(B)

$\therefore$ From equation (*A* and *B*,),

$$T_1 + T_2 = 4800$$

$$2.566\, T_2 + T_2 = 4800$$

or $\quad 3.566\, T_2 = 4800$

or $\quad T_2 = \dfrac{4800}{3.566}$

$$= 1346\, N$$

Then from (*A*), $\quad T_1 = 4800 - 1356$

$$= 3454\, N \textbf{ Ans.}$$

2.15 VELOCITY RATIO

In the open belt drive (Fig. 2.26), if there is no slip between the belt and pulley, the peripheral velocity of each pulley will be equal to the velocity of the belt. Thus if ω_1 and ω_2 w_2 are the angular velocities of the driver and driven pulleys respectively and v the velocity of the belt, then

$$v = r_1\omega_1 = r_2\omega_2$$

Where r_1 and r_2 are the radii of the driver and driven pulleys and the corresponding diameters are D_1 and D_2.

$$\frac{\omega_1}{\omega_2} = \frac{D_2}{D_1} \qquad \text{...(2.67}a\text{)}$$

$$= \frac{r_2}{r_1} \qquad \text{...(2.67}b\text{)}$$

In case the thickness of the belt is taken into account, then the velocity of the center live of the belt is taken and the equation (2.67) reduces to,

$$\frac{\omega_1}{\omega_2} = \frac{r_2 + t/2}{r_2 + t/2}$$

or
$$\frac{\omega_1}{\omega_2} = \frac{2r_2 + t/2}{2r_1 + t/2} = \frac{D_2 + t}{D_1 + t} \qquad \text{...(2.68)}$$

Moreover, the direction of rotation of both the pulleys is same, *i.e* both the pulleys rotate clockwise or anticlockwise.

However is the case of cross belt drive driver and driver pulleys rotates. in the opposite direction. But in this case the angle of lap is same on both the pulleys

$$2\theta = (2\pi - 2\alpha)$$

$$= 2(\pi - \alpha)$$

and
$$\cos\alpha = \frac{r_1 + r_2}{d}$$

$$\alpha = \cos^{-1}\left(\frac{r_1 + r_2}{d}\right)$$

Fig. 2.28

Length of cross-belt is,

$$L = (r_1\theta + r_2\theta) + 2O_2P$$

$$= (r_1 + r_2)\theta + 2\sqrt{d^2 - (r_1 + r_2)^2}$$

$$= (r_1 + r_2)(2\pi - 2\alpha) + 2\sqrt{d^2 - (r_1 + r_2)^2}$$

$$= 2(r_1 + r_2)(\pi - \alpha) + 2\sqrt{d^2 - (r_1 + r_2)^2}$$

$$= 2(r_1 + r_2)\left[\pi - \cos^{-1}\left(\frac{r_1 + r_2}{d}\right)\right] + 2\sqrt{d^2 - (r_1 + r_2)^2} \quad \text{....(2.69)}$$

2.16 BAND BRAKES

Figure 2.28 show as band brake. It consists of a brake drawn over which is wrapped a band at the ends of the band are connected to a lever in such way that when an effort is applied at one end of the lever, the band increases the gripping force thereby applying a braking torque on the drum.

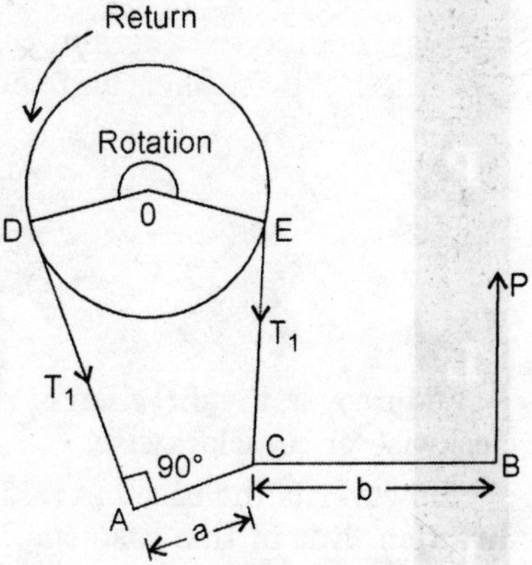

Fig. 2.29

In the arrangement shown in Fig. 2.29 one end of the band is attached to the fulcrum of the lever and the other and to the point A of the lever. The effort is applied at the end B of the lever in order to apply the braking torque on the drum.

$$T_1 = T_2 e^{m\theta} \quad \text{....(2.70)}$$

Taking moments about fulcrum C,

$$T_1 a = Pb \quad \text{...(2.71a)}$$

$$T_1 = \frac{Pb}{a} \quad \text{...(2.71b)}$$

$$T_2 = \frac{T_1}{e^{\mu\theta}}$$

Braking torque = $(T_1 - T_2)r$ where r = radun of drum,

$$T_B = \left(T_1 - \frac{T_1}{e^{\mu\theta}}\right)r$$

$$= T_1\left(1 - \frac{1}{e^{m\theta}}\right)r$$

$$T_B = \frac{Pb}{a}\left(1 - \frac{1}{e^{\mu\theta}}\right)r \quad \text{...(2.70)}$$

EXAMPLE **2.23 :** *In a band brake (Fig. 2.29), the brake drum is of 200 mm diameter the angle of contact with the band is 240° and co-efficient of friction is 0.3. If the lever arms are a = 40 mm and b = 280. and the effort applied at end B is 100N, find the value of braking torque.*

Solution : Ref Fig. 2.29

$$T_1 = T_2\, e^{\mu\theta} = T_2 e^{\frac{0.3 \times 240 \times \pi}{180}}$$

$$= T_2\, e^{1.2566} = 3.5134\, T_2$$

$$T_2 = \frac{T_1}{3.5134}$$

Taking moments about C,

$$T_1 a = P \times b$$

$$T_1 \times 40 = 100 \times 280$$

$$T_1 = \frac{280 \times 100}{40} = 700N$$

$$T_2 = \frac{700}{3.5134} = 199.23N$$

$$T_B = (T_1 - T_2)r$$

$$= (700 - 199.23)\frac{100}{1000}\, Nm$$

$$= 50.77\ Nm \text{ Ans.}$$

PROBLEM FOR PRACTICE

1. Define the terms reaction, co–efficient of friction, limiting friction, friction cone.
2. State whether false or true "Angle of repose is equal to the friction angle". (True)
3. Write down the relationship between co–efficient of friction and friction angle.
4. A body is placed on an inclined rough plane whose inclination is equal to the friction angle. State whether the body will be in equilibrium or move down. (Equilibrium)
5. A body of weight W is placed in a rough inclined plane which makes an angle α with the horizontal. *A* push P acts on the body, along the horizontal direction and the body is just on the point of moving up. Write down the relationship between P and W. The friction angle is. (**Ans.** $P = W \tan (\alpha + \phi)$)
6. State the laws of static friction.
7. What is a wedge? What for is it used? Explain that the rising block must have sliding support so as to have two reactions which are in opposite directions.
8. Is there a limiting value of the wedge angle? Explain when $(a + \phi) \Rightarrow 90°$ will the wedge action be possible?
9. While cutting steel rod with hand hacksaw, the cutting is done during forward stroke but in case of wood it is done on reverse stroke.
10. A body of weight 990 N was placed on rough horizontal plane. The co–efficient of friction being such that a pull of 180 N applied upward at 30° to the rough plane just moved the body to the right. Find μ and determine the push required to be applied to the body at 30° to horizontal so that the body is just on the point of moving towards left side. (**Ans.** M = 0.1732, 220 N)
11. A body of weight W is lying on a rough horizontal place. *A* pull P_1 is applied to the body at an angle of 30° with the horigontal which just moves the body to the right. Again a push P_2 is applied to the body, inclined at 30°, to horizontal to just move it to the right. If $\mu = 0.2$, find the ratio of P_2 to P_1 $\left(\textbf{Ans.}\ \frac{P_2}{P_1} = 1.26\right)$
12. A body weighing 1.6 kN is lying on a rough horizontal plane having $\mu = 0.2$ A pull P of magnitude 1 kN is applied on the body at 30° to horizontal as shown in Fig. 2.30. Find the magnitude of pull Q applied to the left along horizontal direction so the that body is just on the point of moving to the right (v) the body is just on the point of moving to the left. (**Ans.** 0.046 kN, 1.086 kN)

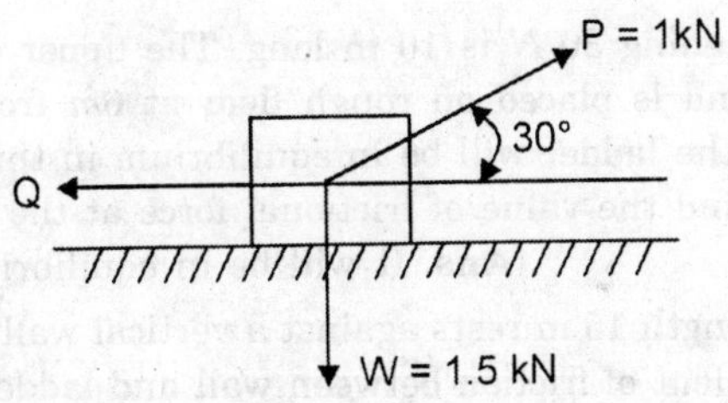

Fig. 2.30

13. Debris from the river bed is being drawn up with the help of drag line in which two trolleys A and B connected with wire rope passing over a smooth pulley are loaded weighing 0.75 *kN* and 0.5 *kN* respectively (Fig. 2.31). The machine applies a pull *P* on the leading trolley at an angle of 30° to the horizontal. If the co–efficient of friction between the trollies and guiding track is 0.2, find the value of *P*. (**Ans.** 0.853 *kN*)

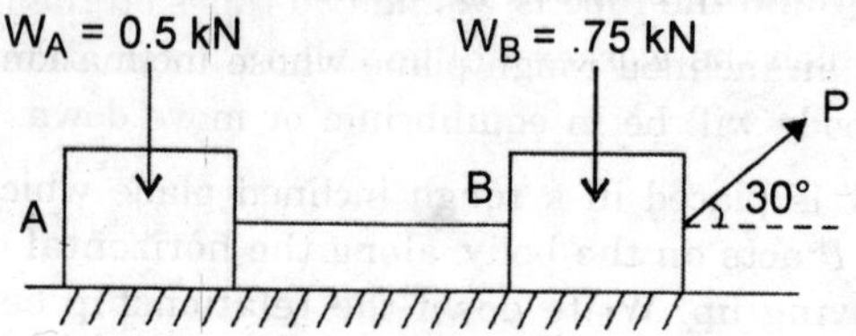

Fig. 2.31

14. A uniform leader of length 13 *m* and weighing 25 *N* is placed against a smooth vertical wall with its lower end 5m from the wall. The co–efficient of friction between ladder and floor is0.3 Show that the ladder will remain in equilibrium in this position. What is the frictional force acting on the ladder at the point of contact between ladder and floor. (*AMIE, 1981*)

(**Ans.** 7.5 *N*)

(**Hint.** $\Sigma Y = 0$ will give normal reaction at the lower and .Then μR will give frictional force available. $\Sigma X = 0$ will give reaction at upper and the for equilibrium. Calculate frictinal force at lower end by setting $\Sigma M = 0$, which will give actual frictional force which will be less than the above calculated value. Thus, this ladder will not slip and stay in equilibrium.

15. Two bodies *A* and *B* weighing 1 *kN* and 0.5 *kN* respectively are connected cord and placed on inclined rough plane as shown in Fig. 2.32, the co–efficient of friction between A, the plane is 0.4 and that between *B* and plane is 0.2. If the motion is just on the point of taking place down the plane, find the tension in the cord and the inclination θ of in plane with the horizontal. (**Ans.** 18.41° and 63.086 *N*)

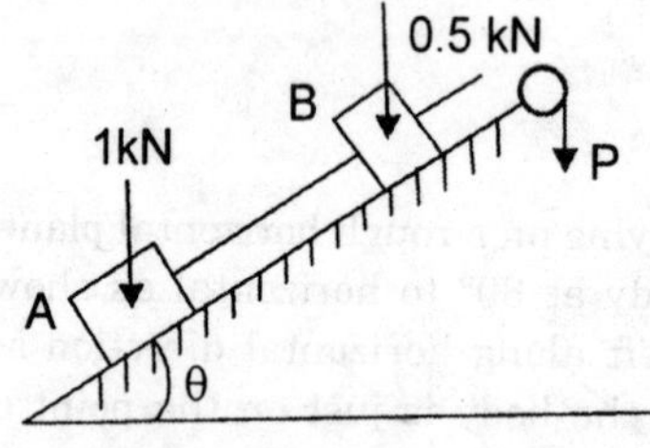

Fig. 2.32

16. A uniform ladder weighing 30 *N* is 10 m long. The upper end is leaning against a smooth vertical wall, lower end is placed on rongh floor at 6*m* from wall. Take μ = 0.45 show by calculations whether the ladder will be in equilibrium in this position or it will slip? In case it is in equilibrium, find the value of frictional force at the lower end of ladder.

(**Ans.** It will be in equilibrium and resultant force = 7.5 *N*)

17. A uniform ladder of length 15 m rests against a vertical wall making an angle of 60° with the horizontal the co–efficient of friction between wall and ladder and ground and ladder are 0.3 and 0.25 respectively. A man weighing 500*N* ascends the leader. How high will he be able to go before ladder slips?

Also, find the weight necessary to be put at the lower end of the ladder so as to be just sufficient to permit the man to go to the top. Assure weight of leader as 850 *N*. (*AMIE 1974*)

(**Ans.** 6.69 *m*, 607.6 *N*)

18. Rope and fixed pulley is used to lower a weight *W* by a worker in a factory. If the worker can apply a pull of 400 *N* and the rope is wound two times on the fixed pulley, Fig. 2.32, find the maximum weight which the worker will be able to lower. Take μ between the pulley and the rope as 0.2. (**Hint.** $\theta = 2 + 2\pi + \pi = 5\pi$)

(**Ans.** 9256 *N*)

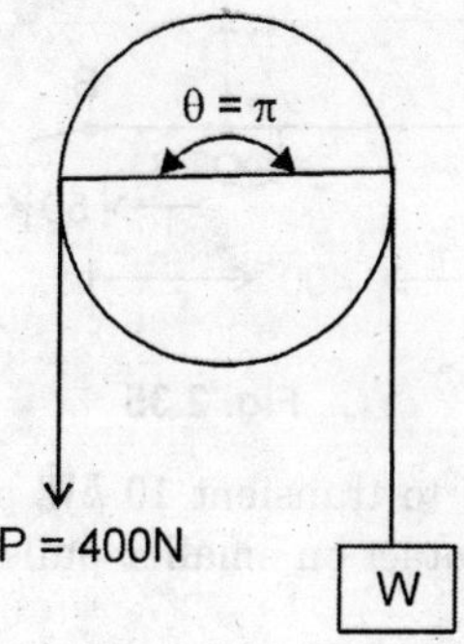

Fig. 2.33

19. A block '*A*' weighing 2.5 *kN* is lying m a rough plane having $\mu = 0.2$ A cord attached to the block '*A*' passes over a fixed pulley B and on the other end is hanging a load of *1 kN*. If μ at the pulley is 0.1, find the value of the pull *P* acting on the block so that (*a*) the block *A* is just on the point of moving to right (*B*) on the point of moving to the left.

(**Ans.** 293.42 *N*, 11.21.58 *N*)

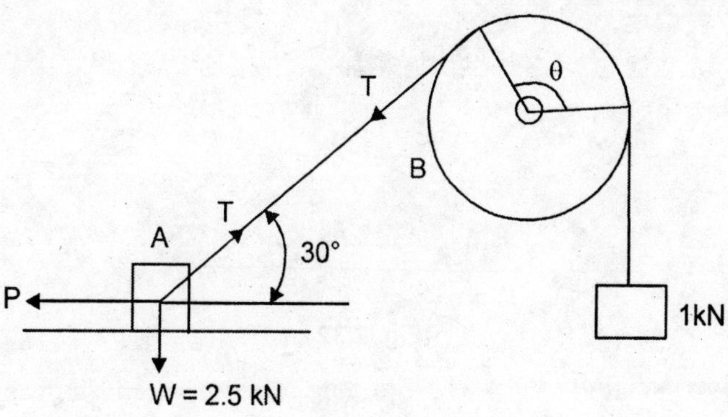

Fig. 2.34

20. Band brake alongwith brake drum are car shown in Fig. 2.35. The tight side of the band is connected to end *C* of the lever *ABC* whose fulcrum is at B. The slack side is attached to the lever at *D*. If *AB* = 400 mm, *BC* = 50 mm and *DB* = 70 *mm*, find the value of braking torque if *P* = 100 *N and* co–efficient of friction between band and drum is 0.2, angle of lap is 270°. Take drum diameter = 200 *mm*. (**Ans.** 10.7 *N-m*)

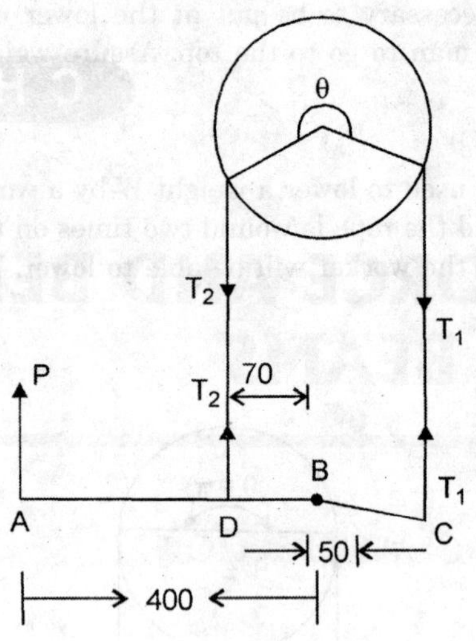

Fig. 2.35

21. An open belt drive is required to transient 10 kW power. If the velocity of the belt is 10 m/s and m = 0.2 and angle of contact on smaller pulley is 13°, calculate the maximum tension in the belt. **(Ans.** T_1 = 2.666 kN)

22. Two parallel shafts are 3m away from each other. Pulleys of diameters 270 mm and 450 mm are keyed on them respectively. They are connected by cross belt drive. The speed of bigger pulley is 500 $r.p.m.$ Find the length of the belt and power. Also, find the tension on the tight side.

CHAPTER 3

SHEARING FORCE AND BENDING MOMENT IN BEAMS

3.1 BEAMS AND THEIR CLASSIFICATION

Any member of a machine or structure under the action of lateral forces in the axial plane is called a beam. The effect of these forces will be to cause bending of the member. Depending upon the end conditions, there are different types of beams. A beam with both the ends rigidly fixed is known as a fixed or encastre beam. If one of the ends is rigidly fixed and the other is absolutely free, it is known as a cantilever beam. If both the supports at the ends are such that the beam is capable of taking any angular position at the end and not the lateral movement, the beam is called to be a freely or simply supported beam ; in actual practice this requirement is met by providing a well lubricated hinge at one end and a support on rollers at the other.

If such a beam has ends extending beyond the supports, it is called an overhanging beam. A beam which is supported at more than two supports is called a continuous beam.

As beams are invariably horizontal, and the external forces are weights, the beams will be taken horizontal and the external forces as vertical, although the same conclusions will hold good in other cases. Members of structures and machines transmit axial loads in addition to lateral forces acting upon them. However, in the present chapter, only transverse forces will be considered.

A concentrated or point load is one which is applied to the beam through a knife edge. A uniformly distributed load is that which is spread evenly over a certain length of the beam so that the load carried by a portion of the beam is proportional to the length of that portion. A distributed load will be called a varying load, when the rate of loading varies from one point to another.

3.1.1 Moment of a Force, Plane of Moment and Axis of Moment

Moment of a force about a point is the product of the force and the perpendicular distance of the point from the force vector. Plane containing the force vector and the point about which moment taken is called Plane of Moment. And the axis passing through the point and normal the plane of moment is called the axis of the moment. Since force is a vector quantity, the moment will also be a vector quantity, as the distance is a scalar quantity. This force vector is represented by an arrowhead, the length of which to some chosen scale represents the magnitude of the force and the arrowhead, indicates the direction of the force. Similarly, to represent a moment vector with an arrowhead right handed screw rule is followed. The length of the arrowhead will represent the magnitude of the moment to some scale. However, before we describe the method to decide the placement of the arrow and the direction to which it is to point, we have to see whether the moment about the given point is clockwise moment or anticlockwise moment, depending upon the location of the point with respect to the direction of the arrowhead representing the force vector. If the point lies on the left side of force vector arrowhead, as seen along the arrowhead, then the moment will be reckoned as anticlockwise. Similarly, if the point in question lies on the right hand side of the force vector, then it will be called the clockwise moment because this moment will tend to rotate the minutes arm of the clock, placed at the point where the moment is taken, in the clockwise direction or vice versa. After having decided whether the moment is clockwise or anticlockwise, we proceed to apply the right hand screw rule. The right handed screw rule states that with respect to a fixed nut placed in the plane of the moment, the direction of axial movement of the right handed screw, when acted upon by the given moment, will be the direction of the arrowhead representing the moment vector. Evidently, the axial movement of the screw with respect to fixed nut which is placed in the plane of the moment, will be normal to in plane of the moment. Thus, if the plane of the moment is vertical and is visible in elevation, the normal will be visible in the plane view and will be normal to plane of the moment. It may be noted that the arrowhead will be normal to the plane of the moment and it will be parallel to the axis of the moment.

3.1.2 Classification of Moment : Bending Moment and Twisting Moment

The next question arises as to whether the moment acting on any section of a beam or shaft is bending moment or twisting moment. When the plane of the moment is normal to the cross–sectional plane of the beam or shaft on which the moment is acting, the moment is called Bending moment, as it will tend to bend the beam or shaft. Alternately, moment will be a Bending Moment, if the axis of the moment is parallel to or lies in the cross–sectional plane of the beam.

Similarly, if the plane of the moment is parallel to or contains the cross–sectional plane of the beam or shaft, it will be called twisting moment, as it tends to twist the shaft. Alternately the moment will be a twisting moment, if the axis of the moment is normal to the cross-sectional plane of the beam or shaft or if the axis of the moment coincides with the longitudinal axis if the shaft.

3.2 SHEARING FORCE AND BENDING MOMENT

Consider a simply supported beam *AB* under the action of vertical forces P_1, P_2, P_3 and P_4 with reactions R_A and R_B shown in the Fig. 3.1.

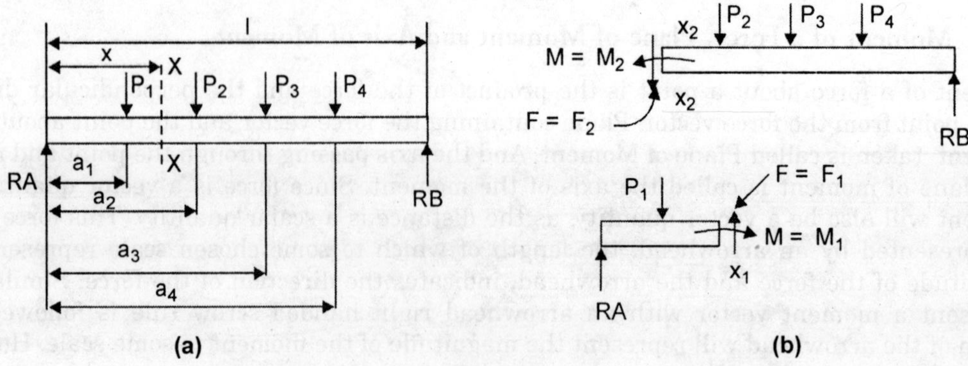

Fig. 3.1

Since the beam is in equilibrium the algebraic sum of all the vertical forces acting on the beam will be zero,

i.e., $$R_A + R_B = P_1 + P_2 + P_3 + P_4 \quad ...(3.1)$$

Again taking a section along XX at a distance x from A and considering the equilibrium of the two portions separately, if F_1 and M_1 are the force and resisting moment along the section $X_1 X_1$ offered by element No.2 and F_2 and M_2 are the force and moment offered by element No. 1 onto element No.2 along the section X_2X_2, then $F_1 = F_2$ (= F say) and also,

$$F_1 = F = -R_A + P_1 \quad ...(3.2)$$

$$F_2 = F = P_2 + P_3 + P_4 - R_B \quad ...(3.3)$$

If $R_A < P_1$ then from equation (3.1) $R_B < (P_2 + P_3 + P_4)$ and the values of F from equations (3.2) and (3.3) will be positive. In other words, the forces F_1 (= F) and F_2 (=F) on the cross–section XX will be as shown in the free body diagrams of the two elements (Fig. 3.1b), that is, the section X_1X_1 of element No.1 will have a tendency to move downwards (in the direction of F_2) and section X_2 X_2 of the element No. 2 will tend to slide upwards with respect to X_1 X_1. Again if $R_A > P_1$, the sliding tendency will be reversed. The force F (= $F_1 = F_2$) is known as the shearing force over the section XX and is defined with the help of equations (3.2) and (3.3) as the algebraic sum of all the external lateral forces on one side of the section. If we start from the left hand side and the algebraic sum is positive (*i.e.* $R_A < P_1$) the direction of shearing force F_2 will be downward and that of F_1 upward, it is said to be a positive shearing force otherwise negative Fig. 3.2.

Once again for the equilibrium of the two elements separately,

$$M = -R_A x + P_1(x - a_1) \quad ...(3.4)$$

$$M = -R_B (1 - x) + P4 (a_4 - x) + P_3 (a_3 - x) + P_2 (a_2 - x) \quad ...(3.5)$$

This moment M is called the bending moment over the section XX, as its axis is perpendicular to the longitudinal axis of beam.

From equations (3.4) and (3.5), it is evident that the bending moment over a section is the algebraic sum of the moments about that section of the forces on one side of the section. If the moment is positive as evaluated from equations (3.4) and (3.5), the directions will be as shown in Fig. 3.1(*b*) and is called a sagging moment (after the shape of a sagging wire held over two vertical poles), as shown in Fig. 3.2.

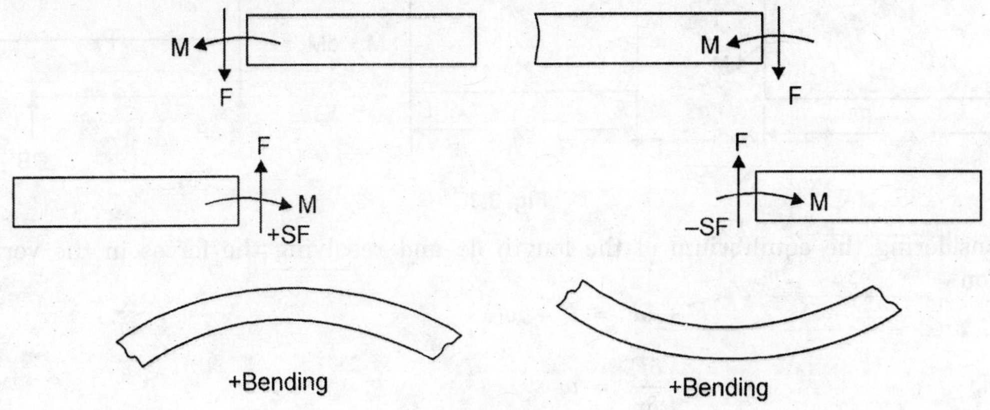

Fig. 3.2

3.3 SHEARING FORCE AND BENDING MOMENT DIAGRAMS

The values of the shearing force and bending moment will go on varying from point to point along the length of the beam. The curves showing the variation of shearing force and bending moment along the length of the beam are known as the shearing force and bending moment diagrams respectively.

The point on the beam where the bending moment changes sign is known as the point of contraflexion. Evidently, at this point the B.M. curve will intersect the base line ; that is, the value of bending moment will be zero.

For finding the sign of the Bending Moment due to any force (Load or reaction) over the section *XX*, following procedure may be conveniently adopted :

Consider the beam clamped at the section *XX* and forget about all other forces except the one in question due to which the moment is to be found (say the reaction (R_A). The tendency of R_A will be to move the point *A* upwards with respect to *X* and the shape of the portion *XA* of the beam will be similar to that of a sagging wire on two electric poles. Thus, the moment (R_A *x*) will be called a sagging moment and rendered negative. Similarly, the moment P_1 $(x - a_1)$ is positive.

3.4 RELATION BETWEEN BENDING MOMENT AND SHEARING FORCE

Consider a small length δx of a beam situated at a distance *x* from some chosen origin. Let the beam by carrying some distributed load which may or may not be uniformly distributed load. But δx being sufficiently small the rate of loading w MN/m run may be taken as constant over the length δx.

The shearing force F and $(F + \delta F)$ alongwith the bending moments *M* and $(M + \delta M)$ on either end of this element are assumed to be, as shown in Fig. 3.3 conforming to the positive signs as per the convention adopted in article 3.2.

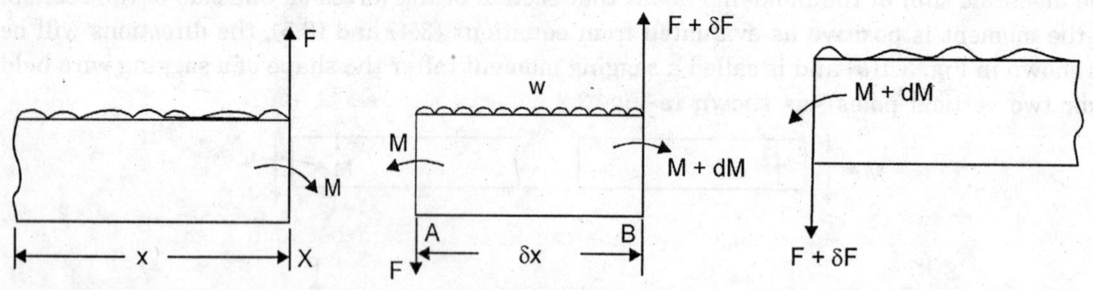

Fig. 3.3

Considering the equilibrium of the length d*x* and resolving the forces in the vertical direction

$$F + \delta F = F + w\delta x$$

i.e.,
$$\frac{\delta F}{\delta x} = w$$

In the limiting case, when $\delta x \to 0$

$$\frac{dF}{dx} = w \qquad ...(3.6)$$

That is, the rate of change of shearing force or the slope of the shearing force curve at XX is numerically equal to the rate of loading at that point.

Again for the equilibrium of the length δx, equating moments of all the forces about any point in the section A,

$$(M + dM) + (w\delta x).\frac{\delta x}{2} = M + (F + \delta F).\delta x$$

Neglecting the product of two small quantities, it gives,

$$\frac{\delta M}{\delta x} = F$$

and in the limiting case when delta $x \to 0$, we have,

$$\frac{dM}{dx} = F \qquad ...(3.7)$$

That is, the rate of change of bending moment or the slope of the bending moment curve at XX is equal to the shearing force at that section.

Evidently from equations (3.6) and (3.7), we have,

$$w = \frac{dF}{dx} = \frac{d^2M}{dx^2} \qquad ...(3.8)$$

Again integrating equations (3.6) and (3.7) between sections x_0 to x,

$(F - F_0)$ = Change in the shearing force between x_0 and x

$$= \int_{x0}^{x} w dx \qquad ...(3.9a)$$

= Total load between x_0 and x ...(3.9b)

and $(M - M_0)$ = Change in bending momentbetween x_0 and x

$$= \int_{x0}^{x} F dx \qquad ...(3.10)$$

= Area of the shearing force diagram between x_0 and x.

3.5 SHEARING FORCE AND BENDING MOMENT DIAGRAMS

3.5.1 (*a*) Beam Simply Supported at its Ends and Carrying a Load at The Mid–Span :

Consider a beam AB supported at its ends and a load W at C. The reactions R_A and R_B will both be equal to $\frac{W}{2}$ as the beam is loaded symmetrically.

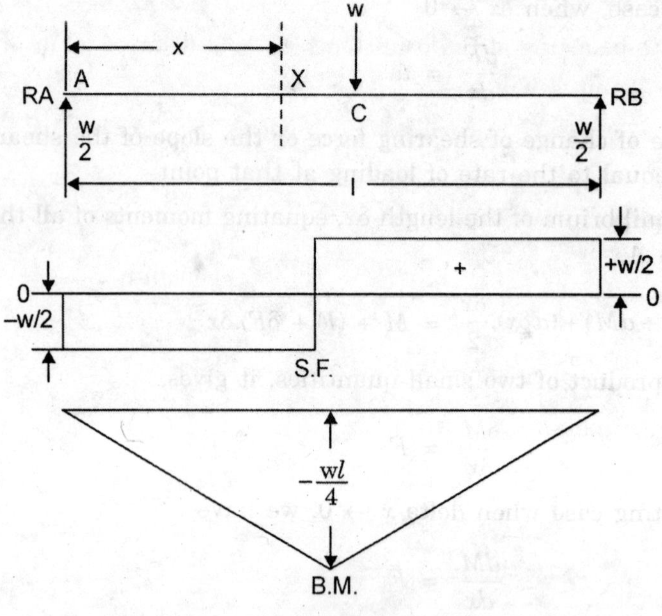

Fig. 3.4

For the section XX lying in between A and C at a distance x from chosen origin at A and according to the adopted sign convention of Fig. 3.2,

$$\text{S.F.} = -R_A = -\frac{W}{2}$$

For the section XX lying in between C and B,

$$\text{S.F.} = -R_A + W$$

$$= -\frac{W}{2} + W = \frac{W}{2}$$

B.M. at the section XX between A and C,

$$= -\frac{W}{2}x,$$

which is straight line giving the value of B.M. as zero at A and $-\frac{Wl}{4}$ at C.

B.M. at any section between C and B,

$$= -\frac{W}{2}x + W\left(x - \frac{1}{2}\right)$$

$$= \left(\frac{Wx}{2} - W\frac{l}{2}\right)$$

which is again a straight line ; variation being $-\frac{Wl}{4}$ and C and zero at B.

The two diagrams are shown in Fig. 3.4.

(*b*) Point Load Not at the Mid Span

Consider the load W to be acting at the point C which is at distance 'a' from A and 'b' from B.

The reactions at A and B will be,

$$R_A = \frac{Wb}{l}; R_B = \frac{Wa}{l}$$

Taking the section XX between A and C,

$$\text{S.F.} = -R_A = -\frac{Wb}{l} = \text{constant}$$

and
$$\text{B.M.} = M = -R_A x + -\frac{Wb}{l}.x$$

which is a straight line variation with,

$$M = 0 \text{ at } x = 0$$

and
$$M = -\frac{Wba}{l} \text{ at } x = a$$

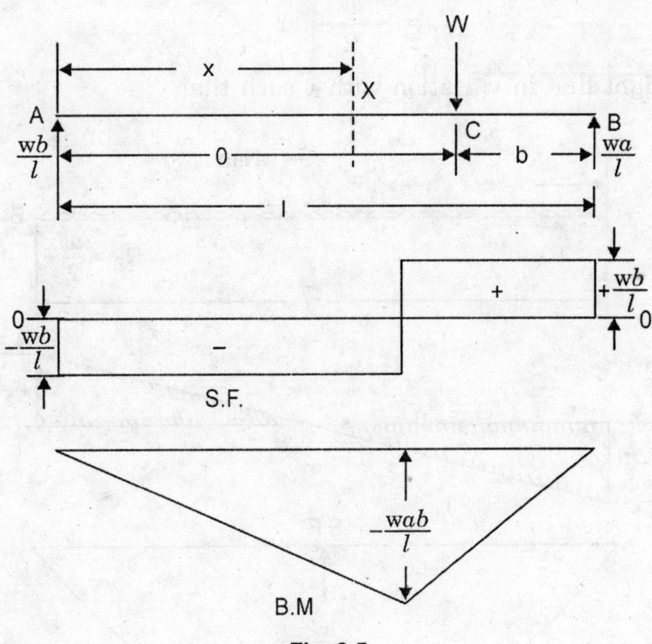

Fig. 3.5

For the section XX between C and B,

$$\text{S.F.} = -R_A + W = -\frac{Wb}{l} + W = \frac{Wa}{l}$$

$$= \text{constant}$$

$$M = -R_A x + W(x-a)$$

$$= -\frac{Wba}{l}x + W(x-a)$$

which is again a straight line such that,

$$M = -\frac{Wba}{l} \quad \text{at } x = a$$

and $$M = 0 \quad \text{at } x = 1 .$$

The shearing force and bending moment diagrams are shown in Fig. 3.5.

3.5.2 Simply Supported Beam With Uniformly Distributed Load

Let the beam *AB* of span l be acted upon by a uniformly distributed load of w Newton per metre length as shown in the Fig. 3.6.

The beam is loaded symmetrically. Therefore,

$$R_A = R_B = \frac{wl}{2}$$

For any section *XX* at a distance x from *A*,

$$\text{S.F.} = -R_A + wx$$

$$R = -\frac{wl}{2} + wx$$

which is a straight line in variation with x such that,

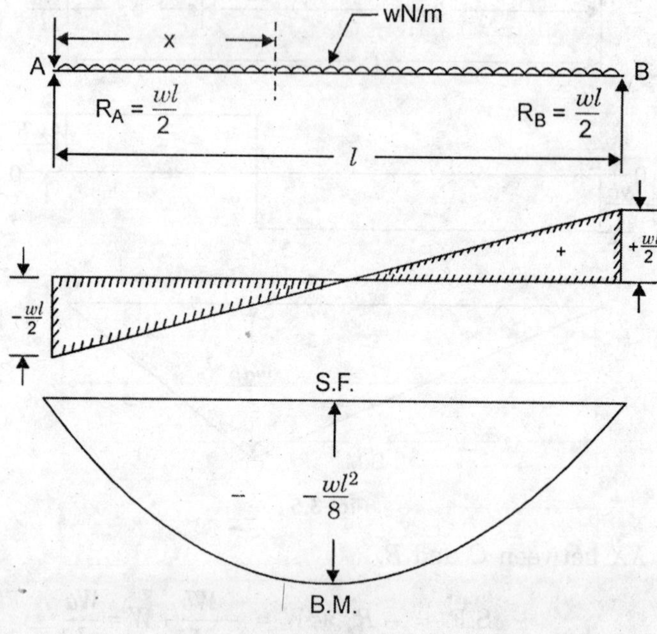

Fig. 3.6

$$\text{S.F.} = -\frac{wl}{2} + wx \quad \text{at } x = 0$$

and $$\text{S.F.} = +\frac{wl}{2} \quad \text{at } x =_1$$

Also, $$M_x = -R_A x + wx\left(\frac{x}{2}\right)$$

$$= -\frac{wl}{2}x + \frac{wx^2}{2}$$

which is parabolic variation with respect to x,

$$M_x = 0$$

at x $=0$ as well as at $x = l$.

The magnitude of M_x is maximum at $x = \frac{l}{2}$, where it is equal to $\left(-\frac{wl^2}{2}\right)$

The magnitude of B.M. can be determined for few more values of x and the curve is then plotted as shown in Fig. 3.6.

3.6 SHEARING FORCE AND BENDING MOMENT DIAGRAMS FOR CANTILEVERS

3.6.1 With Point Load at Free End

Taking a section XX at a distance x from A, the SF at x is given by,

$$\text{S.F.} = W = \text{constant.}$$

And $$\text{B.M.} = M_X = W_x,$$

which is a straight line variation with x such that,

$$M = 0 \quad \text{at } x = 0$$
$$M = W_X \quad \text{at } x = l$$

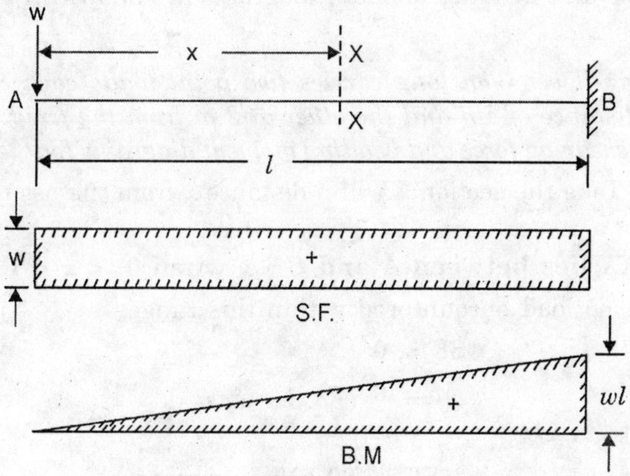

Fig. 3.7

The shearing force and bending moment diagrams are shown in Fig. 3.7.

3.6.2 Load Distributed Over Whole Span

Taking a section xx at a distance x form A, Fig. 3.8 the SF at x is given by,

$$SF = wx$$

White is a straight line variation with SF = 0 at $x = 0$ and SF = wl at $x = l$.

And bending moment M is given by,

$$M = wx \cdot \frac{x}{2} = \frac{wx^2}{2}$$

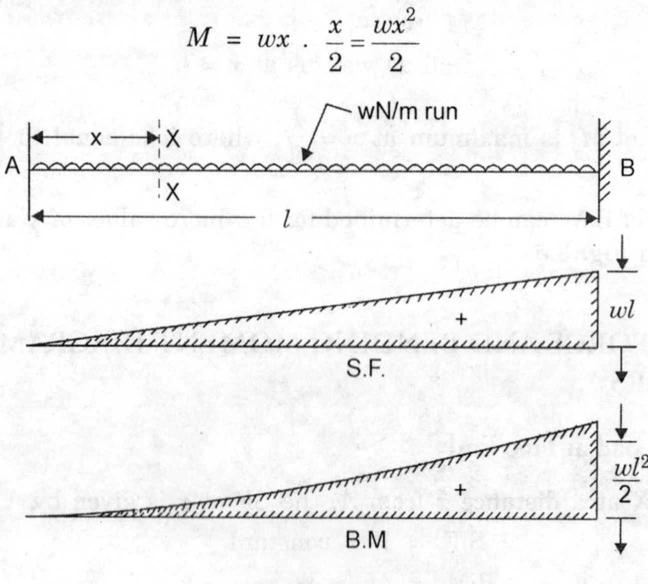

Fig. 3.8

which is parabolic in variation with $M = 0$ at $x = 0$ and $M = \frac{wl^2}{2}$ at $x = l$.

The shearing force and bending moment diagrams are shown in Fig. 3.8.

EXAMPLE **3.1 :** *A cantilever 3 m long carries two point loads each 80 kN, one placed at distance of 1m and the other at 2 m from the fixed end. Draw to scale the shearing force and bending moment diagrams for the cantilever.*

Solution : Take the section XX at a distance x from the assumed origin at A.

(*i*) **When *XX* lies between *A* and *C* *i.e.* when $0 < x < 1$**

As there is no load encountered within this range,

$$SF = 0$$

$$M = 0$$

(*ii*) When $1 < x < 2$,

$$SF = 80 \text{ kN} = \text{constant}$$

$$M = 80(x - 1) \quad \text{(Straight line variation.)}$$

Therefore, $M = 0$ at $x = 1$ m

and $M = 80$ kN-m at $x = 2$ m

(*iii*) When $2 < x < 3$, SF = 80 + 80

= 160 kN = constant

$M = 80\ (x - 1) + 80\ (x - 2)$ Straight line

Therefore, at $x = 2$ m $M = 80$ kN – m

and at $x = 3$ m $M = 80\ x\ (3 - 1) + 80\ (3 - 2) = 240$ kN – m

The shearing force and bending moment diagrams are shown in Fig. 3.9.

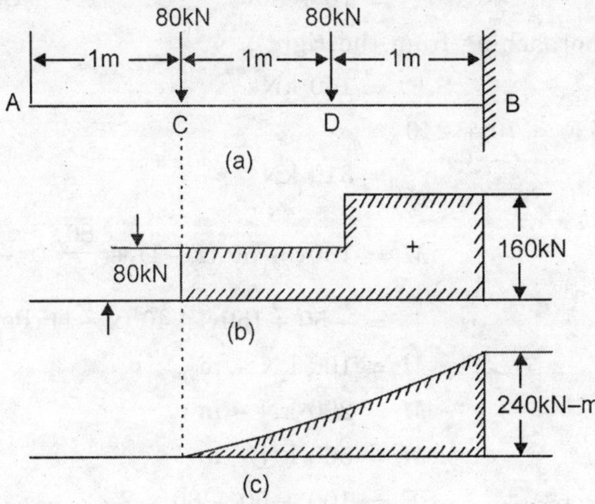

Fig. 3.9

3.2 : *Draw the S.F. and B.M. diagrams for the cantilever as shown in Fig. 3.10.*

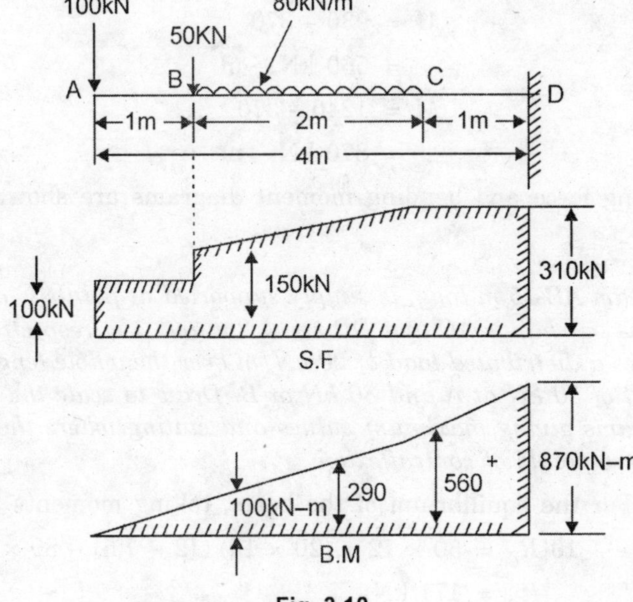

Fig. 3.10

E X A M P L E

Solution : Taking a section XX at a distance x from the origin at A,

(*i*) When $0 < x < 1$, SF = 100 kN = constant

$$M = 100\,x$$

This is a straight line such that at

$$x = 0,\ M = 0$$

and at $x = 1$ m, $M = 100$ kN – m

(*ii*) When $1 < x < 3$, SF = 100 + 50 + 80 (x – 1)

= 70 + 80x Straight line

at x = 1 m (approaching from the right),

S.F. = 150 kN

at x = 3 m, S.F. = 70 + 240

= 310 kN

$$M = 100x + 50\,(x-1) + \frac{80}{2}\,(x-1)^2$$

$= -50 + 150x + 40\,(x-1)^2$ Parabola

at $x = 1$, M = 100 kN – m

at $x = 2$, M = 290 kN – m

at $x = 3$, M = 560 kN – m

(*iii*) When $3 < x < 4$, SF = 100 + 50 + 80 × 2

= 310 kN = constant

$$M = 100x + 50\,(x-1) + (80 \times 2)\,(x-2)$$

$= 310x - 370$ Straight line

at x = 3 M = 930 – 370

= 560 kN – m

at x = 4 M = 1240 – 370

= 870 kN – m

The shearing force and bending moment diagrams are shown in Fig. 3.10.

3.3 : *A beam AB,15 m long, is simply supported at points C and D,10 m apart, the overhangs AC and DB being 2m and 3 m respectively. The beam carries a distributed load of 20 kN/m over the whole length along with a load of 50 kN at A and 80 kN at B. Draw to scale the S.F. and B.M. diagrams giving maximum values and stating where these occur. Also, find the points of contraflexion.*

Solution. For the equilibrium of the beam, taking moments about D.

$$10\,R_C = 50 \times 12 + (20 \times 15)\,(12 - 7.5) - 80 \times 3$$

i.e. R_C = 171 kN

EXAMPLE

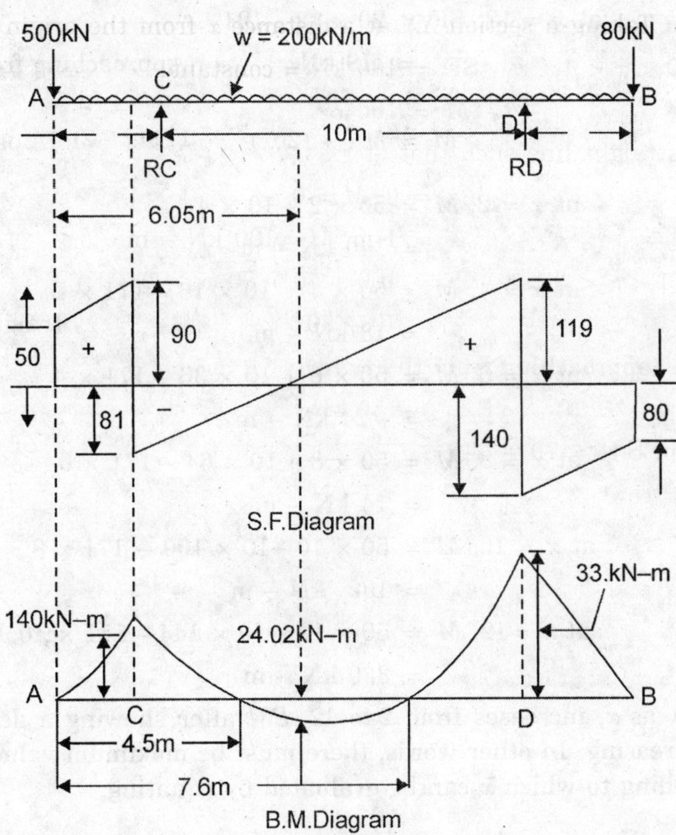

Fig. 3.11

Therefore, $R_D = 50 + (20 \times 15) + 80 - 171$

$= 259$ kN.

Shearing force and B.M. are calculated by taking a section *XX* at a distance x from A as follows :

(*i*) When $0 < x < 2$ SF $= 50 + 20x$ (A straight line)

at $x = 0$, SF $= 50$ kN

at $x = 2$, SF $= 90$ kN (Approaching from left)

$$M = 50x + (20x)\frac{x}{2} \quad \text{(A parabola)}$$

$$= 50x + 10x^2$$

at $x = 0$, $M = 0$

at $x = 1$, $M = 50 + 10 = 60$ kN–m

at $x = 2$, $M = 50 \times 2 + 10 \times 4 = 140$ kN–m

(*ii*) When $2 < x < 12$, SF $= 50 + 20x - 171$

$= 20x - 121$ (A straight line)

at $x = 2$, SF $= 40 - 121 = -81$ kN

(Approaching from right)

$$\text{at } x = 12,\ SF = 20 \times 12 - 121$$
$$= 119 \text{ kN} \qquad \text{(Approaching from left)}$$

And $$M = 50x + (20x)\frac{x}{2} - 171\,(x-2) \quad \text{(A parabola)}$$

$$\text{at } x = 2,\ M = 50 \times 2 + 10 \times 4$$
$$= 140 \text{ kN} - \text{m}$$
$$\text{at } x = 4,\ M = 50 \times 4 + 10 \times 16 - 171 \times 2$$
$$= 18 \text{ kN} - \text{m}$$
$$\text{at } x = 6,\ M = 50 \times 6 + 10 \times 36 - 171 \times 4$$
$$= -24 \text{ kN} - \text{m}$$
$$\text{at } x = 8,\ M = 50 \times 8 + 10 \times 64 - 171 \times 6$$
$$= 14 \text{ kN} - \text{m}$$
$$\text{at } x = 10,\ M = 50 \times 10 + 10 \times 100 - 171 \times 8$$
$$= 132 \text{ kN} - \text{m}$$
$$\text{at } x = 12,\ M = 50 \times 12 + 10 \times 144 - 171 \times 10$$
$$= 330 \text{ kN} - \text{m}$$

Evidently, as x, increases from 2 to 12, BM after showing a decrease, starts increasing. In other words, there must be maximum value of M, corresponding to which x can be evaluated by equating,

$$\frac{dM}{dx} = 0 \text{ or } SF = 0$$

Thus, equating, $$SF = 20x - 121 = 0$$

$$x = \frac{121}{20} = 6.05 \text{ m}$$

For which $M = 50 \times 6.05 + 10\,(6.05)^2 - 171\,(6.05 - 2)$

$$= -24.02 \text{ kN} - \text{m}$$

Also, since the value of M is changing sign, the points of contraflexion with be within this range and the location can be found by equating M to zero and solving for x

That is $\quad 50x + 10x^2 - 171\,(x - 2) = 0$

or $$10x^2 - 121x + 342 = 0$$

Therefore, $$x = \frac{1}{20}\left[121 \pm (121 \times 121 - 4 \times 342 \times 10^{\frac{1}{2}}\right]$$

i.e. $$x = 4.5 \text{ m or } 7.6 \text{ m}$$

Thus, the points of contraflexion are at 4.5 m and 7.6 m from A.

EXAMPLE

E X A M P L E

(*iii*) When $12 < x < 15$,

$$SF = 50 + 20x - 171 - 259$$

$$= 20x - 380 \quad \text{(A straight line)}$$

at $x = 12$, $SF = 20 \times 12 - 380$

$$= -140 \text{ kN} \quad \text{(Approaching from right)}$$

at $x = 15$, $SF = 20 \times 5 - 380$

$$= -80 \text{ kN}$$

And $M = 50x + (20x)\dfrac{x}{2} - 171(x-2) - 259(x-12)$

$$= 50x + 10x^2 - 171(x-2) - 259(x-12)$$

(A parabola)

at $x = 12$, $M = 50 \times 12 + 10 \times 144 - 171 \times 10$

$$= 330 \text{ kN} - \text{m}$$

at $x = 13$, $M = 50 \times 13 + 10 \times 169 - 171 \times 11 - 259 \times 1$

$$= 200 \text{ kN} - \text{m}$$

at $x = 14$, $M = 50 \times 14 + 10 \times 196 - 171 \times 12 - 259 \times 2$

$$= 90 \text{ kN} - \text{m}$$

at $x = 15$, $M = 0$

The complete *SF* and *BM* diagrams are shown in Fig. 3.11, from which the maximum *BM* = 330 kN–m occurring at point *D* and maximum shearing force is 119 kN occurring just on the left of point *D*.

3.4 : *A beam of length l, carrying a uniformly distributed load over its whole length, is symmetrically supported over two simple supports. Find the position of the supports to keep the maximum bending moment on the beam as low as possible.*

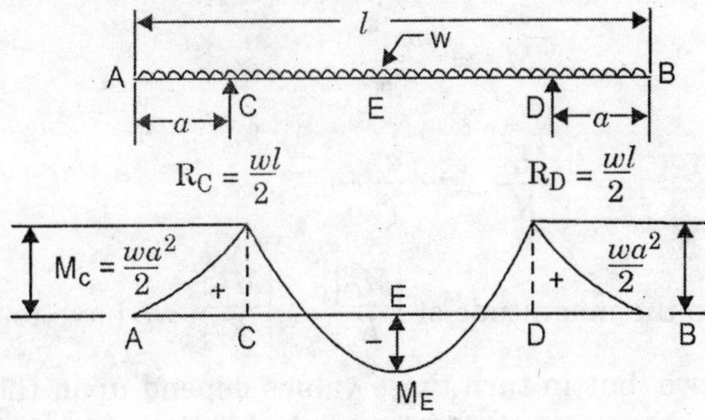

(a)

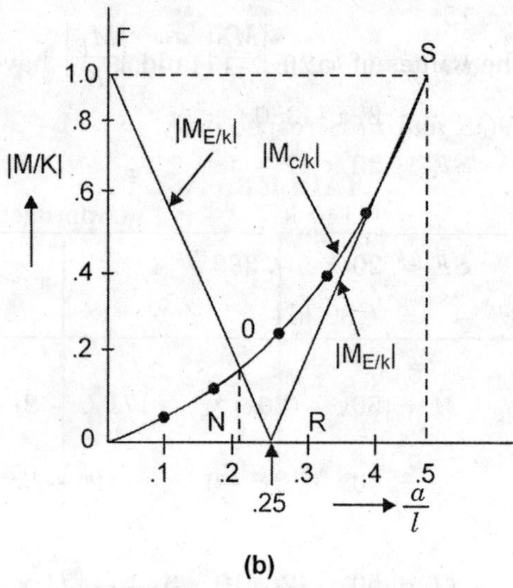

(b)

Fig. 3.12

Solution : Fig. 3.12(*a*) shows the bending moment diagram for some value of '*a*'. For any value of '*a*', the bending moments at the supports and the mid–point E are to be evaluated and the greater of the two will correspond to the maximum bending moment on the beam.

Thus, for the load w kN/m run and the configuration of Fig. 3.12.

$$M_E = \left(\frac{wl}{2}\right)\frac{l}{4} - \frac{wl}{2}\left(\frac{l}{2} - a\right)$$

$$= -\frac{wl^2}{8}\left(1 - 4\frac{a}{l}\right)$$

or $$\frac{M_E}{K} = -1 + 4\frac{a}{l} \quad \text{...(A)}$$

Where, $$K = \frac{wl^2}{8} = \text{constant}$$

and $$M_C = \frac{wa^2}{2}$$

or $$\frac{M_C}{K} = 4\frac{a^2}{l^2} \quad \text{...(B)}$$

Depending upon the magnitude of $\left|\frac{M_C}{K}\right|$ or $\left|\frac{M_E}{K}\right|$, we have to pick up the greater of the two, but in turn these values depend upon the ratio (a/l) which has the extreme values of zero and 0.5 corresponding to the supports being at the ends or merging at the midpoint E respectively.

For this range of the values of (a/l), $\left|\frac{M_C}{K}\right|$ and $\left|\frac{M_E}{K}\right|$ have been plotted in Fig. 3.12(*b*) as *OQS* and *PRS* respectively.

TABLE 3.1

$\frac{a}{l}$	$\frac{M_C}{K}$	$\frac{M_E}{K}$	$\left\lvert\frac{M_E}{K}\right\rvert$
0.0	0.00	– 1.0	1.0
0.1	0.04	– 0.6	0.6
0.2	0.16	– 0.2	0.2
0.25	0.25	0.0	0.0
0.3	0.36	0.2	0.2
0.4	0.64	0.6	0.6
0.5	1.00	1.0	1.0

Evidently, for the values of (*a*/*l*) between *O* and *N*, the maximum value of the B.M. is to be given by the curve *PQ* (equation *B*) and for any value of (*a*/*l*) between *N* and 0.5, the maximum bending movment on the beam will be picked up from the curve *QS* (equation *A*). Thus, the curve *PQS* is the curve which gives the magnitudes of the maximum B.M. for any value of (*a*/*l*). Clearly the point *Q* which is the intersection of the curves of equations (*A*) and (*B*) gives the least value of the maximum *B.M.* on the beam and the corresponding value of (*a*/*l*) can be found by equating the magnitudes of moments, as obtained from equations (*A*) and (*B*). At the same time keeping in view that M_E is actually negative for the portion of the curve *PR* on which lies point *Q*.

Thus, for the greatest B.M. to be as small as possible, we have,

B.M. at supports = Maximum *B.M.* between supports

$$|M_C| = |M_E|$$

$$K.\left(4\frac{a^2}{l^2}\right) = -K\left(-1+4\frac{a}{l}\right)$$

or
$$4\frac{a^2}{l^2}+4\frac{a}{l}-1 = 0$$

Therefore,
$$\frac{a}{l} = \frac{1}{8}[-4 \pm (16+16)^{1/2}]$$

i.e.
$$\frac{a}{l} = 0.207 \text{ or } -1.207$$

The only admissible value is,

$$\frac{a}{l} = 0.207$$

or $$a = 0.207l.$$

3.5 : *A beam AB, 10 m long, is carrying a uniformly distributed load w N/m. The beam is to be supported freely at two piers 6 m apart in such a way that the bending moment on the beam is as small as possible. The piers may not necessarily be placed symmetrical. Find the position of the piers with respect to the ends of the beam.* *(Jaipur University)*

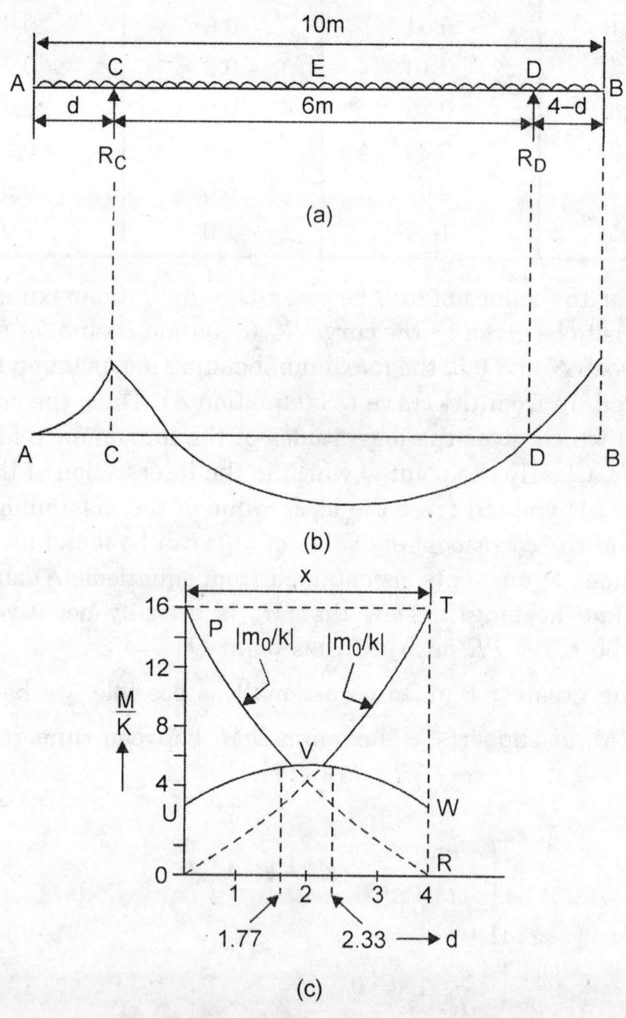

Fig. 3.13

Solution : Let the left hand support C be situated at a distance 'd' from A. The distance of the other support D will then be $(4 - d)$ from B as shown in Fig. 3.13. Then for this configuration taking moments about D.

$$6\,R_C = 10w\,[(6 + d) - 5]$$

EXAMPLE

E X A M P L E

i.e. $$R_C = \frac{5}{3} w (1 + d)$$

Also, $$M_C = \frac{w}{2} d^2$$

and $$M_D = \frac{w}{2} (d - 4)^2$$

Thus, if $$K = \frac{w}{2}, \quad \frac{M_C}{K} = d^2 \qquad \text{...(A)}$$

and $$\frac{M_D}{K} = (d - 4)^2 \qquad \text{...(B)}$$

Also, B.M. at section *XX* which is at a distance *x* from *A* and lying between *C* and *D*.

$$M_x = \frac{wx^2}{2} - \frac{5}{3} w(l + d)(x - d)$$

Now the magnitude of M_x will be maximum when,

$$dM_x/dx = 0$$

That is when $$wx - \frac{5}{3} w (1 + d) = 0$$

or $$x = \frac{5}{3}(1 + d)$$

Thus, the maximum bending moment between the supports is given by M_E as,

$$M_E = \frac{w}{2}\left[\frac{5}{3}(1 + d)\right]^2 - \frac{5}{3} w(1 + d)\left[\frac{5}{3}(1 + d) - d\right]$$

$$= \frac{5}{18} w (1 + d)(d - 5)$$

Therefore, $$\frac{M_E}{K} = \frac{5}{9}(1 + d)(d - 5) \qquad \text{...(C)}$$

From Fig. 3.13(*b*), it is evident that the maximum bending moment will be obtained at one of the three points, viz. *C*, *D* and *E*. Thus, plotting the three curves corresponding to equations (*A*), (*B*) and (*C*), we see from Fig. 3.13(*c*) that for any value of d within the possible range of1m to 4m the maximum bending moment on the beam is obtained from the curve *PYVZT*.

Evidently, the points *Y* and *Z* both give the least values of the maximum B.M. acting on the beam. Points *Y* and *Z* are intersections of the equation (*C*) with equations (*B*) and (*A*) respectively. It may also be pointed out that the actual values of the moments given by equation (*C*) are negative.

Thus, equating $|M_C|$ to M_E and $|M_D|$ to M_E separately we get the location of points Y and Z respectively as under :

$$Kd^2 = -K\left[\frac{5}{3}(1+d)(d-5)\right]$$

i.e. $14d^2 - 20d - 25 = 0$

Then the admissible solution is,

$$d = 2.23 \text{ m} \qquad ...(D)$$

And from equations (B) and (C),

$$K(d-4)^2 = -K\left[\frac{5}{9}(1+d)(d-5)\right]$$

i.e., $14d^2 - 92d + 119 = 0$

Therefore, the admissible solution is

$$d = 1.77 \text{ m} \qquad ...(E)$$

The result of equation (E) could, also, be obtained from equation (D) because the points Y and Z are symmetrically placed about $d = 2.0$ which is evident from the Fig. 3.13(c).

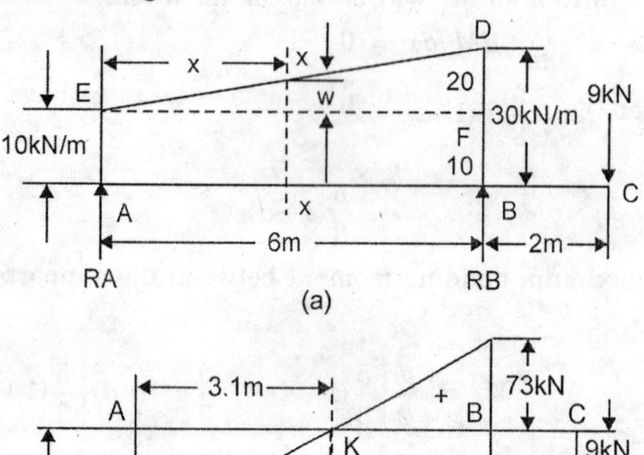

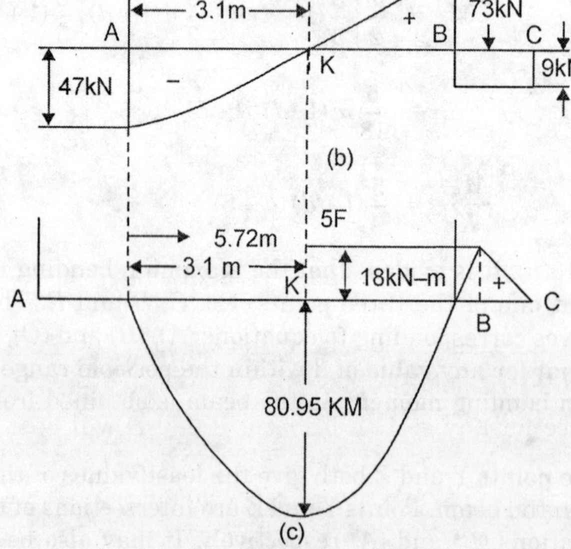

Fig. 3.14

EXAMPLE

Thus, the smallest M_{max} is obtained from equation (E) by substituting 2.23 or 1.77 for d.

$$i.e. \quad \text{Least} \quad |M_{max}| = -\frac{5}{9}\frac{w^2}{2}(1+1.77)(1.77-5)$$

$$= 2.485w \text{ Nm}.$$

3.6 : *An overhanging beam ABC, carrying a varying load and point load as shown in the Fig. 3.14, is freely supported at A and B.*

Draw to scale the shearing force and bending moment diagrams.

Solution : The varying load can be split into a triangular load and a rectangular (uniformly distributed) load as shown in the Figure.

Taking moments about B,

$$6R_A = (10\times 6)\frac{6}{2}+\left(\frac{1}{2}\times 6\times 20\right)\frac{6}{3}-9\times 2$$

$$i.e. \quad R_A = 47 \text{ kN}$$

$$\text{and} \quad R_B = \frac{1}{2}(10+30)\,6-47+9=82 \text{ kN}$$

For the section XX at a distance x from A (since we have from similar triangles $w = \frac{20}{6}x$)

$$SF = R_A + 10x + \frac{1}{2}xw$$

$$= -47 + 10x + \frac{5}{3}x^2 \qquad \text{(a parabola)}$$

at $x = 0$, $SF = -47$

at $x = 2$, $SF = -20.33$

at $x = 4$, $SF = +19.67$

at $x = 6$, $SF = +73$

Also, $SF = 0$ when $\frac{5}{3}.x^2 + 10x - 47 = 0$

or when $x = 3.1$ m

This will also correspond to Max B.M. position.

Shearing force for any section between B and C will be

$$= -47 + \frac{1}{2}(10+30)\times 6 - 82$$

$$= -9 \text{ kN} \qquad \text{(constant)}$$

For the section XX between A and B at a distance x from A,

$$M = -R_A x + (10.x) \times \frac{x}{2} + \left(\frac{1}{2} x \,.\, w\right) \frac{x}{3}$$

$$M = -47x + 5x^2 + \frac{5}{9} x^3$$

at $x = 0$, $M = 0.0$

at $x = 1$, $M = 36.45$

at $x = 2$, $M = -69.56$ kN – m

at $x = 3.1$, $M_{max} = -80.95$ kN – m

at $x = 4$, $M = -72.45$ kN – m

at $x = 6$, $M = +18$ kN–m

For the point of contraflexion,

$$M = 0$$

or $-47x + 5x^2 + \frac{5}{9} x^3 = 0$

Since $x \neq 0$ for the point of contraflexion it reduces to,

$$\frac{5}{9} x^2 + 5x - 47 = 0$$

The admissible solution of this equation is,

$$x = 5.72 \text{ m.}$$

For the section XX between B and C at a distance x from A.

$$M = -R_A x + (10 \times 6)(x-3) + \left(\frac{1}{2} \times 6 \times 20\right)(c-4) - R_B (x-6)$$

i.e. $M = (72 - 9x)$ (A straight line)

at $x = 6$, $M = 18$ kN – m

at $x = 8$, $M = 0.0$ kN – m

The shearing force and bending moment diagrams are drawn in Fig. 3.14.

EXAMPLE

3.7 : *A beam AB, 8 m long, is simply supported at two points A and B 6 m apart. It is loaded with two distributed loads and a point load at C as shown in Fig. 3.15. Draw the shearing force and bending moment diagrams.*

Solution : Taking moments about B,

$$6R_A - 104 - (20 \times 2) \times 1 + (60 \times 2) \times 1 = 0$$

i.e. $R_A = 4$ kN

Therefore, $R_B = 60 \times 2 + 20 + 20 \times 2 - R_A$

$= 176$ kN

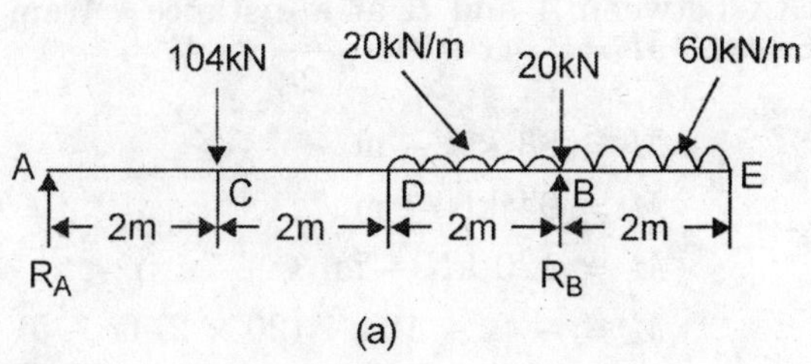

(a)

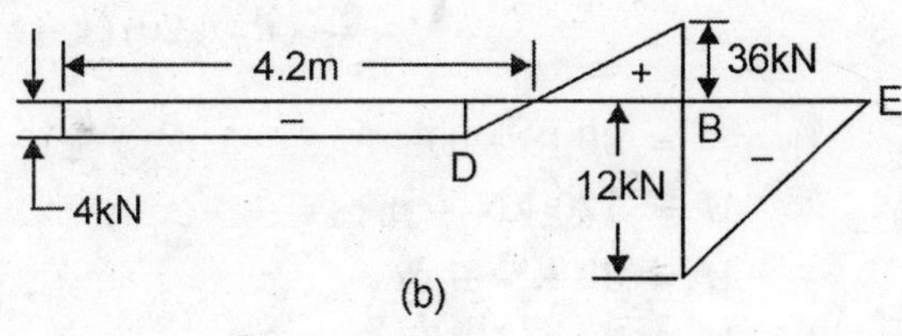

(b)

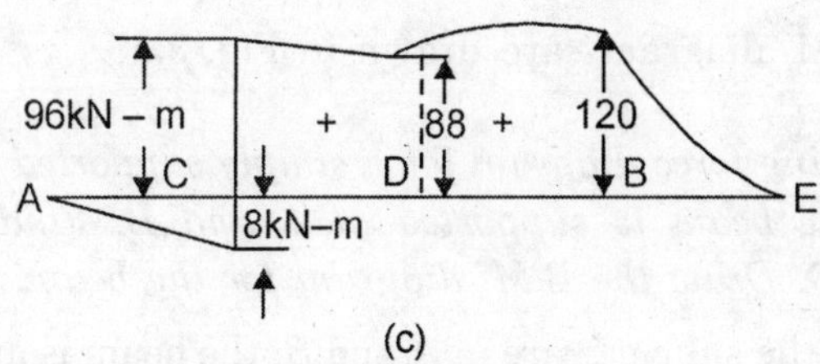

(c)

Fig. 3.15

(*a*) **Shearing Force at any section** bold *XX* **at a distance** bold x **from** bold *A*

(*i*) $0 < x < 4$	$SF = -R_A$		
	$= -4$ kN = constant		
(*ii*) $4 < x < 6$	$SF = -4 + 20(x - 4)$	(A straight line)	
at $x = 4$,	$SF = -4$ kN		
at $x = 4.2$,	$SF = 0$	(Maximum B.M. occurs here)	
at $x = 6$,	$SF = +36$kN	(Approaching from left)	
(*iii*) $6 < x < 8$	$SF = -4 + 20 \times 2 + 20 - 176 + 60\ (x - 6)$		
	$= -120 + 60\ (x - 6$	(A straight line)	
at $x = 6$,	$SF = -120$	(Approaching from right)	
at $x = 8$,	$SF = 0$		

(*b*) **B.M. at any section** ***XX*** **distant** x **from** A

(*i*) $0 < x < 2$	$M = -R_A\,x = -4x$	(A straight line)
at $x = 0$,	$M = 0$	
at $x = 2$,	$M = -8$ kN – m	(Approaching from the left)
(*ii*) $2 < x < 4$	$M = -4x + 104$	(A straight line)
at $x = 2$,	$M = 96$ kN – m	(Approaching from the right)
at $x = 4$,	$M = 88$ kN – m	

(*iii*) $4 < x < 6$ $M = -4x + 104 + \frac{20}{2}(x+4)^2$ (Parabola)

at $x = 4$, $M = 88$ kN – m

at $x = 5$, $M = 94$ kN – m

at $x = 6$, $M = 120$ kN – m

(*iv*) $6 < x < 8$ $M = -4x + 104 + (20 \times 2)(x - 5)$

$$-(R_B - 20)(x-6) + \frac{60}{2}(x-6)^2$$

$= 30(8 - x)^2$ (Parabola)

at $x = 6$, $M = 120$ kN – m

at $x = 7$, $M = 30$ kN – m

at $x = 8$, $M = 0$.

The S.F. and B.M. diagrams are drawn in Fig. 3.15.

3.8 : *The shearing force diagram for a simply supported beam is given in Fig. 3.16. The beam is supported at A and B. Find the distance between B and D. Draw the B.M. diagram for the beam.*

Solution : Since the supports are at *A* and *B*, the beam is an overhanging one. The S.F. remains constant and positive for the portion *CA*. Hence there is a load of 50 kN at the point *C*. At point *A*, there is a vertical drop of *SF* from + 50 to – 40 kN. Thus, the reaction $R_A = 50 - (-40) = 90$ kN.

From points *A* to *B*, the S.F. changes gradually from – 40 kN to 100 kN.

Thus, there will be a uniformly distributed load. If its intensity is *w* kN/m then,

$$w = \frac{dF}{dx} \text{ (Slope of the shearing force line)}$$

$$= \frac{100 - (-40)}{4} = \frac{140}{4} = 35 \text{ kN/m}$$

Again shearing force drops suddenly from + 100 kN to – 80 kN at the point *B*. That is, there is a reaction $R_B = 100 - (-80) = 180$ kN at the support *B*. Further the S.F. remains constant at 80 kN for the portion *BD* which signifies that there will be a point load of 80 kN at the point *D*.

The load diagram is shown in Fig. 3.16(*b*).

The value of '*a*' is found by taking moments about *B*. That is

$-50 \times 5 + 90 \times 4 - (35 \times 4) \times 2 + 170 \times a = 0$

i.e. $a = 1$ m.

E X A M P L E

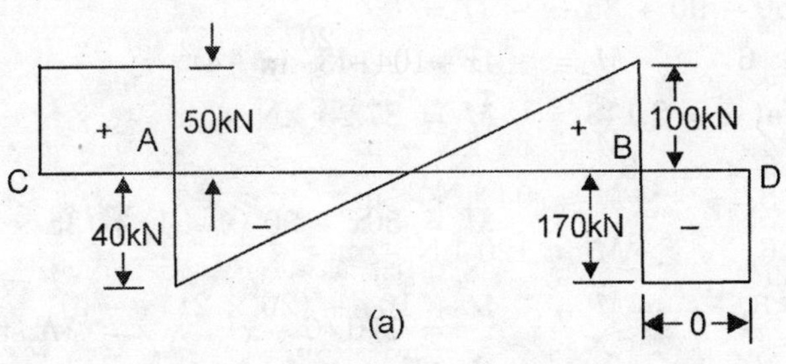

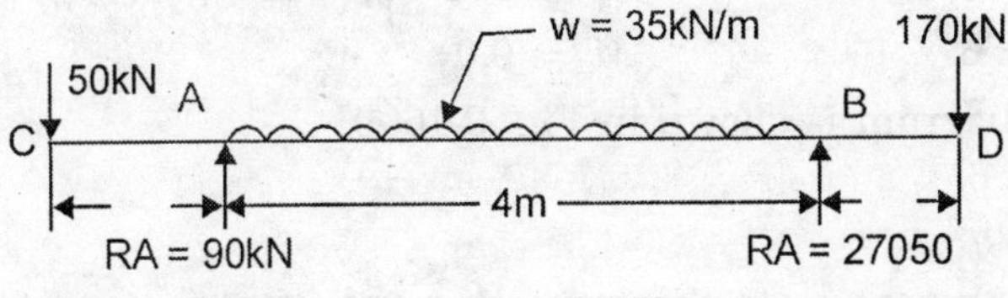

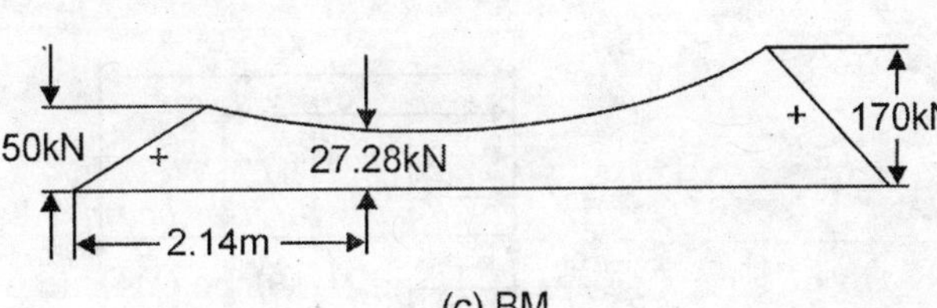

Fig. 3.16

For the B.M. diagram take a section *XX* at a distance x from the point *C*.

(*i*) When $0 < x < 1$

$$M = 50x \quad \text{(A straight line)}$$

at $x = 0$, $M = 0$

at $x = 1$, $M = 50$ kN − m

(*ii*) When $1 < x < 5$

$$M = 50x - 90\ (x - 1) + \frac{35}{2}(x-1)^2$$

at $x = 1$, $M = 50.0$ kN − m

at $x = 2$, $M = 27.5$ kN − m

at $x = 3$, $M = 40.0$ kN − m

at $x = 4$, $M = 87.5$ kN − m

at $x = 5$, $M = 770.0$ kN − m

Evidently, the value of B.M. decreases within the range $x = 1$ to 2 m and then starts rising. To find the location of least B.M. within this range set,

$$\frac{dM}{dx} = 0$$

i.e. $50 - 90 + 35 (x - 1) = 0$

or $x = 2.143$ m

and at $x = 2.143$ $M = 27.28$ kN – m

(*iii*) When $5 < x < 6$

$$M = 50x - 90 (x - 1) + (35 \times 4) (x - 3) - 270 (x - 5)$$

$$= 170 (6 - x) \quad \text{(A straight line)}$$

at $x = 5$, $M = 170$ kN – m

at $x = 6$, $M = 0.0$

The B.M. diagram is shown in Fig. 3.16(*c*).

EXAMPLE

3.7 GRAPHICAL METHOD (LINK POLYGON METHOD)

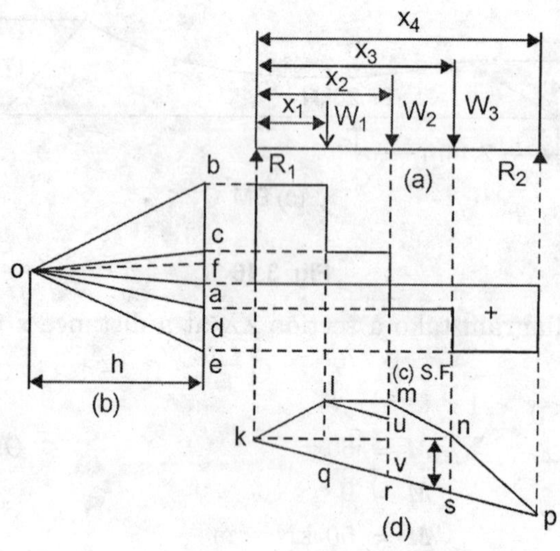

Fig. 3.17

Fig.3.17(*a*) shows the beam carrying the point loads for which it is desired to draw the S.F. and B.M. diagrams. For this to start with, the spaces are named using the Bow's notation e.g. *A*, *B*, *C*.... *E* etc. as shown in the Fig.. Choose a pole *O* after drawing the load line bcde (Fig. 3.17*b*) such that to some scale $bc = W_1$, $cd = W_2$ and $de = W_3$. Join *O* to *b*, *c*, *d* and *e*. This is called the polar diagram.

Choose any point *k* on the line of action of R_1, draw *kt* parallel to *Ob* in the space '*B*', *lm* parallel to *Oc* in the space '*C*', mn parallel to *Od* in the space '*D*' and similarly *np* parallel to *Oe* in the space '*E*'. Join *kp*. Draw *Oa* parallel to *kp*. Then *ab* and *ea* represent the reactions R_1 and R_2 and *k l m n p* the bending moment diagram on base *kp*, the bending moment *M* being proportional to vertical ordinates like *ql*, *rm* etc.

Proof : In Fig. 3.17(*b*), W_1 is represented by *bc* and it acts through the point l (Fig. 3.17*d*). Since *kl* is parallel to *Ob* and *lm* is parallel to *Oc*, W_1 can be replaced by forces *bO* along *lk* and *Oc* along *lm* (law of triangle of force). Similarly, W_2 at m is equivalent to two forces represented by cO along ml and Od along mn. *In* the same way W_3 at n is replaced by dO along nm and Oe along np. Evidently, all these forces nullify each other except bO along lk and Oe along np. The two forces must be balanced by R_1 and R_2 as the beam is in static equilibrium. This is only possible, if R_1 at k is replaced by Ob along kl and aO along pk, because R_2 at p is equivalent to eO along pn and Oa along kp. Thus, R_1 is given by ab and R_2 by ea.

– Now coming to the bending moments since the triangles Oba and klq are similar,

$$lq = ba.kq/Oa \quad ...(3.11)$$

Drawing Of and kt perpendicular to be and lq respectively we have from the similar triangles Oaf and kqt,

$$kqlOa = x_1/h \quad ...(3.12)$$

Thus, substituting equation (3.12) in equation (3.11),

$$lq = ba \;.\; x_1/h \quad ...(3.13)$$

But as proved above, ba represents R_1. Therefore, the product $ba.\ x_1$ (= $R_1 \cdot x_1$) is proportional to the bending moment at the point where W_1 acts. Hence, from equation (3.13), for a given position of the pole O, lq will represent to a certain scale the bending moment at a distance x_1 from the left hand support.

Again draw lu parallel to kp to meet mr in u, and the perpendicular qv will meet mr in v. Evidently triangles cOa and mlu are similar and

$$mu = ca \;.\; lu/\mathbf{Oa} \quad ...(3.14)$$

Also, since triangles qvr and Ofa are similar,

$$lu = qr = Oa \;.\; qv/Of$$

or

$$lu/Oa = qv/Of \quad ...(3.15)$$

Thus, from equations (3.14) and (3.15),

$$mu = ca \;.(x_2 - x_1)/h \quad ...(3.16)$$

Therefore,

$$\begin{aligned} mr &= mu + ur \\ &= mu + lq \\ &= ca \;.\; (x_2 - x_1)/h + ba \;.\; x_1/h \qquad \text{From (3.13)} \\ &= (ba - bc)\;(x_2 - x_1)/h + ba \;.\; x_1/h \\ &= \frac{1}{h}[(R_1 - W_1)(x_2 - x_1) + R_1 x_1] \end{aligned}$$

i.e.,

$$mr = \frac{1}{h}[R_1 x_1 - W_1\,(x_2 - x_1)] \quad ...(3.17)$$

Thus, for a given position of the pole O, for which h is fixed, equations (3.13) and (3.17) give that the ordinates ql and rm of the diagram 3.17d represent to some scale the bending moments at these points. Similarly, it can be shown that at any other point the ordinate of the diagram in Fig. 3.17(d) will give the bending moment at that point according to the same scale, which is determined as follows:

Let the scale of load in polar diagram be 1 mm = S_F Newton, that of the length in Fig. 3.17(a) as 1 mm = S_L metre and of the bending moment in Fig. 3.17(d) as 1 mm = S_M Nm. Then from equation 3.13 the length,

$$ql = \left(\frac{R_1}{S_F} \cdot \frac{x_1}{S_L}\right) \Big/ h \qquad ...(3.18)$$

Also, length $\qquad ql = M_1/S_M \qquad ...(3.19)$

Thus, since $M_1 = R_1 x_1$, we have from equations (3.18) and (3.19),

$$S_M = h \,.\, S_F \; S_L \qquad ...(3.20a)$$

Thus, for any point if the ordinate in Fig. 3.17(*d*) is *y* mm then B.M.= $yS_M = y \,.\, h \,.\, SF \,.\, S_L$ Nm, where h is measured in mm.

For drawing the S.F. diagram, horizontal lines are projected from the points *b, c, d,e* and a across from the polar diagram. The S.F. diagram is obtained with the horizontal line from '*a*' as the base line.

This method can also be used for distributed load by dividing it into a large number of very small point loads concentrated at the centre of gravity of each small strip.

EXAMPLE

3.9. Draw the *SF* and *BM* diagram of the cantilever shown in adjoining figure 3.18 (*a*) choosing the free end of the cantilever *D*, as origin and measuring *x* positive towards left.

For the range $0 < x < \dfrac{l}{2}$,

$$F_x = \text{weight of the portion } CX$$

$$= wx = \frac{2W}{l} x \text{ st. line)}$$

Therefore, $(F_x)_{x=0} = 0$

At $x = \dfrac{l}{2}$,

$$F = \frac{2W}{l}\frac{l}{2} = W$$

For the range $\dfrac{l}{2} < x < l$

$$F_n = w\frac{l}{2} + W = 2W$$

$$= \frac{2W}{l} \cdot \frac{1}{2} + W = W + W$$

(constant)

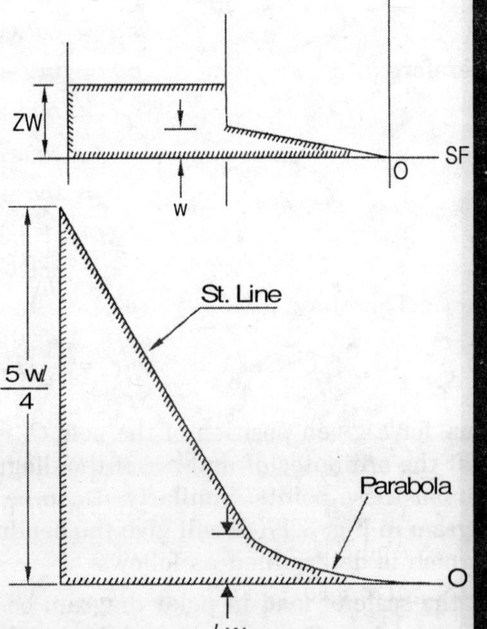

Fig. 3.18

For *BM* diagram

For the range $0 < x < \dfrac{l}{2}$

$$M_x = \frac{wx^2}{2} = \frac{2W}{l}\frac{x^2}{2} = \frac{Wx^2}{l} \qquad (A)$$

EXAMPLE

At $x = 0$ $\quad M_x = 0$

$$x = \frac{l}{2} \qquad M_B = \frac{w}{2}\left(\frac{l}{4}\right)^2 = \frac{w}{l}\frac{l^2}{16} = \frac{Wl}{16}$$

For range $\frac{l}{2} < x < l$,

$$M_x = \left(\frac{wl}{2}\right)\left(x - \frac{l}{4}\right) + W\left(x - \frac{l}{2}\right)$$

$$= \left(\frac{2W}{l}\frac{l}{2}\right)\left(x - \frac{l}{4}\right) + W\left(x - \frac{l}{2}\right)$$

$$= W\left(x - \frac{l}{4}\right) + w\left(x - \frac{l}{2}\right)$$

$$= 2Wx - \frac{3l}{4}W \text{ (st. line)}$$

$$\text{At } x = \frac{l}{2} \qquad M_B = 2W\frac{l}{2} - \frac{3l}{4}W = \frac{Wl}{4}$$

$$x = l \qquad M_A = 2Wl - \frac{3l}{4}W = \frac{5W}{4}l$$

3.7. Draw the *SF* and *BM* diagrams of the cantilever shown in the figure 3.19 (*a*).

Solution. Take origin at point *C* and measure distance x toward left as positive.

Then for the range $0 < x < 1.5$ m

SF $\quad F = 500 + wx = 500 + 100x \quad$ (st. line)

BM $\quad M_x = 500x + \frac{wx^2}{2} = 500x + 50x^2 \quad$ (parabola)

Therefore, $\quad$ at $x = 0$

$(F)_C = 500$

$M_C = 0$

at $x = 0.5$m $\quad (F)_{0.5} = 500 + 100 \times .5 = 550$ kN

$x = 1.0 \quad (F)_{1.0} = 500 + 100 \times 1 = 600$ kN

$x = 1.5 \quad (F_B) = 500 + 100 \times 1.5 = 650$ kN

And *BM* $\quad$ at $x = 0 \quad (M)_0 = 0$

at $x = .5 \quad M_{0.5} = 500 \times .5 + 50 \times (.5)^2$

$= 250 + 12.5 = 282.5$ kN-m $\quad x = 1.0$

$(M)_{1.0} = 500 \times 1.0 + 50\,(1)^2 = 550$ kN-m$x = 1.5$

$$(M)_{1.5} = M_B = 500 + 1.5 + 50 \times (1.5)2 = 750 + 112.5$$
$$= 862.5 \text{ kN-m}$$

For the range $1.5 < x < 2$

SF $\quad F = 500 + 100 + 1.5 = 650 \text{ kN} \quad$ (constant)

BM $\quad M = 500x + w\ (1.5)\ (x - 0.75)$

$$= 500x + 150\ (x - 0.75) \quad \text{(straignt. line)}$$

Therefore, $\quad M_{1.5} = 500 + 1.5 + 150\ (1.5 - 0.75)$

or $\quad M_B = 750 + 150 \times 0.75$

$$= 750 + 112.5 = 862.5\ k\text{N-m}$$

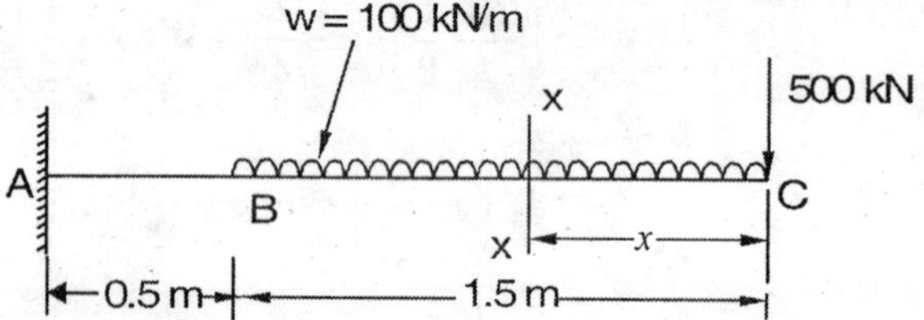

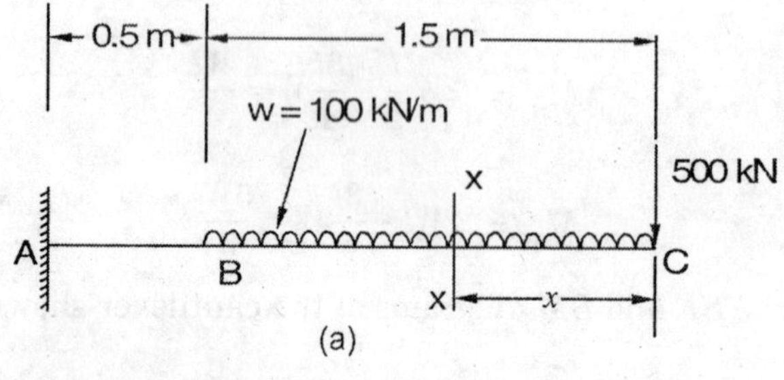

(a)

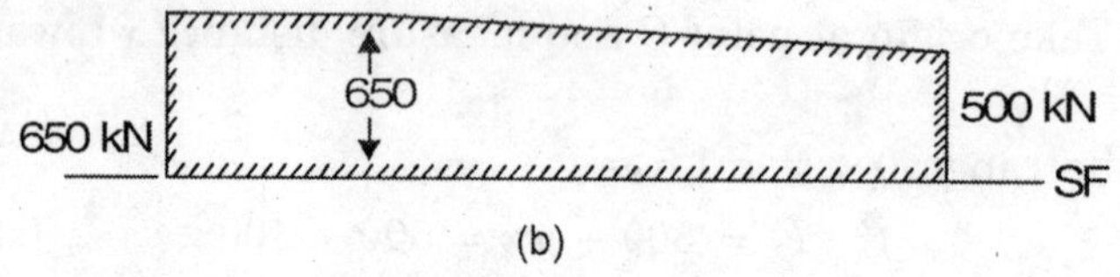

(b)

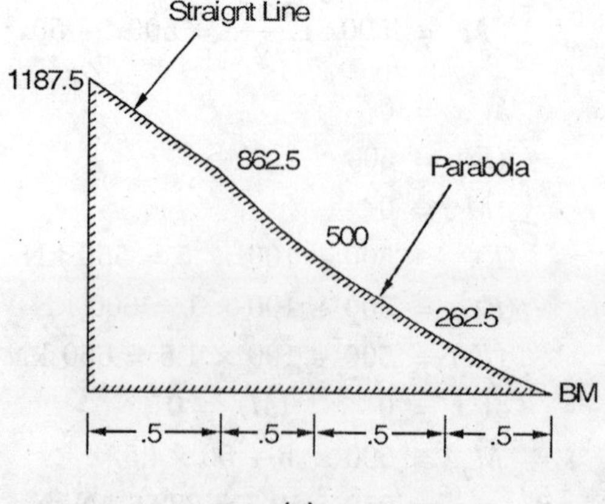

(c)

Fig. 3.19

Similarly, at $x = 2.0$ $M_A = 500 \times 2 + 150\,(2 - .75)$

$= 1000 + 150 \times 1.25$

$= 1187.5$ kN-m

The SF and *BM* diagrams are as shown in figure 3.19

3.10. Draw *SF* and *BM* diagram for the cantilever shown in figure (3.20)

Solution. (Take origin at *D* and measure *x* from *D*, left words positive)

For Range $0 < x < 0.5$)

$F = 0,\ M = 0$

because there is no load sustained by portion *CD*

For the range $0.5 < x < 1.5$ m

$$F_x = w(x - 0.5)$$

At $x = .5$ $F_C = 0$

At $x = 1.5$ $F_B = w(1.5 - .5) = 500 \times 1$

$= 500$ kN-m

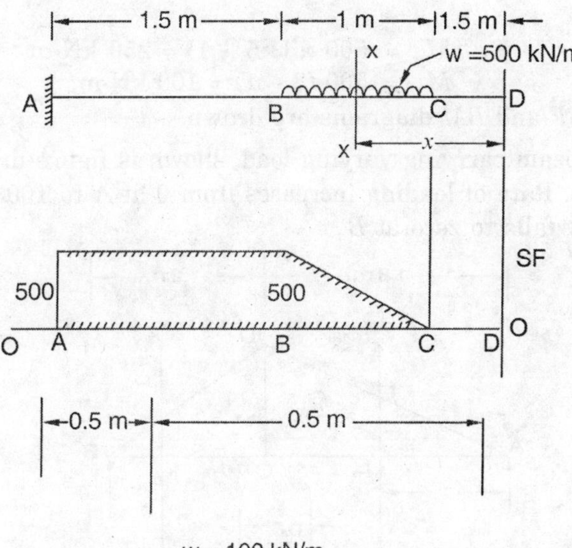

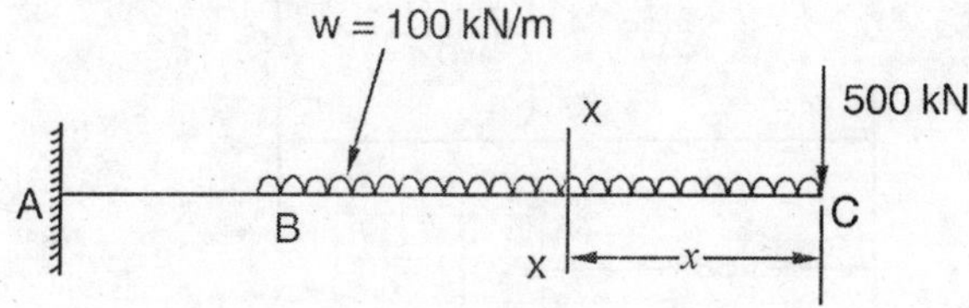

Fig. 3.20

$$M_x = \frac{w}{2}(x - 0.5)^2 \qquad \text{(parabola)}$$

At $x = 0.6$m $M_C = 0$

At $x = 0.6$m $(M)_{.6} = \dfrac{500}{2}(0.6 - 0.5)^2 = \dfrac{500 \times .01}{2}$

$= 2.5$ kN m

At $x = 0.7$ $(M)_{.6} = 250 \times (0.7 - .5)^2$

$= 250 \times 0.04$

$= 10.0$ kN-m

At $x = 0.8$ $(M)_{.8} = 250 \times (.8 - .5)^3$

$= 22.5$ kN-m

At $x = 1.0$ $(M)_{1.0} = 250 \times (1.2 - .5)^2$

$= 250 \times 0.49$

$= 122.5$ kN m

at $x = 1.5$ $(M)_B = 250\ (1.5 - .5)$

$= 250.0$ kN-m

For Range $1.5 < x < 3.0F = w \times 1$

$= 500$ kN constant

$M = (w \times 1)\ (x - 1)$

(because the CG of u/d load is at mid pint of BC is it is at 1.0 m from D)]

$= 500\ (x - 1)$ (st. line)

Therefore, at at $x = 1.5$

at $x = 3$

$M_B = 500 \times (1.5 - 1) = 250$ kN-m

$M_A = 500\ (3 - 1) = 100$ kN-m

Accordingly SF and BM diagrams are drawn.

3.11. For the beam carrying varying load, shown is figure draw SF and MB diagrams. Rate of loading increases from 0 at A to 100 kN-m at C and gradually falls to zero at B.

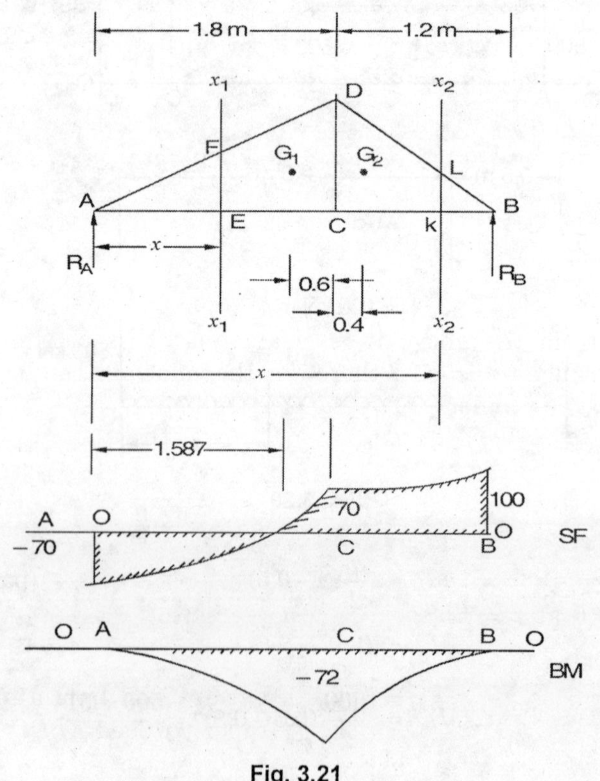

Fig. 3.21

E X A M P L E

EXAMPLE

Distance $AC = 1.8$ m

and $BC = 1.2$ m

Solution : Splitting the load into two triangular loads ACD and DCB. Taking moments about B for the static equilibrium of the beam, we get,

$$R_A\ (1.8+1.2)-\frac{1}{2}(1.8)100\,(0.6+1.2) = -\frac{1}{2}(.12)\,100\,(0.8)=0$$

$$3R_A - 90 \times 1.8 - 48 = 0$$

$$R_A = \frac{210}{3} = 70 \text{ kN}$$

For the range $0 < x < 1.8$ m with origin at A,

$$F_x = -R_A + \text{triangular weight } AEF$$

$$= -70 + \frac{1}{2}x(EF)$$

$$= -70+\frac{1}{2}x\left[\frac{CD}{1.8}\times x\right]$$

$$= -70+\frac{1}{2}x^2\left[\frac{100}{1.8}\right]$$

$$= -70 + 27.7777\,x^2 \qquad \text{(parabola)}$$

Therefore, at

$x = 0$ $\quad F_A = -70$ kN

$x = 0.8$ $\quad (F)_{.8} = -70 + 27.7777 \times (.8)^2$

$= -70 + 17.7777$

$= -52.2223$ kN

$x = 1.0$ $\quad (F)_{1.0} = -70 + 27.7777$

$= -42.2223$ kN

$x = 1.8$ $\quad (F)_C = -70 + 27.777 \times (1.8)^2$

$= 19.9997$ kN

Evidently, from the above it is clear that SF is changing its sign within the range $0 < x < 1.8$ therefore, the value of x can be calculated, where SF = 0 as under

$$-70 + 27.7777\,x^2 = 0$$

or $$x = \sqrt{\frac{70}{27.7777}} = \sqrt{2.5200}$$

or $$x = 1.587 \text{ m}$$

Thus, $SF = 0$ at $x = 1.587$

For the range $1.8 < x < 3.0$ m,

F_x = Total downward load from a to k

i.e., $F_x = -R_A$ + Trapezium load $KLDC$

and triangular load *ADC*

Now $\frac{KL}{KB} = \frac{CD}{1.2}$ (From similar Δs *BKL* and *BCD*] or

$$KL = (KB)\frac{CD}{1.2} = (3-x)\frac{100}{1.2}$$

Also, $CK = AK - AC = (x - 1.8)$

Therefore, $$F_x = -70 + \frac{1}{2}(1.8)\times 100 + \frac{1}{2}\left[CD + (3-x)\frac{100}{1.2}\right](x-1.8)$$

$$= -70 + 90 + \frac{1}{2}[100 + 50\,(x-1.8)\;\;41.66\,(3-x)\,(x-1.8)$$

$\therefore$ At $x = 1.8$, $F_C = 20 + \frac{1}{2}(100 + 0 + 0] = 70$ kN

At $x = 2$m, $(F)_{2.0} = 20 + \frac{1}{2}[100 + 50\times .2 + 41.66\,(3.2)\,(0.2)$

$$= 20 + \frac{1}{2}[110 + 8.332] = 79.166 \text{ kN}$$

At $x = 2.5$, $(F)_{2.5} = 20 + \frac{1}{2}[100 + 50\,(2.5 - 1.8) + 41.66\,(0.5)\,(0.7)]$

$$= 20 + \frac{1}{2}[135 + 14.581] = 94.79 \text{ kN}$$

At $x = 3.0$, $F_B = 20 + \frac{1}{2}[100 + 50\times 1.2 + 41.66\times 0\times 1.2]$

$$= 20 + \frac{1}{2}[160] = 100 \text{ kN}$$

For BM diagram

For range $0 < x < 1.8$ m,

$$M_x = -R_A x + \frac{1}{2}(AE)\,(EF)\left(\frac{1}{3}x\right)$$

$$= -70x + \frac{1}{2}x\left(\frac{CD}{1.8}\times x\right)\frac{x}{3}$$

$$= -70x + \frac{50}{1.8}x^2\left(\frac{x}{3}\right)$$

$$= -70x + 9.259x^3$$

At $x = 0$, $M_A = 0$

At $x = 0.8$, $(M)_{0.8} = -70 \times .8 + 9.259 \times (.8)^3$

EXAMPLE

$= -56 + 4.746 = -51.2594$ kN-m

At $x = 1$, $(M)1.0 = -70 \times 1 + 9.259$

$= -68.74$ kN-m

At $x = 1.8$, $(M)_C = -70 \times 1.8 + 9.259 \times (1.8)3$

$= -126 + 53.9985$

$= -72$ kN-m

At $x = 1.187$ $(M)_{max} = -70 \times 1.587 + 9.259\,(1.587)^3$

$= 111.09 + 37.008$

$= -74.82$ kN-m

For range $1.8 < x < 3$,

$$M_X = R_A x + \frac{1}{2} AC \times CD \times (x - 1.2)$$

$$+\left(\frac{1}{2} NL \times ND\right)\left(\frac{2}{3} CK\right) + \frac{1}{2}(LK \times CK)\left(\frac{1}{2} CK\right)$$

$$= -70x + \frac{1}{2}(1.8)(100)(x-1.2)$$

$$+\frac{1}{2}(x-1.8)\left[100 - \frac{100}{1.2}(3-x)\right]$$

$$\times\left(\frac{2}{3}\right)(x-1.8) + \frac{1}{2}\left[(3-x)\frac{100}{1.2}\right][x-1.8]^2$$

$$= -70x + 90(x-1.2) + \frac{1}{2}(x-1.8)[100]\left[1 - \frac{1}{1.2}(3-x)\right]$$

$$\times \frac{2}{3}(x-1.8) + \frac{1}{2} \times \left[\frac{100}{1.2}(3-x)\right](x-1.8)^2$$

$\therefore$ At $x = 1.8$, $M_C = -70 \times 1.8 + 90 \times 0.6 + 0 + 0 + 0$

$= -126 + 54 = -72$ kN-m

At $x = 2$, $$(M)_{x=2} = -70 \times 2 + 90(2-1.2) + \frac{1}{2}(2-1.8)[100]\left[1 - \frac{1}{1.2}\right]$$

$$\times \frac{2}{3}(2-1.8) + \frac{50}{1.2}(3-2)(2-1.8)^2$$

$$= -140 + 72 + (0.2 \times 50)\left(\frac{1.2-1}{1.2}\right) \times \frac{2}{3} \times 0.2 + \frac{50}{1.2}(0.2)^2$$

$= -68 + .2222 + 1.6666$

$= -66.1112$ kN-m

At $x = 2.5$, $$(M)_{2.5} = -70 \times 2.5 + 90\,(2.5 - 1.2) + \frac{1}{2}(2.5 - 1.8)\,100$$

$$\times \left[1 - \frac{(3-2.5)}{1.2}\right] \times \frac{2}{3}(2.5 - 1.8) + \frac{50}{1.2}(3 - 2.5)(2.5 - 1.8)^2$$

EXAMPLE

$$= -175 + 117 + 9.52778 + 10.2083$$

$$= -38.2639 \text{ kN-m}$$

At $x = 3$, $\quad (M)_B = -70 \times 3 + 90(3 - 1.2) + \frac{1}{2}(3 - 1.5)$

$$\times\ 100\,(1.0)\frac{2}{3}(3 - 1.8) + \frac{50}{1.2}(0)\,(3 - 1.8)^2$$

$$= -210 + 162 + 48 = 0$$

Thus, the *SF* and *BM* are obtained by plotting these points.

3.12. A beam 5m long is simply supported at its ends. It carries a u/d load of 25 kN-m over *AC* and a point load at *D* 1m from end *B* draw *SF* and *BM* diagrams for this beam.

Solution. For finding the reactions take moments about point B for the equilibrium of beam.

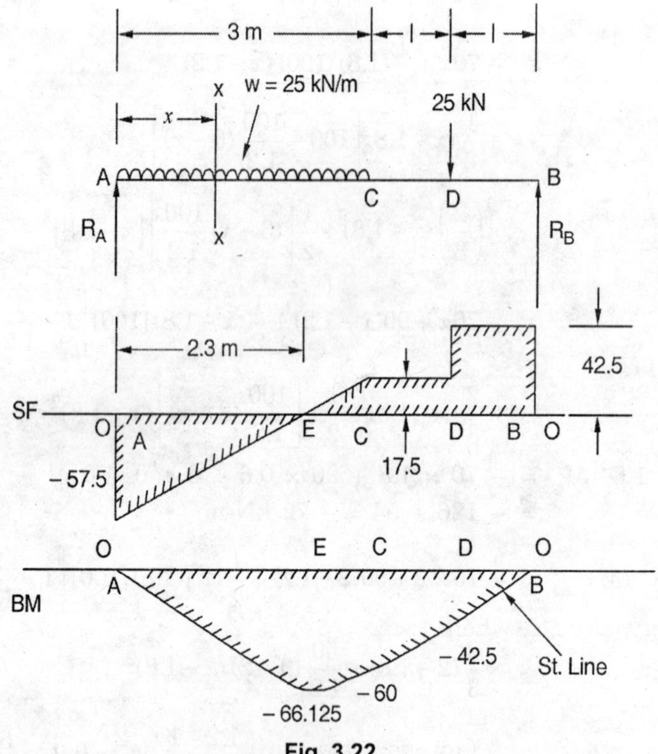

Fig. 3.22

$$5R_A - (25 \times 3)\,(3.5) - 25 \times 1 = 0$$

or $\quad R_A = \frac{1}{5}[262.5 + 25] = 57.5 \text{ kN}$

$$R_B = \text{Total load} - R_A$$

$$= 25 \times 3 + 25 - 57.5 = 42.5 \text{ kN}$$

Take origin at A and measure x from A.

For the segment AC, where $0 < x < 3$

$$F_x = -R_A + wx = -57.5 + 25 \times x \text{ st line}$$

$$M_x = -R_A x + 25\frac{x^2}{2} \qquad \text{(parabola)}$$

$$(F)_0 = -57.5 \text{ kN}$$

$$(F)_{x=3} = F_C = -57.5 + 25 \times 3 = 17.5 \text{ kN}$$

It changes sign and $F_C = 0$ when $x = \dfrac{57.5}{25} = 2.3$ m

$$(M)_A = 0 \text{ at } x = 0$$

At $x = 1$, $\quad (M)_{x=1} = -57.5 \times 1 + 12.5 \times (1)^2$

$$= -45 \text{ kN-m}$$

At $x = 2$, $\quad (M)_{x=2} = -57.5 \times 2 + 12.5 \times (2)^2$

$$= -115 + 50 = -65$$

At $x > 2.3$, $\quad (M)_{2.3} = -57.5 \times 2.3 + 12.5\,(2.3)^2 = -66.125$

At $x = 3$, $\quad M_C = -57.5 + 3 + 12.5 \times (3)^2$

$$= 172.5 + 112.5 = -60$$

For the segment CD when $3 < x < 4$,

$$F_x = -R_A + 25 \times 3 = -57.5 + 75 = 17.5 \text{ constant}$$

$$M_x = -R_A x + (25 \times 3)\,(x - 1.5)$$

$$= -57.5x + 75\,(x - 1.5) \text{ st. line}$$

At $x = 3$, $\quad M_C = -57.5 \times 3 + 75(3 - 1.5)$

$$= -172.5 + 25 \times 15 = -60 \text{ kN m}$$

At $x = 4$, $\quad M_D = -57.5 \times 4 + 75(4 - 1.5)$

$$= -230.0 + 187.5 = -42.5 \text{ kN-m}$$

For the segment DB when $4 < x < 5$

$$SF = -R_A + 25 \times 3 + 25 = 57.5 + 100$$

$$= 42.5$$

$$M_x = -R_A x + (25 \times 3)\,(x - 1.5) + 25\,(x - 4) \text{ st line}$$

$$(M_x)_{x>4} = M_D = -57.5 \times 4 + 75\,(4 - 1.5) + 0$$

$$= -42.5 \text{ kN-m}$$

At $x = 5$, $\quad (M_B) = -57.5 \times 5 \times 75\,(5 - 1.5) + 25\,(5.4)$

$$= 0 - 287.5 + 262.5 + 25 = 0$$

With the above information the SF and BM diagrams are drawn as shown.

REVIEW QUESTIONS

1. Explain the terms "Moment of a force about point A" and "Plane of the Moment".
2. State the right handed screw rule for drawing the "moment vector." How is the vector represented ?
3. Differentiate between "Banding Moment" and "Twisting Moment".
4. Write down the equation showing the relationship between Bending Moment shearing force and Rate of Loading at a particular point.
5. Define Point of Contraflexion.

PROBLEM FOR PRACTICE

1. A beam, 10 m long, is simply supported over a span of 5 m, the overhangs on the left and right hand supports being 3 m and 2 m respectively. The left and right hand overhangs carry uniformly distributed loads of 60 kN/m and 105 kN/m respectively. The latter load also extends upto left hand support. Draw to scale the shearing force and bending moment diagrams for the beam. Calculate the value of the least B.M. between the supports and find the position where it occurs.
(**Ans.** 41.85 kN–m at 3.9 m from the left hand support)
2. An overhanging beam *ABC*, 9 m long, is simply supported at points *A* and *B*, 7 m apart. The beam carries a load of 225 kN uniformly distributed over its whole length together with three point loads of 120 kN, 70 kN and 150 kN at 1 m, 4 m and 6 m respectively from the point *C*.
Draw to scale the S.F. and B.M. diagrams giving the maximum value of B.M.
(**Ans.** 394.3 kN–m at 3 m from *A*)
3. A wooden beam 5 m long and 0.4 m × 0.4 m in section, floating in horizontal position is loaded with a point load at the centre just sufficient to immerse it completely. If the mass of 1 cubic metre of wood is 800 kg, draw and S.F. and B.M. diagrams for the beam. What are the maximum values of S.F. and B.M. and where do they occur?
(**Ans.** 1570 N and 981 Nm both at the centre}
4. If the beam in question 3 were to be loaded with two weights placed symmetrically at distance '*a*' from the ends so that the maximum bending moment to which the beam is subjected is as small as possible, find the value of each weight and its position. What is then the magnitude of the maximum bending moment? (**Ans.** 785 N ; 1.035 m from each end ; 169 Nm)
5. A concrete pile, of length L and uniform cross–section, is to be lifted with the help of a crane by putting two slings symmetrically around it so that it always stays horizontal. Find the most suitable positions for putting the slings, if it is assumed that the pile is most adversely affected by the bending moments caused by self–weight and not by shearing forces.
(**Ans.** 0.207 *L* from each end)
6. A horizontal beam *AB*, 5 m long is supported freely at the ends. It carries a vertical load of 15 kN at a distance of 4.0 m from *B*. In addition there acts, at a section 3 m from *B*, clockwise couple of magnitude 5 kN–m on the beam. The plane of the couple being vertical and passing through the axis of the beam, draw the B.M. and S.F. diagrams. Where does the maximum B.M. occur ? Find its value. (**Ans.** 3 m from *B* ; 12 kN–m)
7. A beam *AB*, 7 m long, is simply supported at its ends. It carries a load of 30 kN, applied through a *L*–shaped bracket *CDE* which is fixed to the beam at C, distant 2 m from *A*, such that *CDE* lies in the vertical plane passing through the axis of the beam and the portion *DE* is towards the side of point *B*. If *DE* = 0.5 m. Draw the S.F. and B.M. diagrams for the beam *AB* giving the values of maximum S.F. and B.M. $\left(\textbf{Ans. } \frac{135}{7} \text{ kN}; \frac{375}{7} \text{ kN} - \text{m just to the right of C}\right)$

8. A cantilever, 2 m long, carries a load varying from zero at the free and to 50 kN/m run at the fixed end. Draw the S.F. and B.M. diagrams. What are the values of maximum S.F. and bending moment ? (**Ans.** 50 kN ; 33.33 kN–m both at the fixed end)

9. An overhanging beam *ABCD*, 18 m long, is simply supported at *A* and *C* 12 m apart. It carries a load of 15 kN concentrated at *B* which is at a distance of 4 m from *A*. In addition there acts a distributed load with intensity varying from zero at *A* and *D* to 15 kN/m run at *C*.

Draw the S.F. and B.M. diagrams and calculate the position and magnitude of maximum B.M. Find from diagram the position of the point of inflexion.

(**Ans.** 5.3 m from A ; 121.6 kN–m ; 10.55 m)

10. The shearing force diagram, for a beam, simply supported at *A* and *B* is given in Fig. 3.23. Draw the B.M. diagram for the same.

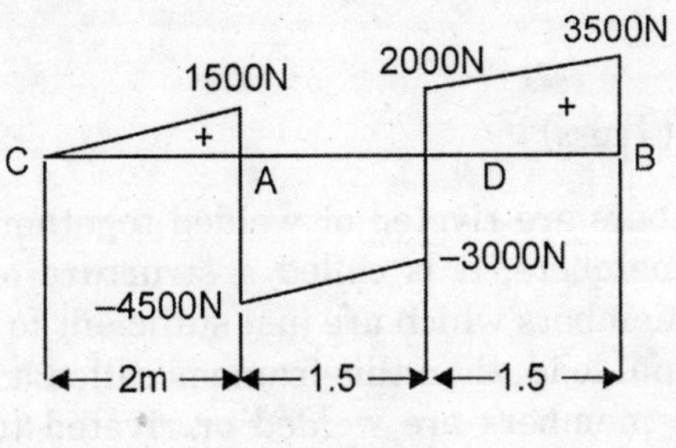

Fig. 3.23

11. For an overhanging beam *ABCD*, simply supported at *B* and *C*, the S.F. diagram is given in Fig. 3.24. Draw the B.M. diagram.

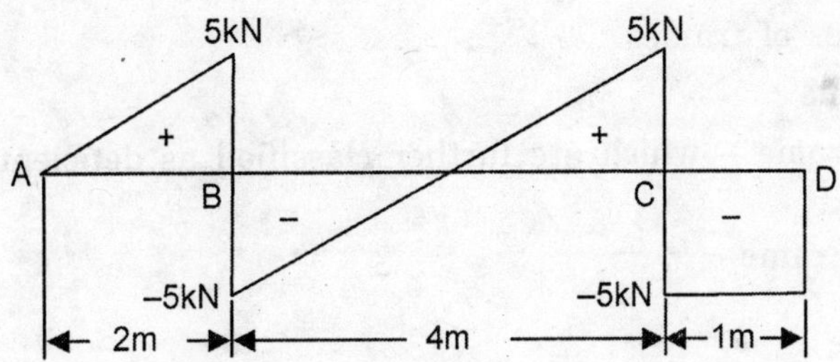

Fig. 3.24

❑❑❑

CHAPTER 4

TRUSS

4.1 INTRODUCTION

4.1.1 Structure or Frame (Truss)

When several bars or members are riveted or welded together such that there is no relative motion between any two members, it is called a structure or frame (or Truss). When the frame is composed of such members which are just sufficient to keep the frame in equilibrium, when it is supporting external load, then this frame is called a perfect frame. Despite the fact that in actual practice the members are welded or riveted together at their joints but for calculation purposes, it is not only customery but universally accepted that the joints are hinged or pin–joints.

4.1.2 Types of Frame

There are three types of frames :

(i) Perfect Frame

(ii) Imperfect Frame – which are further classified as deficient frame and redundant frame.

(iii) Redundent Frame

4.1.2.1 Perfect Frame

A perfect frame is one which consists of just sufficient number of members so as to keep the frame in equilibrium, when it is supporting an external load. Evidently, the simplest perfect frame in a triangle an shown in Fig. 4.1 which consists of 3 bars or members *viz. AB, BC* and *CA* and it has 3 joints at the points *A, B* and *C*. The forces carried or sustained by each member can be analysed, if we are given the load coming on the frame.

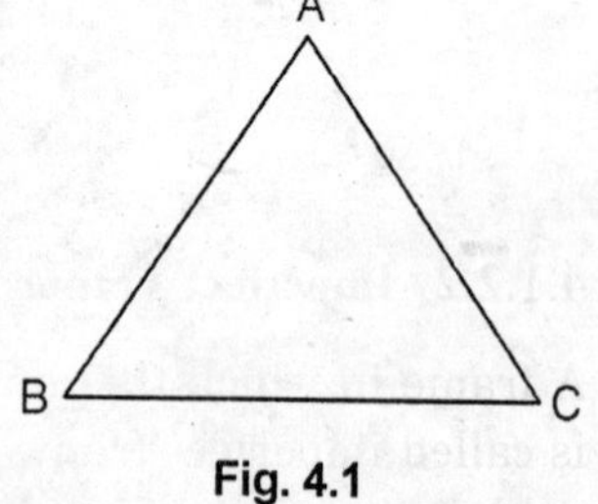

Fig. 4.1

Further let two members *AD* and *CD* and a joint *D* he added to the triangular frame *ABC* to get another perfect frame *ABCD*. It is shown in Fig. 4.2 (*a*). This frame also can easily be analysed by the condition of equilibrium. Similarly, if we add a set of two members *DE* and *CE* like the one shown in Fig. (4.2*b*) it results in a perfect frame. From the above discussion we conclude the following relationship between the number of members and number of joints in a perfect frame or truss:

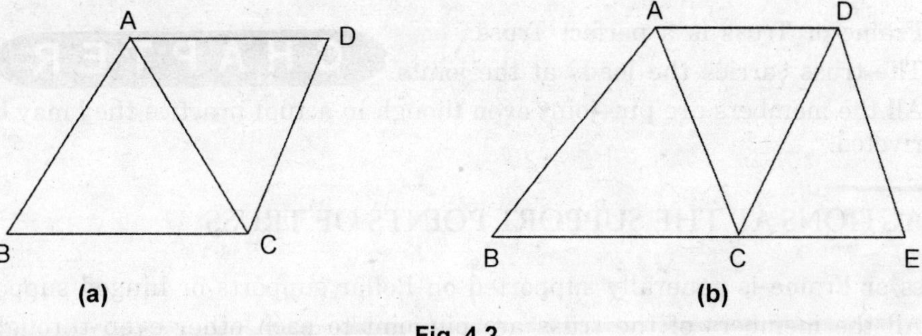

Fig. 4.2

$$n = 2j - 3 \quad \text{...(4.1)}$$

Where, n = number of members

j = number of joints

Important Note

It is very important to note that equation (4.1) gives only a necessary but not sufficient condition of a perfect truss. For example, two trusses shown in Fig. (4.3*a*) and Fig. (4.3*b*) have the same number of members and same number of joints. The truss shown at Fig. (4.3*a*) is perfect but the one shown at Fig. (4.3*b*) will not retain its shape, if loaded at joint 2. Thus, we conclude that only necessary and sufficient condition of a perfect difficient frame.

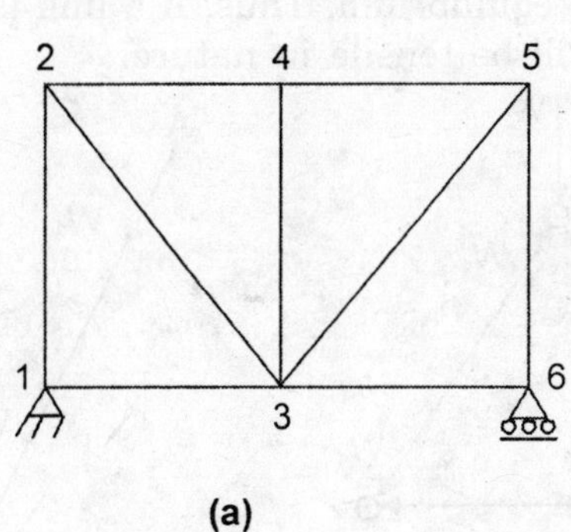

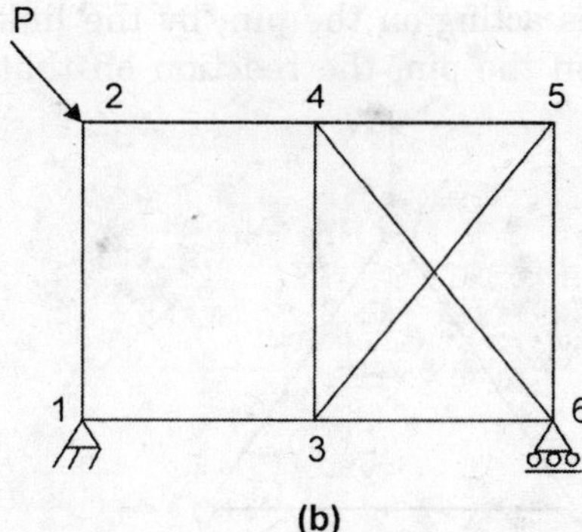

Fig. 4.3

4.1.2.2 Imperfect Frames

A frame in which this number of joints and number of members do not satisfy equation (4.7) is called imperfect frame or truss. If the number of members is less than $(2j - 3)$ it is known as deficient frame.

On the other hand, if the number of members in a frame or truss are more than $(2j - 3)$, it is known as Redundant Frame.

4.2 FORCES IN VARIOUS MEMBERS OF A TRUSS

Before we proceed to find the forces carried by various members of a truss following assumptions are made :

(*i*) Frame or Truss is a perfect Truss.

(*ii*) The truss carries the loads at the joints.

(*iii*) All the members are pin–joint even though in actual practice they may be welded or riveted.

4.3 REACTIONS AT THE SUPPORT POINTS OF TRUSS

The Truss or Frame is generally supported on Rollar supports or hinged supports:

(*iv*) All the members of the truss are pin-joint to each other even through in actual practice they are not pin joined.

(*v*) The weight of the individual members of the truss are negligible.

(*vi*) Cross-section of the individual members is uniform. However, if the cross-section of an individual members varies the centre of gravity of the section is assumed to be located along the same longitudinal line.

(*vii*) For the reason that the ends of the members of the truss are pin-connected, the method to evaluate the forces in various members in popularly known as : Analysis of pin-connected plane trusses.

4.3.1 Nature of Forces in Members

As per the assumptions made, at a joint where the load is applied, various members are rivetted in the form of a pin–joint. In other words, if we consider the equilibrium of the joint the forces acting on the pin, by the links will be in equilibrium. Thus, if a link is applying a push on the pin, the reaction on that member will be tensile in nature.

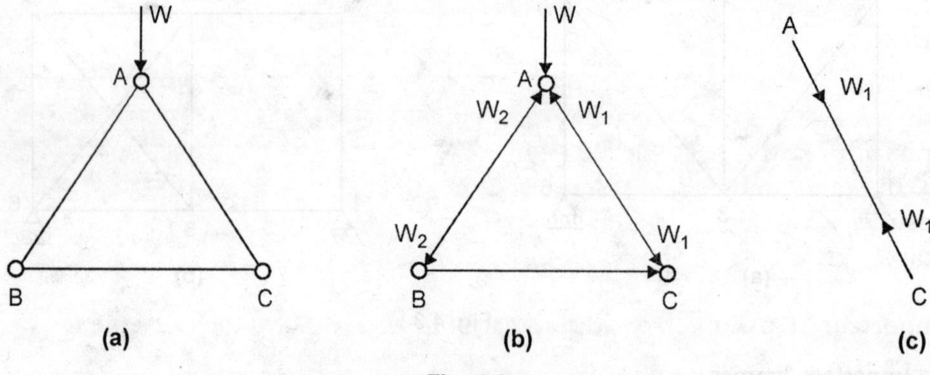

Fig. 4.4

For example, if we consider the simplest perfect truss *ABC* with a load in acting at *A*. Then for the equilibrium of pin–joint *A* (refer to Fig. 4.4) the push offered by the links *BA* and *BA* are W_1 and W_2. As a reaction on these links links *AB* and *CA,* the forces set up in the links will be as marked in Fig. 4.4 (*c*). Thus, we conclude that in the analysis of frame, we mark the forces on the joints instead of forces in the links. A very important point to note is that a compressive force in a particular member is represented in the figure by two arrow heads pointing away from each other as in Fig. (4.4*b*). Similarly, tensile force will be represented by arrowheads pointing towards each other. As a matter of fact the markings on the members represent the internal reactive forces developed which are opposite in direction to the applied forces. We now proceed to discuss the available methods of analysing the forces in various members of a perfect truss.

4.4 METHODS OF ANALYSIS

There are three methods for analysis of pin-joint frames :

1. Method of joints
2. Method of section
3. Graphical method

However, first two methods being included in syllabus of the UP Tech University, will be discussed here.

4.4.1 Method of Joints

The load acting at a joint alongwith the forces in the members ends of which are assumed to be riveted there constitute a system of concurrent forces. Thus, for the equilibrium of the joint, the forces can be resolved along two arbitrarily chosen convenient perpendicular directions and independently equated to zero and the solution of these simultaneous equation will give the values of unknown forces in two member.

Thus, if there are j member of joints, there will be $2j$ member of equations which can be formed. Further, since there will be in general three members at a joint, we will have $(m + 3)$ unknowns corresponding to 'm' member of members in a frame. In Mathematical terms the problem can be solved, if there are as many equations as are there unknown variables since in this case,

$$2j = m + 3$$

or

$$m = 2j - 3$$

which is the same equation as (4.1).

Thus, the truss analysis is determinate. If $m > (2j - 3)$, there the member of unknowns is more than the member of equations, then the redundant frame will be indeterminate. On the contrary if $m < (2j - 3)$, then the number of equations will be more than the number of unknons. In mathematical terms more than one solutions can satisfy the equations and it will result in "Instabilty" of the structure. In other words, a deficient frame or truss is not stable.

4.4.2 Important Steps to Follow while Analysing Forces by Method of Joints

Step 1. By considering the equilibrium of the truss as a whole, calculate the reactions at the supports. In the case of cantilever type frame usually the reaction at the support may not be required to be calculated.

Step 2. Locate a joint, if anywhere two forces are colinear, then the third force will be zero. Ref. Fig. 4.5 resolving the forces perpendicular to F_1 or F_2 and equating & zero we get,

$$F_3 \sin \theta = 0$$

or

$$F_3 = 0$$

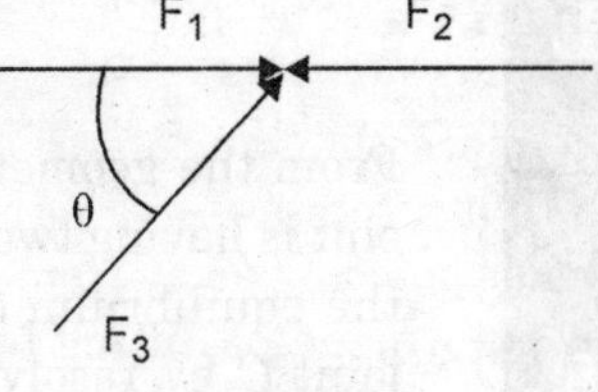

Fig. 4.5

Step 3. Next important step is to locate a joint, where there are only two unknowns. That is in that joint the forces in only two members are unknown and in other members as well the external load, if any acting at that joint is known. These forces can then be determined by considering the equilibrium of the joint for which two equation are formed by setting ΣX and ΣY separately equal to zero. Then the directions of the forces are marked in members meeting at that joint. It is then followed by marking the forces in opposite direction in these individual members at their respective other ends which form the next joint to be analysed. Here, again two unknown will be found by considering the equilibrium of this joint and so on till the forces in all the members are known. We now proceed to illustrate the method of joints by taking a few examples.

EXAMPLE

4.1 : *Determine the forces in a triangular shaped truss shown in Fig. (4.6) which is simply supported at points A and B.*

Solution :

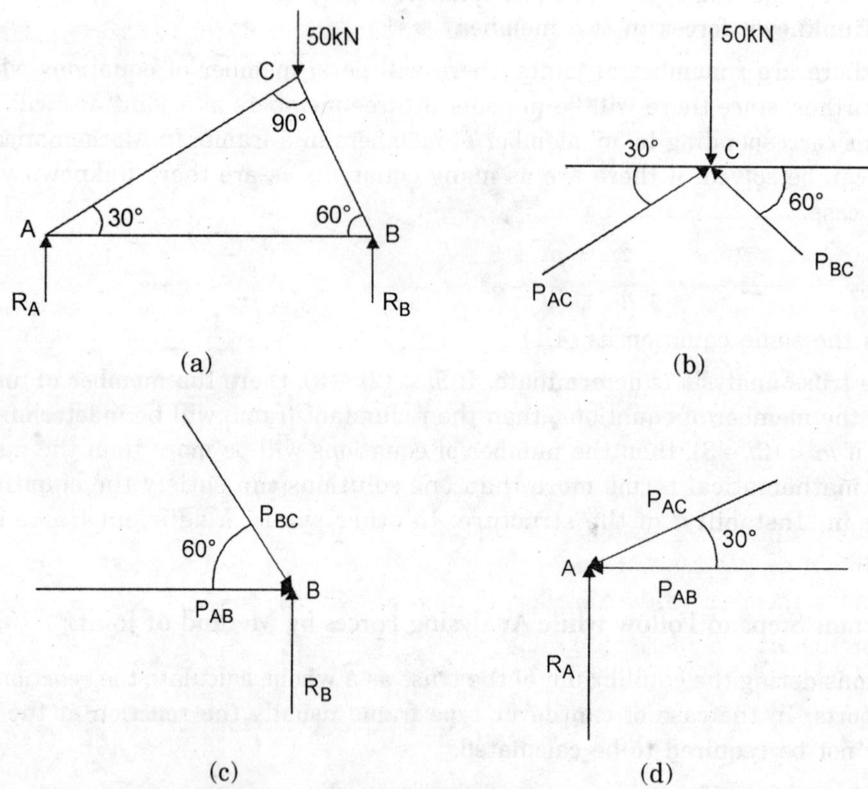

Fig. 4.6

From the geometry of the frame, we see that $\angle ACB = 90°$. Further, the oint is having two unknown forces viz. P_{BC} and P_{AC}. Therefore, considering the equilibrium of the joint C, we have from the Free Body diagram of joint C by resolving the forces in the horizontal and vertical directions we get,

For $\Sigma X = 0$, we have,

$$P_{BC} \cos 60 = P_{AC} \cos 30$$

$$\frac{1}{2} P_{BC} = \frac{\sqrt{3}}{2} P_{AC}$$

or

$$P_{BC} = \sqrt{3}\, P_{AC} \qquad ...(A)$$

For $\Sigma Y = 0$, we get,

$$P_{BC} \sin 60 + P_{AC} \sin 30 = 50 \text{ kN}$$

or

$$\frac{\sqrt{3}}{2} P_{BC} + \frac{1}{2} P_{AC} = 50$$

Substituting value from equation *A*,

$$\frac{\sqrt{3}}{2}\left(\frac{\sqrt{3}}{2} P_{AC}\right) + \frac{1}{2} P_{AC} = 50$$

or

$$3P_{AC} + P_{AC} = 50 \times 2$$

$$4P_{AC} = 100$$

or

$$P_{AC} = 25 \text{ kN}$$

And from (*A*),

$$P_{BC} = \sqrt{3} \times 25 = 43.3 \text{ kN} \qquad ...(B)$$

From the Free Body diagram of joint *C* (Fig. 4.6*b*), we see both these are push. The force P_{BC} at the joint *B* is shown in Fig. (4.6*c*) *i.e.*, the arrowhead will be in the opposite direction to that at the joint *C*. The arrowheads represent the internal reactions set up in the individual members. Thus, the member *BC* will be in compression.

The reactions at supports *A* and *B* can be determined by considering the equilibrium of the joints *A* and *B* separately. In the present case at both joints *A* and *B* there are total three concurrent forces, we can also apply Lami's theorem while considering their equilibrium.

For the joint *B* (as in Fig. 4.6*c*),

$$\frac{P_{BC}}{\sin 90} = \frac{P_B}{\sin 60} = \frac{P_{AB}}{\sin (180 + 30)}$$

From (*B*),

$$\frac{25\sqrt{3}}{1.0} = \frac{2P_B}{\sqrt{3}} = \frac{P_{AB}}{\left(-\frac{1}{2}\right)}$$

i.e.,

$$P_B = \frac{25\sqrt{3} \times \sqrt{3}}{2} = 37.5 \text{ kN}$$

EXAMPLE

$$P_{AB} = \left(-\frac{1}{2}\right) 25\sqrt{3} = -21.65 \text{ kN} \qquad ...(C)$$

Negative sign of P_{AB} shows that the actual direction of the arrowhead is opposite to that assumed in Free Body diagram as given in Fig. (2.5*c*). Similarly, considering the equilibrium of joint A (Fig. 4.6*d*) and applying Lami's theorem,

$$\frac{P_A}{\sin 30} = \frac{P_{CA}}{\sin 90} = \frac{P_{AB}}{\sin(180+60)}$$

or
$$\frac{2P_A}{1.0} = \frac{P_{CA}}{1.0} = \frac{2P_{AB}}{(-\sqrt{3})}$$

or
$$2P_A = \frac{25.0}{1.0} = -\frac{2}{\sqrt{3}} P_{AB}$$

or
$$P_A = \frac{25}{2} = 12.5 \text{ kN}$$

and
$$P_{AB} = \frac{25 \times \sqrt{3}}{2}$$

$$= -21.65 \text{ kN (same as in } (C))$$

As a check for the equilibrium of the entire from

$$P = P_A + P_B$$
$$= 12.5 + 37.5$$
$$= 50 \text{ kN (given load)}$$

4.2 : *Find the forces in all the members of the truss shown in Fig. (4.7).*

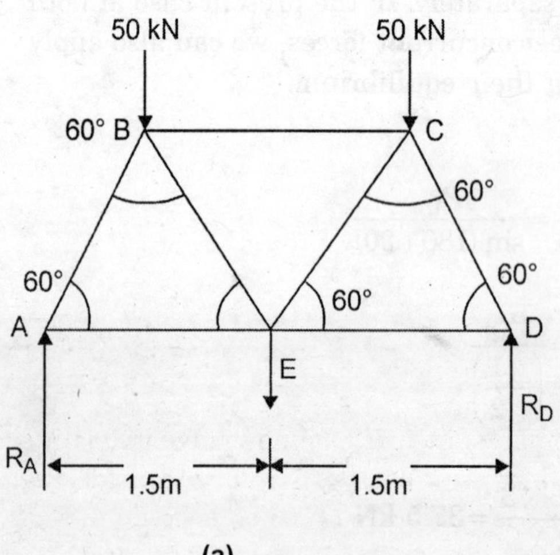

(a)

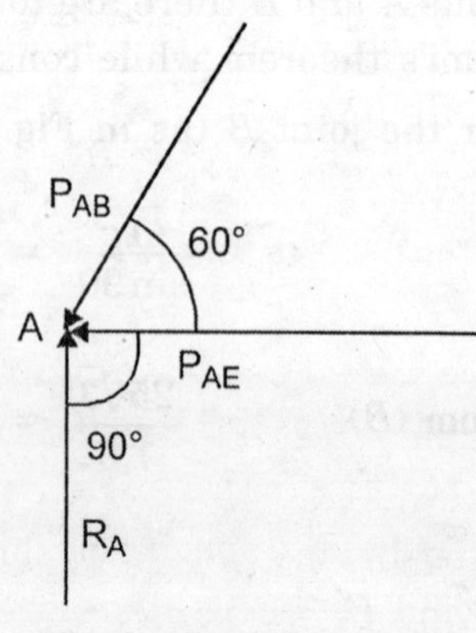

(b)

EXAMPLE

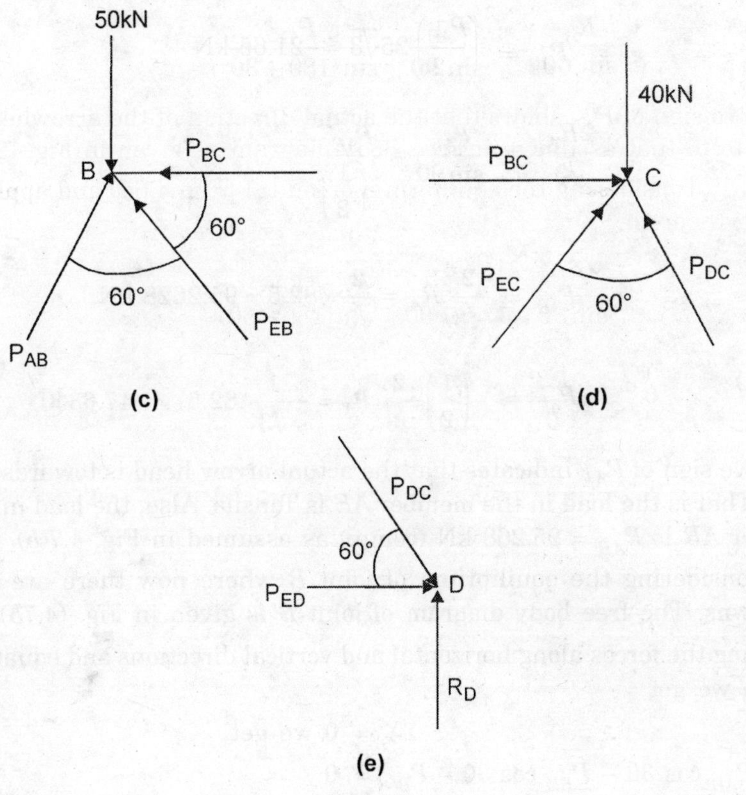

Fig. 4.7

Solution : In the truss shown in Fig. (4.6), there is no such joint where only two unknown forces are required to be found. We then proceed to find the reactions R_B and R_D by considering lie equilibrium of the entire frame.

Therefore, taking sum of moments actual the point A and equating it to zero we get,

$$50(AF) + 40(AG) + 70(AE) = R_D(AB)$$

i.e., $50 \times (1.5 \cos 60) + 40\,(1.5 + 1.5 \cos 60) + 70\,(1.5) = 3.0\, R_D$

or $3R_D = 50 \times 1.5 \times \frac{1}{2} + 40 \times 1.5\left(1 + \frac{1}{2}\right) + 70(1.5)$

$= 1.5\,[25 + 60 + 70] = 1.5 \times 155$

$R_D = 77.5$ kN ...(*A*)

Then, $R_A = (50 + 40 + 70) - 77.5$

$= 160 - 77.5 = 82.5$ kN ...(*B*)

After having calculated the reactions R_A over R_D we now have two joints *A* and *D,* where there are two unknowns each. Therefore, considering the equilibrium of joint *A*, the free body diagram of which is given in Fig. (4.7*b*) and applying Lami's theorem.

EXAMPLE

$$\frac{R_A}{\sin 60} = \frac{P_{AB}}{\sin 90} = \frac{P_{AE}}{\sin(180+30°)}$$

$$\therefore \quad \frac{2R_A}{\sqrt{3}} = \frac{P_{AB}}{\sin 90} = \frac{P_{AE}}{-\left(\frac{1}{2}\right)}$$

$$\therefore \quad P_{AB} = \frac{2}{\sqrt{3}} R_A = \frac{2}{\sqrt{3}} \times 82.5 = 95.2628 \text{ kN}$$

and
$$P_{AE} = -\left(\frac{1}{2}\right)\frac{2}{\sqrt{3}} P_A = -\frac{1}{\sqrt{3}}(82.5) = -47.63 \text{ kN}$$

Negative sign of P_{AE} indicates that the actual arrow head is towards the right. That is the load in the member *AE* is Tensile. Also, the load in the member *AB* is P_{AB} = 95.268 kN (comp.) as assumed in Fig. 4.7(*b*).

Now considering the equilibrium of joint *B* where now there are two unknowns. The free body diagram of joint *B* is given in Fig. (4.7*b*).

Resolving the forces along horizontal and vertical directions and equating to zero we get,

$\Sigma X = 0$ we get,

$$P_{AB} \cos 30 - P_{BE} \cos 30 - P_{BC} = 0$$

$$(95.2628)\frac{\sqrt{3}}{2} - P_{BE}\frac{\sqrt{3}}{2} - P_{BC} = 0 \qquad ...(C)$$

For $\Sigma Y = 0$ we get,

$$P_{AB} \cos 30 + P_{BE} \cos 30 = 50$$

$$(P_{AB} + P_{BE}) = 50 \times \frac{2}{\sqrt{3}}$$

$$P_{BE} = \frac{100}{\sqrt{3}} - P_{AB} = \frac{100}{\sqrt{3}} - 95.2628$$

$$= 57.7350 - 95.2628$$

$$= -37.528 \text{ kN}$$

Thus, The force acting along *BE is* 37.5278 *kN* (Tensile) from (*C*) and (*D*),

$$P_{BC} = 95.2628\left(\frac{\sqrt{3}}{2}\right) + 37.5228 \times \frac{\sqrt{3}}{2}$$

$$= 132.79 \times \frac{\sqrt{3}}{2} = 115.0 \ kN$$

Again considering the equilibrium of joint D. The free body diagram of this joint is shown in Fig. 4.6 d,

Since there are only the forces, we get by applying lamer, theorem.

$$\frac{R_D}{\sin 60} = \frac{P_{CD}}{\sin 90} = \frac{P_{ED}}{\sin(180+30)}$$

$$\frac{2R_D}{\sqrt{3}} = \frac{P_{CD}}{1.0} = \frac{P_{CD}}{\left(-\frac{1}{2}\right)}$$

From (A) R_D = 77.5 kN,

$$\therefore \quad P_{CD} = \frac{2}{\sqrt{3}} \times 77.5 = 86.49 \ kN$$

$$P_{ED} = -\frac{1}{2}\left(\frac{2}{\sqrt{3}} \times 77.5\right) = -44.75 \ kN \qquad \text{(Tensile)}$$

Again for the equilibrium of joint C. The force body diagram of joint C is given is Fig. (4.6.e). Resolving along vertical direction,

$$40 = P_{CE} \cos 30 + P_{CD} \cos 30$$

$$P_{CE} \cos 30° = 40 - P_{CD} \cos 30$$

$$= 40 - (-44.75)\frac{\sqrt{3}}{2}$$

or
$$P_{CE} = \frac{2}{\sqrt{3}} 40 + 44.75 \times \frac{\sqrt{3}}{2} \cdot \frac{2}{\sqrt{3}}$$

$$= \frac{80}{\sqrt{3}} + 44.75$$

$$= 46.188 + 44.75 = 90.938 \ kN \qquad \text{(comp)}$$

4.3 : *A concubine truss is shown is Fig. 4.8, find the forces in all the members of the truss.*

Solution: Conceding the equilibrium of joint E (Fig. 4.8 b) by Lami's theorem,

$$\frac{10}{\sin 60} = \frac{P_{DE}}{\sin 90} = \frac{P_{CE}}{\sin(180+30)}$$

$$\frac{10 \times 2}{\sqrt{3}} = \frac{P_{DE}}{1.0} = \frac{P_{CE}}{-\left(\frac{1}{2}\right)}$$

$$\therefore \quad P_{DE} = \frac{20}{\sqrt{3}} = 1.1547 \ kN \text{ (comp)} \qquad ...(A)$$

$$P_{CE} = \frac{20}{\sqrt{3}}\left(-\frac{1}{2}\right) = -\frac{70}{\sqrt{3}}$$

$$= -0.57735 \ kN \text{ (Tensile)} \qquad ...(B)$$

EXAMPLE

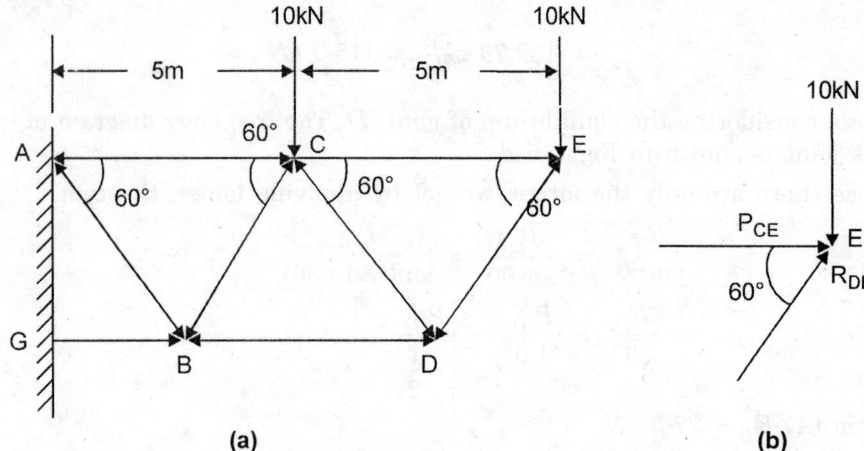

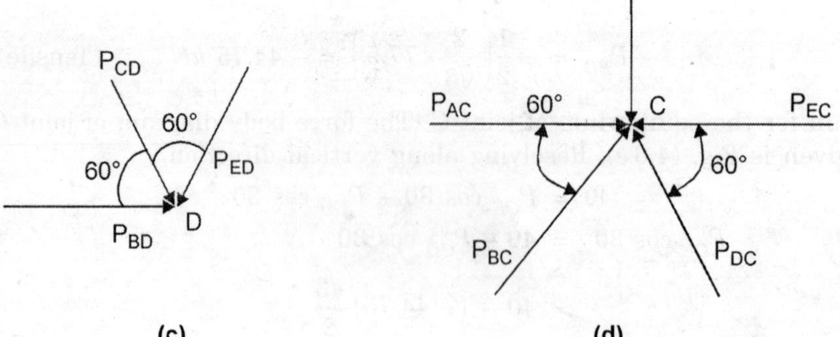

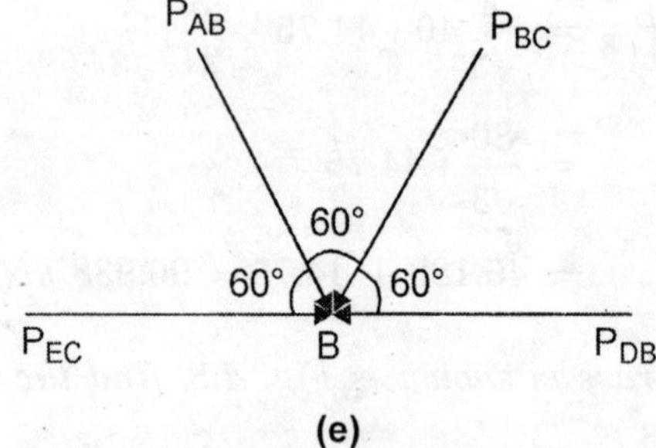

Fig. 4.8

For the joint D by Lami's theorem on (Fig. 4.8 c),

$$\frac{P_{DE}}{\sin 60} = \frac{P_{BD}}{\sin 60} = \frac{P_{CD}}{\sin(180+60°)} \qquad ...(C)$$

$$\therefore \qquad P_{BC} = \frac{2P_{DE}}{\sqrt{3}}\left(-\frac{\sqrt{3}}{2}\right) = -P_{DE}$$

$$= -\frac{20}{\sqrt{3}} = -\ 1.1547\ kN$$

$$P_{DE} = P_{BE} = \frac{2P_{CD}}{-\sqrt{3}}$$

From A, $P_{DE} = P_{BD}$

$$= 1.1547\ kN$$

$\therefore$ $P_{CD} = -\frac{\sqrt{3}}{2} \times 1.1547$

$$= -10\ kN$$

For joit C (Fig. 4.8 d),

$$\Sigma X = 0$$

and $\Sigma Y = 0$

is $P_{AC} - P_{CE} + P_{BE} \cos 60 - P_{DC} \cos 60 = 0$

From (B) and P_{AC} $\left(-\frac{10}{\sqrt{3}}\right)\frac{1}{2} - (-10)\frac{1}{2} + \left(-\frac{20}{\sqrt{3}}\right)\frac{1}{2}$,

$$P_{DC} = -\frac{5}{\sqrt{3}} - 5 + \frac{10}{\sqrt{3}}$$

$$= -\left(5 + 2.8867 + \frac{10}{\sqrt{3}}\right)$$

$$= -7.8867\ kN + 5.773$$

$$= -2.1132$$

For the joint B (Free body diagram is in Fig. (4.8 e)

$$\Sigma X = 0$$

$$\Sigma Y = 0$$

$$P_{AB} \cos 30 + P_{BC} \cos 30 = 0$$

$$P_{AB}\frac{\sqrt{3}}{2} - \frac{20}{\sqrt{3}} \times \frac{\sqrt{3}}{2} = 0$$

$$P_{AB} = \frac{20}{\sqrt{3}}$$

$$= 1.1547\ kN$$

4.4 : *Find the forces in all the members of the frame shown in Fig. 4.9.*

Solution: At the joint C the forces P_{BC} and P_{DC} are unknown, therefore considering the equilibrium of joint C for which free body diagram is given in Fig. 4.9 b and applying Lami's theorem,

EXAMPLE

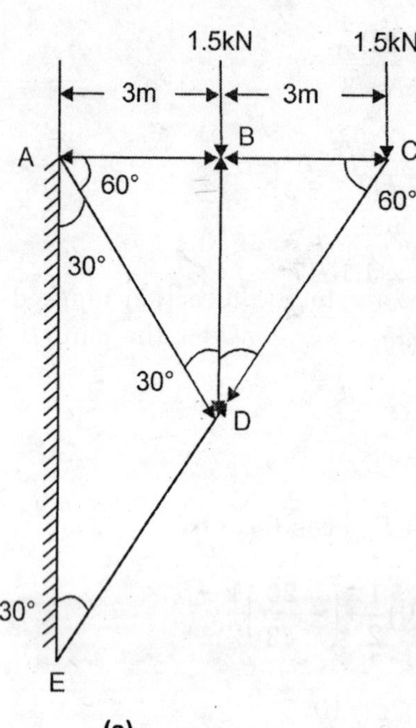

(a)

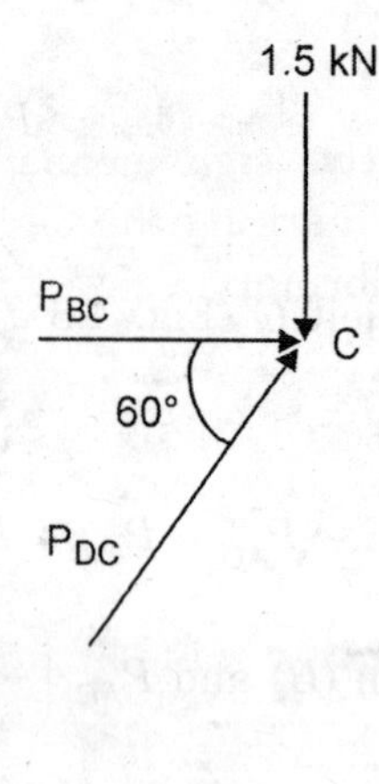

(b)

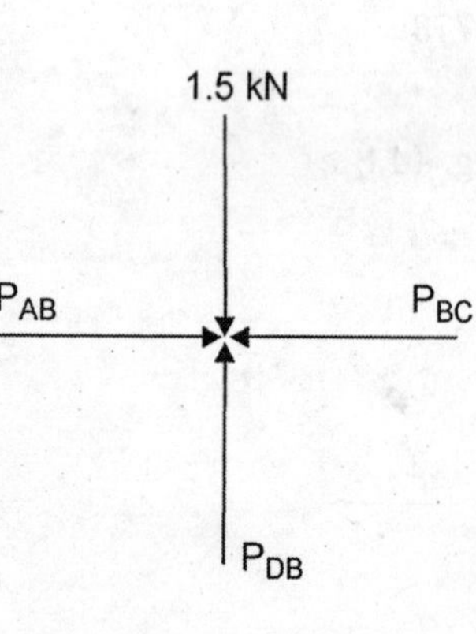

(c)

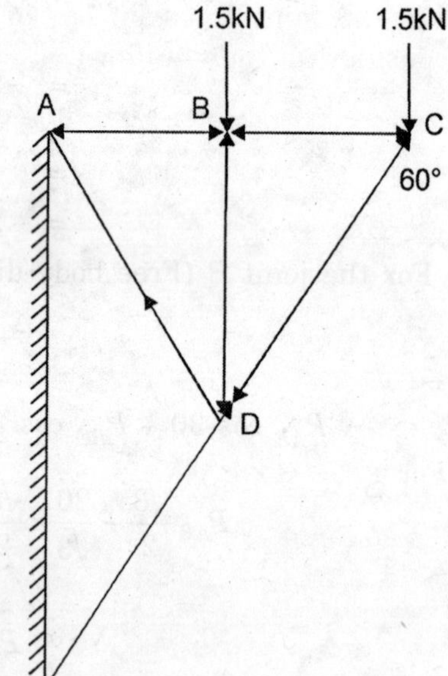

(d)

Fig. 4.9

$$\frac{1.5}{\sin 60} = \frac{P_{DC}}{\sin 90} = \frac{P_{BC}}{\sin(180+30)}$$

$$\frac{2(1.5)}{\sqrt{3}} = P_{DC} = \frac{P_{BC}}{-\frac{1}{2}}$$

i.e., $P_{DC} = \sqrt{3}$

$= 1.732\ kN$

and $P_{BC} = -\frac{1}{2} \times \frac{2(1.5)}{\sqrt{3}}$

$= -0.866\ kN$

Negative sign means it is in apposite to the direction marked on *FB* after marking the forced P_{DC} and P_{BC}, we consider the joint *B*. For its equilibrium,

$\Sigma X = 0$

and $\Sigma Y = 0$

For $\Sigma X = 0$ we get,

$P_{AB} = P_{BC} = 0$

i.e, $P_{AB} = P_{BC} = -0.866\ kN$

for $\Sigma Y = 0$ we have,

$1.5 = P_{BD}$

or $P_{DB} = 1.5\ kN$

For the joint *D*, again out of four forces two are unknown, therefore considering the equilibrium of joint *D*

For $\Sigma X = 0$

We get,$- P_{DC} \sin 30 + P_{ED} \sin 30 + P_{AD} \cos 60 = 0$

$$-P_{DC}\left(\frac{1}{2}\right) + P_{ED}\left(\frac{1}{2}\right) + P_{AD}\left(\frac{1}{2}\right) = 0$$

or $- P_{DC} + P_{ED} + P_{AD} = 0$

$$-\sqrt{3} + P_{ED} + P_{AD} = 0 \qquad \text{...(B)}$$

For $\Sigma Y = 0$ we get,

$- P_{BD} - P_{DC} \cos 30 + P_{DE} \cos 30 - P_{AD} \cos 30 = 0$

$$-1.5 - \sqrt{3}\left(\frac{\sqrt{3}}{2}\right) + P_{DE}\frac{\sqrt{3}}{2} - P_{AD}\frac{\sqrt{3}}{2} = 0$$

$$-3 + P_{DE}\frac{\sqrt{3}}{2} - P_{AD}\frac{\sqrt{3}}{2} = 0$$

$$P_{DE} - P_{AD} = \frac{2}{\sqrt{3}} \times 3 = 2\sqrt{3}$$

i.e, $P_{DE} = 2\sqrt{3} + P_{AD}$ ...(*C*)

EXAMPLE

Form (*B*) and (*C*),

$$-\sqrt{3}+\left(2\sqrt{3}+P_{AD}\right)+P_{DA} = 0$$

or
$$2P_{AD} = -\sqrt{3}$$

$$P_{AD} = -\frac{\sqrt{3}}{2}$$

$$= -0.866\ kN$$

and
$$P_{ED} = 2\sqrt{3}-\frac{\sqrt{3}}{2}=\frac{3\sqrt{3}}{2}kN$$

Final types of forces are marked in Fig. (4.9 *d*)

4.5 : *Find the forces in all the number of the truss shown in Fig. 4.10*

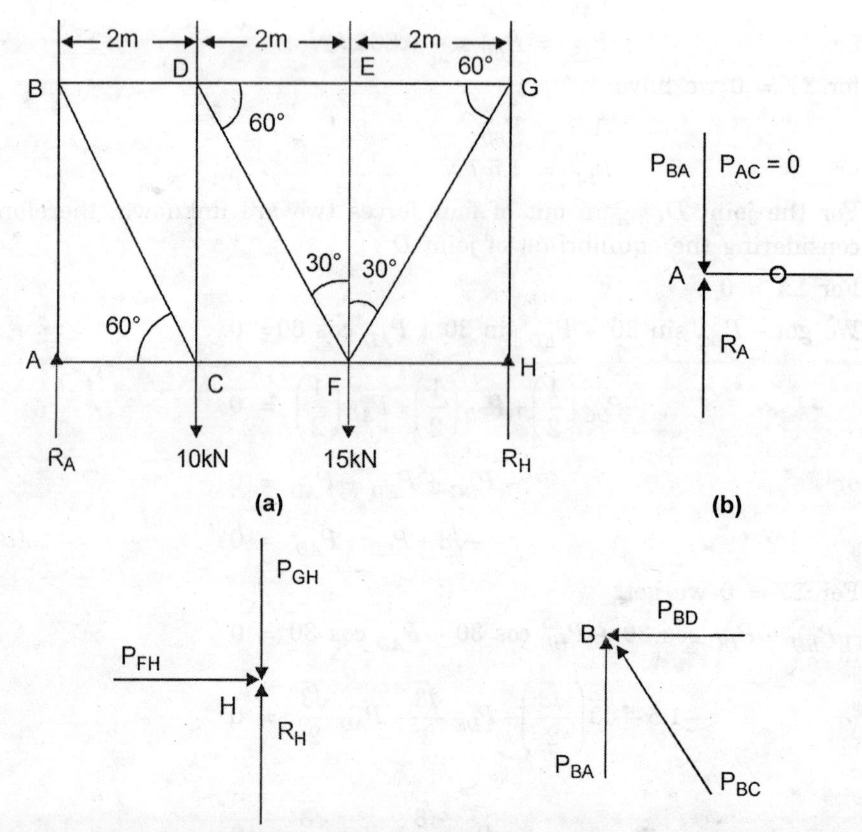

Fig. 4.10

Solution : For finding reaction on simple supports *A* and *H* take moments about *H* for the equilibrium of entire frame,

$$R_A \times 6 = 10 \times 4 + 15 \times 2$$

$$= 40 + 30 = 70$$

E X A M P L E

$$P_A = \frac{70}{6} = kN$$

and $$P_B = 5+10-\frac{70}{6} = 15-\frac{70}{6}$$

$$= \frac{80}{6} kN$$

At Joint A there are three forces, out of the three, two forces P_A and P_{BA} are colinear therefore $P_{AC} = 0$ ...(A)

Similarly, at joint H, P_{GH} and R_A are collinear. Therefore,

$$P_{FH} = 0 \quad \text{...(B)}$$

and $P_{GH} = R_H$ On similar grounds at joint E as P_{GE} and P_{ED} are colinear

$$P_{EF} = 0 \quad \text{...(C)}$$

For the equilibrium of A since $P_{AC} = 0$

$$P_A = P_{BA} = \frac{70}{6} kN$$

$$= 11.667\ kN$$

At joint B from Fig (4.10 *d*),

As $$\Sigma Y = 0,$$

$$P_{AB} + P_{BC} \cos 30 = 0$$

$$\frac{70}{6} + P_{BC}\frac{\sqrt{3}}{2} = 0$$

$$P_{BC} = -\frac{70}{6} \times \frac{2}{\sqrt{3}}$$

$$= -\frac{20}{3\sqrt{3}}$$

$$= -13.47 kN$$

For $$\Sigma X = 0$$

$$P_{BD} = P_{BC} \cos 60$$

$$= 13.47 \times \frac{1}{2}$$

$$= 6.735\ kN$$

4.6 : *A truss is shown in Fig. (4.11 a)). Find the forces in all the members of the truss and indicate it is tension or compression.*

(U.P. Tech-Univ 2000-2001)

EXAMPLE

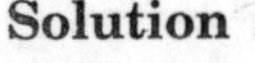
Solution :

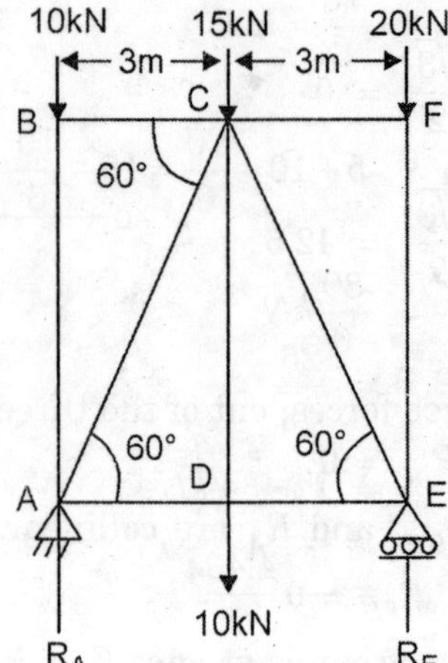

Fig. 4.11 (a)

The support at A is hinged and that at E is on rollers. Therefore, R_E will be vertical. Further since all the loads are verticle the reaction R_A will also be vertical.

To find the reactions R_E taking moments about A and $M_A = 0$

$$15 \times BC + 10 \times AD + 20 \times AE = R_E \times AE$$

$$15 \times 3 + 10 \times 3 + 20 \times 6 = R_E \times 6$$

$$\therefore \qquad R_E = \frac{75+120}{6} = \frac{195}{6}$$

$$= 32.5 \; kN$$

Fig. 4.11 (b)

Then $\qquad R_A = 10 + 15 + 10 + 20 - 32.5$

$$= 22.5 \; kN$$

At the joint B, two forces are unknown, therefore for its equilibrium are have two forces which are colinear. 10 kN and P_{BA}

Therefore, third force,

$$P_{BC} = 0$$

or $\qquad P_{BA} = 10 \; kN \qquad \text{...(A)}$

Joint $\boldsymbol{A}$ Here, again there are two unknown forces

For $\qquad \Sigma X = 0$ we get

$$-P_{AC} \cos 60 - P_{AD} = 0$$

or $\qquad \frac{1}{2} P_{AC} = -P_{AD} \qquad \text{...(B)}$

For $\qquad \Sigma Y = 0$

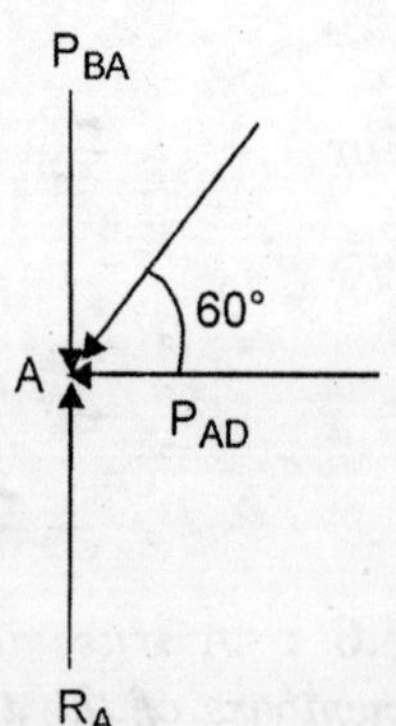

Fig. 4.11 (c)

$$R_A - P_{BA} - P_{AC}\cos 30 = 0$$

$$22.5 - 10 - P_{AC}\frac{\sqrt{3}}{2} = 0$$

$$P_{AC}\frac{\sqrt{3}}{2} = 12.5$$

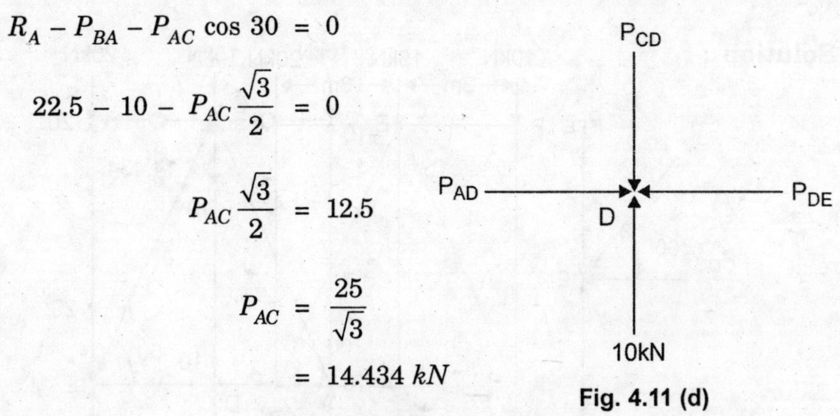

Fig. 4.11 (d)

or $$P_{AC} = \frac{25}{\sqrt{3}}$$

$$= 14.434\ kN$$

∴ From (*B*) and (*C*) $$P_{AD} = -\frac{14.434}{2}$$

$$= -7.217\ kN$$

Joint *D* for equilibrium of joint *D*

$$\Sigma X = 0$$

$$P_{AD} = P_{DE}$$

$$P_{DF} = -7.217\ kN$$

(– ve sign indicates P_{DF} is tensid, as in Free Body diagram it is marked comp.

For $$\Sigma Y = 0 \text{ we have}$$

$$P_{CD} = 10\ kN \qquad ...(E)$$

Joint *F* Here, there are only three forces out of which two viz 20 *kN* and P_{RE} are colinear. Therefore, the third force $P_{CF} = 0$

and $$P_{EF} = 20\ kN \qquad ...(F)$$

Joint *E* For equilibrium,

$$P_{FE} + P_{CF}\cos 30 = R_E$$

For equ. *F* $$20 + P_{CE}\left(\frac{\sqrt{3}}{2}\right) = 32.5$$

$$P_{CE}\left(\frac{\sqrt{3}}{2}\right) = 12.5$$

$$P_{CF} = \frac{2}{\sqrt{3}} \times 12.5$$

$$= 14.434\ kN$$

Fig. 4.11 (e)

Thus, the final forces are as shown in Fig. 4.11 (*g*)

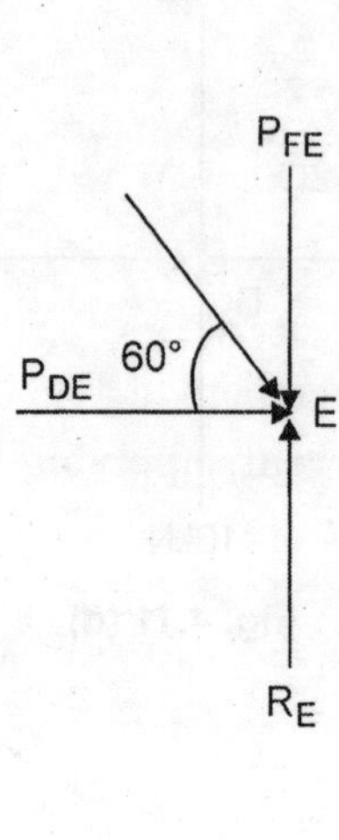

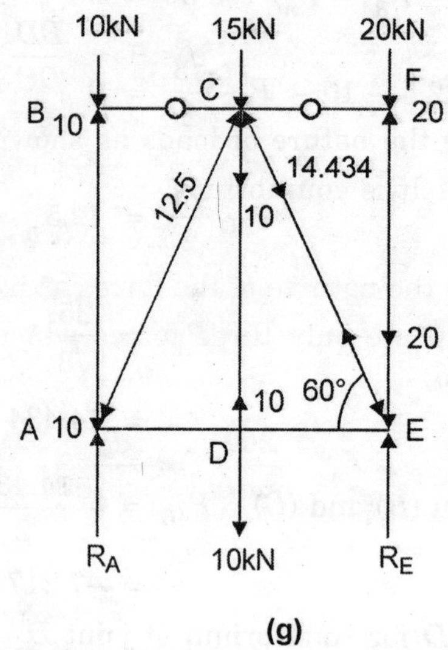

(f) (g)

Fig. 4.11

4.7 : *Determine the support reactions and nature and magnitude of forces in the members of the truss shown in Fig. 4.12 a.*(U.P. Tech the . 2001-2002)

Solution : From Δ *CBD*,

$$\tan\theta = \frac{BC}{DB}$$

$$= \frac{2}{4} - \frac{1}{2}$$

$$CD = \sqrt{4^2 + 2^2} = \sqrt{20}$$

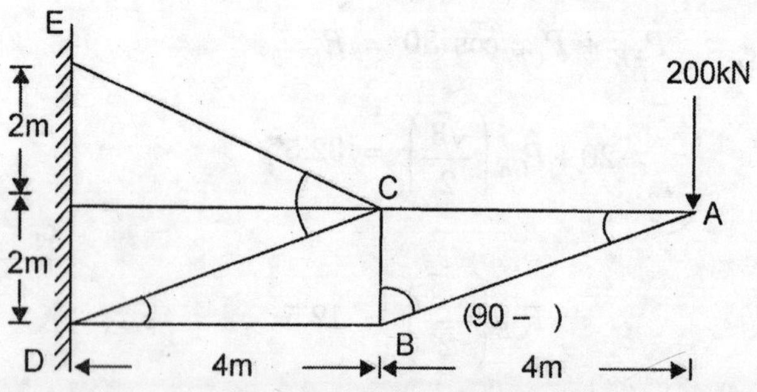

Fig. 4.12 (a)

Also ,

$$\sin\theta = \frac{BC}{CD}$$

$$= \frac{2}{\sqrt{20}}$$

Also $\cos\theta = \dfrac{BD}{DC} = \dfrac{4}{\sqrt{20}}$

Assume the nature of loads as shown in Fig. 4.12(a).

Joint *A* It is equilibrium,

For $\Sigma X = 0$ we have

Assume the naturze of the forces as indicated in members at Fig. (4.12 *b*)

As there are only three forces which are concurrent, applying Lami's theorem,

$$\frac{200}{\sin\theta} = \frac{P_{BA}}{\sin 90}$$

$$= \frac{P_{CA}}{\sin(180+90-\theta)}$$

$$\frac{200}{\sin\theta} = P_{BA} = \frac{P_{CA}}{\cos\theta}$$

or $$\frac{200\times\sqrt{20}}{2} = P_{BA} = -P_{CA}\frac{\sqrt{50}}{4}$$

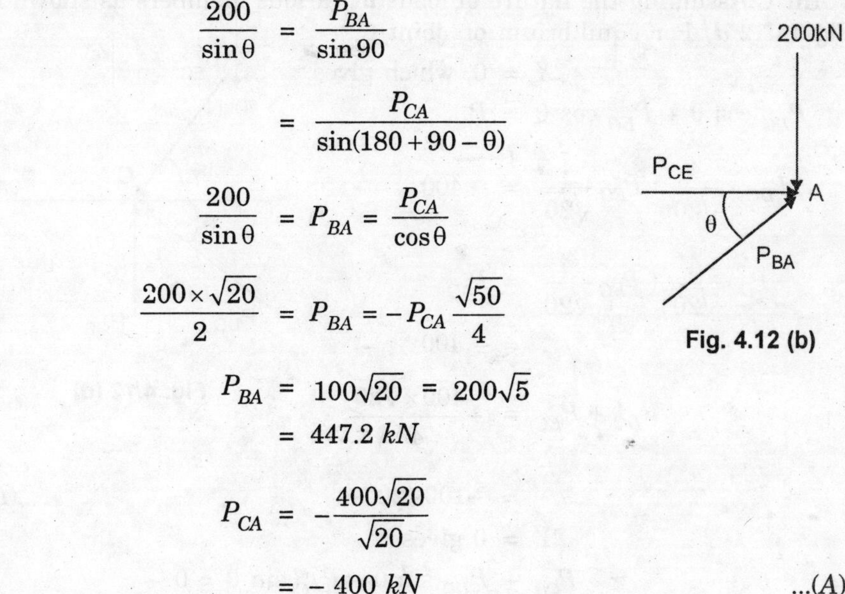

Fig. 4.12 (b)

$\therefore$ $$P_{BA} = 100\sqrt{20} = 200\sqrt{5}$$

$$= 447.2\ kN$$

$$P_{CA} = -\frac{400\sqrt{20}}{\sqrt{20}}$$

$$= -\ 400\ kN \qquad \text{...}(A)$$

Joint *B* (Fig 4.12 *c*) Assuming the nature of loads as shown

There are three forces, therefore applying Lami's theorem,

$$\frac{P_{BA}}{\sin 90} = \frac{P_{DB}}{\sin(90-\theta)}$$

$$= \frac{P_{CB}}{\sin(180+\theta)}$$

$$\frac{P_{BA}}{1.0} = \frac{P_{DB}}{\cos\theta} = \frac{P_{CB}}{-\sin\theta}$$

Fig. 4.12 (c)

$$P_{BA} = \frac{P_{DB}}{4}\sqrt{20}$$

$$= -\frac{P_{CB}\sqrt{20}}{2} \qquad \text{...}(C)$$

Form (A), (B) and (C),

$$200\sqrt{5} = P_{DB}\frac{\sqrt{20}}{4} = -P_{CB}\sqrt{5}$$

$$P_{DB} = \frac{4\times 200\sqrt{5}}{\sqrt{20}} = 400\ kN$$

$$P_{CB} = -\frac{200\sqrt{5}}{\sqrt{5}} = -\ 200\ kN$$

Joint *C* Assuming the nature of loads in various members as shown in Fig. (4.12 *d*) For equilibrium of joint *C*,

$$\Sigma X = 0, \text{ which gives,}$$

$$P_{DC}\cos\theta + P_{EC}\cos\theta = P_{AC}$$

$$P_{DC}\times\frac{4}{\sqrt{20}} + P_{EC}\frac{4}{\sqrt{20}} = -\ 400$$

$$P_{DC}\frac{4}{\sqrt{20}} + P_{EC}\frac{4}{\sqrt{20}} = P_{AC}$$

$$= -\ 400$$

$$P_{DC} + P_{EC} = \frac{-400\times\sqrt{20}}{4}$$

$$= -100\sqrt{20} \qquad ...(D)$$

Fig. 4.12 (d)

$$\Sigma Y = 0 \text{ gives}$$

$$P_{CB} + P_{DC}\sin\theta - P_{EC}\sin\theta = 0$$

or

$$P_{CB} = -\frac{2}{\sqrt{20}}P_{DC} + P_{EC}\frac{2}{\sqrt{20}}$$

$$\frac{\sqrt{20}}{2}P_{CB} = -\ P_{DC} + P_{EC} \qquad ...(E)$$

Solving eq. (D) and (E),

$$-\frac{\sqrt{20}}{2}\times 200 = -\ P_{DC} + P_{EC} \qquad ...(F)$$

∴ From equation (D), and (F),

$$P_{DC} + P_{EC} = -100\sqrt{20}$$

and

$$-\ P_{DC} + P_{EC} = -100\sqrt{20}$$

By adding,

$$2P_{EC} = -200\sqrt{20}$$

$$P_{EC} = -100\sqrt{20}\ kN$$

and $$P_{DC} = -100\sqrt{20} + P_{EC} = 0$$

$$= -100\sqrt{20} + 100\sqrt{2} = 0$$

Joint *D* Assessing the nature of forces as shown in Fig(4.12 *e*)

$$\Sigma X = 0 \text{ gives}$$

$$P_D = P_{BD}$$

$$= 400\ KN$$

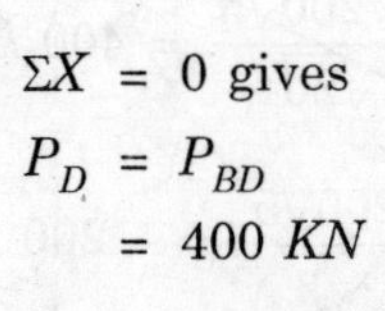

Fig. 4.12 (e)

Joint *E* $$P_{EC} = -100\sqrt{20}$$

Let *H* and *V* be the horizontal and vertical components of reaction at *E*

The joint *E* is in equilibrium under the action of 3 forces, *V* and P_{EC}

Then $$\Sigma X = 0 \text{ gives}$$

$$H = P_{EC} \cos\theta$$

$$= -100\sqrt{20} \times \frac{4}{\sqrt{20}}$$

$$= -400\ KN$$

And $$\Sigma Y = 0 \text{ give}$$

$$V = P_{EC} \sin\theta$$

$$= -100\sqrt{20} \times \frac{2}{\sqrt{20}}$$

$$= -200\ KN$$

Fig. 4.12 (f)

Then, the final result is as shown in Fig. (4.12 g)

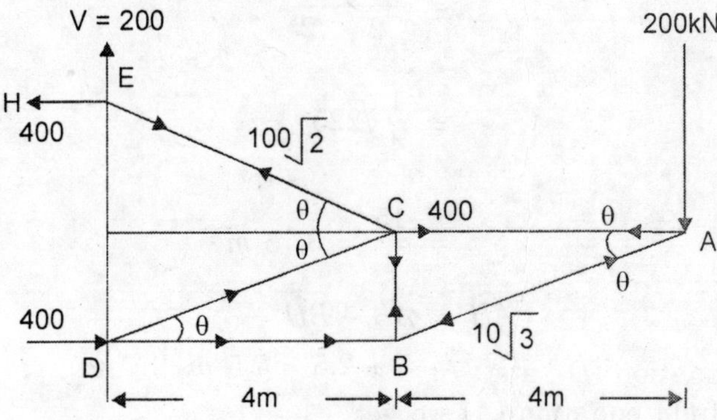

Fig. 4.12 (g)

4.8 : *A frame shown on Fig (4.13 a) carries a load of 20 kN at the joint D and 24 kN at C . It in simply supported at points A and E. Find the load carried by each member.*

E X A M P L E

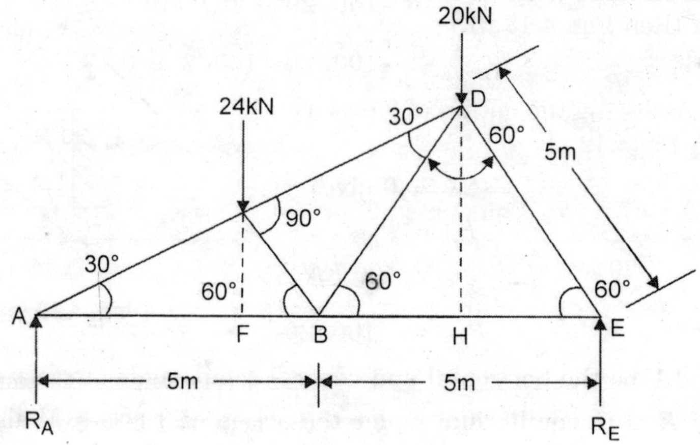

Fig. 4.13 (a)

Solution : Let us first find the reactions at the supports A and E

Taking moments about A we get,

$$10\,RE = 24 \times AF + 20\,AH \qquad ...(A)$$

Now $$AD^2 + DE^2 = AE^2$$

$$AD^2 + (5)^2 = (10)^2$$

$$AD = \sqrt{75}$$

Then $$AC = \frac{1}{2}AD = \frac{\sqrt{75}}{2}$$

Therefore, from Δ ACF, $$AF = AC\cos 30$$

$$= \frac{\sqrt{75}}{2}\cdot\frac{\sqrt{3}}{2}$$

$$= \frac{1}{4}\sqrt{225}$$

$$= \frac{15}{4} = 3.75\ m$$

and $$AH = AB + BH$$

$$= 5 + 2.5 = 7.5\ m$$

Therefore, from equation (A) we get,

$$10\,R_E = 24 \times 3.75 + 20 \times 7.5$$

or $$R_E = \frac{1}{10}[90 + 150] = 24\ kN$$

Therefore, $$R_A = (24 + 20) - 24 = 20\ kN$$

Joint *A* Assume loads in *AC* and *AB* are compressive, then by Lami's theorem (Reb Fig. 4.13 *b*)

$$\frac{R_A}{\sin 30} = \frac{P_{AC}}{\sin 90}$$

$$= \frac{P_{AB}}{\sin(180+60)}$$

$$\frac{20\times 2}{1} = \frac{P_{AC}}{1.0}$$

$$= \frac{2P_{AB}}{-\sqrt{3}}$$

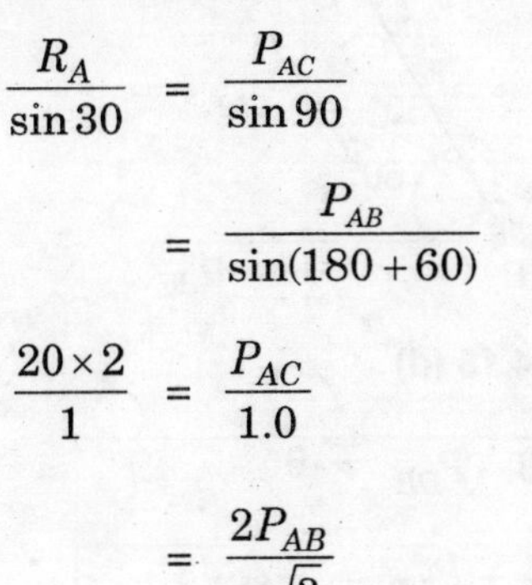

Fig. 4.13 (b)

$$\therefore \quad P_{AC} = 40\ kN$$

and $$P_{AB} = -\frac{\sqrt{3}}{2}\times 40$$

$$= -20\sqrt{3} \text{ (–ve. Therefore, Tensite)} \qquad ...(B)$$

Joint *C* For equilibrium of joint *C*, $\Sigma X = 0$ and $\Sigma Y = 0$ Take ΣX along *AC* and ΣY normal to *AC* for $\Sigma X = 0$ we get Fig. 4.13 *c*

AsruncP_{AC}, P_{CD} and P_{CB} comp.

Fig. 4.13 (c)

$$P_{AC} - P_{CD} - 24\cos 60 = 0$$

$$40 - P_{CD} - 24\times\frac{1}{2} = 0$$

or $$P_{CD} = -40 + 12$$

$$= -28\ kN \text{ (Tensite)} \qquad ...(C)$$

For $\Sigma Y = 0$ $$P_{CB} = 24\sin 60$$

$$= 24\frac{\sqrt{3}}{2}$$

$$= 12\sqrt{3} \text{ comp} \qquad ...(D)$$

Joint B For equilibrium of joint B,

$$\Sigma X = 0$$

and $$\Sigma Y = 0$$

For $\Sigma X = 0$,

$$P_{AB} - P_{BE} + P_{CB}\cos 60 - P_{DB}\cos 60 = 0$$

$$\left(-20\sqrt{3}\right) - P_{BE} + 12\sqrt{3}\times\frac{1}{2} - P_{DB}\frac{1}{2} = 0$$

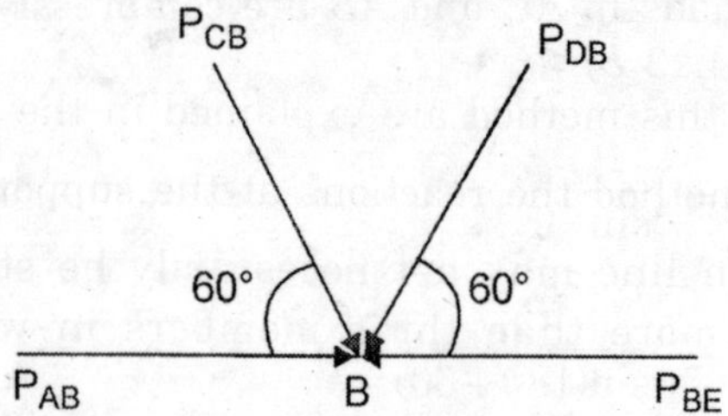

Fi.g 4.13 (d)

$$-40\sqrt{3} - 2P_{BE} + 12\sqrt{3} - P_{DB} = 0$$

or $$2P_{BE} + P_{DB} = -28\sqrt{3} \qquad ....(E)$$

ΣY gives $$P_{CB} \cos 30 + P_{DB} \cos 30 = 0$$

$$\frac{1}{2}(P_{CB} + P_{DB}) = 0$$

or $$P_{CB} = -P_{DB} \qquad ...(F)$$

From (*E*), and (*D*), $$P_{DB} = -P_{CB}$$

$$= -12\sqrt{3} \text{ kN} \qquad ...(G)$$

$\therefore$ From (*E*) and (*G*),

i.e., $$P_{CB} = 12\sqrt{3}$$

$$2P_{BE} - 12\sqrt{3} = -28\sqrt{3}$$

or $$P_{BE} = -8\sqrt{3} \ kN \qquad ...(H)$$

Joint *E* $\Sigma Y = 0$ gives

$$P_{DE} = (\cos 30) = R_E$$

$$P_{DE} \times \frac{\sqrt{3}}{2} = 24$$

$$P_{DE} = \frac{48}{\sqrt{3}}$$

$$= 16\sqrt{3} \ kN \qquad ...(J)$$

Fig. 4.13 (e)

$\Sigma X = 0$ gives $$P_{BE} + P_{DE} \cos 60 = 0$$

$$P_{BE} = -\frac{1}{2} P_{DE}$$

or $$P_{DE} = -2 \times \left(-5\sqrt{3}\right)$$

$$= 16\sqrt{3} \ kN \text{ (same as in Eqn. } J)$$

Final loads are shown in Fig. (4.13 *f*)

4.4.3 Method of Sections

The steps to be taken is this method are explained in the following paragraphs:

Step 1. By the usual method the reactions at the support points are determined.

Step 2. Draw a section line may not necessarily be straight but the section line must not cut more than three members in which the forces are to the determined.

Step 3. After taking section as stated above the equilibrium of one of these two parts obtained by sectioning is considered and the unknown forces acting on the members cut by the section line are calculated by solving the system of their non-concurrent forces acting on either part of the truss. While taking moments the point of intersection of two unknown forcers is preferably chosen. In this method, it is an essential condition that the section line must not pass through more than their members in which forces are not known.

The main difference in the method of joints and method of section is that the former involves the analysis of concurrent forces system and the latter deals with analysis of non-concurrent forces system.

Thus, we conclude that the method of section is preferred over the method of joints under the following circumstances:

(*i*) In the case of a large truss in which the forces acting only in a few members are required to be calculated.

(*ii*) In those cases where the method of joints fails to proceed with the analysis, as no joint is available, where there are only two unknown forces.

EXAMPLE

4.9 : *In the truss given in Fig. (4.14 a), find the forces in the element CD, BD and BE by two method of sections.*

Solution : As found in example (4.8)

$$R_A = 20\ kN$$

and $$R_E = 24\ kN$$

Taking the section along the curve 1, 2, 3 and drawing the free body diagram of the left portion. Let the force in the element *CD* be P_{CD} and in the element *BD* be P_{BD} and that in element *BE* be P_{BE}

Then considering the equilibrium of the left portion of which free body diagram is given in Fig. (4.14 *b*) and taking moments about *B*, we get,

$$R_A \times 5 + P_{CD}\,(CB) - 24 \times FB = 0 \qquad ...(A)$$

Now from ΔACB, $$CB = AB \sin 30 = 5 \times \frac{1}{2}$$

$$= \frac{5}{2} = 2.5$$

(a)

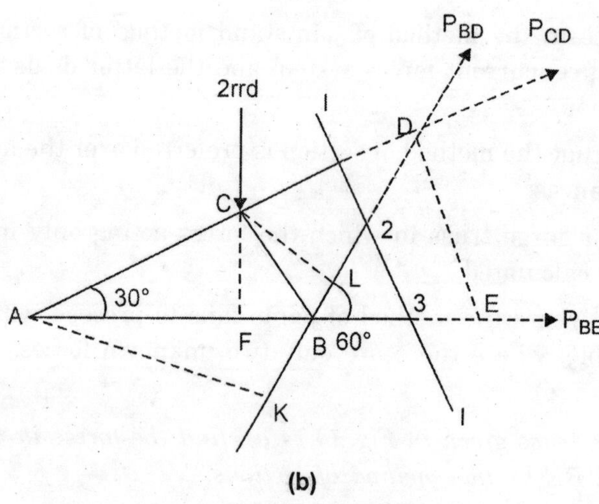

(b)

Fig. 4.14

Also, from $\Delta C\ FB$, $\quad FB = CB \sin 30$

$$= \frac{5}{2} \times \frac{1}{2} = 1.25$$

Also, R_A as in example 4.8 in calculated,

as $\quad R_A = 20\ kN$

Therefore, substituting values in eqn. A

$$20 \times 5 + P_{CD}(2.5) - 24 \times 1.25 = 0$$

or $\quad 2.5\ P_{CD} = -100 + 30 = -70$

$$P_{CD} = -\frac{70}{2.5}$$

$$= -28\ kN$$

Again for evaluating P_{BD} take moments about A and $\Sigma M_A = 0$

$$24\,AF - P_{BD}\,AK = 0 \qquad ...(B)$$

Now from $\Delta\,AFC$, $\quad AF = AC \cos 30$ (From $\Delta\,ABC$)

$$= AB \cos 30 \,(\cos 30)$$

$$= 5\times\frac{\sqrt{3}}{2}\cdot\frac{\sqrt{3}}{2} = \frac{15}{4}$$

And from $\Delta\,ABK$, $\quad AK = AB \sin 60 = 5\times\frac{\sqrt{3}}{2}$

Therefore, substituting values in eqn. B,

$$24\times\frac{15}{4} - P_{BD}\frac{5\sqrt{3}}{2} = 0$$

or

$$P_{BD} = \frac{90\times 2}{5\sqrt{3}} = \frac{36}{\sqrt{3}} = 12\sqrt{3}$$

For calculating P_{BE} take moments about point C

And $\quad M_C = 0$

$$-R_A\,AF + P_{BD}\times CL + P_{BE}\times CF = 0 \qquad ...(C)$$

From $\Delta\,ABC$, $\quad AC = AB \cos 30$

and $\quad AF = AC \cos 30 = AB(\cos 30)^2 = 5\times\frac{3}{4}$

From $\Delta\,BLC$, $\quad CL = BC \sin 60$ $\quad (\therefore < CBL = 60°)$

$$= AB \sin 30 \sin 60$$

$$= 5\times\frac{1}{2}\times\frac{\sqrt{3}}{2} - \frac{5\sqrt{3}}{4}$$

From $\Delta\,CFB$, $\quad CF = BC \sin 60 = AB \sin 30 \sin 60$

$$= \frac{5\sqrt{3}}{4}$$

$\therefore$ From eqn: C $\quad -20\times\frac{15}{4} + P_{BD}\frac{5\sqrt{3}}{4} + P_{BE}\cdot\frac{5\sqrt{3}}{4} = 0$

or

$$P_{BE}\frac{5\sqrt{3}}{4} = 75 - P_{BD}\frac{5\sqrt{3}}{4}$$

$$= 75 - 12\sqrt{3}\times\frac{5\sqrt{3}}{4}$$

$$= 75 - 45 = 30$$

$$P_{BE} = \frac{30\times 4}{5\sqrt{3}} = \frac{26}{\sqrt{3}} = 8\sqrt{3}\ kN \qquad \text{(Tensile)}$$

EXAMPLE

4.10 : *Find the forces in the members AB and AC of the truss shown in Fig. (4.15) using method of sections.* **(U.P. Tech. Un. 2002-2003)**

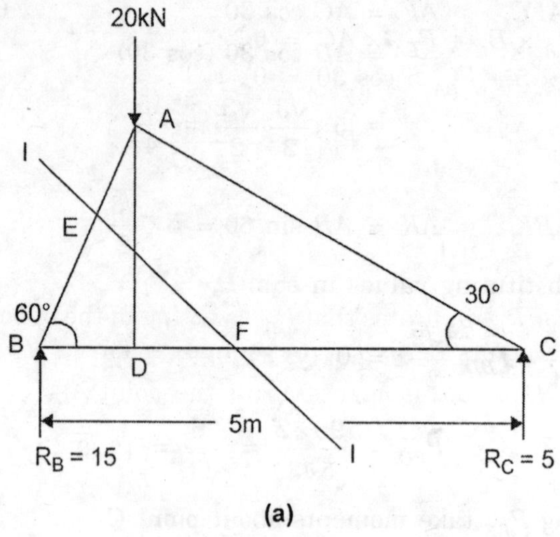

(a)

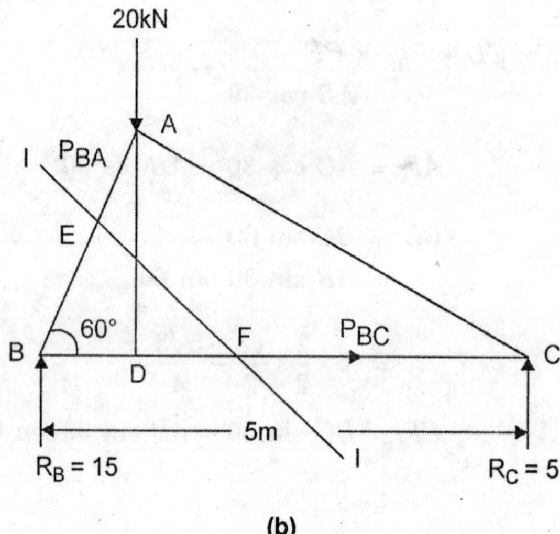

(b)

Fig. 4.15

Solution : Let the reaction at the support points be R_B and R_C. Then taking moments about B give,

$$M_B = 0$$

$$5R_B = 20\ DB = 20 \times 2.5 \cos 60$$

$$= 20 \times \frac{2.5}{2} = 25$$

$$\therefore \qquad R_C = \frac{25}{5} = 5\ kN$$

and

$$R_B = 20 - 5 = 15\ kN$$

Take section 1 – 1 as shown in Fig. 4.15 and the frec body diagram of the left portion of the truss is shown in Fig. 4.15 b. Then taking moments about C gives,

$$R_B \times B_C + P_{BA} \times AC = 0$$
$$15 \times 5 + P_{BA}\ 5 \cos 30 = 0$$
$$P_{BA} = -\frac{15 \times 5 \times 2}{5 \times \sqrt{3}}$$
$$= -\frac{30}{\sqrt{3}} = -10\sqrt{3}$$
$$= -17.32\ kN$$

Negative sign for P_{BA} indicates that P_{BA} is acting in the opposite direction *i.e,* towards the points B . is P_{BA} in comparing force.

For evaluation P_{BC} take moments about the point A,

$$R_B(BD) - P_{BC} \times AD = 0$$

or
$$P_{BC} = 15 \times \frac{BD}{AD}$$
$$= 15 \times \frac{AB \cos 60}{AB \sin 60}$$
$$= 15 \times \frac{1}{2} \times \frac{2}{\sqrt{3}}$$
$$= \frac{15}{\sqrt{3}} = 5\sqrt{3}\ kN$$
$$= 8.66\ kN$$

4.11 : *Determine the nature and magnitude of forces in the members of the truss shown in Fig. 4.16 by method of sections. (U.P. Tech. Un.)*

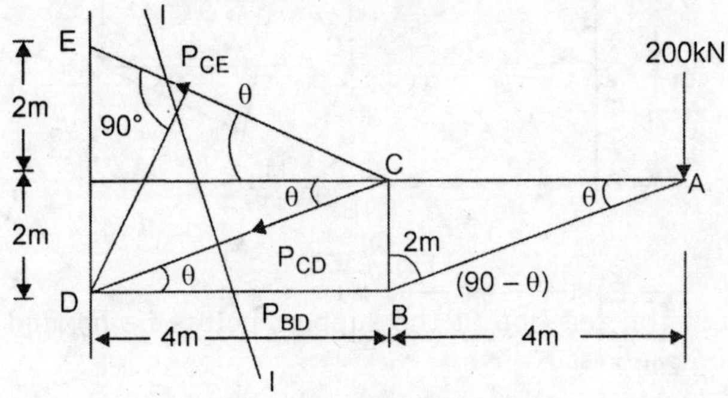

Fig. 4.16 (a)

Solution : Take section 1–1 and let the forces be P_{CE} P_{CD} and P_{BD} (Fig. 4.16 *b*)

Taking moments about C (point of intersection of forces P_{CE} and P_{CD}) considering the equilibrium of the right hand portion.

$$2 \times P_{BD} + 200 \times CA = 0$$
$$P_{BD} = -100 \times CA$$
$$= -400\ kN$$

E X A M P L E

– Ve means apposite to assumed P_{DB} direction. That is tensile

Again taking moments about D (Point of intersection of forces P_{CD} and P_{DB})

$$P_{CE}(DK) + 200 \times 8 = 0$$

$$P_{CE} = \frac{1600}{DK}$$

From D BCD we have,

$$CD = \sqrt{(BD)^2 + (BC)^2} = \sqrt{(4)^2 + (2)^2} = \sqrt{20}$$

And $$\sin\theta = \frac{2}{\sqrt{20}}$$

$$\cos\theta = \frac{4}{\sqrt{20}}$$

From triangle DCK, $$DK = CD \sin 2\theta = 2CD \sin\theta \cos\theta$$

$$= 2 \times \sqrt{20} \times \frac{2}{\sqrt{20}} \times \frac{4}{\sqrt{20}}$$

$$= \frac{16}{\sqrt{20}}$$

$\therefore$ $$P_{CE} = \frac{1600}{16} \times \sqrt{20} = 100\sqrt{20} \ kN$$

Take the section and Let P_{AC} and P_{AB} be tensile forces in members AC and AB respectively

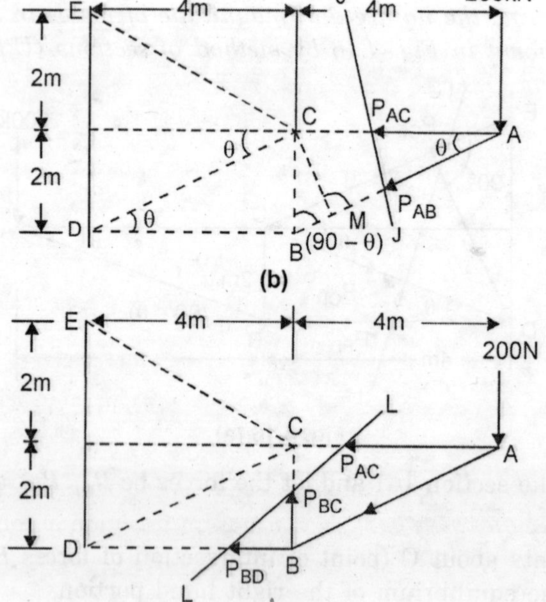

(b)

(c)

Fig. 4.16

Taking moments about C,

$$200 \times 4 + P_{AB}\, CM = 0$$

or

$$P_{AB} = -\frac{800}{CM}$$

$$= -\frac{800}{CA \sin\theta}$$

$$= -\frac{800}{4} \times \frac{\sqrt{20}}{2}$$

$$= -100\sqrt{20}\ kN$$

$$= -\ 447.2\ kN$$

Negative sign indicates that P_{AB} is compressive

For calculating P_{AC} take moments about the point B,

$$200 \times 4 - P_{AC} \times (CB) = 0$$

$$800 - (P_{AC})(2) = 0$$

$$P_{AC} = 400\ kN$$

For force in CB. *Take* a section $L - L$ Let P_{BC} be tensile in CB. Take moments about D,

$$P_{BC} \times 4 + P_{BD}\, 2 = 0$$

$$P_{BC} = -\frac{1}{2} P_{BD}$$

$$= -\left(\frac{-400}{2}\right) = +\ 200\ kN$$

4.12 : *Determine the nature and magnitudes of forces in members CD, GD, and GF of the truss shown in Fig. 4.17 by the method of sections.*

Solution : As the truss is symmetrically loaded the reactions R_A and R_B will be equal and

$$R_A = R_B$$

$$= \frac{1}{2}[10 + 15 + 10 + 15 + 10]$$

$$= 30\ kN$$

Let the forces in the members CD, GD and GF be P_{CD}, P_{GD} and P_{GF} *be* compressive in nature and they are marked accordingly in Fig. 4.17. Then take a section 1-1 and considering the equilibrium of the left hand portion Taking moments about A (Point of intersection of forces P_{CD} and P_{GF})

$$15 \times AK + P_{GD}\, AL = 0$$

$$15 \times AC \cos 30 + P_{GD}\, AG \sin 60 = 0$$

$$15 \times (AG \cos 30) \cos 30 + P_{GD}\, (AG \sin 60) = 0$$

EXAMPLE

$$15 \times \left(\frac{\sqrt{3}}{2}\right)^2 = -P_{GD}\frac{\sqrt{3}}{2}$$

or $$P_{GD} = \frac{-15 \times \sqrt{3}}{2}\ kN.$$

$$= -\ 12.99\ kN$$

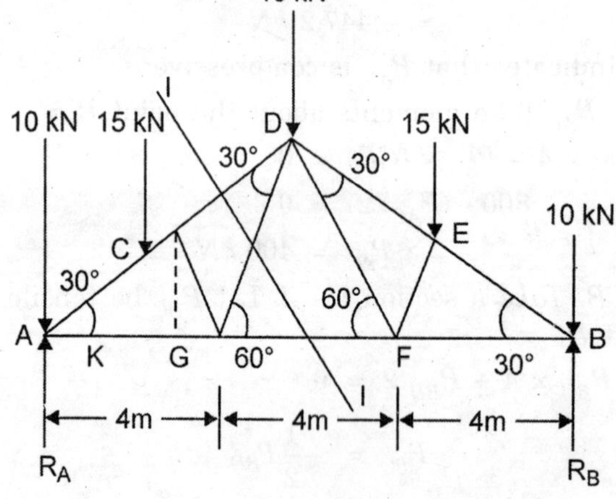

(a)

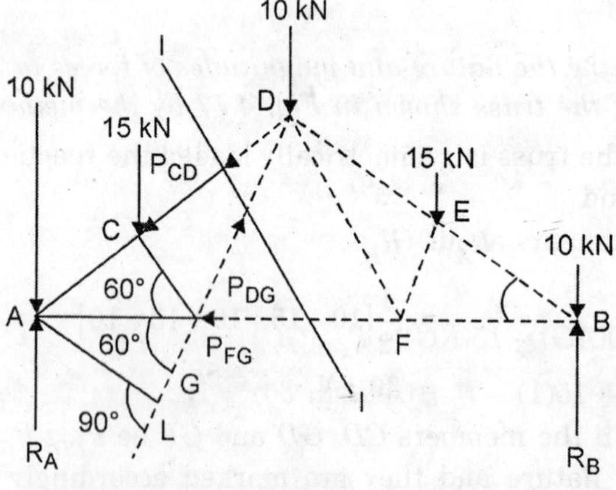

(b)

Fig. 4.17

Since it has come out to be negative P_{GD} is tensile.

(Point of intersection of forces P_{CD} *and* P_{GD}) is D

Taking moments about D we get,

$$R_A(AM) - 10(AM) - 15(KM) + P_{GF} \times DM = 0$$

$$R_A 6 - 10 \times 6 - 15(KG + GM) + P_{GF} \times DM = 0 \quad ...(A)$$

Now $KG = CG \cos 60$

or $KG = AG \sin 30 \cos 60$

$$= 4 \times \frac{1}{2} \times \frac{1}{2} = 1$$

$$GM = GD \cos 60$$

($\therefore$ GD is side of an equilateral Δ GDF)

$$= 4 \times \frac{1}{2} = 2m \quad (\therefore AG = DG)$$

and $DM = AD \sin 30$

$$= 2AC \sin 30$$

$$= 2(AG \cos 30) \sin 30$$

$$= 2 \times 4 \times \frac{\sqrt{3}}{2} \times \frac{1}{2} = 2\sqrt{3}$$

$\therefore$ Eqn. (A) reduces to,

$$30 \times 6 - 10 \times 6 - 15(1 + 2) + P_{GF} 2\sqrt{3} = 0$$

$$180 - 60 - 45 + P_{GF} 2\sqrt{3} = 0$$

$$P_{FG} = -\frac{75}{2\sqrt{3}}$$

$$= -2.165\ kN$$

Again taking moments about G (point of intersection of forces P_{GD} and P_{GP}

$$R_A(AG) - 10(AG) - 15(KG) - P_{CD}(GC) = 0$$

$$30(4) - 10 \times 4 - 15(1) - P_{CD}\ (AG \sin 30) = 0$$

$$120 - 40 - 15 - P_{CD}\left(4 \times \frac{1}{2}\right) = 0$$

$$65 - 2P_{CD} = 0$$

or $P_{CD} = \frac{65}{2}$

$$= 32.5\ kN$$

EXAMPLE

PROBLEM FOR PRACTICE

1. Define and explain the terms perfect frame, imperfect frame or deficient frame and reduldant frame. **(U.P.Tech 2002-2003)**
2. What is a truss ? Write down the assumptions made while finding the forces in the members of a frame.
3. Write down the steps taken while finding the forces in various members, by the method of sections.
4. Write down the steps to be followed while determining the forces in the members of a truss by the method of joints.
5. A truss is shown in Fig. 4.18. Find the forces in all the members of the truss and indicate whether it is in tension or compression. **(UPTU. 2000-2001)**

(**Ans.** P_{BC} = 0, P_{AC} = 14.43 *kN* P_{AD} = 7.215 *kN* P_{FC} = 0, P_{FE} = 20 *kN* P_{ED} = 7.215 *kN*))

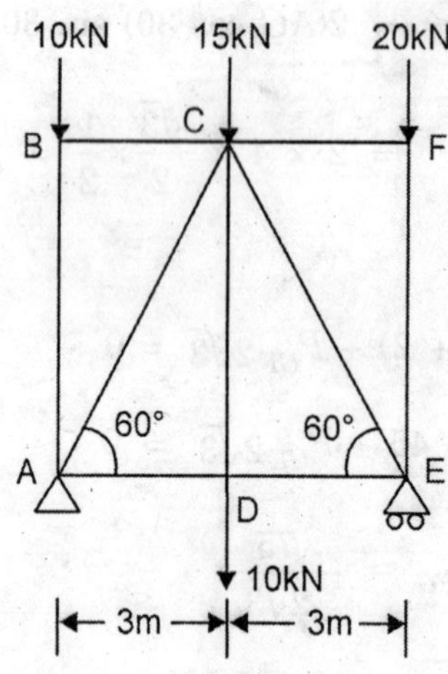

Fig. 418

6. A truss is shown in Fig. 4.19, where $\theta = \tan^{-1}\left(\frac{3}{4}\right)$ Determine the forces in the members *DG GF* and *FH*.

(**Ans.** P_{DG} = 416.67 *kN* P_{GF} = 500 *N* P_{FH} = 333.33 *N*))

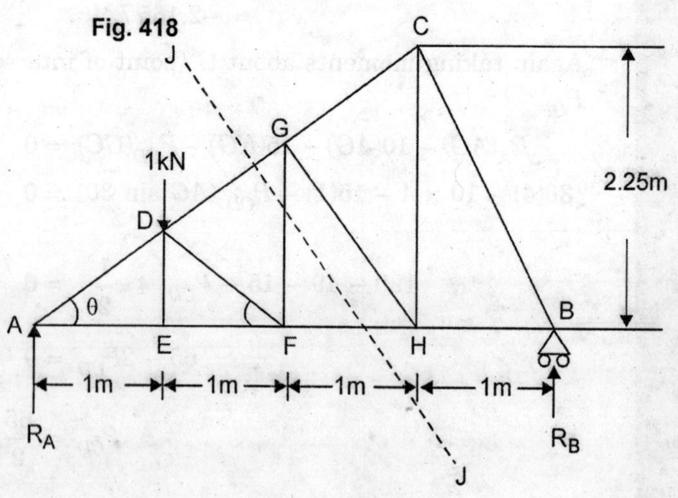

Fig. 4.19

7. A cantilever truss in loaded as shown in Fig. 4.20. Find the forces in all the members.

(**Ans.** P_{CE} = 11.55 kN
P_{CB} = 11.55 N
P_{CD} = – 23.1 kN
P_{BD} = 23.1 kN
P_{ED} = – 23.1 kN
P_{BA} = 40.2, P_{EA} = 46.2, P_{EP} = – 69.3))kN

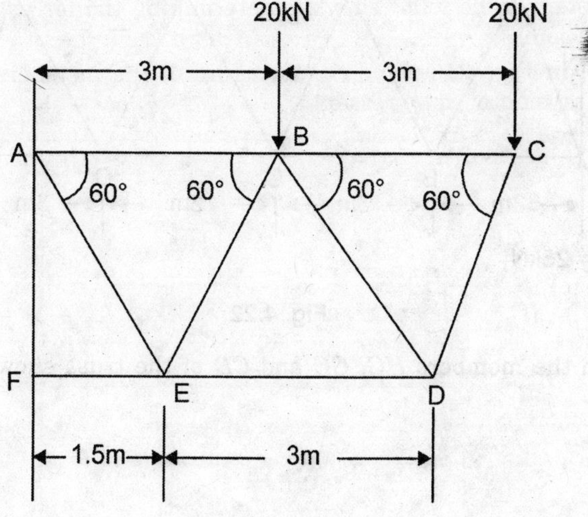

Fig. 4.20

8. Find the forces in the members *AB, AF, AH, BC* and *FC* of the truss shown in Fig. 4.21

(**Ans.** F_{AB} = 92.617 kN Comp
F_{BF} = 21.213 kN comp
F_{BC} = 71.404 kN, F_{FC} = 40.98 kN (Tension)
(Comp) F_{AG} = 30 kN) (Tension))

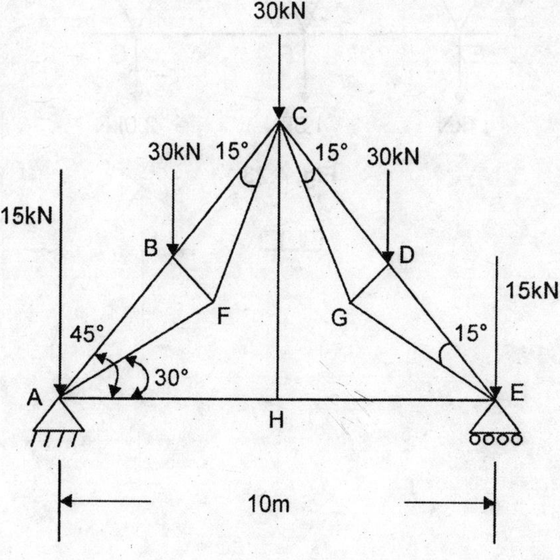

Fig. 4.21

9. Find the load in the members *CD, DF* and *FG* of the turns shown in Fig. 4.23

(**Ans.** F_{CD} = 2.145 *kN* (comp)
F_{DF} = 0.7 *kN* (Tension)
F_{FG} = 1.75 *kN* (Tension))

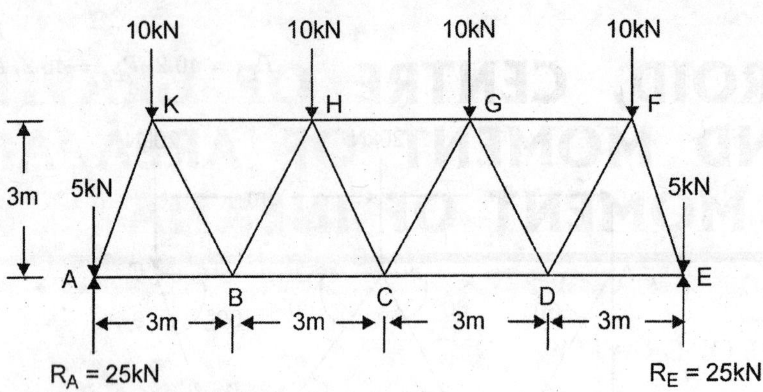

Fig. 4.22

10. Find the load in the members *HG, GC* and *CD* of the truss shown in Fig. 4.22.

(**Ans.** F_{EL} = 20 *kN* (comp)
F_{CG} = 0
F_{AM} = 20 *kN* (Tension))

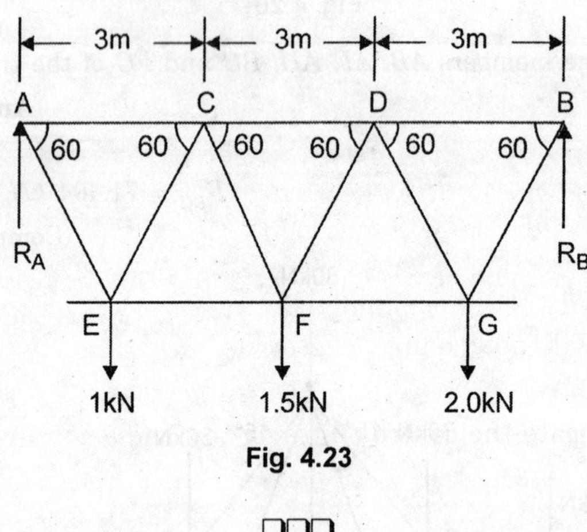

Fig. 4.23

□□□

CHAPTER 5

CENTROID, CENTRE OF GRAVITY, SECOND MOMENT OF AREA AND MASS MOMENT OF INERTIA

5.1 CENTRE OF GRAVITY

A solid body is built-up of a large number of very small particles. Each and every particle is attracted by the earth and the force of attraction is towards the centre of earth. Since the size of the body is very small as compared to the radius of earth, these individual forces acting on the particles of which the body is made of, will be parallel to each other. The force on each particle will be proportional to the mass of that particle. If we find the resultant of all these individual small forces, it will give the total weight of the body *i.e.*, the force with which it is being attracted by the earth and it is directed towards the centre of earth. Let us imagine a supporting force which is equal and opposite to the resultant force of attraction of the earth, then the body will be in equilibrium. In other wards, it is meant to say that the resultant upward force is passing through the centre of gravity. Since the body can move or rotate again the resultant force of attraction lay earth on the body will pass through the C.G. and its magnitude will be equal to the weight of the body. This force of attraction will again represent the total weight of the body and its line of action will pass through the centre of gravity. The point of intersection of these two resultant forces in two positions will give the location of centre of gravity (C.G. or supply G.). Since the centre of gravity is one and only one of a solid body, it will not change is location irrespective of the rotation or angular position of the solid body.

Now we proceed to locate the centre of gravity by using Varignon's theorem for different simple cases.

5.1.1 C.G. of a Straight Wire of Length l

Let AB be the wire of uniform section and w be the weight per unit length of wire. Take A as the origin and consider a small length dx situated at a distance x from A.

Fig. 5.1

The weight of the element CD of length dx will be,

$$dw = w\ dx$$

Total weight of wire AB

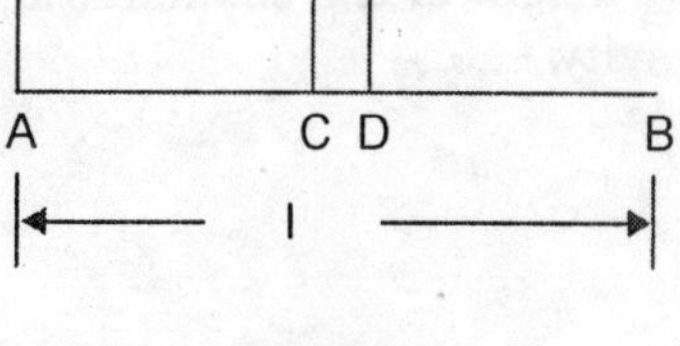

$$W = \int_0^P w\,dx = wl \qquad ...(5.1)$$

Let x be the distance of C.G. of wire AB through which the line of action of the resultant force of attraction by earth will pass. Thus, taking moments about 'A' of the various small forces acting on individual particles,

$$M = \int_0^l (w\,dx)\,x$$

$$= \left(w\frac{x^2}{2}\right)_0^l = \boldsymbol{wl^2}$$

$$= \frac{1}{2}W(l)$$

From Varignon's theorem, this moment is equal to the moment of the resultant force of attraction of earth which will be equal to W and its location is assumed to be at distance $\bar{x}$ from A,

i.e., $$W\bar{x} = \frac{W}{2}l$$

$$\bar{x} = \frac{l}{2} \qquad ...(5.3)$$

5.1.2 Corollary

In case we consider a line only. Theoretically, the line does not have mass but still we take a thin wire of very small diameter. Its centre of gravity will be at a distance of $\frac{l}{2}$ from one end. However, in the case of line we use the word centroid rather than the centre of gravity which term is used when the force of attraction of earth is considered.

5.1.3 Centre of Gravity of Plane Lamina of Uniform Thickness

As the plane lamina has uniform thickness t and we take ρ as the density of the material.

Let dA be the elemental area around the point (x, y) and its thickness is t along z axis.

$$dv = \rho\, t(dA)$$

Weight of the element dw, and if 'g' acceleration due to gravity,

$$dw = (\rho\, dv)\, g$$

$$w = \int \rho g\, dv = \int \rho g\, dA t$$

$$= \rho g t \int dA$$

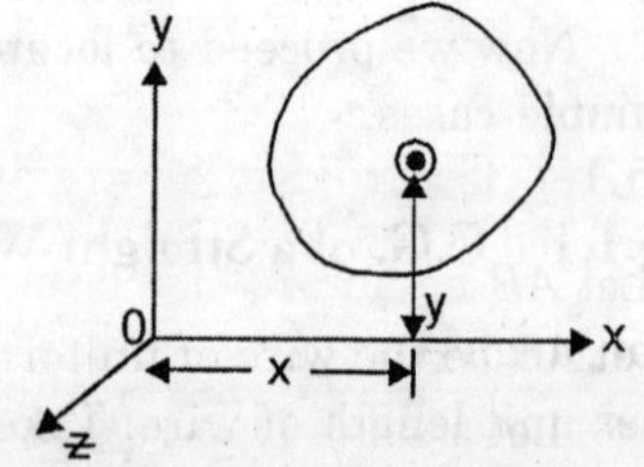

Fig. 5.2

$$= \rho\, g t\, A \qquad ...(5.3)$$

Where, $A = \int dA$

$= \text{Total area}$

Then from Varignon's theorem taking moment about *X-X*-axis,

$$W\bar{x} = \oint y\, dw$$

$$= \int y \rho g t\, dA$$

$$(\rho g t A)\bar{y} = \rho g t \int y\, dA$$

$$\bar{y} = \frac{\int y\, dA}{\int dA} \qquad ...(5.4)$$

Similarly, if we take moments about $\bar{y}$-axis and $\bar{x}$ is the distance of C.G. from $\bar{y}$-axis,

$$(\rho g t A)\bar{x} = \frac{\int x\, dA}{\int dA} \qquad ...(5.5)$$

Along the *x*-direction the thickness in uniform and equal to '*t*' the C.G. will be at their thickness. Thus, the co–ordinates of the C.G. are fully known as $\left(\bar{x}, \bar{y} \text{ and } \bar{z} = \frac{f}{z}\right)$.

Corollary

If a plane lamina of almost zero thickness is considered, we will get the co–ordinates of the point which is known as centroid of the plane figure.

Thus, $$\bar{x} = \frac{\int x\, dA}{\int dA} \qquad ...(5.4)$$

$$\bar{y} = \frac{\int y\, dA}{\int dA} \qquad ...(5.5)$$

5.1.4 Centroid of an are of *A* Circle

Let *AB* be an arc of a circle of radius *R* and subtending an angle 2α at the centre.

Let a small element *CD* of the arc which subtends an angle *d*θ at the centre be at angular position θ from the central line *OX* which divides the angle 2α.

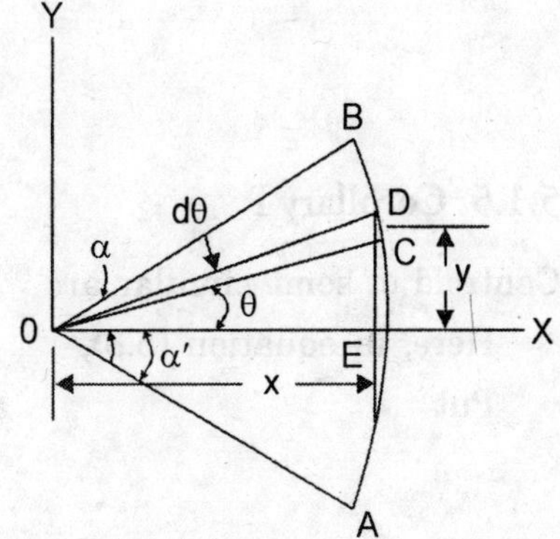

Fig. 5.3

If $(\bar{x}, \bar{y})$ are the co-ordinates of the centroid (C.G.) of the arc *AB*, we have using Varignon's theorem.

Moment of the resultant must be equal to the sum of

moments of weights of elemental lengths about Y and X axes respectively.

$$\bar{x} \times m\int_{-\propto}^{+\propto} R\,d\theta = \int_{-\propto}^{+\propto} m\,dl\,R\cos\theta$$

$$= \int_{-\propto}^{+\propto} m(R\,d\theta)\,R\cos\theta$$

or $$m\bar{x}R[\theta]_{-\propto}^{\propto} = mR^2\int_{-\propto}^{\propto}\cos\theta\,d\theta \qquad ...(5.6a)$$

$$mR\bar{x}\; z\propto = mR^2[\sin\theta]_{-\propto}^{\propto}$$

$$= m\;R^2(\sin^{\propto} + \sin\propto)$$

$$= 2m\;R^2\;\sin\propto$$

$$\bar{x} = \frac{R\sin\propto}{\propto} \qquad ...(5.6b)$$

For distance of C.G. from x-axis. Let it be $\bar{y}$. Then taking moments what X-axis and using Varignon's theorem,

$$\bar{y}\int_{\propto}^{\propto} mR\,d\theta = \int_{-\propto}^{\propto} y(R\,d\theta)\,m$$

$$\bar{y}_m R(\theta)_{-\propto}^{\propto} = \int_{-\propto}^{\propto}[m(R\,d\theta)]R\sin\theta$$

$$= mR^2\int_{-\propto}^{\propto}\sin\theta\,d\theta$$

$$= mR^2[-\cos\theta]_{-\propto}^{\propto}$$

$$\bar{y}mR\;2\propto = mR^2[-\cos\propto + \cos\propto]$$

i.e., $$\bar{y} = 0 \qquad ...(5.7)$$

Thus, the C.G. lies on the central axis and is at distance,

$$\bar{x} = \frac{R\sin\propto}{\propto} \qquad ...(5.8)$$

5.1.5 Corollary I

Centroid of semi-circular arc

Here, in equation (5.8),

Put $$z\propto = \pi$$

$$\propto = \pi/2 \qquad ...(5.8)$$

$\therefore$ $$\bar{x} = R\frac{\sin\pi/2}{\pi/2} = \frac{2R}{\pi} \qquad ...(5.9)$$

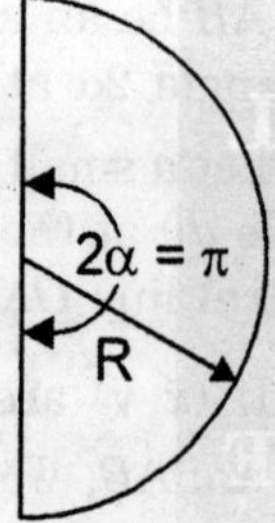

Fig. 5.4

5.1.6 Corollary II

Centroid of a equation of circular

Put $2\alpha = \pi/2$

or $\alpha = \pi/4$

Therefore, $\bar{x} = R\dfrac{\sin \pi/4}{\pi/4} = \dfrac{4R}{\pi} \times \dfrac{1}{\sqrt{2}}$

$$= \frac{2\sqrt{2}}{\pi} R \qquad \text{...(5.10)}$$

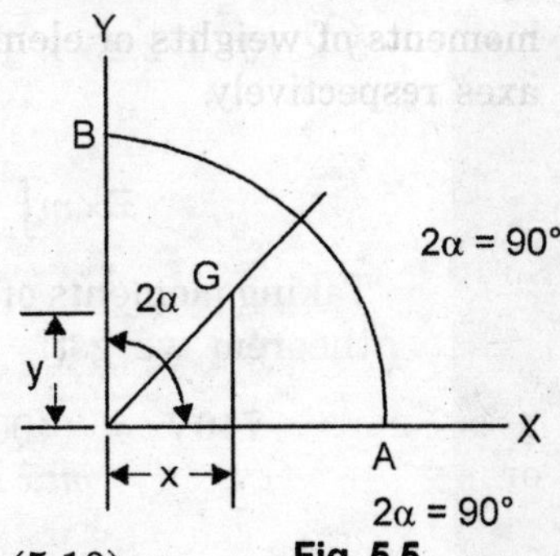

Fig. 5.5

EXAMPLE **5.1 :** *Determine the co-ordinates of the centroid of the lines shown in fig. 5.6.*

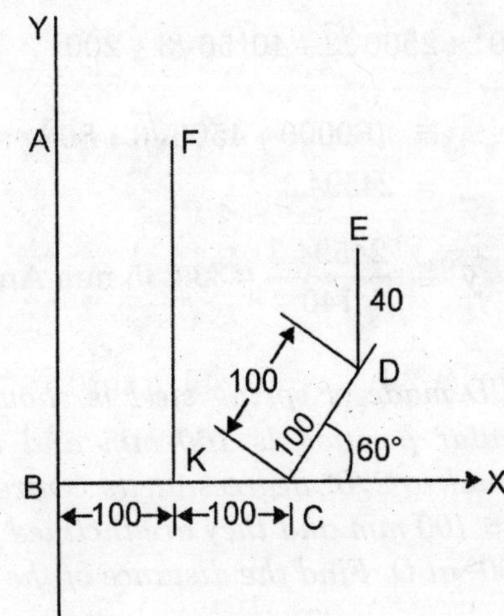

Fig. 5.6

Solution : Taking the centroids of the respective lengths at their mid -points and taking moments of these individual length segments about Y axis, we get by applying Varignon's theorem.

$$AB \times 0 + (FK \times BK) + \left(BC \times \frac{1}{2} BC\right) + CD\left(BC + \frac{DC}{2}\cos 60\right) + ED(BC + CD\cos 60)$$

$$= \bar{x} \times 740$$

or $$0 + 400 \times 100 + \left(200 \times \frac{200}{2}\right) + 100\left(200 + \frac{100}{2} \times \frac{1}{2}\right) + 40\left(200 + 100\frac{1}{2}\right)$$

$$= \bar{x} \times 740$$

or $$4 \times 10^4 + 1 \times 10^4 + 100\ (225) + 40\ (250)$$

$$= 740\bar{x}$$

$5 \times 10^4 + 2.25 \times 10^4 + 10000 = 740\overline{x}$

$$\overline{x} = \frac{8.25\times10^4}{740} = 1.115 \times 10^2 = 111.5$$

Taking moments of length segments about X-axis and applying varignon's theorem we get,

$$740\overline{y} = (400\times200)+(400\times200)+(BC\times0)+$$

$$\left(CD\times\frac{CD}{2}\sin60\right)+DE\left(CD\sin60+\frac{DE}{2}\right)$$

$$= 8\times10^4+8\times10^4+0+\left(100\times50\times\frac{\sqrt{3}}{2}\right)+40\left(100\frac{\sqrt{3}}{2}+\frac{1}{2}\times40\right)$$

$$= 16\times10^4+2500\sqrt{3}+40\left(50\sqrt{3}+200\right)$$

$$= 160000+4500\sqrt{3}+800$$

$$= 24594.3$$

$$\therefore \qquad \overline{y} = \frac{24594.3}{740} = 33.235 \text{ mm } \textbf{Ans.}$$

5.2 : *A clip E ABCD made of spring steel is shown in Fig. 5.7. The radius R of the circular position is 100 mm and the curved portion subtends an angle equal to 270° degrees at its centre. The straight legs EA and DC are each = 100 mm and they are inclined at 45° OY line and subtend an angle of 90° at O. Find the distance of the centroid of the clip wire from XX line.*

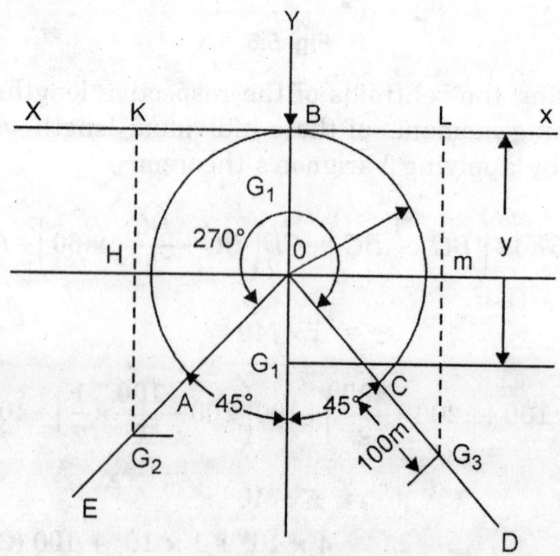

Fig. 5.7

E X A M P L E

E X A M P L E

Solution : Let G_1 be the centroid of the curved position *ABC*. We have from equation 5.8,

$$OG_1 = \frac{R\sin\alpha}{\alpha}$$

$$R = 100 \text{ mm}$$

Where, $2\alpha = 270°$

$$= \frac{3\pi}{2} \text{ radius}$$

or $\alpha = \frac{3\pi}{4}$

$$\therefore \quad OG_1 = \frac{100 \times \sin\left(\frac{3\pi}{4}\right)}{\left(\frac{3\pi}{4}\right)}$$

$$= 100 \times \frac{1}{\sqrt{2}} \times \frac{4}{3\pi} = 30.01 \text{ mm}$$

Also, length of arc,

$$ABC = l_1 = R(2\alpha)$$

$$= R \times \frac{3\pi}{2}$$

$$= 150\,\pi = 471.24 \text{ mm}$$

Geometrically,

If G_2 and G_3 and C.G. of *EA* and *DC* respectively,

$$AG_2 = \frac{100}{2} = 50 \text{ mm}$$

$$OG_2 = OA + 50 = 100 \times 50 = 150$$

$$\therefore \quad G_2H = OG_2 \sin 45° = \frac{150}{\sqrt{2}}$$

$$G_2K = G_2H + HK = G_2H + OB$$

$$= \frac{150}{\sqrt{2}} + 100 = 206.06 \text{ mm}$$

$$= y_2 \text{ (say)}$$

$$= G_2L = y_3 \text{ (say)}$$

Taking moments about the line $X - X$, if $\bar{y}$ is the distance of C.G. of the clip wire from XX

Total length $l = \text{arc}\ (ABC + AE + CD)$

$$= l_1 + l_2 + l_3$$

$$= R\ (2\alpha) + 100 + 100$$

$$= 100\left(\frac{3\pi}{2}\right) + 200$$

$$l = 671.24 \text{ mm}$$

$\therefore$ $$l\bar{y} = l_1 y_1 + l_2 y_2 + l_3 y_3 \qquad \text{...(B)}$$

where, y_1 = distance of G_1 from $XX = G_1B$

y_2 = distance of G_2 from $XX = G_2K$

y_3 = distance of G_3 from $XX = G_3L$

Now as her Varignon's theorem,

$$l\bar{y} = l\bar{y}_1 + l_2\bar{y}_2 + l_3\bar{y}_3$$

And $$l_1 = l_1 l\bar{y}_1 + l_2\bar{y}_2 100\frac{3\pi}{2} = 150\pi = 471.24$$

$$l_2 = l_3 = 100$$

$$y_1 = G_1B = OB - OG_1 = 100 - 30.01 = 69.09 \text{ mm}$$

$$y_2 = G_2K = 206.06 \text{ mm}$$

Therefore, substituting values in equation (*B*) we get,

$$l\bar{y} = 471.24 \times 69.09 + 2(100)\ (206.06)$$

or $$671.24\bar{y} = 32939.7 + 41212$$

$$= 74151.7$$

$$\bar{y} = \frac{74151.7}{671.24} = 110.47 \text{ mm} \quad \textbf{Ans.}$$

EXAMPLE

5.2 LOCATION OF CENTRE OF GRAVITY AND CENTROID

We now proceed to develop the method to locate the centre of gravity of a body whose mass is M and density of the material in ρ (say). The solid body can be considered to be built up of large number of elemental masses of weights W_1, W_2 ... which are located at points $(x_i, y_i, z_i)_{i = 1 \text{ to } n}$ with respect to certain chosen axis of references. Let the co-ordinates of the centre of gravity be $(\bar{x}, \bar{y}, \bar{z})$. If W is the total weight of the body. Then W being the resultant of the elemental weights ,

$$W = W_1 + W_2 + W_3$$

or $$W = \Sigma W_i \qquad \text{...(5.11)}$$

Taking moments about axes of reference in turn and applying Varignon's theorem we get,

$$Mg\bar{x} = \Sigma g\ m_i\ x_i$$

or
$$M\bar{x} = \oint x\,dm$$

It ρ is the density and V the volume, it gives,

$$\rho V\bar{x} = \oint \rho x\,dV$$

or
$$V\bar{x} = \oint x\,dV \qquad ...(5.11a)$$

Similarly for y-and z axis,

$$V\bar{y} = \oint y\,dV \qquad ...(5.11b)$$

$$V\bar{z} = \oint z\,dv \qquad ...(5.11c)$$

Also,
$$V = \int dv$$

Then equation (5.11) reduce to,

$$\bar{x} = \frac{\oint x\,dV}{\oint dV} \qquad ...(5.12a)$$

$$\bar{y} = \frac{\oint y\,dV}{\oint dV} \qquad ...(5.12b)$$

$$\bar{z} = \frac{\oint z\,dV}{\oint dV} \qquad ...(5.12c)$$

$(\bar{x}, \bar{y}, \bar{z})$ also, the co-ordinates of the centre of gravity of the total mass whose density is uniform.

Further, if we have a body consisting of plane laminas of uniform thickness of plane areas A_1, A_2 ... then equation (5.12) reduce to the following equation:

For figures consisting of plane areas as,

$$\bar{x} = \frac{\oint x\,dA}{\oint dA} \qquad ...(5.13a)$$

$$\bar{y} = \frac{\oint y\,dA}{\oint dA} \qquad ...(5.13b)$$

Thus, centre of gravity refers to the mass of the three dimensional body while a centroid pertains to plane areas.

5.2.1 Difference between Centre of Gravity and Centroid

From the analysis done in article 5.2, it is concluded that:

(a) The term centre of gravity applies is the bodies having mass so that the resultant force of attraction due to the gravitational pull of earth will pass through the centre of gravity irrespective of the mevement of the body.

(b) On the other hand centroid is associated with lines, plane areas having no mass and the moment of area about any axis through the centroid is zero.

5.3 AXIS OF SYMMETRY AND LOCATION OF CENTROID

In case, the area has an axis of symmetry, the centroid will be on it. The proof of this statement is given below:

Let the y-axis be the axis of symmetry of the given area. Consider any two elemental areas A_1 and A_2 which are at the same distance from y-axis *i.e.*, x_1 and x_2 and $x_1 = x_2 = a$. The entire area on the left of y-axis is of the same shape as that on the right of y-axis. Therefore for every elemental area on the left side we will have similar elemental area on the right side of y-axis. The moment of such similar elemental area about y-axis will cancel each other when the summation of $\Sigma x_i A_i$ is taken.

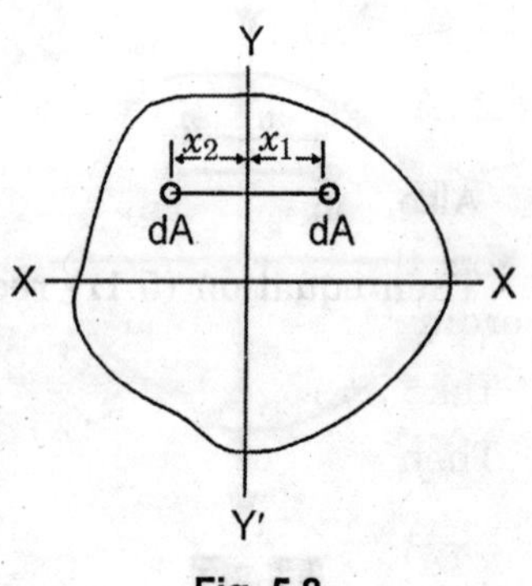

Fig. 5.8

Thus, $\oint x\, dA = 0$

Therefore, from equation (5.13a) we conclude that,

$$\bar{x} = \frac{\oint x\, dA}{\oint dA} = 0$$

From this it is concluded that the centroid will be on Y-axis. Similarly, if X-axis is also the axis of symmetry, the centroid will also be on x-axis which leads to the conclusion if both X-X and Y-Y are the axis of symmetry, then the origin, the point of intersection of axis of reference is the centroid of the section.

5.4 AREAS AND CENTROID OF PLANE FIGURES

5.4.1 Rectangle

Fig. 5.9, shows a rectangle $ABCD$ is shown of width 'b' and depth 'd'. If we take the X-X and Y-Y axes midway between AB and DC and between AD and BC respectively, then these will be axes of symmetry. Therefore, the origin O is the centroid of the section.

For finding the area of rectangle $ABCD$, take an element of width dx at a distance x from the Y-axis. The length of the element is d.

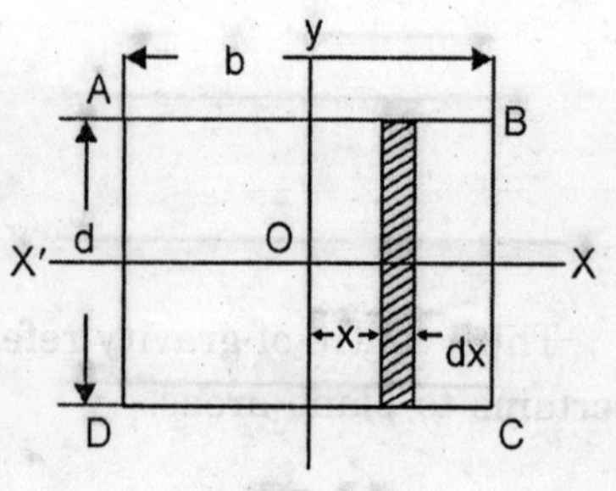

Fig. 5.9

$\therefore$ Elemental area $= dA = d(dx)$

Integrating, we get the total area,

$$A = \int_{-\frac{b}{2}}^{b/2} d(dx)$$

$$= d\int_{-\frac{b}{2}}^{b/2} dx = d[x]_{-b/2}^{b/2}$$

$$= d\left(\frac{b}{2}+\frac{b}{2}\right) = bd \qquad \text{...(5.14)}$$

5.4.2 Area and Centroid of Triangle

Fig. 5.10 shows a triangle, whose base *BC* of length *b* and height *h*.

Let the point *B* be the origin and take the *X*-axis along *BC* and *OY* as the *Y*-axis. Let the co-ordinates of the centroid be $(\bar{x}, \bar{y})$.

Take an elemental area at height *y* of depth *dy* parallel to *BC* or *X*-axis.

Then we have from similar triangles *ADE* and *ABC*,

$$\frac{DE}{BC} = \frac{AF}{AH}$$

$$= \frac{h-y}{h}$$

Fig. 5.10

That

$$DE = (b)\frac{(h-y)}{h}$$

$$= \left(b - \frac{by}{h}\right) \qquad \text{...(5.15)}$$

Therefore, from equation (5.13),

$$\bar{y} = \frac{\oint y\,dA}{\oint dA} = \frac{\int y(DE)\,dy}{\int dA} \qquad \text{...(5.16)}$$

$$\int_0^h y\,dA = \int_0^h y\left(b - \frac{by}{h}\right)dy$$

$$= \left[\frac{y^2}{2}b - \frac{b}{h}\frac{y^3}{3}\right]_0^h$$

$$= \frac{bh^2}{2} - \frac{1}{3}\frac{bh^3}{h}$$

$$= \frac{bh^2}{2} - \frac{bh^2}{3} = \frac{bh^2}{6} \qquad ...(5.17)$$

Also,
$$A = \int dA = \int_0^h (DE)\, dy$$

$$= \int_0^h \left(b - \frac{by}{h}\right) dy$$

$$= \left[by - \frac{by^2}{2h}\right]_0^h = bh - \frac{bh^2}{2h}$$

$$= \frac{bh}{2} \qquad ...(5.18)$$

$\therefore$ From equation (5.16) and (5.17) and (5.18), we get,

$$\bar{y} = \frac{bh^2}{6} \times \frac{2}{bh} = \frac{h}{3} \qquad ...(5.19)$$

From (5.19) it is concluded that the centroid of a triangle is at $\frac{2^{rd}}{3}$ of its height from the base.

Similarly, if the side AB is taken as base and the perpendicular distance of C from AB as height, the centroid will lie at $\frac{1}{3}$ of height of C from base AB. The intersection of these will give the location of centroid G.

5.4.3 Centroid of a Sector of a Circle

Fig. 5.11 shows a sector AOB subtending an angle 2α at the centre. Let the axis be taken along the bisector of the angle AOB and Y axis OY perpendicular to OX.

An elemental area $(r\, d\theta)\, dr$ which lies at an angular position θ from OX is shown,

$$dA = (r\, d\theta)\, dr$$

Then
$$A = \int_{\theta=-\alpha}^{\alpha} \int_{r=0}^{R} dA$$

$$= \int_{\theta=-\alpha}^{\alpha} \int_{r=0}^{R} (r\, dr)\, d\theta$$

$$= \int_{\theta=-\alpha}^{\alpha} \left[\frac{r^2}{2}\right]_0^R d\theta$$

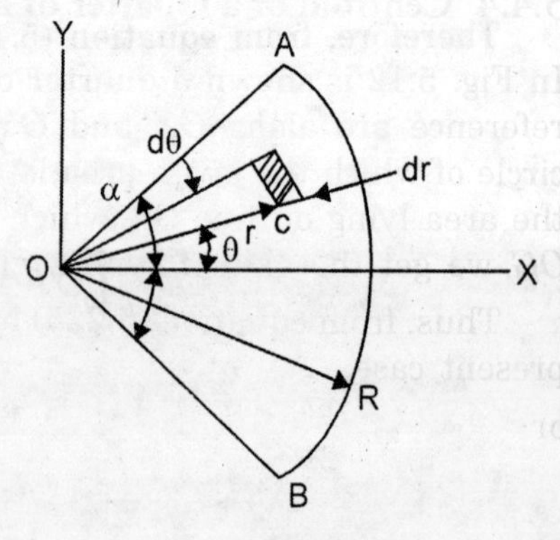

Fig. 5.11

$$= \left[\frac{R^2}{2}\theta\right]_{-\propto}^{\propto} = \frac{R^2}{2}[\propto + \propto] = R^2 \propto \qquad ...(5.20)$$

and if $\overline{x}$ is the distance of centroid from 0, along OX the axis of symmetry

$$\overline{x} = \frac{\int x\,dA}{\int dA}$$

$$= \frac{\int \text{moment about } y\text{-axis}}{\text{Area}}$$

$$\int \overline{x}\,dA = \int_{\theta=-\propto}^{\propto}\int_{r=0}^{R}(r\,d\theta\,dr)r\cos\theta$$

$$= \int_{\theta=-\propto}^{\propto}\left[\int_{r=0}^{R} r^2\,dr\right]\cos\theta\,d\theta$$

$$= \frac{R^3}{3}\int_{-\propto}^{\propto}\cos\theta\,d\theta = \frac{R^3}{3}[\sin\theta]_{-\propto}^{\propto}$$

$$= \frac{R^3}{3}2\sin\propto = \frac{2}{3}R^3\sin\propto \qquad ...(5.21)$$

From equation (5.20) and (5.21),

$$\overline{x} = \left(\frac{2}{3}R^3\sin\propto\right)\frac{1}{R^2\propto}$$

$$= \frac{2}{3}\frac{R\sin\propto}{\propto} \qquad ...(5.22a)$$

Since OX axis divides the angle $2\propto$ subtended by the arc AB at the centre, this is the axis of symmetry for the area and the centroid is at a distance of $\overline{x}$ given by equation (5.22*a*)

i.e.,
$$OG = \frac{2}{3}\frac{R\sin\alpha}{a} \qquad ...(5.22b)$$

5.4.4 Centroid of a Quarter of a Circle of Radius R

In Fig. 5.12 is shown a quarter of circular area CD. The axis of reference are along OX and OY, where OX the centre of the circle of which the given area is a part. Let G be the centroid of the area lying on line OA which bisects the angle COD Joining OG we get the axis of symmetry.

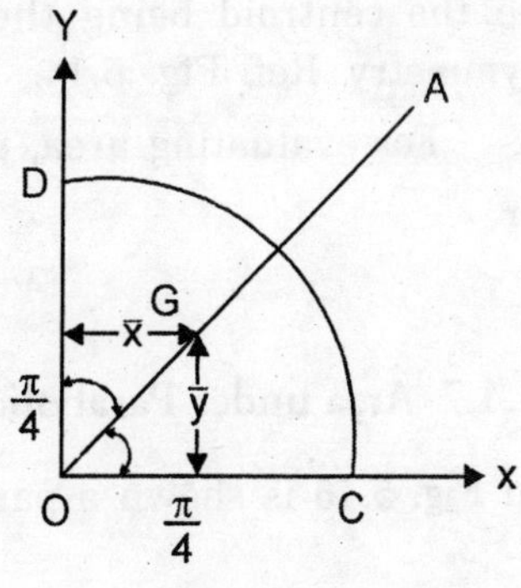

Fig. 5.12

Thus, from equation (5.22*b*) by putting $2\propto = \pi/2$ we get in the present case,

or
$$\propto = 45° = \pi/4$$

$$OG = \frac{2}{3}\frac{R\sin\left(\frac{\pi}{4}\right)}{\pi/4}$$

$$= \frac{8R}{3\pi}\frac{1}{\sqrt{2}} = \frac{8}{3}\frac{R}{\pi\sqrt{2}}$$

Thus, if $(\bar{x}, \bar{y})$ we the co-ordinates of G with respect to the axis OX and OY.

$$\bar{x} = OG \cos\frac{\pi}{4} = \frac{8R}{3\pi\sqrt{2}} \cdot \frac{1}{\sqrt{2}} = \frac{4R}{3\pi} \quad ...(5.23a)$$

Also,
$$\bar{y} = OG \sin\frac{\pi}{4} = \frac{8R}{3\pi}\frac{1}{(\sqrt{2})}\frac{1}{\sqrt{2}} = \frac{4R}{3\pi} \quad ...(5.23b)$$

And from equation (5.20),

$$A = R^2 \propto = R^2\,\pi/4 = \pi R^2/4$$

5.4.5 Centroid of A Semi–circular *AREA*

In Fig. 5.13 is shown a semi–circular area ABC or radius R. How the arc ABC subtends angle of π of the centre 0. Also, G lies x Y-axis, as it is an axis of symmetry.

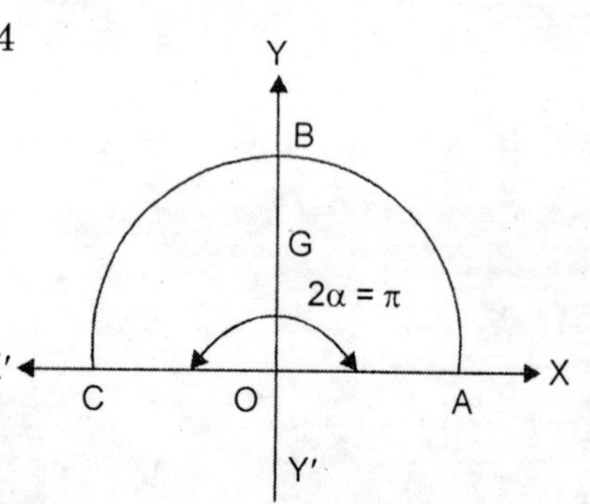

Fig. 5.13

Thus, $2\propto = \pi$ or $\propto = \pi/2$

∴ From equation (5.22),

$$OG = \frac{2}{3}R\frac{\sin\frac{\pi}{2}}{\pi/2} = \frac{4R}{3\pi} \quad ...(5.25)$$

Also, from equation (5.20) area of semi–circle A dy putting $\propto = A_2$,

We get,
$$A = \frac{\pi}{2}R^2 \quad ...(5.26)$$

5.4.6 Centroid and Area of Circle

In the case of a circular area both X and Y axes are the axes of symmetry. Therefore, the centre O of the circle itself will be the centroid being the put of intersection of areas of symmetry. Ref. Fig. 5.14.

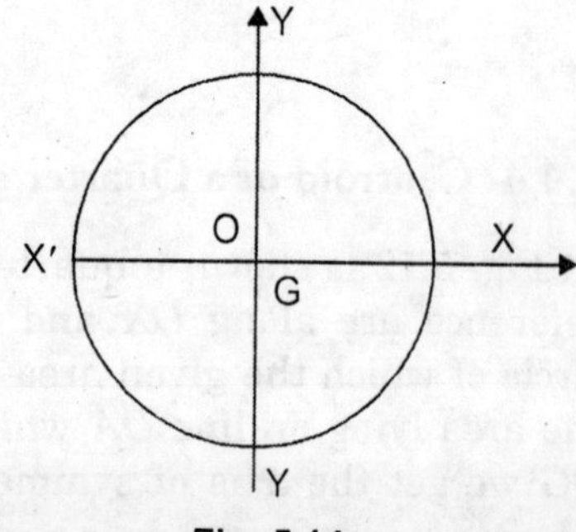

Fig. 5.14

For evaluating area, put $2\propto = 2\pi$ in equation (5.22)

or
$$\propto = \pi$$

$$A = R^2\propto = \pi R^2 \quad ...(5.27)$$

5.4.7 Area under Parabolic Curve $y = kx^2$ and its Centroid

In Fig. 5.15 is shown a parabolic curve OA of equation

$$y = kx^2 \quad ...(5.28)$$

Elemental area of width (dx) and length y is taken at a distance x from the origin

i.e.,
$$dA = y\ dx$$
$$= kx^2\ dx \qquad \text{...(5.29)}$$

It is desired to find the area under the curve OA, where the co–ordinates of A are $(x_1\ y_1)$. Integrating equation (5.28) we get μx area under the curve OA as,

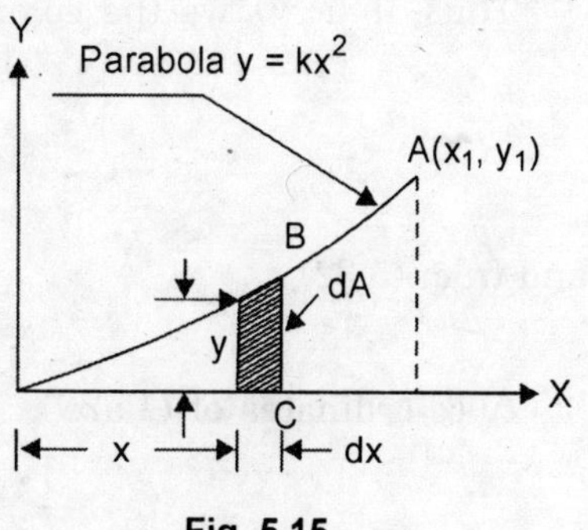

Fig. 5.15

$$A = \int_{x^2 a}^{x_1} kx^2 dx$$

$$= k\left[\frac{x^3}{3}\right]_0^{x_1} = \frac{kx_1^3}{3} \qquad \text{...(5.30)}$$

If $(\bar{x}, \bar{y})$ are the co-ordinates of centroid G, then taking moments about Y-axis give,

$$A\bar{x} = \text{moment of area about } Y\text{-axis}$$

$$= \int_0^{x_1} x\,dA = \int_0^{x_1} (x)(y\,dx) = \int_0^{x_1} (x)(kx^2)\,dx$$

or
$$A\bar{x} = k\int_0^{x_1} x^3 dx = k\left[\frac{x^4}{4}\right]_0^{x_1} = \frac{kx_1^4}{4} \qquad \text{...(5.31)}$$

Therefore, Fig. (5.29), (5.30),

$$\bar{x} = \frac{kx_1^4}{4} \times \frac{3}{kx_1^3} = \frac{3}{4}x_1 \qquad \text{...(5.32)}$$

Again taking moments of elemental area about X-axis we get,

$$A\bar{y} = \int_{x=0}^{x_1} dA \cdot \left(\frac{y}{2}\right)$$

$$= \int_{x=0}^{x_1} (y\,dx)\frac{y}{2} = \int_{x=0}^{x_1} kx^2 \cdot \frac{kx^2}{2} dx$$

$$= \int_0^{x_1} \frac{k^2 x^4}{2} dx = \frac{k^2}{2}\left[\frac{x^5}{5}\right]_0^{x_1}$$

$$= \frac{k^2}{10} x_1^5$$

$$\therefore \quad \bar{y} = \frac{k^2}{10} x_1^5 \times \frac{3}{kx_1^3} = \frac{3}{10} k\, x_1^2 \qquad \text{...(5.33)}$$

Since the point $(x_1\ y_1)$ lies on parabola we have Fig. (5.28),

$$y_1 = kx_1^2 \quad i.e. \quad k = \frac{y_1}{x_1^2} \qquad ...(5.34a)$$

$$\therefore \qquad \bar{y} = \frac{3}{10}\frac{y_1}{x_1^2}x_1^2 = \frac{3}{10}y_1 \qquad ...(5.34b)$$

and from (5.32), $\bar{x} = \frac{3}{4}x_1$

$\therefore$ co-ordinates of G are $\left[\frac{3}{4}x_1, \frac{3}{10}y_1\right]$

And Area A $= \frac{kx_1^3}{3} = \frac{y_1}{x_1^2}\cdot\frac{x_1^3}{3} = \frac{x_1y_1}{3}$ ...(5.35)

5.4.8 Area under Parabolic curve $y^2 = kx$ and its Centroid

In Fig. 5.39 is shown the parabolic curve,

$$y^2 = kx \qquad ...(5.36)$$

Elemental area $dA = y\ dx$

$$= \sqrt{k}\,x^{\frac{1}{2}}\,dx$$

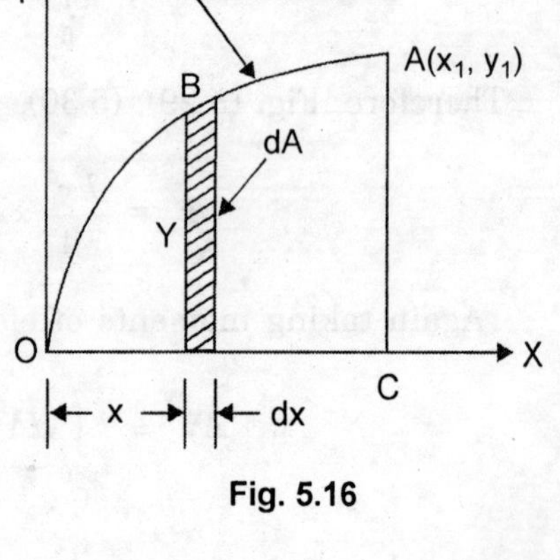

Fig. 5.16

Therefore, Area under the curve OBA when the co-ordinates of point A are (x_1, y_1)

$$A = \int_{x=0}^{x_1} dA = \int_{x=0}^{x_1} y\,dx$$

$$= \int_{x=0}^{x_1} \sqrt{k}\ x^{\frac{1}{2}}\,dx = \sqrt{k}\left[\frac{2x^{3/2}}{3}\right]_0^{x_1}$$

$$A = \frac{2}{3}\sqrt{k}\,x_1^{3/2} \qquad ...(5.37a)$$

Since point A (x_1, y_1) lies on curve therefore from equation (5.36),

$$y_1^2 = kx_1$$

or $$k = \frac{y_1^2}{x_1}$$

Then equation (5.37a) reduces to,

$$A = \frac{2}{3}\sqrt{\frac{y_1^2}{x_1}}\,x_1^{3/2}$$

$$= \frac{2}{3}\frac{y_1}{x^{1/2}}x_1^{3/2} = \frac{2}{3}x_1 y_1 \qquad \text{...(5.37}b\text{)}$$

If G is the centroid of the area under the curve OBA, the co-ordinates of A being (x_1, y_1) then moment about Y-axis gives,

$$A\bar{x} = \int_{x=0}^{x_{1x}} x\,dA = \int_{x=0}^{x_1} x\,y\,dx$$

$$= \int_{x=0}^{x_1} x\sqrt{k}\,x^{\frac{1}{2}}\,dx$$

$$= \sqrt{k}\int_{x=0}^{x_1} x^{3/2}\,dx = \sqrt{k}\left[\frac{2x^{5/2}}{5}\right]_0^{x_1}$$

$$= \frac{2}{5}\sqrt{k}\,x_1^{5/2}$$

Substituting value of A from equation (5.37a),

$$\therefore \quad \bar{x} = \frac{\frac{2}{5}\sqrt{k}\,x_1^{5/2}}{\frac{2}{3}\sqrt{k}\,x_1^{3/2}} = \frac{3}{5}x_1 \qquad \text{...(5.38)}$$

Similarly taking moments about X-axis we get,

$$A\bar{y} = \int_{x=0}^{x_1}\frac{y}{2}\,dA = \frac{1}{2}\int_{x=0}^{x_1} y\cdot(y\,dx)$$

$$= \frac{1}{2}\int_{x=0}^{x_1} y^2\,dx \quad \text{from equation (5.36)}$$

$$= \frac{1}{2}\int_{x=0}^{x_1} kx\,dx$$

or

$$A\bar{y} = \frac{1}{2}k\cdot\left(\frac{x^2}{2}\right)_0^{x_1} = \frac{1}{4}k\,x_1^2$$

$$\therefore \quad \bar{y} = \frac{\frac{1}{4}k\,x_1^2}{\frac{2}{5}\sqrt{k}\,x_1^{3/2}} = \frac{5\sqrt{k}\sqrt{x_1}}{8} \qquad \text{...(5.38)}$$

Also

$$k = \frac{y_1^2}{x_1}$$

$$\sqrt{k} = \frac{y_1}{\sqrt{x_1}} \qquad ...(5.39)$$

Therefore, from (5.38) and (5.39),

$$\bar{y} = \frac{5}{8}\frac{y_1}{\sqrt{x_1}}\sqrt{x_1} = \frac{5}{8}y_1$$

Therefore, the centroid G has co-ordinates,

$$(\bar{x}, \bar{y}) \text{ as } \left(\frac{3}{5}x_1, \frac{5}{8}y_1\right) \qquad ...(5.40)$$

and

$$A = \frac{2}{3}x, y_1 \qquad ...(5.41)$$

5.4.9 Area and Centroid of an Ellipse

Let the ellipse be represented by the equation,

$$\frac{x^2}{a^2} + \frac{y^2}{b^2} = 1 \qquad ...(5.42)$$

The curve for this equation is drawn in Fig. (5.17).

Since the area is symmetrical about both X and Y-axes, the centroid will be the origin O.

Let an elemental area be the strip AB of width dx at a distance x from origin O.

Then $$dA = y\,dx$$

But from equation (5.42), $$y = \frac{b}{a}(a^2 - x^2)^{\frac{1}{2}}$$

$$\therefore \quad dA = \frac{b}{a}(a^2 - x^2)^{\frac{1}{2}}\,dx$$

and total area $$A = 4\int_{x=0}^{a} dA$$

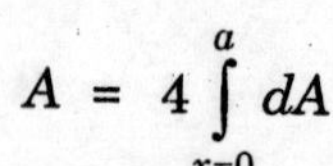

Fig. 5.17

$$= 4\int_{x=0}^{a} \frac{b}{a}(a^2 - x^2)^{\frac{1}{2}}\,dx$$

Substitute
$$x = a \sin \theta$$
$$dx = a \cos \theta \, d\theta$$

$$\therefore \quad A = \frac{4b}{a}\int_{0}^{\pi/2} (a^2 - x^2 \sin^2 \theta)^{\frac{1}{2}}\, a \cos\theta \, d\theta$$

$$= \frac{4b}{a} \int_0^{\pi/2} a^2 \cos^2 \theta \, d\theta$$

$$= \frac{4ba}{2} \int_0^{\pi/2} 2\cos^2\theta \, d\theta$$

$$= 2ba \int_0^{\pi/2} (1 + \cos 2\theta) \, d\theta$$

$$= 2ba \left[\theta + \frac{\sin 2\theta}{2} \right]_0^{\pi/2}$$

$$= 2ba \left(\frac{\pi}{2} \right) = \pi ab \qquad \text{...(5.43)}$$

5.4.10 Area and Centroid of Quarter Ellipse,

Let the curve be defined by equation,

$$\frac{x^2}{a^2} + \frac{y^2}{b^2} = 1$$

Since the curve is neither symmetrical about X-axis nor Y-axis the centroid of this area will not lie at O.

Let the elemental area CD be of width dx lying at a distance of x from OY.

Then the area,

$$dA = y \, dx$$

Fig. 5.18

$$A = \int_{x=0}^{a} y \, dx = \frac{1}{4} \text{ (area of entire ellipse as found in equation 5.43)}$$

$$= \frac{\pi ab}{4} \qquad \text{...(5.44)}$$

Moments about x-axis gives $\bar{y}$, so that

$$A\bar{y} = \int_{x=0}^{a} (y \, dx) \frac{y}{2} = \frac{1}{2} \int_{x=0}^{a} y^2 \, dx$$

But

$$\frac{y^2}{b^2} = 1 - \frac{x^2}{a^2}$$

$$y^2 = \frac{b^2(a^2 - x^2)}{a^2}$$

$$\therefore \quad A\,\bar{y} = \frac{1}{2}\int_{x=0}^{a} \frac{b^2}{a^2}(a^2 - x^2)\,dx$$

$$= \frac{1}{2}\int_{0}^{a}\left(b^2 - \frac{b^2}{a^2}x^2\right)dx$$

$$= \frac{1}{2}\left(b^2 x - \frac{b^2}{a^2}\frac{x^3}{3}\right)_0^a$$

$$= \frac{1}{2}\left(b^2 a - \frac{b^2}{3a^2}a^3\right) = \frac{1}{2}\frac{2}{3}ab^2 = \frac{b^2}{3}a$$

$$\therefore \quad \bar{y} = \frac{ab^2}{3}\cdot\frac{1}{A} = \frac{ab^2}{3}\cdot\frac{4}{\pi ab} = \frac{4b}{3\pi} \quad \textbf{Ans.} \qquad ...(5.45)$$

For finding $\bar{x}$ take moments about y-axis,

$$\bar{x}\,A = \int_{x=0}^{a} (y\,dx)\,x$$

Now

$$y^2 = b^2\left(1 - \frac{x^2}{a^2}\right)$$

$$= \frac{b^2}{a^2}(a^2 - x^2)$$

or

$$y = \frac{b}{a}(a^2 - x^2)^{\frac{1}{2}}$$

$$\therefore \quad \bar{x}\,A = \int_{x=0}^{a} \frac{b}{a}(a^2 - x^2)^{\frac{1}{2}}\,x\,dx$$

Put

$$x = a\sin\theta$$

$$dx = a\cos\theta\,d\theta$$

$$\therefore \quad \bar{x}\,A = \int_{0}^{\pi/2} \frac{b}{a}(a\cos\theta)(a\sin\theta)(a\cos\theta)\,d\theta$$

$$= a^2 b\int_{0}^{\pi/2} \cos^2\theta\,\sin\theta\,d\theta$$

$$= -a^2 b\int_{0}^{\pi/2} \cos^2\theta(-\sin\theta)\,d\theta$$

$$= -a^2 b\left(\frac{\cos^3\theta}{3}\right)_0^{\pi/2}$$

$$= -\frac{a^2 b}{3}[0-1] = \frac{a^2 b}{3}$$

$$\therefore \quad \bar{x} = \frac{a^2 b}{3}\times\frac{1}{A} = \frac{a^2 b}{3}\frac{4}{\pi ab} = \frac{4a}{3\pi} \quad \text{...(5.46a)}$$

Therefore, the co–ordinates of the centroid are,

$$(\bar{x}, \bar{y}) \rightarrow \left(\frac{4a}{3\pi}, \frac{4b}{3\pi}\right) \quad \text{...(5.46b)}$$

and

$$A = \frac{\pi ab}{4} \quad \text{...(5.47)}$$

5.4.11 Centroid and Area of Semi-elliptic Area

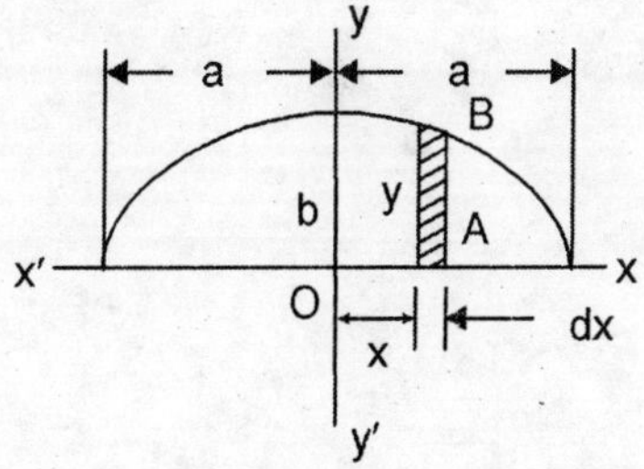

Fig. 5.18

$$\frac{x^2}{a^2}+\frac{y^2}{b^2} = 1$$

or

$$y^2 = \left(1-\frac{x^2}{a^2}\right)b^2$$

As the Y-axis is the axis of symmetry the C.G. will be on it

For finding the value of $\bar{y}$ take moments about X-axis. For this elemental area ($y\ dx$) as shown in Fig. 5.19.

$$\bar{y}\int_{x=-a}^{a} y\,dx = \int_{x=-a/1}^{a} (y\,dx)\left(\frac{y}{2}\right)$$

$$2\bar{y}\int_0^a\left(1-\frac{x^2}{a^2}\right)^{\frac{1}{2}} b\,dx = \frac{1}{2}\int_{x=-a}^{a} y^2\,dx$$

$$\bar{y}\,\frac{2b}{a}\int_0^a\sqrt{a^2-x^2}\,dx = \frac{1}{2}\times 2\int_0^a y^2\,dx$$

or

$$\frac{2b}{a}\bar{y}\int_0^a\sqrt{a^2-x^2}\,dx = \int_0^a b^2\left(1-\frac{x^2}{a^2}\right)dx$$

$$= \frac{b^2}{a^2}\int_0^a (a^2-x^2)\,dx$$

$$= \frac{b^2}{a^2}\left[a^2x - \frac{x^3}{3}\right]_0^a = \frac{2}{3}\frac{b^2a^3}{a^2} = \frac{2}{3}ab^2 \quad ...(5.48)$$

For LHS put
$$x = a \sin \theta$$
$$dx = a \cos \theta \, d\theta$$

$$\text{LHS} = \frac{2b}{a}\bar{y}\int_0^{\pi/2}\sqrt{a^2 - a^2\sin^2\theta}\,(a\cos\theta\, d\theta)$$

$$= \frac{2b}{a}\bar{y}\int_0^{\pi/2}(a\cos)(a\cos\theta)\,d\theta$$

$$= 2ab\,\bar{y}\times\frac{1}{2}\int_0^{\pi/2}2\cos^2\theta\, d\theta$$

$$= \frac{2ab}{2}\bar{y}\int_0^{\pi/2}(1+\cos 2\theta)\,d\theta$$

$$= ab\,\bar{y}\left(\theta + \frac{\sin 2\theta}{2}\right)_0^{\pi/2}$$

$$= \frac{\pi ab\,\bar{y}}{2}$$

Above expression, then reduces to,

$$\frac{\pi ab\,\bar{y}}{2} = \frac{2}{3}ab^2$$

$$\therefore \quad \bar{y} = \frac{2}{3}\frac{ab^2\times 2}{\pi ab} = \frac{4b}{3\pi} \quad ...(5.49)$$

Thus, the centroid of semi-ellipse lie at a distance $\frac{4b}{3\pi}$ from the central axes (*i.e.*, *OX*) and on the *Y*-axis.

5.5 CENTRE OF GRAVITY OF SOLIDS

Centre of gravity or in short (C.G.) is the point, where the Resultant Earth's gravitational force of attraction is passing. The method to be followed for locating the C.G. will be same as that followed while finding the co-ordinates of centroids of plane figures by involving the Varignon's theorem *i.e.*, by taking moments of the individual forces about the three axes of reference in turn.

Table 5.1. Areas and Location of Centroid of Common Plane Figures.

	Figure	Area	Centroid $(\bar{x}, \bar{y})$
Rectangle		$b \times d$	(0, 0)
Triangle		$\frac{1}{2} bh$	$\bar{y}$ of $G = \frac{h}{3}$
Semi–circle		$\frac{\pi}{2} R^2$	$\bar{y} = \frac{4R}{3\pi}$
Quarter Circle		$\frac{\pi R^2}{4}$	$\bar{x} = \frac{4R}{3\pi}$ $\bar{y} = \frac{4R}{3\pi}$
Parabola $y = kx^2$		$\frac{x_1 y_1}{3}$	$(\bar{x}, \bar{y})$ $\left(\frac{3}{4} x_1, \frac{3}{10} y_1\right)$
Parabola $x = ky^2$		$\frac{2}{3} x_1 y_1$	$\left(\frac{3}{5} x_1, \frac{5}{8} y_1\right)$
Ellipse		πab	(0, 0)
Quarter of Ellipse		Area $= \frac{\pi ab}{4}$	centroid $\left(\frac{4a}{3\pi}, \frac{4b}{3\pi}\right)$
Semi–ellipse		$\frac{\pi ab}{2}$	$\left(\frac{4b}{3\pi}\right)$

5.5.1 C.G. of a Solid Sphere

Since the sphere is symmetrical about the three axes of reference OX, OY and OZ. Thus, their common point *i.e.*, O, the origin is C.G. of the solid sphere. (Fig. 5.20)

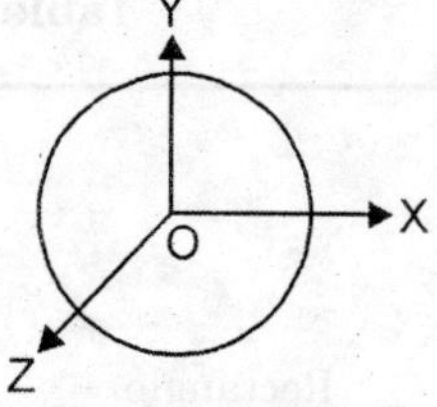

Fig. 5.20

5.5.2 Centre of Gravity of a Right Circular Cone

In Fig. 5.21 is shown a right-circular curve of height h and circular base of radius r.

Evidently AO in the axis of symmetry and the C.G. will be x the axis *i.e.*, OY. Chose the axis of reference OX, OY and OZ with origin at O, the centre of base circle.

Let an elemental volume in the form of a thin circular disc DFE of radius DF and thickness dy be lying at a height of $FO = y$ from the base circle

Then
$$dV = \pi(DF)^2\, dy \quad \text{...(5.50)}$$

From Fig. (5.32) symmetrically we have,

$$\frac{DF}{BO} = \frac{AF}{AO}$$

or
$$DF = \frac{AF}{AO}(BO)$$

$$= \frac{(h-y)}{h}r$$

or
$$DF = r\left(\frac{h-y}{h}\right)$$

$$= r\left(1-\frac{y}{h}\right) \quad \text{...(5.51)}$$

Fig. 5.21

Thus, from equation (5.50) and (5.51),

$$\therefore \quad dV = \pi\left(\frac{r^2}{h^2}(h-y)^2\right)dy$$

$$= \pi r^2\left[\left(1-\frac{y}{h}\right)^2\right]dy$$

$$V = \int_{y=0}^{h} \pi r^2\left\{1+\frac{y^2}{h^2}-\frac{2y}{h}\right\}dy$$

$$= \pi r^2\left[y+\frac{y^3}{3h^2}-\frac{2y^2}{2h}\right]_0^h$$

$$= \pi r^2 \left[h + \frac{h^3}{3h^2} - \frac{h^2}{h} \right]$$

$$= \pi r^2 \left[h + \frac{h}{3} - h \right]$$

$$= \frac{\pi r^2 h}{3} \qquad \text{...(5.52)}$$

If the density of the cone material is ρ, then

Weight of cone $= \rho \times V$

$$W = \frac{\pi}{3} r^2 h \rho$$

Taking moments about X-axis,

$$\bar{h} W = \int_{y=0}^{h} (\rho \, dV) \, y$$

$$= \int_{y=0}^{h} \rho y \{ \pi (DF)^2 dy \}$$

$$= \pi\rho \int_{y=0}^{h} y \left\{ r^2 \left(1 - \frac{y}{h} \right)^2 dy \right.$$

$$= \pi r^2 \rho \int_0^h y \left(1 + \frac{y^2}{h^2} - \frac{2y}{h} \right) dy$$

$$= \pi r^2 \rho \int_0^h \left(y + \frac{y^3}{h^2} - \frac{2y^2}{h} \right) dy$$

$$= \pi r^2 \rho \left[\frac{y^2}{2} + \frac{y^4}{4h^2} - \frac{2y^3}{3h} \right]_0^h$$

$$= \pi r^2 \rho \left[\frac{h^2}{2} + \frac{h^2}{4} - \frac{2}{3} h^2 \right]$$

$$= \rho \frac{\pi r^2}{12} h^2 [6 + 3 - 8]$$

or
$$\bar{h}\frac{\pi}{3}r^2 h\rho = \frac{\pi r^2 h^2}{12}\rho$$

$$\bar{h} = \frac{h}{4} \qquad ...(5.53)$$

That is, the C.G. lies on the axis of cone and it is at a distance of $\frac{h}{4}$ from the base.

5.5.3 Volume of a Solid Sphere of Radius r

In Fig. 5.22 section of the sphere by $X - Z$ plane is shown.

ABC is disc of radius x and it is at a distance of z from the diametral plane *OXY* and of thickness dz.

$$dV = \pi x^2 \, dz$$
$$= \pi(r^2 - z^2)\, dz$$

$$V = 2\int_{z=0}^{r} \pi(r^2 - z^2)dz$$

$$= 2\pi \int_{z=0}^{r} (r^2 - z^2)dz$$

$$= 2\pi\left[r^2 z - \frac{z^3}{3}\right]_0^r$$

$$= 2\pi\left[r^3 - \frac{r^3}{3}\right]$$

$$= 2\times\frac{2\pi r^3}{3} = \frac{4}{3}\pi r^3 \qquad ...(5.54)$$

Fig. 5.22. Section of sphere in x – z plane.

Since the sphere is symmetrical about the three axes, the C.G. coincides with 0.

5.5.4 To Find the Location of the C.G. of a Solid Hemisphere

In Fig. 5.23 a section by ~~zx~~ plane of the hemisphere is shown. *ABC* is a disc of thickness dz at a distance of z from the base of the hemisphere,

$$BC^2 = x^2 = r^2 - z^2$$
$$dV = \pi x^2 \, dz$$
$$= \pi(r^2 - z^2)\, dz$$

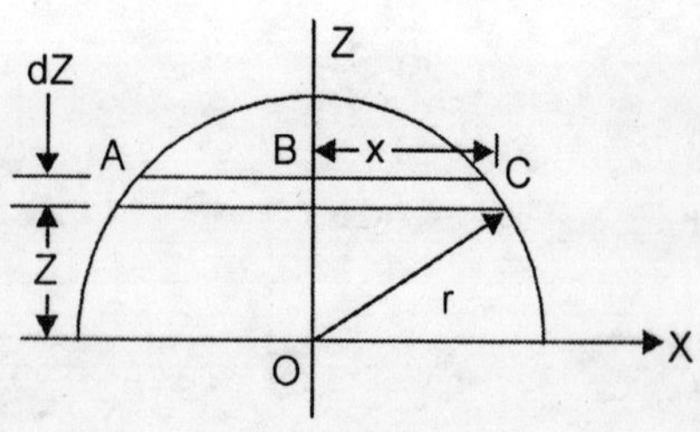

Fig. 5.23

$$V = \int_{z=0}^{r} \pi(r^2 - z^2)\,dz$$

$$= \pi\left[r^2 z - \frac{z^3}{3}\right]_0^r$$

$$= \pi\left[r^3 - \frac{r^3}{3}\right] = \frac{2}{3}\pi r^3 \qquad \text{...(5.55)}$$

For finding the location of C.G. which will be lying on the axis of symmetry OZ. If $\bar{z}$ is the height of the C.G. from the base, then using Varignon's theorem and taking moments about OX we get,

$$(\rho V)\bar{z} = \int_{z=0}^{r} (\rho\, dV)\, z$$

$$= \int_{z=0}^{r} \rho \pi x^2 z\, dz$$

$$= \int_{z=0}^{r} \pi \rho (r^2 - z^2) z\, dz$$

$$= \pi\rho \int_{z=0}^{r} (r^2 z - z^3)\, dz$$

$$= \pi\rho\left[\frac{r^2 z^2}{2} - \frac{z^4}{4}\right]_0^r$$

Substitute the value of V from equation 5.55,

or $$\bar{z}\left(\frac{2}{3}\pi r^3 \rho\right) = \frac{\pi\rho}{4}[2r^4 - r^4]$$

or $$= \frac{\pi\rho r^4}{4}$$

or $$\bar{z} = \frac{\pi\rho r^4}{4} \times \frac{3}{2\pi r^3 \rho}$$

$$\bar{z} = \frac{3r}{8} \qquad \text{...(5.56)}$$

CENTROID OF COMPOSITE AREAS AND C.G. OF COMPOSITE BODIES

5.1 : *Find the location of centroid of a T-section shown in Fig. 5.24.*

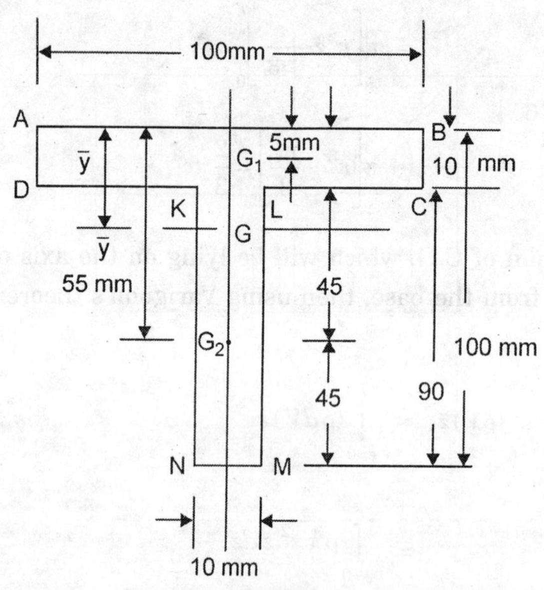

Fig. 5.24

Solution : The composite section consists of two rectangular areas *ABCD* and *KLMN*.

Let these areas be represented by A_1 and A_2 with their centroids G_1 and G_2 such that the distances of G_1 and G_2 from the reference line *AB* are y_1 and y_2. Let the distance of centroid *G* of the composite section be $\bar{y}$ from *AB*. Then taking moments of the respective areas about *AB*, we will get by using Varignon's theorem.

These $\quad A_1 = 100 \times 10 = 1000 \text{ mm}^2$

$$y_1 = \frac{1}{2} \times 10 = 5 \text{ mm}$$

$$A_2 = 90 \times 10 = 900 \text{ mm}^2$$

$$y_2 = 100 - 45 = 55 \text{ mm}$$

Then $\quad (A_1 + A_2)\bar{y} = A_1 y_1 + A_2 y_2$

$$(1000 + 900)\bar{y} = 1000 \times 5 + 900 \times 55$$

$$1900\bar{y} = 5000 + 99500$$

$$= 54500$$

$$\bar{y} = \frac{54500}{1800} = 28.68 \text{ mm}$$

E X A M P L E

EXAMPLE

5.2 : *A plate ABCD is 600 mm wide and 400 mm deep A hole of diameter 15 mm is drilled at a point G_2 which is 60 mm from AB and 60 mm from AD. Find the location of the centroid of the plate.*

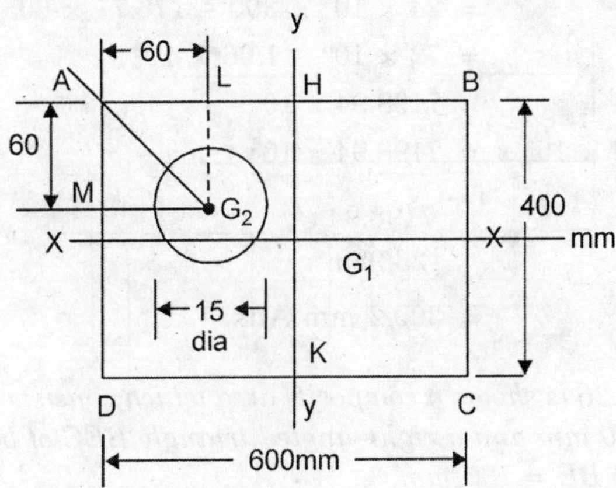

Fig. 5.25

Solution : The centroid of the plate when there is no hole will be G_1 the point of intersection of the axes of symmetry *XX* and *YY*.

Then $A_1 = b \times d$

where, $b = 600$

$d = 400$

$= 600 \times 400$

$= 24 \times 10^4 \text{ mm}^2$

The C.G. of the hole will be at G_2 and its area is,

$$A_2 = \frac{\pi}{4} \times (15.0)^2 = 176.71 \text{ mm}^2$$

$\therefore$ Resultant area,

$$A_1 - A_2 = 240000 - 176.71$$

$$= 23.98 \times 10^4 \text{ mm}^2$$

Let the centroid of the final area be at a distance $\bar{y}$ from *AB*.

Taking moments about *AB*,

$$(A_1 - A_2)\bar{y} = A_1 \times (HG_1) - A_2(LG_2)$$

$$= 24 \times 10^4 \times 200 - 176.71 \times 60$$

$$= 48 \times 10^6 - 1.06 \times 10^4$$

or $23.98 \times 10^4 \ \bar{y} = 47.989 \times 10^6 \text{ mm}^3$

$$\therefore \quad \bar{y} = \frac{47.989 \times 10^6}{23.98 \times 10^4} = 200.1\text{mm } \textbf{Ans.}$$

If $\overline{x}$ is the distance of G from AD, then taking moment about AD gives,

$$(A_1 - A_2)\overline{x} = A_1 \times (NG_1) - A_2 (M\,G_2)$$

$$= 24 \times 10^4 \times 300 - 176.71 \times 60$$

$$= 72 \times 10^6 - 1.06 \times 10^4$$

$$= 7198.94 \times 10^4$$

or $$23.98 \times 10^4\, x = 7198.94 \times 10^4$$

$$\overline{x} = \frac{7198.94}{23.989}$$

$$= 300.2 \text{ mm } \textbf{Ans.}$$

5.3 : *In Fig. 5.26 is shown a composite area which consists of a rectangle ABCD 600 × 300 mm and a right–angled triangle BEC of base BC = 300 mm and height BE = 120 mm.*

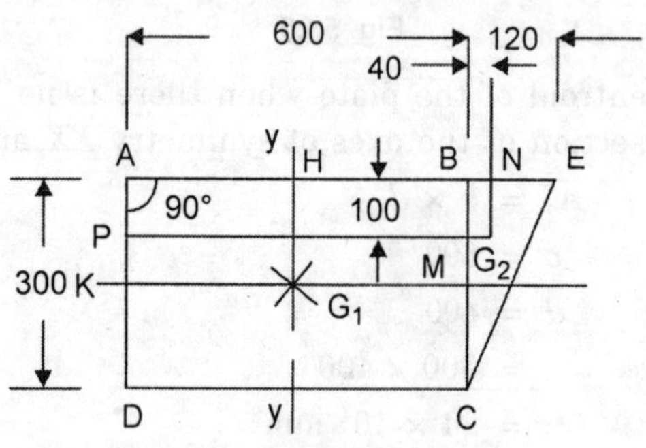

Fig. 5.26

Find the distances of the centroid of the given figure from AD and AB.

Solution :

$$A_1 = \text{Area of rectangle } ABCD$$

$$= 600 \times 300$$

$$= 18 \times 10^4 \text{ mm}^2$$

$$G_1H = 150 \text{ mm}$$

$$KG_1 = 300 \text{ mm}$$

$$A_2 = \text{Area of triangle } BEC$$

$$= \frac{1}{2} \times 300 \times 120 = 18000 \text{ mm}^2$$

$$MG_2 = \frac{1}{3}(BE) = \frac{1}{3} \times 120 = 40 \text{ mm}$$

$\therefore$ $$PG_2 = 600 + 40 = 640 \text{ mm}$$

and $$NG_2 = \frac{1}{3}(BC) = \frac{1}{3} \times 300 = 100 \text{ mm}$$

Therefore, total area = $(A_1 + A_2)$

$= 18 \times 10^4 + 18000$

$= 198 \times 10^3 \text{ mm}^2$

Taking moments of areas about *AD* and applying Varignon's theorem, if $\bar{x}$ is the distance of *G* from *AD*.

$$(A_1 + A_2)\bar{x} = A_1(KG_1) + A_2(PG_2)$$

$$= 18 \times 10^4 \times \frac{600}{2} + 18000 \times 640$$

or $$198 \times 10^3 \bar{x} = 54 \times 10^6 + 11.526 \times 11^6$$

i.e., $$198 \times 10^3 \bar{x} = 65.526 \times 10^6$$

$$\bar{x} = \frac{65.526 \times 10^6}{198 \times 10^3}$$

$$= 33.94 \text{ mm}$$

Again if $\bar{y}$ in the distance of *G* from *AB* taking moments about *AB* of areas,

$$(A_1 + A_2)\bar{y} = A_1(G_1H) + A_2(G_2N)$$

$$= 18 \times 10^4 (150) + 18000 \times 100$$

$$= 27 \times 10^6 + 18 \times 10^5$$

$$198 \times 10^3 \bar{y} = 288 \times 10^5 \text{ mm}^3$$

$$\therefore \quad \bar{y} = \frac{288 \times 10^5}{198 \times 10^3}$$

$$= 145.95 \text{ mm } \textbf{Ans.}$$

EXAMPLE 5.4 : *In Fig. 5.27 is shown a trapezium of height 'h' and base of width 'b'. Find the height of the centroid of the trapezium from AB.*

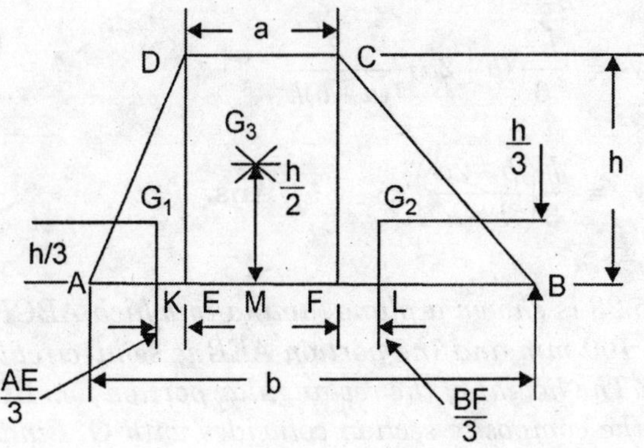

Fig. 5.27

Solution : Given data

$AB = b$

$CD = a$

Height $DE = h$

The given area can be split into 3 areas *viz* triangle *ADE*, Rectangle *DCFE* and triangle *CFB* having their centroids at G_1, G_3 and G_2 respectively such what $G_1$and G_2 are at heights $\frac{h}{3}$ from *AB* and G_3 is at height $\frac{h}{2}$ from *AB*.

Let the centroid *G* of the trapezium be at height $\bar{y}$ from *AB* taking moments of areas *ADE, DEFC* and *CFB* about the base *AB* we get,

$$(\text{Area } ABCD)\,\bar{y} = (\text{Area } AED)\,(G_1K) + (\text{Area } DEFC)\,(G_3A) + (\text{Area } CFB)\,(G_2L)$$

$$\left[\frac{1}{2}(a+b)h\right]\bar{y} = \left[\frac{1}{2}h(AE)\right]\frac{h}{3} + (ah)\frac{h}{2} + \left[\frac{1}{2}h(FB)\right]\frac{h}{3}$$

$$= \frac{h^2}{6}(AE+FB) + \frac{ah^2}{2}$$

$$= \frac{h^2}{6}(b-a) + \frac{ah^2}{2}$$

$$= \frac{h^2}{6}[b-a+3a]$$

$$= \frac{h^2}{6}(b+2a)$$

Therefore, $$\bar{y} = \frac{h^2}{6}(b+2a)\cdot\frac{2}{(a+b)h}$$

or $$\bar{y} = \frac{h}{3}\left(\frac{b+2a}{b+a}\right)... \quad \textbf{Ans.}$$

5.5 : *In Fig. 5.28 is shown a plane lamina in which ABCD is rectangle with side AB = 100 mm and the portion AEB is semi–circular area with AB as diameter. The height of the rectangular portion i.e., BC is such that the centroid of the composite section coincides with O. Find the value of h.*

EXAMPLE

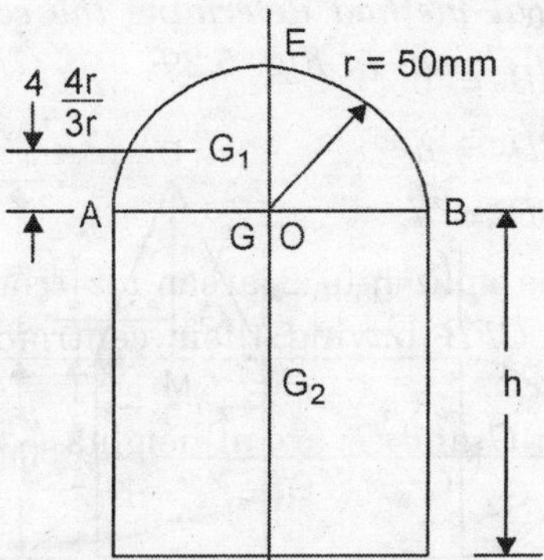

Fig. 5.28

Solution : Given data Radius of semi–circular area,

$$r = \frac{100}{2} = 50 \text{ mm}$$

$\therefore$ Area of semi–circular portion,

$$A_1 = \frac{1}{2}(\pi r^2) = \frac{1}{2}\pi(50)^2$$

$$= 3928.99 \text{ mm}^2$$

From equation (5.25), we have the distance of centroid of semi–circular area from O.

$$OG_1 = \frac{4r}{3\pi} = \frac{4 \times 50}{3 \times \pi} = 21.22 \text{ mm}$$

For this rectangular portion *ABCD*,

$$A_2 = AB \times h = 100\,h$$

and $$OG_2 = \frac{1}{2}h$$

If we take *AB* as the axis of reference and since the centroid of the composite section is to be on it the moments of area A_1 and A_2 which are a opposite sides of *AB* must balance,

i.e., $$A_1(OG_1) = A_2(OG_2)$$

or $$(3926.99 \times 21.22) = (100\,h)\frac{4}{2}$$

$$= 50\,h^2$$

$$\therefore \quad h^2 = \frac{3928.99 \times 21.22}{50} = 1660.614$$

$$\therefore \quad h = 40.824 \text{ mm}$$ **Ans.**

5.6 : *Using analytical method determine the centre of gravity of the plane uniform lamina shown in Fig. 5.29.* *(U.P. Tech Un. 2001-02)*

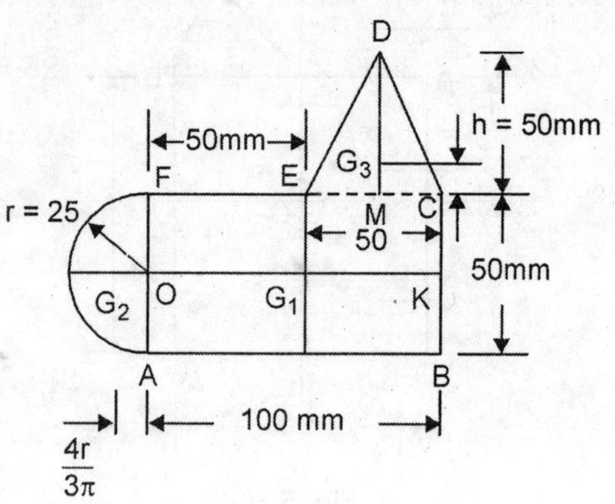

Fig. 5.29

Solution : The lamina contains 3 areas *viz.* rectangular portion A_1 semi–circular portion A_2 and triangular portion A_3.

Given Data $\quad r = \dfrac{50}{2} = 25$ mm

$$A_1 = 100 \times 50 = 5000 \text{ mm}^2$$

$$AG_1 = 50$$

$$A_2 = \frac{1}{2}(\pi r^2) = \frac{\pi(25)^2}{2} = 981.748 \text{ mm}^2$$

$$OG_2 = \frac{4R}{3\pi} = \frac{4 \times 25}{3\pi} = 10.01 \text{ mm}$$

$$G_2K = 10.01 + 100 = 110.01 \text{ mm}$$

$$A_3 = \frac{1}{2}(CE)\,DM$$

Now $\quad CE = \text{CF} - EF = 100 - 50 = 50$

and $\quad A_3 = \dfrac{1}{2}(CE) \times \text{height of } \Delta CDE$

$$= \frac{1}{2} \times 50 \times 50 = 1250 \text{ mm}^2$$

$$G_3M = \frac{1}{3} \times \text{height of } \Delta DEC = \frac{1}{3} \times 25 = \frac{25}{3}$$

$$= 8.3333 \text{ mm}$$

EXAMPLE

Taking moments about AB, if $\bar{y}$ is the distance of *C.G from* AB

$$\bar{y}(A_1 + A_2 + A_3) = A_1 \times 25 + A_2 \times 25 + A_3\,(8.3333 + 50)$$

$$\bar{y}(500 + 981.748 + 1250) = 5000 \times 25 + 981.748 \times 25 + 1280 \times 58.3333$$

$$= 125000 + 24543.7 + 72916.625$$

$$7231.748\,\bar{y} = 222460.32$$

$$\therefore \quad \bar{y} = \frac{222460.32}{7231.748}$$

$$= \mathbf{30.76}\ \text{mm}$$

Now take moments about BC, if $\bar{x}$ is the distance of CG from BC

$$\bar{x}(5000 + 981.748 + 1250) = A_2 \times (G_2K) + A_3(MC) + A_1 \times (G_1K)$$

$$= (981.748 \times 110.01) + (1250 \times 25) + (5000 \times 50)$$

$$7231.746\,\bar{x} = 108002.1 + 31250 + 250000$$

$$= 389252.1$$

$$\bar{x} = \frac{389252.1}{7231.746}$$

$$= 53.825 \text{ mm } \textbf{Ans.}$$

5.7 : *Determine the location of centroid of the I-section shown in Fig. 5.30.*

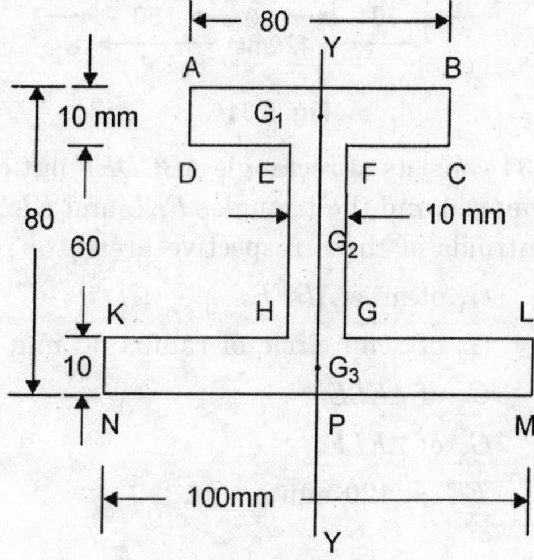

Fig. 5.30

Solution : Since YY is the axis of symmetry the centroid will be on it. Therefore, it is required to find the distance of G from the line NM.

Given data A_1 = Area $ABCD$ = 80 × 10 = 800 mm^2

A_2 = Area $EFGH$ = 60 × 10 = 600 mm^2

A_3 = Area $KLMN$ = 100 × 10 = 1000 mm^2

Let $\bar{y}$ be the distance of CG from NM. Then moments about NM give,

$$(A_1 + A_2 + A_3)\bar{y} = A_1(G_1P) + A_2(G_2P) + A_3(G_3P)$$

$$(800 + 600 + 1000)\bar{y} = 800\ (80 - 5) + 600\ (10 + 30) + 1000 \times 5$$

or

$$2400\bar{y} = 60000 + 24000 + 5000$$

$$= 89000$$

$$\therefore \quad \bar{y} = \frac{89000}{2400} = 37.08 \text{ mm}$$

5.8 : *Find the co-ordinates of the centroid of the plane area shown in Fig. 5.31, if the axes of reference are taken along FA and AD.*

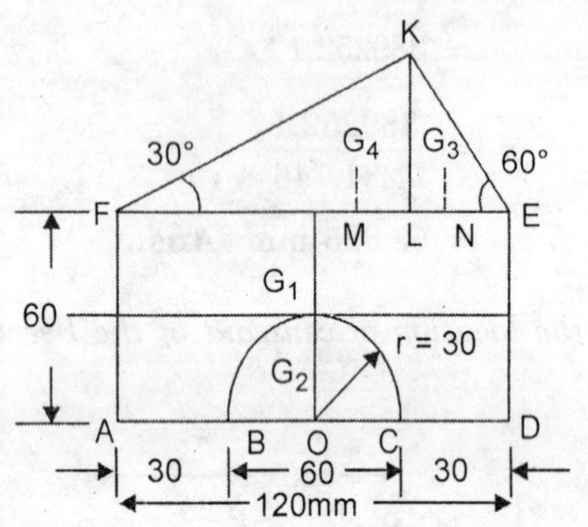

Fig. 5.31

Solution : Fig. 5.31 consists of rectangle $ABCDEF$ out of which a semi-circular area is removed and the triangles FKL and KLE. Let G_1, G_2, G_3 and G_3 be the centroids of these respective areas,

G_1 of area $ADEF$

G_2 of semi circle of radius 30 mm

G_3 of ΔKLE

G_4 of ΔKLF

Given data FE = 120 mm

$$FK = FE\cos 30° = 120 \times \frac{\sqrt{3}}{2} = 103.92$$

and

$$KE = EF\sin 30° = 120 \times \frac{1}{2} = 60$$

EXAMPLE

$$\therefore \quad FL = FK\cos 30 = 103.92 \times \frac{\sqrt{3}}{2} = 89.99 \text{ mm}$$

$$KL = FK\sin 30 = 103.92 \times \frac{1}{2} = 51.96$$

$$\therefore \quad G_4M = G_3N = \frac{1}{3} \times KL = \frac{51.96}{3} = 17.32$$

and $$ML = \frac{1}{3}FL = \frac{1}{3} \times 89.99 = 29.9967$$

$$FM = FL - ML$$
$$= 89.99 - 29.9967 = 59.99$$

$$LN = \frac{1}{3}LE$$

$$= \frac{1}{3}[120 - FL]$$

$$= \frac{1}{3}(120 - 89.99)$$

$$= (10.003) \text{ mm}$$

$$OG_1 = 60/2 = 30 \text{ mm}$$

$$OG_2 = \frac{4r}{3\pi} = \frac{4 \times 30}{3 \times \pi} = 12.73 \text{ mm}$$

For finding $\bar{x}$ take moments about AF,

$$(A_1 - A_2 + A_3 + A_4)\bar{x} = A_1x_1 - A_2x_2 + A_3x_3 + A_4x_4 \quad ...(A)$$

Now $$A_1 = 120 \times 60 = 7200 \text{ mm}^2$$

$$A_2 = \frac{1}{2}\pi r^2 = \frac{1}{2} \times \pi \times 30^2 = 1413.72$$

$$A_3 = \Delta KLE = \frac{1}{2} \times LE \times KL$$

$$= \frac{1}{2}(120 - FL) \times 51.96$$

$$= \frac{1}{2}(120 - 89.99) \times 51.96$$

$$= 779.66 \text{ mm}^2$$

$$A_4 = \text{Area } FKL$$

$$= \frac{1}{2} FL \times KL$$

$$= \frac{1}{2}(89.99)51.96$$

$$= 2337.94 \text{ mm}^2$$

$$x_1 = \frac{1}{2} \times 120 = 60$$

$$x_2 = 60$$

$$x_3 = FN = FL + LN$$

$$= 89.99 + 100.003$$

$$= 99.993 \text{ m}$$

$$x_4 = FM = 59.993$$

Also $$y_1 = G_1O = \frac{1}{2} \times 60 = 30$$

$$y_2 = OG_2 = 12.73$$

$$y_3 = 60 + G_3N$$

$$= 60 + 17.32 = 77.32$$

$$y_4 = 60 + GM$$

$$= 60 + 17.32 = 77.32$$

Then from equation (*A*),

$$\bar{x}(7200 - 1413.72 + 779.66 + 2337.94)$$

$$= 7200 \times 60 - 1413.72 \times 60 + 779.66 \times 99.99 + 2337.94 \times 59.993$$

i.e., $$8903.8\bar{x} = 432000 - 848232 + 77911.4 + 140112.74$$

$$= 495200.9$$

$$\bar{x} = \frac{495200.9}{8903.8}$$

$$= 55.61 \text{ mm}$$

Taking moments about *ABD*,

$$(A_1 - A_2 + A_3 + A_4)\bar{y} = A_1y_1 - A_2y_2 + A_3y_3 + A_4y_4$$

$$8903.8\bar{y} = 7200 \times 30 - 1413.72 \times 12.73 + 779.66 \times 77.32 + 2337.94 \times 77.32$$

$$= 216000 - 17996.6 + 60283.3 + 18076.9$$

$$= 276363.6$$

$$\bar{y} = 31.039 \text{ mm}$$

EXAMPLE

5.6 SECOND MOMENT OF PLANE *AREA*

5.6.1 (*AREA* Moment of Inertia)

Definitions and theorem of perpendicular axes

Consider an elemental area dA Fig. (3.32 a) at a distance r from the line AB. Then the term rdA is called the first moment of area about AB just similar to the term moment of a force. The first moment of the entire area will be $\oint r\,dA$. Similarly, the term $\oint r^2\,dA$ is known as the second moment of area. However, it is a popularly called moment of inertia even through it is a misnomer as the term moment of intertia is to be associated with the mass of the body. Therefore, in order to avoid confusion, it is called 'second moment of area" or "Area moment of Inertia.

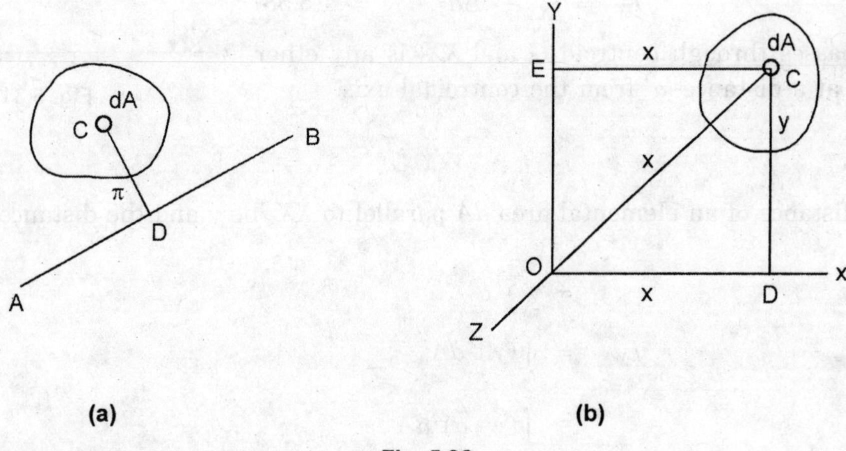

Fig. 5.32

Now consider the area dA with reference to axes OX and OY. That is dA has co-ordinates (x, y).

Then $\oint x^2\,dA$ is called the second moment of area about y-axis and is represented by I_{yy}

Similarly, $\oint y^2\,dA$ is known as second moment of area about x-axis and is represented by I_{xx} when the area lies in XY plane.

However, if dA is situated at a distance r from z-axis which is normal to plane in which the area lies, then the term $\oint r^2\,dA$ is called the "Polar second moment area" or "Area Polar Moment of Inertia".

Evidently, $$r^2 = x^2 + y^2$$

$$\therefore \qquad \oint r^2 dA = \oint (x^2 + y^2)\,dA$$

$$I_{ZZ} = \oint x^2 dA + \oint y^2 dA$$

$$= I_{yy} + I_{xx}$$

i.e.,
$$I_{ZZ} = I_{XX} + I_{YY} \quad ...(5.57)$$

Equation (5.57) is known as theorem of perpendicular axes

5.6.2 Theorem of Parallel Axes

Statement

Second moment of area about an axis in the plane of the area is equal to the sum of second moment of area about parallel centroidal axis and the product of area and square of the distance between the two parallel axes. Mathematically, the statement is,

$$I_{XX'} = I_{XX} + Aa^2 \quad ...(5.58)$$

where *GG′* passes through centroid *G* and *XX′* is any other parallel axis at a distance *'a'* from the centroidal axis.

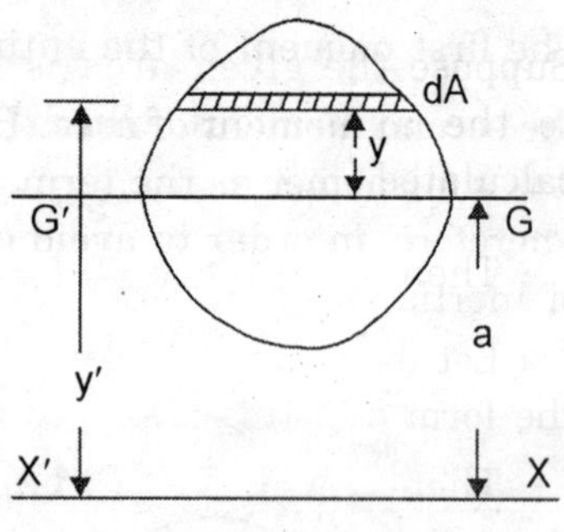

Fig. 5.31

Proof

Let the distance of an elemental area *dA* parallel to *XX* be *y* and the distance from *XX′* is *y′*.

Evidently,
$$y' = y + a$$

Then
$$I_{XX'} = \oint (y')^2 dA$$

$$= \oint (y+a)^2 dA$$

$$= \oint [y^2 + a^2 + 2aydA]$$

or
$$I_{XX'} = \oint y^2 dA + a^2 \int dA + 2a \int dA\, y \quad ...(5.59)$$

The term $\oint y dA$ is nothing but first moment of area about *Ge′* which is postioning through centroid *G* that its distance is zero from *CG*.

Thus,
$$\oint y dA = 0$$

Therefore, equation (5.59) reduces to,

$$I_{XX'} = \oint y^2 \cdot dA + a^2 A$$

$$= I_{XX} + a^2 A \quad ...(5.60a)$$

or
$$= I_{GG} + a^2 A. \text{ Hence, the Proof} \quad ...(5.60b)$$

While using the expression (5.60 *a*) sometimes the students commit a serious mistake and they use I_{XX} for any axis other than centroidal axis. This will be wrong. Therefore, the form (5.60 *b*) is better as I_{GG} will automatically indicate that it is second moment of area about an axis passing through *CG*.

5.7 RADIUS OF GYRATION

Suppose the given area is split up into small strips parallel to the axis about which second moment of are is to be calculated.

Then $$I = A_1r_1^2 + A_2r_2^2 + A_3r_3^2 \quad ...(5.61)$$

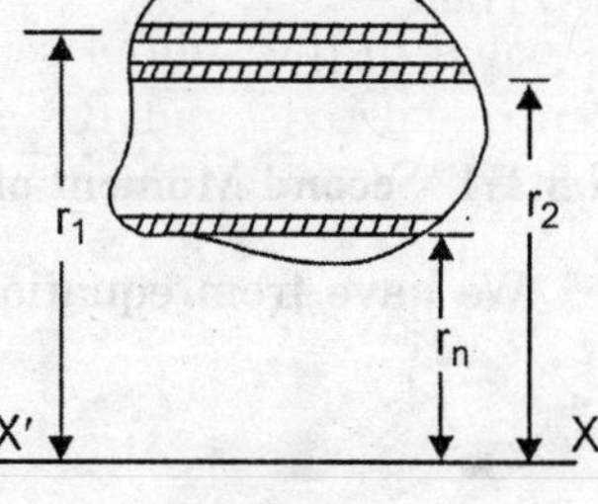

Fig. 5.34

Let us assume that whole (mass) or area is concentrated in the form of a this strip of area *A* at distance *k*

Then $$I = Ak^2 \quad ...(5.62)$$

Where, $$A = A_1 + A_2 + A_3. = \sum_{i=1}^{n} A_1$$

Thus, equating (5.61) and (5.62),

$$k = \sqrt{\frac{I}{A}} \quad ...(5.63)$$

This *k* is known as radius of gyration. The units of *I* will be mm^4 and or *A* = mm^2. Evidently *k* will be in mm.

5.8 SECOND MOMENT OF AREA OF STANDARD CASES (MOMENT OF INERTIA)

5.8.1 Rectangle

In Fig. 5.45 is shown a rectangle *ABCD* of length '*l*' and depth '*d*'. It has two axes of symmetry *XX, YY* with 0 considering with the centroid *a*.

We now proceed to find the second moment of area about *XX* axis. Take a strip *EF* of depth *dy* at a height '*y*' from the *x*-axis.

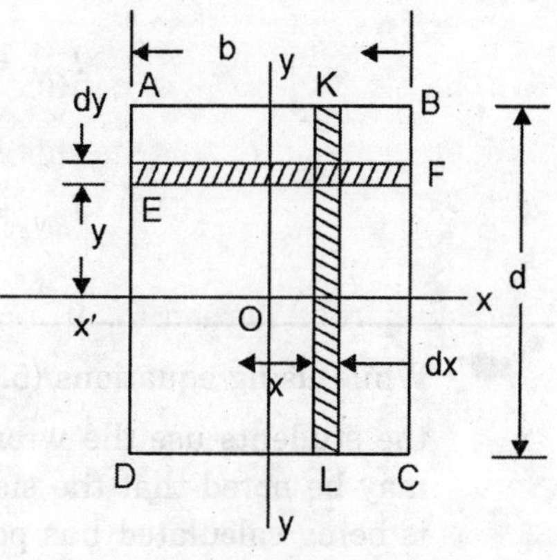

Fig. 5.35

Then $$d\,I_{xx} = dA\,y^2$$

or $$I_{XX} = \int_{d/2}^{-d/2} dA\,y^2 = \int_{-d/2}^{d/2} y^2 (b\,dy)$$

$$= \left[\frac{by^3}{3}\right]_{-\frac{d}{2}}^{d/2} = \frac{b}{3}\left[\frac{d^3}{8} - \left(-\frac{d^3}{8}\right)\right]$$

$$= \frac{bd^3}{3 \times 4} = \frac{bd^3}{12} \qquad \text{...(5.64}a\text{)}$$

For calculating I_{yy} take elemental strip of width dx parallel to y-xis at a distance x.

Then $$I_{yy} = \int_{-h/2}^{h/2} x^2\, dA = \int_{-h/2}^{h/2} x^2\, d(dx) = d\left[\frac{2^3}{3}\right]_{-h/2}^{h/2} = \frac{dh^3}{12}$$

5.8.1.1 Second Moment of Area of a Rectangle about its sides

We have from equation (5.61),

$$I_{XX} = \frac{BD^3}{12}$$

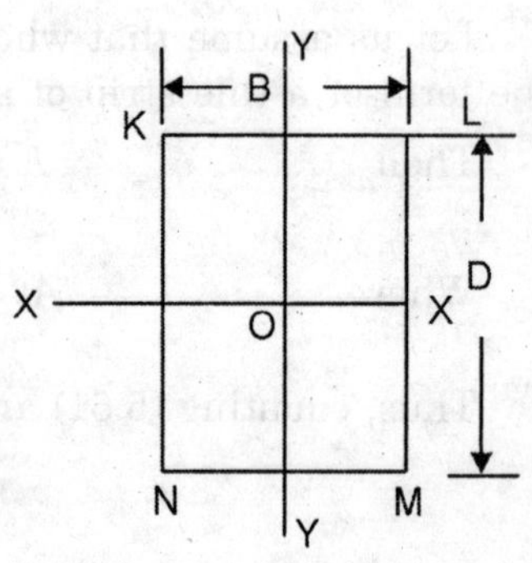

Fig. 5.36 (a)

Using theorem of parallel axes,

$$I_{NM} = I_{XX} + A\left(\frac{1}{2}LM\right)^2$$

$$= \frac{BD^3}{12} + (BD)\left(\frac{1}{2}D\right)^2$$

$$= \frac{BD^3}{12} + \frac{BD^3}{4} = \frac{BD^3}{3} \qquad \text{...(5.64}b\text{)}$$

Similarly, $$I_{YY} = \frac{DB^3}{12}$$

$$I_{KN} = I_{YY} + A\left(\frac{1}{2}B\right)^2 = I_{yy} + BD\left(\frac{B}{2}\right)^2$$

$$I_{KN} = \frac{DB^3}{12} + \frac{B^3D}{4} = \frac{B^3D}{3} \qquad \text{...(5.66)}$$

☛ While using equations (5.64) and (5.65) one has to be very careful. Most of the time the students use the wrong value for I_{XX} and I_{YY} as the expressions are simular. It may be noted that the side parallel to the axis about which area moment of inertia is being calculated has power 3.

5.8.2 Second Moment of Square Area

In equation (5.64) and (5.63),

If we put $b = d = a$ (say)

Then $I_{XX} = I_{YY} = \dfrac{a^4}{12}$

If I_{zz} is moment of inertia about *zz* axis, then from the theorem of perpendicular axes, we have,

Fig. 5.36 (b)

$$I_{ZZ} = I_{XX} + I_{YY}$$

$$= 2I_{XX} \qquad [\because I_{XX} = I_{YY}] \qquad ...(5.67)$$

Further for a square section, the diagonals are perpendicular to each other

$$\therefore \quad I_{ZZ} = I_{DB} + I_{CA}$$

$$= 2I_{DB} = 2I_{CA} \qquad ...(5.68)$$

$\therefore$ From (5.67) and (5.68),

We get, $I_{XX} + I_{YY} = I_{ZZ}$

or $2I_{XX} = I_{ZZ} = 2I_{CA}$

or $I_{XX} = I_{YY} = I_{AC} = I_{BD}$

$$= \frac{a^4}{12} \qquad ...(5.69)$$

5.8.3 Second Moment of Area of a Triangle about its Base

Fig. 5.37 shows a triangle *ABC* whose base *BC* is of length '*l*' and height of the vertex *A* is '*h*'.

Geometrically $AD = h$

$DF = y$

$AF = (h - y)$

EFK is an elemental area *dA* of thickness *dy* at a distance *y* from the base *BC*.

From similar triangles *ABC* and *AEK*,

$$\frac{EK}{BC} = \frac{AF}{AD}$$

$$\therefore \quad EK = \left(\frac{h-y}{h}\right) \times BC$$

$$= b\left(1 - \frac{y}{h}\right)$$

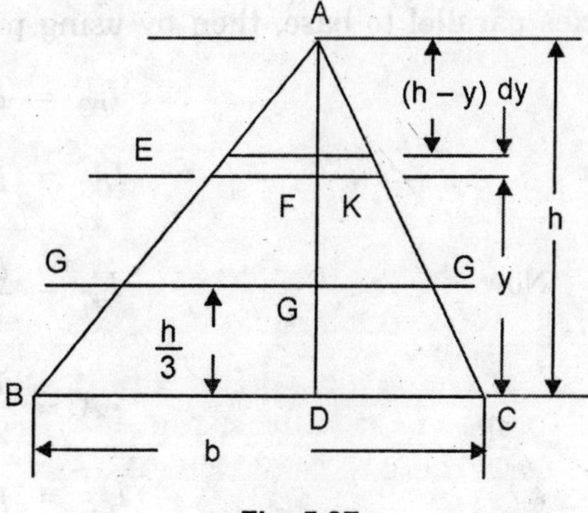

Fig. 5.37

$$\therefore \quad I_{BC} = \int_{y=0}^{h} y^2 dA$$

$$= \int_{y=0}^{h} y^2 \cdot (EK)\, dy$$

$$= \int_0^h y^2 \left[b\left(1 - \frac{y}{h}\right) \right] dy$$

$$= b \int_{y=0}^{h} \left(y^2 - \frac{y^3}{h} \right) dy$$

$$= h\left[\frac{y^3}{3} - \frac{y^4}{4h} \right]_0^h$$

i.e.,

$$I_{BC} = b\left[\frac{h^3}{3} - \frac{h^4}{4h} \right]$$

$$= bh^3\left(\frac{1}{3} - \frac{1}{4} \right)$$

$$= \frac{bh^3}{12} \quad ...(5.70)$$

5.8.3.1 Corollary

It has been proved earlier in equation (5.19). The centroid of a triangle is at a height of $\frac{h}{3}$ from the base and the area of the triangle is $\frac{1}{2}bh$.

Therefore, of I_{GG} which is called the second moment of triangular area about centroidal axes parallel to base, then by using principle of parallel axes, we have

$$I_{BC} = I_{GG} + A(\overline{y})^2$$

or

$$I_{GG} = I_{BC} - A(\overline{y})^2$$

Now

$$I_{BC} = \frac{bh^3}{12} \quad ...(5.70)$$

$$A = \frac{1}{2}bh$$

$$\therefore \quad I_{GG} = I_{BC} - A(\overline{y})^2$$

or $$I_{GG} = \frac{bh^3}{12} - \left(\frac{bh}{2}\right)(\bar{y})^2$$

$$= \frac{bh^3}{12} - \frac{bh}{2}\cdot\left(\frac{h}{3}\right)^2$$

$$= \frac{bh^3}{12} - \frac{bh^3}{18} = bh^3\left(\frac{b}{12+18}\right)$$

or $$I_{GG} = \frac{bh^3}{36} \qquad \text{...(5.71)}$$

5.8.4 Second Moment of Area of Hollow Rectangular Section

Fig. 5.51 shows a hollow rectangular section which is symmetrical about both the *XX* and *YY* axis.

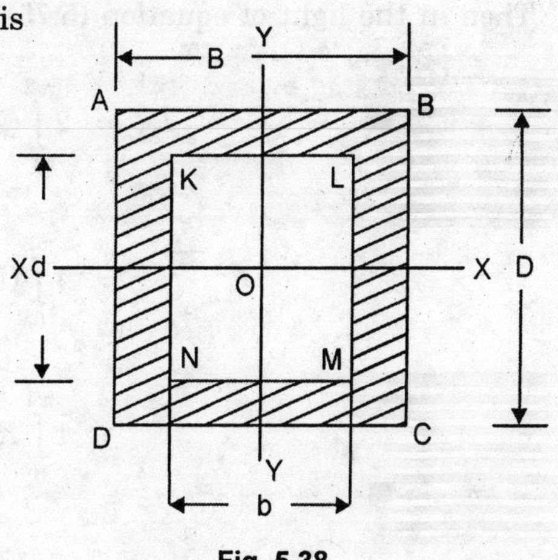

Fig. 5.38

$$I_{XX} \text{ of solid rectangle } ABCD = \frac{BD^3}{12}$$

$$I_{XX} \text{ of solid rectangle } KLMN = \frac{bd^3}{12}$$

$\therefore$ I_{XX} of the given hollow section

$$= \frac{BD^3}{12} - \frac{bd^3}{12} \qquad \text{...(5.72)}$$

Similarly,

$$I_{YY} \text{ of the hollow section } = \frac{DB^3}{12} - \frac{db^3}{12} \qquad \text{...(5.73)}$$

5.8.5 Second Moment of Area of Plane Circular Section

First Method

Fig. 5.39 shows a circular section of radius *R*. Since the area is symmetrical about any diameter whether along *X*-axis or along *Y*-axis. The origin itself will be the *CG* of the section. Take an elemental area *ABC* parallel to *X*-axis of thickness *dy* and at a distance of *y* from *X*-axis. Then

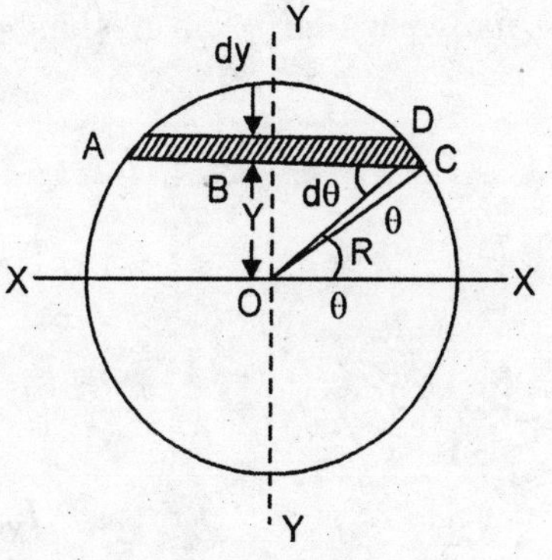

Fig. 5.39

$$I_{xx} = \int_{y=-d/2}^{d/2} y^2 dA$$

$$= \int_{y=-\frac{d}{2}}^{d/2} (AC)\,dy\,y^2$$

$$I_{XX} = \int_{y=-\frac{d}{2}}^{d/2} 2(BC)\,dy\,y^2 \qquad ...(5.74)$$

If we assume that the radial line OC is inclined at an angle θ with X-axis,

$$y = R \sin \theta \qquad ...(5.75a)$$

$$dy = R \cos \theta \, d\theta \qquad ...(5.75b)$$

Also $BC = R \cos \theta$

Then in the light of equation (5.75), equation (5.74) reduces to,

$$I_{XX} = 2\int_{0}^{\pi/2} (2BC)\,y^2 dy \quad (\because \text{ section is symmetrical about } Y\text{-axis})$$

$$= 4\int_{0}^{\pi/2} (\mathrm{R}\cos\theta)(R^2 \sin^2\theta)(R\cos\theta)\,d\theta$$

$$= 4\int_{0}^{\pi/1} R^4 \sin^2\theta\cos^2\theta\,d\theta$$

$$= R^4 \int_{\theta=0}^{\pi/2} (4\sin^2\theta\cos^2\theta)\,d\theta$$

$$= \frac{R^4}{2}\int_{0}^{\pi 2} 2(\sin 2\theta)^2 d\theta$$

$$= \frac{R^4}{2}\int_{\theta=0}^{\pi/2} (1-\cos 4\theta)\,d\theta$$

$$= \frac{R^4}{2}\left[\theta + \frac{\sin 4\theta}{4}\right]_0^{\pi/2} = \frac{\pi R^4}{4}$$

$$I_{XX} = \frac{\pi}{4}\left(\frac{D}{2}\right)^4 = \frac{\pi D^4}{64} \qquad ...(5.76)$$

Second Method

In this method we take elemental area in the form of a ring of radial thickness *dr* situated at a distance of *r* from *ZZ*-axis.

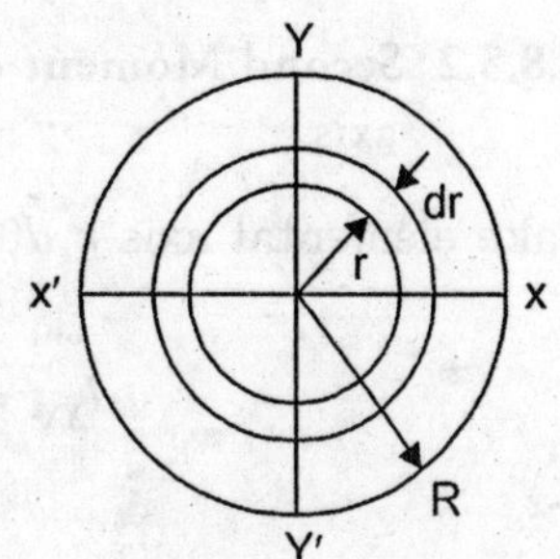

Fig. 5.40

Then $$I_{ZZ} = \int_0^R r^2 dA = \int_{r=0}^R r^2 (2\pi r\, dr)$$

$$= 2\pi \int_{r=0}^R r^3\, dr$$

$$\therefore \quad I_{ZZ} = 2\pi \left[\frac{r^4}{4}\right]_0^R = \frac{2\pi}{4} R^4 = \frac{\pi R^2}{2}$$

But we know from the theorem of perpendicular axes that,

$$I_{ZZ} = I_{XX} + I_{YY}$$

Also, $$I_{XX} = I_{YY}$$

$$\therefore \quad I_{ZZ} = 2I_{XX}$$

or $$I_{XX} = \frac{1}{2}(I_{ZZ})$$

$$= \frac{1}{2}\left(\frac{\pi R^4}{2}\right) = \frac{\pi R^4}{4}$$

$$= \frac{\pi D^4}{64} \qquad \text{...(5.78)}$$

☛ Evidently the second method is much simpler and less time consuming. Therefore, it is suggested that the students follow this method in the examination unless it is stated otherwise in the question.

5.8.5.1 Corollary

In case of a hollow plane circular section as shown in Fig. 5.41.

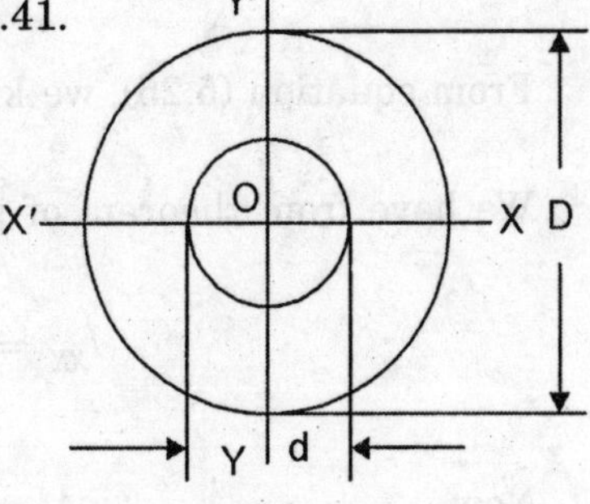

Fig. 5.41

$$I_{XX} = I_{YY} = \frac{\pi D^4}{64} - \frac{\pi d^4}{64}$$

and $$I_{ZZ} = I_{XX} + I_{YY}$$

$$= 2\, I_{XX}$$

$$= 2\left[\frac{\pi D^4}{64} - \frac{\pi d^4}{64}\right]$$

$$I_{ZZ} = \frac{\pi}{32}(D^4 - d^4) \qquad \text{...(5.79)}$$

5.8.5.2 Second Moment of Area of a Semi–circular Area about base and Centroidal axis

Take elemental axis $r\,d\theta\,dr$ as shows in Fig. 5.42

$$I_{XX} = \int_{r=0}^{R}\int_{\theta=0}^{\pi} (r\,d\theta\,dr)(r\sin\theta)^2$$

$$= \int_{r=0}^{R}\int_{\theta=0}^{\pi} r^3 \sin^2\theta\,d\theta\,dr$$

$$= \int_{r=0}^{R}\left[\frac{r^3}{2}\int_{0}^{\pi}(2\sin^2\theta)\,d\theta\right]dr$$

$$= \int_{r=0}^{R}\frac{r^3}{2}\int_{0}^{\pi}(1-\cos 2\theta)\,d\theta\,dr$$

$$= \int_{r=0}^{R}\frac{r^3}{2}\left[\theta - \frac{\sin 2\theta}{2}\right]_0^{\pi} dr$$

$$= \int_{0}^{R}\frac{r^3}{2}\,[\pi - 0 + 0]\,dr$$

$$= \frac{\pi}{2}\left[\frac{r^4}{4}\right]_0^{R} = \frac{\pi R^4}{8}$$

$$= \frac{\pi}{8}\left(\frac{D}{2}\right)^4 = \frac{\pi D^4}{128} \qquad \text{...(5.80)}$$

Fig. 5.42

From equation (5.25), we know that the CG of semicircle is at a distance $\frac{4R}{3\pi}$ from XX.

We have from theorem of parallel axes,

$$I_{XX} = I_{GG} + A\left(\frac{4R}{3\pi}\right)^2$$

Now $$A = \frac{\pi R^2}{2}$$

$$\therefore \qquad I_{GG} = I_{XX} - A\left(\frac{4R}{3\pi}\right)^2$$

$$= \frac{\pi R^4}{8} - \frac{\pi R^2}{2} \times \frac{16R^2}{9\pi^2}$$

$$= \frac{\pi R^4}{8} - \pi R^4 \cdot \frac{8}{9\pi}$$

$$I_{GG} = \frac{\pi R^4}{8}\left[1 - \frac{64}{9\pi^2}\right] \qquad \text{...(5.81a)}$$

$$= \frac{\pi R^4}{8}\left[1 - \frac{64}{9\pi^2}\right]$$

$$= \left[\frac{\pi D^4}{128} - \frac{D^4}{18\pi}\right] \qquad \text{...(5.81b)}$$

5.8.5.3 Area Moment of Inertia of Quarter Circle

In Fig. 5.56, an elemental area is shown,

$$dA = (r\,d\theta)\,dr$$

located at an angle θ with OX. Therefore,

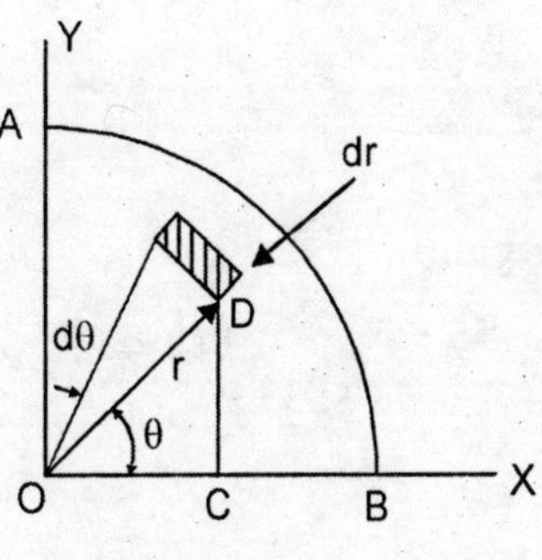

Fig. 5.43

$$I_{XX} = \int_{r=0}^{R}\int_{\theta=0}^{\pi/2} (r\,d\theta)\,dr\,(r\sin\theta)^2$$

$$= \int_{r=0}^{R} r^3 \int_{\theta=0}^{\pi/2} \sin^2\theta\, d\theta\, dr$$

$$= \int_{r=0}^{R} \frac{r^3}{2} \int_{0}^{\pi/2} 2\sin^2\theta\, d\theta\, dr$$

or

$$I_{XX} = \int_{r=0}^{R} \frac{r^3}{2} \int_{0}^{\pi/2} (1 - \cos 2\theta)\, d\theta\, dr$$

$$= \int_{r=0}^{R} \frac{r^3}{2}\left[\theta - \frac{\sin 2\theta}{2}\right]_0^{\pi/2} dr$$

$$= \int_{r=0}^{R} \frac{r^3}{2}\left[\frac{\pi}{2} - 0 + 0\right] dr$$

$$= \int_{r=0}^{R} \frac{\pi r^3}{4} dr = \frac{\pi}{16}\left[r^4\right]_0^R$$

$$I_{XX} = \frac{\pi R^4}{16} \qquad \text{...(5.82}a\text{)}$$

$$= \frac{\pi}{16} \times \left(\frac{D}{2}\right)^4 = \frac{\pi D^4}{256} \qquad \text{...(5.82}b\text{)}$$

For finding I_{GG} parallel to X-X,

We have $A = \dfrac{\pi R^2}{4}$

Distance of G from X-axis $\bar{y} = \dfrac{4R}{3\pi}$

Now $I_{XX} = I_{GG} + A(\bar{y})^2$

$\therefore$ $I_{GG} = I_{XX} - A(\bar{y})^2$

$$= \frac{\pi R^4}{16} - \frac{\pi R^2}{4}\left(\frac{4R}{3\pi}\right)^2$$

$$= \frac{\pi R^4}{16} - \pi R^4 \frac{4}{9\pi^2}$$

or $I_{GG} = \dfrac{\pi R^4}{16} - \pi R^4 \times \dfrac{4}{9\pi^2}$

or $I_{GG} = \dfrac{\pi R^4}{16}\left[1 - \dfrac{64}{9\pi^2}\right]$ ...(5.83*a*)

$$= \frac{\pi D^4}{256} - \frac{\pi D^4}{256} \cdot \frac{64}{9\pi^2}$$

$$I_{GG} = \left(\frac{\pi D^4}{256} - \frac{D^4}{36\pi}\right) \qquad \text{...(5.83}b\text{)}$$

Table 5.2. Second Moment of Area of Some Standard Plane Figures

Figure	I
Rectangle (B, D, X–X′, y–y′)	$I_{XX} = \dfrac{BD^3}{12}$ $I_{YY} = \dfrac{DB^3}{12}$
Hollow rectangle (B, D, b, d, O, x–x′, y–y′)	$I_{XX} = \dfrac{BD^3}{12} - \dfrac{bd^3}{12}$ $I_{XX} = \dfrac{DB^3}{12} - \dfrac{db^3}{12}$
Hollow circle (d, D, O, X–X, Y–Y)	$I_{XX} - I_{YY} = \dfrac{\pi D^4}{64} - \dfrac{\pi D^4}{64}$ $I_{ZZ} = \dfrac{\pi D^4}{32} - \dfrac{\pi d^4}{32}$
Triangle (A, B, C, b, h, G–G, CG, h/3)	$I_{BC} = \dfrac{bh^3}{12}$ $I_{GG} = \dfrac{bh^3}{36}$
Semi circle (R, O, G–G, X–X, $\frac{4R}{3\pi}$)	$I_{XX} = \dfrac{\pi D^4}{128}$ $I_{GG} = \left(\dfrac{\pi D^4}{128} - \dfrac{D^4}{18\pi}\right)$
Quarter circle (R, G, X, $\frac{4R}{3\pi}$, $\frac{4R}{3\pi}$)	$I_{XX} = \dfrac{\pi D^4}{256}$ $I_{GG} = \left(\dfrac{\pi D^4}{256} - \dfrac{D^4}{36\pi}\right)$

5.9 AREA MOMENT OF INERTIA OF COMPOSITE SECTIONS

5.9.1 Area Moment of Inertia of T-section

5.9 : *Find the centroidal moment of Inertia of the T-section shown in Fig. 5.44 about horizontal centroidal aixs.*

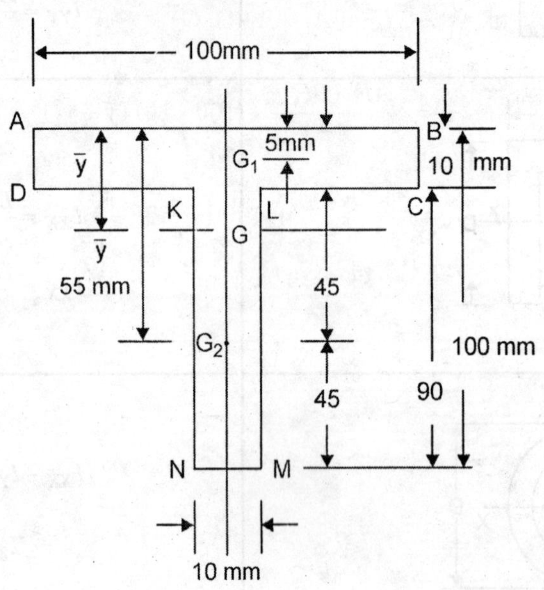

Fig. 5.44

Solution : Given composite section consists of the rectangular areas namely, *ABCD* and *KLMN*. Let the areas of these rectangles be A_1 and A_2 with their centroids at G_1 and G_2 respectively. Assume that the centroid of the whole composite area be of point *G* which is at a distance of $\bar{y}$ from *AB*.

Then taking moments about *AB* of the respective areas we have,

$$A_1 = 100 \times 10 = 1000 \text{ mm}^2$$

$$y_1 = \frac{1}{2} \times 10 = 5 \text{ mm}$$

$$A_2 = 90 \times 10 = 900 \text{ mm}^2$$

$$y_2 = 100 - 45 = 55 \text{ mm}$$

Total area $A = A_1 + A_2$

$$= 1000 + 900 = 1900 \text{ mm}^2$$

$$A\bar{y} = A_1 y_1 + A_2 y_2$$

or $$1900\bar{y} = 1000 \times 5 + 900 \times 55$$

EXAMPLE

$$1900\,\bar{y} = 5000 + 49500$$

or $$\bar{y} = \frac{54500}{1900} = 28.68 \text{ mm}$$

Applying the theorem of parallel axis,

$$I_{GG} = I_{GG} \text{ of } A_1 + I_{GG} \text{ of } A_2 \qquad ...(A)$$

But $$I_{GG} \text{ of } A_1 = I_{G_1} \text{of } A_1 + A_1 (G_1 G)^2$$

$$= \frac{AB \cdot (BC)^3}{12} + (100 \times 10)\,(\bar{y} - y_1)^2$$

$$= \frac{100 \times 10^3}{12} + 1000(28.68 - 5)^2$$

$$= 8.33 \times 10^3 + 560.74 \times 10^3$$

$$= 569.07 \times 10^3 \text{ mm}^4$$

And $$I_{GG} \text{ of } A_2 = I_{G_2G_2} \text{of } A_2 + A_2 (GG_2)^2$$

$$= \frac{10 \times 90^3}{12} + (90 \times 10)(y_2 - \bar{y})^2$$

$$= 607.5 \times 10^3 + 900\,(55 - 28.68)^2$$

$$= 607.5 \times 10^3 + 623.468 \times 10^3$$

$$= 1230.968 \times 10^3 \qquad ...(B)$$

Therefore, Adding (A) and (B) we get,

$$I_{GG} = 569.07 \times 10^3 + 1230.968 \times 10^3$$

$$= 1800.0 \times 10^3 \text{ mm}^4 \textbf{ Ans.}$$

EXAMPLE **5.10 :** *Find the area moment of inertia of the section shown in Fig. 5.45 about centroidal axis parallel to AB.*

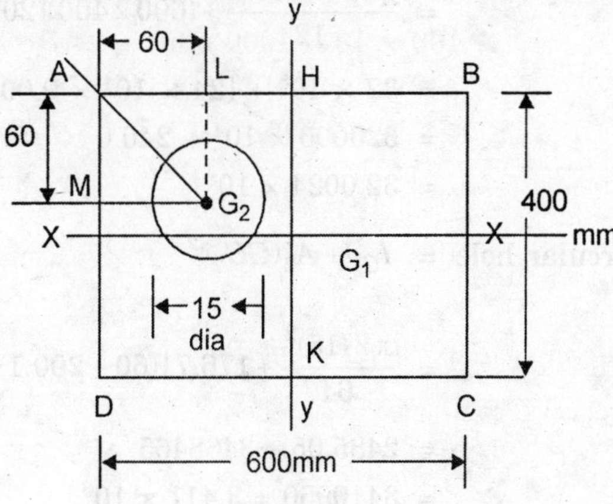

Fig. 5.45

Solution : Given data: Delta size 600 × 400 mm, R_a of hole = 180 mm

A_1 = Area of whole rectangle when there is no hole

$= b \times d$

Where, $b = 600$ mm

$d = 400$ mm

$= 600 \times 400$

$= 24 \times 10^4 \text{ mm}^2$

Let G be the location of section without hole, it will be of the mid point G_2, the CG of circular hole above will be at its centre.

$$A_2 = \frac{\pi}{4}(15)^2 = 176.71 \text{mm}^2$$

Not area $A = A_1 - A_2 = 240000 - 176.71$

$= 23.98 \times 10^4 \text{ mm}^2$

If the centroid G is at distance of $\bar{y}$ from the line AB, taking the moments about AB,

$$A\bar{y} = A_1 \times HG_1 - A_2(LG_2)$$

$= 24 \times 10^4 \times 200 - 176.71 \times 60$

$= 48 \times 10^6 - 1.06 \times 10^4$

or $(23.98 \times 10^4)\,\bar{y} = 47/98 \times 10^6$ ...(A)

$$\therefore \quad \bar{y} = \frac{47.989 \times 10^6}{23.98 \times 10^4}$$

$= 200.1$ mm

$$I_{GG} = I_G \text{ of rectangle } ABCD - R_G \text{ of circular area}$$

Now I_{GG} of $ABCD = I_{G_1} + A_1(y_{G_1} - \bar{y})^2$

$$= \frac{600 \times (400)^3}{12} + (600 \times 400)[200 - 200.1]^2$$

$= 32 \times 10^8 + [24 \times 10^4 \times 0.001]$

$= 320000 \times 10^4 + 240.0$

$= 32.0024 \times 10^8$

After I_{GG} of circular hole $= I_{G_2} + A_2(GG_2)^2$

$$= \frac{\pi \times (15)^2}{64} + 176.71(60 - 200.1)^2$$

$= 2485.05 + 3468465$

$= 3410950 = 3.411 \times 10^6$

E X A M P L E

Thus, of given are $\quad I_G = I_G\ ABCD - I_{GG}$ of hole

$$= 32.0024 \times 10 - 3.411 \times 10^6$$

$$= 3200.24 \times 10^8 \text{ mm}^4 \quad \textbf{Ans.}$$

5.11 : *Find the area moment of inertia about centroidal axis parallel to AB inertia of the composite section shown in Fig. 5.46.*

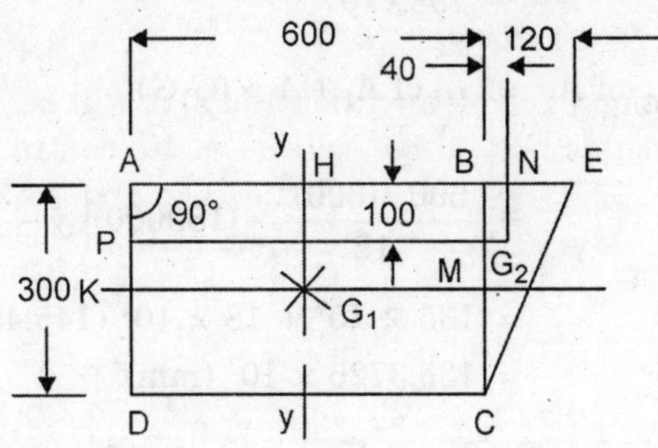

Fig. 5.46

Solution : Given data

$$AB = 600,$$

$$AD = 300,$$

$$G_1K = \frac{1}{2}300 = 150$$

$$A_1 = AB \times AD$$

$$= 600 \times 300 - 180000 \text{ mm}^2$$

$$BE = 129,\ G_2N$$

$$= \frac{1}{3} \times 300 = 100$$

$$A_2 = \text{Area of } DBEC$$

$$= \frac{1}{2} \times 300 \times 120$$

$$= 1800 \text{ mm}^2$$

$$A = A_1 + A_2$$

$$= 18 \times 10^4 + 18000$$

$$= 198 \times 10^3 \text{ mm}^2$$

$$G_2N = \frac{1}{3} \times BC = \frac{1}{3} \times 3N = 100$$

Taking moments about AB,

$$198 \times 10^3\, \overline{y} = A_1 y_1 + A_2 y_2$$

$$= A_1(G_1K) + A_2(G_2N)$$
$$= (18 \times 10^4)\ 150 + 18000 \times 100$$
$$= 27 \times 10^6 + 18 \times 10^5$$
$$= 288 \times 10^5$$

$$\therefore \quad \bar{y} = \frac{288 \times 10^5}{198 \times 10^3} = 145.45 \text{ mm}$$

Now $\quad I_G \text{ of } A_1 = I_{G_1} \text{ of } A_1 + A_1 \times (G_1G)^2$

$$= \frac{600 \times (300)^3}{12} + (180000)\left(\bar{y} - \frac{300}{2}\right)^2$$
$$= 135 \times 10^7 + 18 \times 10^4\ (145.45 - 150)^2$$
$$= 135.3726 \times 10^7 \text{ (mm)}^4$$

$$I_G \text{ of } A_2 = I_{G_2} + A_2\left(\bar{y} - \frac{1}{3} \times 300\right)^2$$
$$= \frac{BE \times BC^3}{12} + (18000)(45.45 - 100)^2$$
$$= 27 \times 10^7 + 3.718 \times 10^7$$
$$= 30.718 \times 10^7 \text{ (mm)}^4$$

$\therefore$ Total $\quad I_G = I_G \text{ of } A_1 + I_G \text{ of } A_2$
$$= 135.3726 \times 10^7 + 30.718 \times 10^7$$
$$= 136.09 \times 10^7 \textbf{ Ans.}$$

AREA MOMENT OF INERTIA OF TRAPEZIUM

5.12 : *A trapezium ABCD is shown in Fig. 5.47. Find the area moment of inertia of the trapezium about the base AB, also, about the centroidal axis parallel to AB.*

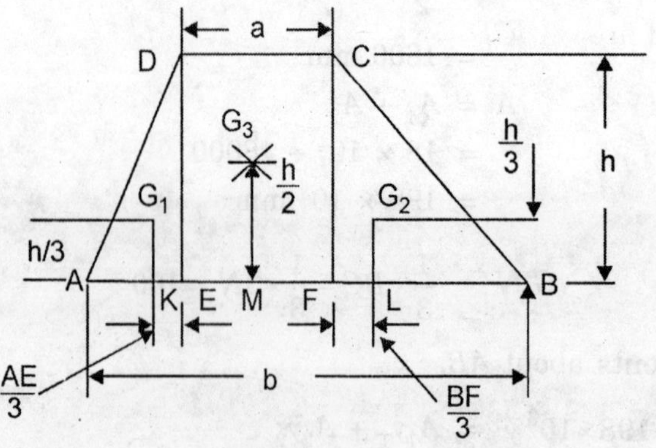

Fig. 5.47

E X A M P L E

EXAMPLE

In example 5.4, the centroid of this very trapezium has been found. The results obtained there are reproduced below. Total area of the trapezium has been split into three areas consisting of the triangle *DAE*, rectangle *DCFE* and trianlge *CFB*. The results obtained there are as below,

$$A_1 = \text{Area of } \Delta DAE = \frac{1}{2}h(AE)$$

$$\text{Height of centroid } G_1 \text{ of area } \Delta DAE = \frac{h}{3}$$

$$A_2 = \text{Area of } \Delta CFB = \frac{1}{2}h(FB)$$

$$\text{Height of centroid } G_2 \text{of area } \Delta\ CFB = \frac{1}{3}h$$

From *AB*, A_3 = Area of rectangle *DCFE* = *ah*

$$\text{Height of centroid } G_3 \text{ for } AB = \frac{h}{2}$$

With the help of equation (5.70) we have for areas A_1 and A_2,

$$I_{AB} \text{ of area } A_1 = \frac{(AE)h^3}{12} \qquad ...(A)$$

$$I_{AB} \text{ of area } A_2 = \frac{(FB)\times h^3}{12} \qquad ...(B)$$

And with the help of equation (5.64*b*),

$$I_{AB} \text{ of area } A_3 = \frac{ah^3}{3} \qquad ...(C)$$

I_{AB} of the whole trapezium by adding (*A*), (*B*) and (*C*),

$$I_{AB} \text{ of area } ABCD = \frac{(AE)h^3}{12} + \frac{(FB)h^3}{12} + \frac{ah^3}{3}$$

$$= \frac{h^3}{12}(AE + FB) + \frac{ah^3}{3}$$

But $AE + FB + a = b$

$\therefore$ $AE + FB = (b - a)$

$$\therefore \quad I_{AB} = \frac{h^3}{12}(b-a) + \frac{ah^3}{3}$$

$$= \frac{h^3}{12}(4a + b - a)$$

$$= \frac{h^3}{12}(3a + b) \quad ...(D)$$

Again from the theorem of parallel axis,

$$I_{AB} = I_G + A(\bar{y})^2$$

Now from example (5.4), the distances of *C.G.* from *AB* is $\bar{y}$,

$$\bar{y} = \frac{h}{3}\left(\frac{b + 2a}{b + a}\right) \quad ...(E)$$

and Area of trapezium $= \left(\frac{a + b}{2}\right)h$, ...(*F*)

$$I_{GG} = I_{AB} - A(\bar{y})^2$$

$$= \frac{h^3}{12}(3a + b) - \frac{1}{2}(a + b)h\left[\frac{h}{3}\frac{(b + 2a)}{(a + b)}\right]^2$$

$$= \frac{h^3}{12}(3a + b) - \frac{h^3}{18}\frac{(b + 2a)^2}{(b + a)}$$

$$= \frac{h^3}{36}\left[3(3a + b) - \frac{2(b + 2a)^2}{(b + a)}\right]$$

$$= \frac{h^3}{36(b + a)}[3(3a + b)(b + a) - 2(b + 2a)^2]$$

or
$$I_{GG} = \frac{h^3}{36(b + a)}[9ab + 9a^2 + 3b^2 + 3ab - 2b^2 - 8a^2 - 8ab]$$

$$= \frac{h^3}{36(b + a)}[4ab + b^2 + a^2]$$

$$= \frac{h^3}{36(b + a)}[(b + a)^2 + 2ab]$$

$$I_{GG} = \frac{h^3}{36}\left[(a + b) + \frac{2ab}{(a + b)}\right] \quad ...(G)$$

EXAMPLE

5.10 PRODUCT OF INERTIA

In Fig. 5.48 is shown an area A of which an elemental area dA is situated at a point P whose co-ordinates with respect to X-Y areas are (x, y). In other words its distance from X-axis is y and that from Y-axis is x. Then the term $xy\,dA$ is called the product of inertia of elemental area dA and $\int xy\,dA$ is known as the product of inertia of the whole area and is represented by I_{xy}.

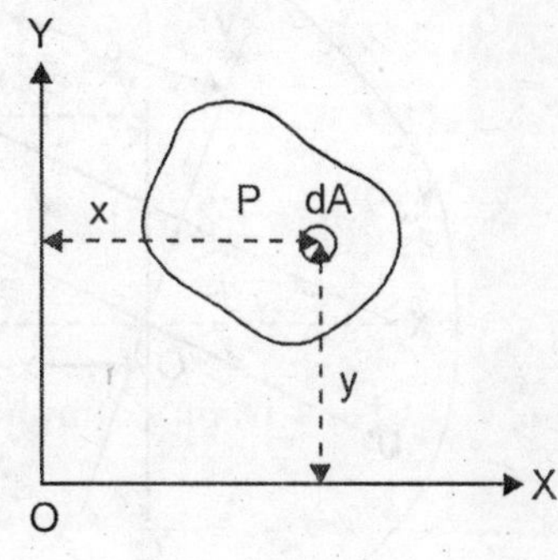

Fig. 5.48

That is $$I_{xy} = \int xy\,dA \qquad ...(5.84)$$

Mathematically, it is expressed as,

$$Ax_Gy_G = \int xy\,dA = \sum_{i=1}^{n} x_i y_i A_i$$

$$= x_1y_1A_1 + x_2y_2A_2 + ... + x_ny_n\,A_n \qquad ...(5.85)$$

where (x_1, y_1) is CG of area A_1, (x_2, y_2) that of area A_2 and so on.

It may be noted that the product of inertia may be positive, negative or even zero.

In case the area is symmetrical about one or both the areas. Then two equal elemental area dA on be located at a distance x, $-x$ (of symmetry axis X-axis) such that the product inertia will have the times $\int xy\,dA$ and $\int (x)y\,dA$ each integrated for half μ_x area. These times will cancel each other and,

$$I_{xy} = 0$$

In addition the product of inertia with respect to centroidal axis will also be zero. Let UU' and VV' be new set of axes which pass through the centroid of the area G. Then if $I_{uv} = 0$, the Product of inertia I_{uv} is called the Principal Product moment of inertia. The product inertia with respect to centroidal axes will also be zero.

The value of I_{uv} will depend upon the orientation of new set of axes with respect to the original axes *i.e.*, OX, OY.

Let us suppose that the new set of axis UU' and VV' are inclined at an angle θ with the original axes OX, OY. As θ will change, so will the values of I_{uv}. At some stage when $I_{uv} = 0$, then at that angle θ, the new axes will be called the Principal Axes and I_{uv} will be called the principal moment of Inertia.

We now proceed to determine the principal moment of inertia and other corresponding location of the axes UU and VV.

Let (x, y) be the co-ordinates of an elemental area dA with respect to axes XOX' and YOY' (Fig. 5.62) with origin at CG. Let UOU' and VOV' be another set of axes inclined at θ (anticlockwise) with the original axes OX, OY. Let the same elemental area dA have (u, v) co-ordinates with respect to OU, OV axes. (Fig 5.59)

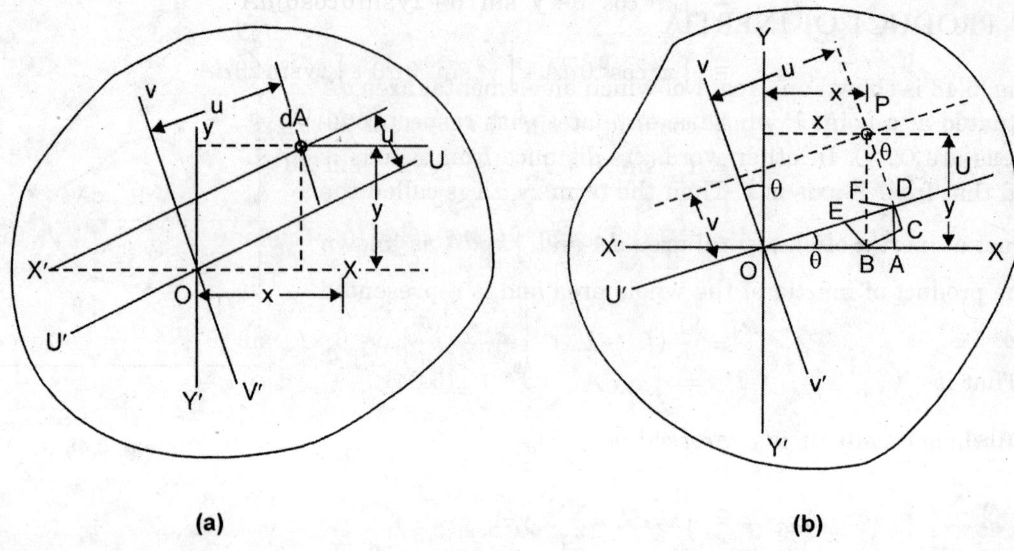

Fig. 5.59

Then Geometrically, $OB = x,\ BP = y,\ OD = u,\ PD = v$

The $u = GD = OD = u = GE + ED$

$$= GB \cos\theta + BC$$

$$= GB \cos\theta + PB \sin\theta$$

$$u = x\cos\theta + y\sin\theta \qquad \text{...(5.86)}$$

Also, $PD = v = PC - CD$

$$= y\cos\theta - BE$$

$$= y\cos\theta - x\sin\theta \qquad \text{...(5.87)}$$

Now $I_{UU} = \int v^2 dA$

$$= \int (y\cos\theta - x\sin\theta)^2\, dA$$

$$= \int (y^2\cos^2\theta + x^2\sin^2\theta - 2xy\sin\theta\cos\theta)\, dA$$

$$= \cos\theta \int y^2 dA + \sin^2\theta \int x^2 dA - \sin 2\theta \int xy\, dA$$

$$I_{UU} = I_{xx}\cos^2\theta + I_{yy}\sin^2\theta - I_{XY}\sin 2\theta$$

$$= \frac{I_{xx}}{2}(1+\cos 2\theta) + \frac{I_{yy}}{2}(1-\cos 2\theta) - I_{XY}\sin 2\theta$$

$$= \frac{1}{2}(I_{xx}+I_{yy}) + \frac{I_{xx}-I_{yy}}{2}\cos 2\theta - I_{xx}\sin 2\theta \qquad \text{...(5.88b)}$$

Similarly $I_{VV} = \int u^2 dA$

$$= \int (x\cos\theta + y\sin\theta)^2\, dA\ \text{z}$$

$$= \int [x^2 \cos^2\theta + y^2 \sin^2\theta + xy \sin\theta \cos\theta] dA$$

$$= \int x^2 \cos^2\theta\, dA + \int y^2 \sin^2\theta\, d\theta + \int xy \sin 2\theta\, dA$$

$$= \cos^2\theta\ I_{yy} + \sin^2\theta\ I_{xx} + (\sin 2\theta)\ I_{xy}$$

or $$I_{VV} = I_{xx} \sin^2\theta + I_{yy} \cos^2\theta + I_{xy} \sin 2\theta$$

$$= \frac{I_{xx}}{2}(1 - \cos 2\theta) + \frac{I_{yy}}{2}(1 + \cos^2\theta) + I_{xy} \sin^2\theta \qquad \text{...(5.89a)}$$

$$= \frac{1}{2}(I_{xx} + I_{yy}) - \left(\frac{I_{xx} - I_{yy}}{2}\right)\cos^2\theta - I_{yy} \sin^2\theta$$

Also, $$I_{uv} = \int uv\, dA \qquad \text{...(5.89b)}$$

$$= \int (x\cos\theta + y\sin\theta)(y\cos\theta - x\sin\theta) dA$$

$$= \int [xy\cos^2\theta - x^2 \sin\theta\cos\theta + y^2 \sin\theta\cos\theta - xy\sin^2\theta] dA$$

$$= \int [xy(\cos^2\theta - \sin^2\theta) + (y^2 - x^2)\sin\theta\cos\theta) dA$$

$$= \int [xy\cos 2\theta) dA + \int \left(\frac{1}{2}\sin 2\theta\right)(y^2 - x^2) dA$$

or $$I_{uv} = \cos 2\theta \int xy\, dA + \frac{\sin 2\theta}{2}\left\{\int y^2 dA - \int x^2 dA\right\}$$

$$I_{uv} = I_{xy} \cos 2\theta + \left(\frac{I_{xx} - I_{yy}}{2}\right)\sin 2\theta \qquad \text{...(5.90)}$$

For finding principal moment of inertia put,

$$I_{uv} = 0$$

∴ From equation (5.90) θ for Principal moment of Inertia, we get by equating I_{uv} to zero.

$$0 = I_{xy} \cos 2\theta + \frac{I_{xx} - I_{yy}}{2} \sin 2\theta$$

or $$\tan 2\theta = \frac{2I_{xy}}{I_{yy} - I_{xx}} \qquad \text{...(5.91)}$$

Therefore, if θ_1 and θ_2 are two values of θ from equation (5.91),

$$\theta_1 = \frac{1}{2}\tan^{-1}\left[\frac{2I_{xy}}{I_{yy} - I_{xx}}\right] \qquad \text{...(5.92a)}$$

$$\theta_2 = \frac{\pi}{2} + \theta_1$$

Corresponding to their values of θ as θ_1 and θ_2, the values of the Principal Moment of Inertia can be obtained from the equations (5.88) and (5.89).

For value of $\theta = \theta_1$, we have from equation (5.92*a*) after constructing the basic trignometric triangle as in Fig. 15.63.

$$\sin 2\theta_1 = \frac{I_{xy}}{\sqrt{\left(\frac{I_{yy} - I_{xx}}{2}\right)^2 + I_{xy}^2}} \quad \text{...(5.93a)}$$

$$\cos 2\theta_1 = \frac{\frac{1}{2}(I_{yy} - I_{xx})}{\sqrt{\left(\frac{I_{yy} - I_{xx}}{2}\right)^2 + I_{xy}^2}} \quad \text{...(5.93b)}$$

Fig. 5.50

Therefore, substituting values from equations (5.93*a*) and (5.93*b*) into equation (5.88) and (5.89), we get for $\theta = \theta_1$.

$$(I_{uu})_{\theta_1} = \frac{1}{2} I_{xx}(2\cos^2\theta_1) + \frac{1}{2} I_{yy}(2\sin^2\theta_1) - I_{xy}\sin 2\theta_1$$

$$= \frac{1}{2} I_{xx}(1 + \cos 2\theta_1) + \frac{1}{2} I_{yy}(1 - \cos 2\theta_1) - I_{xy}\sin 2\theta_1$$

$$= \left(\frac{I_{xx} + I_{yy}}{2}\right) + \left(\frac{I_{xx} - I_{yy}}{2}\right)\cos 2\theta_1 - I_{xy}\sin 2\theta_1$$

$$= \left(\frac{I_{xx} + I_{yy}}{2}\right) + \frac{\left(\frac{I_{xx} - I_{yy}}{2}\right)\left(\frac{I_{yy} - I_{xx}}{2}\right)}{\sqrt{\left(\frac{I_{yy} - I_{xx}}{2}\right)^2 + I_{xy}^2}} - I_{xy}\frac{I_{xy}}{\sqrt{\left(\frac{I_{yy} - I_{zz}}{2}\right)^2 + I_{xy}^2}}$$

$$= \frac{1}{2}(I_{xx} + I_{yy}) - \frac{\left(\frac{I_{yy} - I_{xx}}{2}\right)^2}{\sqrt{\left(\frac{I_{yy} - I_{xx}}{2}\right)^2 + I_{xy}^2}} - \frac{I_{xy}^2}{\sqrt{\left(\frac{I_{yy} - I_{xx}}{2}\right)^2 + I_{xy}^2}}$$

$$(I_{uu})_{\theta_1} = \frac{1}{2}(I_{yy} + I_{xx}) - \sqrt{\left(\frac{I_{yy} - I_{xx}}{2}\right)^2 + I_{xy}^2} \quad \text{...(5.94)}$$

Again substituting values from equation (5.93*a*) and (5.93*b*) into equation (5.89) we get,

$$(I_{VV})_{\theta_1} = \frac{1}{2}(I_{xx} + I_{yy}) - \left(\frac{I_{xx} - I_{yy}}{2}\right)\cos 2\theta_1 + I_{xy}\sin 2\theta_1$$

$$= \frac{1}{2}(I_{xx} + I_{yy}) - \frac{\left(\frac{I_{xx} - I_{yy}}{2}\right)\frac{1}{2}(I_{yy} - I_{xx})}{\sqrt{\left(\frac{I_{yy} - I_{xx}}{2}\right)^2 + I_{xy}^2}} + I_{xy}\frac{I_{xy}}{\sqrt{\left(\frac{I_{yy} - I_{xx}}{2}\right)^2 + I_{xy}^2}}$$

$$= \frac{1}{2}(I_{xx} + I_{yy}) + \frac{\left(\frac{I_{yy} - I_{xx}}{2}\right)^2 + I_{xy}^2}{\sqrt{\left(\frac{I_{yy} - I_{xx}}{2}\right)^2 \times I_{xy}^2}}$$

i.e.,
$$(I_{vv})_{\theta_1} = \frac{1}{2}(I_{xx} + I_{yy}) + \sqrt{\left(\frac{I_{yy} - I_{xx}}{2}\right)^2 \times I_{xy}^2} \qquad \text{...(5.95)}$$

☛ Since the value of θ as θ_1 is the solution of the equation (5.91) which was obtained by equating $I_{uv} = 0$, the corresponding values of $(I_{UU})_{\theta_1}$ and $(I_{VV})\theta_1$ refer to principal moment of inertia and angles θ_1 and θ_2 refer the directions of Principal Axes.

5.10.1 Maximum and Minimum Values of I_{UU} and I_{VV}

For maximum or minimum values of I_{uu} and I_{vv} we must have,

$$\frac{dI_{uu}}{d\theta} = \frac{dI_{vv}}{d\theta} = 0$$

From equation (5.88*b*) and we get after differentiation to *r* to 'θ'.

$$-\left(\frac{I_{xx} - I_{yy}}{2}\right)(2\sin 2\theta) - 2I_{xy}\cos 2\theta = 0$$

$$\tan 2\theta = \frac{I_{xy}}{\frac{1}{2}(I_{yy} - I_{xx})} \qquad \text{...(5.96)}$$

This equation is same as equation (5.91). This leads to the conclusion that. Principal moment of inertia are also maximum and minimum. A perusal of equations (5.94) and (5.95) shows that $(I_{vv})_{\theta_1}$ is greater than $(I_{uu})_{\theta_1}$. Thus, $(I_{vv})_{\theta_1}$ will be maximum and $(I_{uu})_{\theta_1}$ will be minimum value of the moment of inertia.

Also,
$$\theta_2 = \frac{\pi}{2} + \theta_1$$

or
$$2\theta_2 = \pi + 2\theta_1$$

$\therefore$ $\sin 2\theta_2 = \sin(\pi + 2\theta_1) = -\sin 2\theta_1$ ...(5.97*a*)

and $\cos 2\theta_2 = \cos(\pi + 2\theta_1) = -\cos 2\theta_1$ ...(5.97*b*)

Therefore, substituting these values in equation (5.89) we get,

$$(I_{vv})_{\theta_2} = I_{xx}\sin^2\theta_2 + I_{yy}\cos^2\theta_2 + I_{xy}\sin 2\theta_2$$

$$= \frac{1}{2}I_{xx}(2\sin^2\theta_2) + \frac{1}{2}I_{yy}(2\cos^2\theta_2) + I_{xy}\sin 2\theta_2$$

$$= \frac{1}{2}I_{xx}(1-\cos 2\theta_2) + \frac{1}{2}I_{yy}(1+\cos 2\theta_2) + I_{xy}\sin 2\theta_2$$

$$= \frac{1}{2}(I_{xx}+I_{yy}) - \frac{1}{2}(I_{xx}-I_{yy})\cos 2\theta_2 + I_{xy}\sin 2\theta_2$$

$$= \frac{1}{2}(I_{xx}+I_{yy}) + \frac{1}{2}(I_{xx}-I_{yy})\cos 2\theta_1 - I_{xy}\sin 2\theta_1 \quad \text{(From 5.95)}$$

$$= \frac{1}{2}(I_{xx}+I_{yy}) + \frac{1}{2}(I_{xx}-I_{yy})\frac{(I_{yy}-I_{xx})/2}{\sqrt{\left(\frac{I_{yy}-I_{xx}}{2}\right)^2 + I_{xy}^2}} - I_{xy}\frac{I_{xy}}{\sqrt{\left(\frac{I_{yy}-I_{xx}}{2}\right)^2 + I_{xy}^2}}$$

or
$$(I_{VV})_{\theta_2} = \frac{1}{2}(I_{xx}\times I_{yy}) - \sqrt{\left(\frac{I_{yy}-I_{xx}}{2}\right)^2 + I_{xy}^2} \quad \text{...(5.98)}$$

Similarly, from equations (5.88*a*) and (5.97),

$$(I_{uu})_{\theta_2} = \frac{1}{2}(I_{xx}+I_{yy}) + \left(\frac{I_{xx}-I_{yy}}{2}\right)\cos 2\theta_2 - I_{xy}\sin 2\theta_2$$

or
$$(I_{uu})_{\theta_2} = \frac{1}{2}(I_{xx}+I_{yy}) + \left(\frac{I_{xx}-I_{yy}}{2}\right)\cos 2\theta_1 - I_{xy}(-\sin 2\theta_1)$$

$$= \frac{1}{2}(I_{xx}+I_{yy}) - \left(\frac{I_{xx}-I_{yy}}{2}\right)\frac{\left(\frac{I_{yy}-I_{xx}}{2}\right)}{\sqrt{\left(\frac{I_{yy}-I_{xx}}{2}\right)^2 + I_{xy}^2}} + \frac{I_{xy}\times I_{xy}}{\sqrt{\left(\frac{I_{yy}-I_{xx}}{2}\right)^2 + I_{xy}^2}}$$

i.e.,
$$(I_{uu})_{\theta_2} = \frac{1}{2}(I_{xx}+I_{yy}) + \sqrt{\left(\frac{I_{yy}-I_{xx}}{2}\right)^2 + I_{xy}^2} \quad \text{...(5.98)}$$

From equations (5.94), (5.95), (5.97) and (5.98) we see that,

$$(I_{uu})_{\theta_2} = (I_{vv})_{\theta_1} \quad ...(5.99a)$$

and

$$(I_{vv})_{\theta_2} = (I_{uu})_{\theta_1} \quad ...(5.99b)$$

Also,

$$(I_{uu})_{\theta_1} + (I_{vv})_{\theta_1} = (I_{uu})_{\theta_2} + (I_{vv})_{\theta_2} = (I_{xx} + I_{yy}) \quad ...(5.100)$$

5.11 THEOREM OF PARALLEL AXES FOR PRODUCT OF INERTIA

In the Fig. 5.51 is shown an area whose product of inertia about *XX* and *yy* axes which are passing through the centre of gravity *G* are known.

An elemental Area *dA* has co-ordinates (x, y) with respect to the original areas *OX, OY*. passing through *CG*.

$$\therefore \quad I_{XX} = \oint dA\, y^2$$

$$I_{YY} = \oint dA\, x^2$$

$$I_{XY} = \oint dA\, xy$$

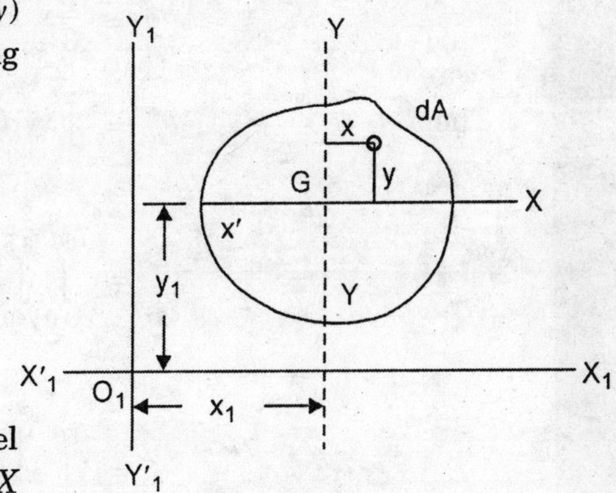

Fig. 5.51

Let X_1X_1 and Y_1Y_1 be another set of axes parallel to the given set *i.e.*, $X_1'O_1X_1'$ is parallel to $X'OX$ and $Y_1O_1Y_1'$ is parallel to YOY' such that the co-ordinates of O will respect to new origin O_1 are say (x_1, y_1). Therefore, the co-ordinates of *dA* with respect to $X_1^1O_1X_1$ and $Y_1^1O_1Y_1$ axes are $(x_1 + x)$ and $(y_1 + y)$.

$$\therefore \quad I_{X_1Y_1} = \oint (x_1 + x)(y + y_1)\, dA$$

$$= \int (x_1 y + x_1 y_1 + xy + xy_1)\, dA$$

$$= x_1 \int y\, dA + x_1 y_1 \int dA + \int xy\, dA + y_1 \int x\, dA \quad ...(5.101)$$

Since $\int x dA$ and $\int y(dA)$ are first moments of area about centroidal axes *OY* and *OX* respectively. They must be zero. Then equation (5.101) reduces to,

$$I_{X_1Y_1} = \oint xy\, dA + x_1 y_1 A$$

$$= I_{XY} + x_1 y_1 A \quad ...(5.102)$$

5.13 : *Find the product moment of inertia of the right angled triangle OAB about XOY axis as shown in Fig. 5.52.*

Solution : Let the elemental area $dA = dxdy$ be taken at any point P whose co-ordinates are (x, y)

Then $$I_{XY} = \oint xy\,dA$$

Now the slope of the OB is

$$\frac{40}{60} = \frac{2}{3}$$

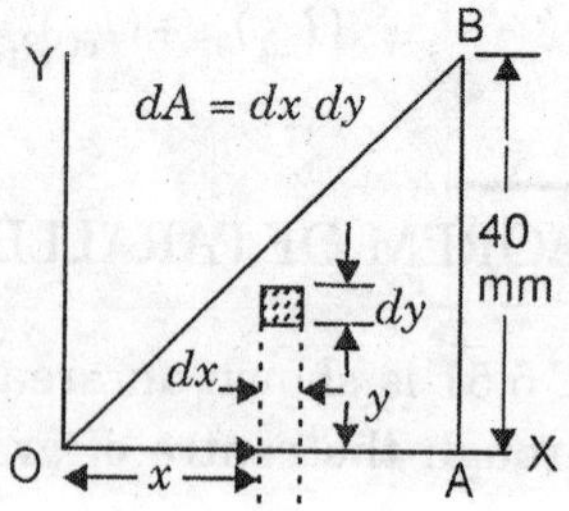

Fig. 5.52

Therefore, equation of OB is,

$$y = \frac{2}{3}x$$

Then $$I_{xy} = \oint xy\,dx\,dy$$

$$= \int_{x=0}^{60}\int_{y=0}^{\frac{2}{3}x} xy\,dy\,dx$$

$$= \int_{x=0}^{60} x\left[\frac{y^2}{2}\right]_{y=0}^{\frac{2}{3}x} dx$$

or $$I_{xy} = \int_{x=0}^{60} \frac{x}{2}\left[\frac{4x^2}{9} - 0\right] dx$$

$$= \int_{x=0}^{60}\left[\frac{2}{9}x^3\right] dx = \frac{2}{9}\left[\frac{x^4}{4}\right]_0^{60}$$

$$= \frac{1}{18}[60]^4$$

$$= 72 \times 10^4 \text{ mm}^4$$

5.14 : *Find the product moment of inertia of a quadrant OAB of a circle about the perpendicular axes OX and OY shown in Fig. 5.16. Take the radius of quadrant as r.*

Solution : Let the elemental area $dx\,dy$ be located at p having co-ordinates (x, y)

Equation of curve AB is,

E X A M P L E

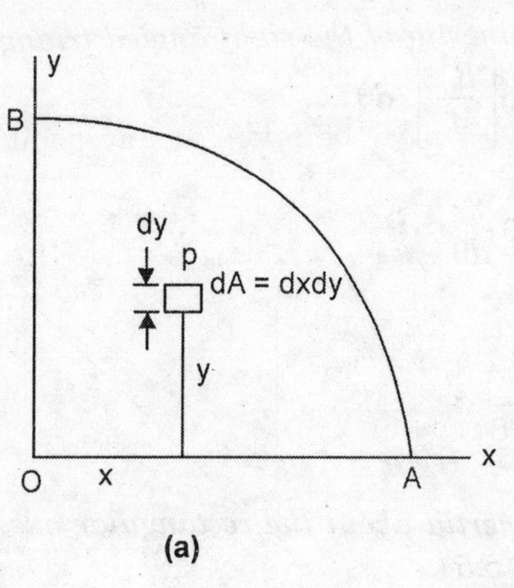

(a)

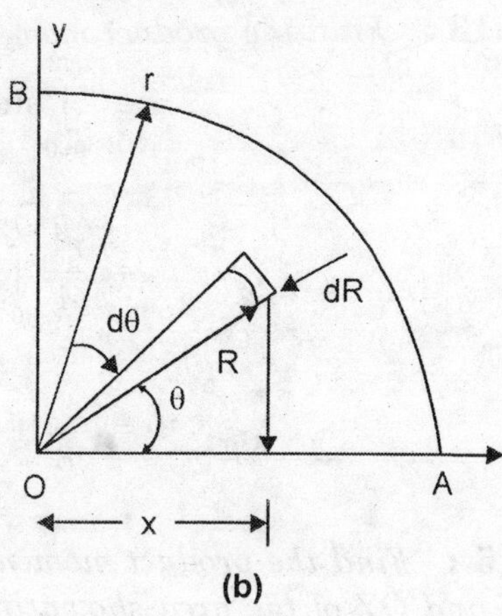

(b)

Fig. 5.53

$$x^2 + y^2 = r^2$$

or $$y = \sqrt{r^2 - x^2}$$

$\therefore$ $$I_{xy} = \int_{x=0}^{r} \left\{ \int_{y=0}^{\sqrt{r^2-x^2}} xy\,dy \right\} dx = \int_{x=0}^{r} \left\{ \left[\frac{xy^2}{2} \right]_0^{\sqrt{r^2-x^2}} \right\} dx$$

or $$I_{xy} = \int_{x=0}^{r} \frac{x}{2}(r^2 - x^2)\,dx$$

$$= \frac{1}{2}\left[r^2 \frac{x^2}{2} - \frac{x^4}{4} \right]_0^r$$

$$= \frac{1}{2}\left(\frac{r^4}{2} - \frac{r^4}{4} \right) = \frac{r^4}{8}$$

Alternatively in polar co-ordinates,

$$I_{xy} = \iint xy\,dA$$

From Fig. (5.61*b*), $dA = R\,d\theta\,dR$

$$x = R\cos\theta,\ y = R\sin\theta$$

Then $$I_{xy} = \int_{\theta=0}^{\pi/2} \int_{R=0}^{r} (R\cos\theta)(R\sin\theta)\,R\,dR\,d\theta$$

$$= \int_{\theta=0}^{\pi/2} \left[\int_{R=0}^{r} (\sin\theta\cos\theta) R^3\,dR \right] d\theta$$

$$= \int_{\theta=0}^{\pi/2} \sin\theta\cos\theta \left[\frac{R^4}{4}\right]_0^r d\theta$$

$$= \frac{r^4}{4}\int_0^{\pi/2} \frac{\sin 2\theta}{2}\, d\theta$$

$$= \frac{r^4}{8}[1-0] = \frac{r^4}{8}$$

5.15 : *Find the product moment of inertia about the rectangular axes OX and OY of the area shown in Fig. 5.54.*

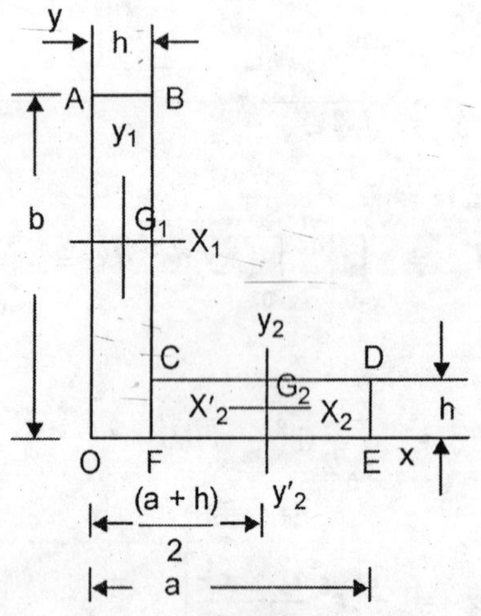

Fig. 5.54

Solution : Let G_1 and G_2 be the centroids of the areas *ABFO* and *CDEF* respectively having the corresponding areas A_1 and A_2. Let the co-ordinates of G_1 and G_2 with respect to *XOY* axis be (x_1, y_1) and (x_2, y_2)

Then for the area *ABFO* from equation (5.102) *i.e.*, theorem of parallel axes for product moment of inertia,

$$I_{xy} = I_{x_1y_1} + A_1x_1y_1$$

Now $I_{x_1y_1}$ of area *ABFO* = 0 since G_1X_1 and G_1V_1 are the principal axes of the area *ABRO* being axes of symmetry.

And $$A_1 = b \cdot h; x_1 = \frac{h}{2} \text{ and } y_1 = \frac{b}{2}$$

Therefore, $I_{\overline{xy}}$ of area

$$ABFO = 0 + (bh)\left(\frac{h}{2}\right)\left(\frac{b}{2}\right) = \frac{b^2h^2}{4}.$$

EXAMPLE

Again for the Area *CDEF*

$$I_{xy} = I_{x_2y_2} + A_2x_2y_2$$

Here also, $I_{x_2y_2} = 0$ as G_2X_2 and G_2Y_2 are axes of symmetry.

and $$A_2 = (a - h)h$$

$$x_2 = OF + \frac{1}{2}(FE) = h + \frac{a-h}{2} = \frac{(a+h)}{2}$$

$$y_2 = \frac{h}{2}$$

Therefore, I_{xy} of area,

$$CDEF = 0 + (a-h)h\left[\frac{a+h}{2}\right]\left(\frac{h}{2}\right)$$

$$= (a^2 - h^2)\frac{h^2}{4}$$

$\therefore$ Total product of inertia of entire area

$$I_{xy} = \frac{b^2h^2}{4} + \frac{h^2}{4}(a^2h^2)$$

$$= \frac{h^2}{4}(b^2 + a^2 - h^2) \text{ **Ans.**}$$

5.16 : *Find the centroidal principal moment of inertia of an equal angle sections shown in Fig. 5.55.*

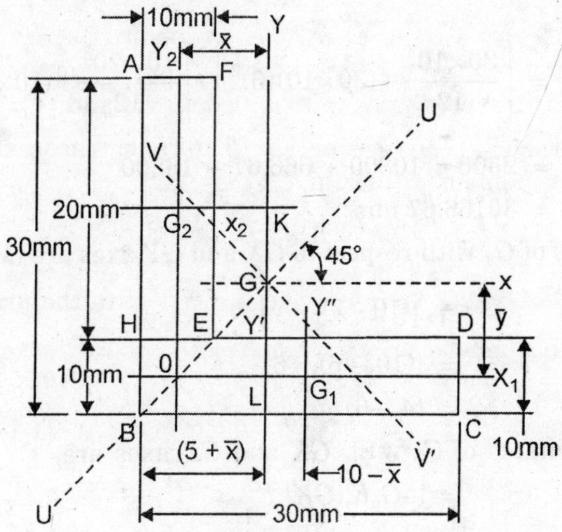

Fig. 5.53

Solution : Let $(\bar{x}, \bar{y})$ be the co-ordinates of the centroid G, with respect to the rectangular axis OX_1 and OY_2. Since this is an equal angle,

$$\bar{x} = \bar{y}$$

For finding the value of $\bar{y}$, take moments of areas about the line OX_1, the centroidal axis of area *BCDH*.

$\therefore$ $(30 \times 10 + 20 \times 10)\,\bar{y}$ = moment of area *HDCB* about OX_1 + moment of area *AFEH*

That is $$500\bar{y} = (30 \times 10 \times 0) + (20 \times 10 \times 15)$$

$$= 3000$$

$$\therefore \quad \bar{y} = \frac{3000}{500} = 6 \text{ mm}$$

Also, because of equal angle specifications,

$$I_{xx} = I_{yy}$$

$$= \left[I_{G_1X_1} \text{ of } HDCB + A_{HDCB} \times (\bar{y})^2\right]$$

$$+ \left[I_{G_2X_2} \text{ of } AFEH + A_{AFEH} \times (KG)^2\right]$$

Now $$KG = KM - MG$$

$$= \frac{1}{2}(AH + HB) - \bar{y}$$

$$= \frac{1}{2}(20 + 10) - 6 = 9$$

$$I_{xx} = I_{yy} = \left[\frac{30 \times 10^3}{12} + (30 \times 10)(6)^2\right] + \left[\frac{10 \times 20^3}{12} + (10 \times 20)(9)^2\right]$$

$$= 2500 + 10800 + 666.67 + 16200$$

$$= 30166.67 \text{ mm}^4$$

Now co-ordinates of G_1 with respect to *GX* and *GY* axes are $(MG_1, -GM)$

$$= [(10 - \bar{x}), -\bar{y}]$$

$$= [(10 - 6), -6]$$

$$= (4, -6)$$

Similarly, co-ordinates of G_2 w.r.t. *GX* and *GY* axes are,

$$= [-G_2K, GK]$$

$$= [-\bar{x}, (KM - MG)]$$

EXAMPLE

$$= [-\bar{x}, 9]$$

$$= [-6, 9]$$

$$\therefore \quad I_{xy} \text{ of area } HDCB = I_{X_1Y_1} \text{ of Area } HDCB + A_{HDCB}(4)(-6)$$

$$= 0 + (30 \times 10)(-24)$$

(Because OX_1, is the axis of symmetry $\therefore$ ($I_{x_1y_1} = 0$)

$$= -7200 \text{ mm}^4$$

Now I_{XY} of area

$$AFEH = I_{x_2y_2} \text{ of } A_{AFEH} + A_{AFEH}(-6)(9)$$

$$= 0 + 10 \times 20\ (-54)$$

$$= -10800 \text{ mm}^4$$

Total $\quad I_{XY} = -7200 - 10800$

$$= -18000 \text{ mm}^4$$

If θ is the inclination of principal axes GU an GV with GX passing through G then,

$$\tan 2\theta = \frac{2I_{XY}}{I_{XX} - I_{YY}} = \infty \ (\because I_{XX} = I_{YY})$$

i.e., $\quad 2\theta_1 = 90°$ or $\theta_1 = 45°$

and $\quad \theta_2 = 90 + 45 = 135°$

Thus, the Principal axes GU and GV are inclined with GX at 45° and 135°.

For $\theta = 45°$ $\quad I_{UU} = \frac{1}{2}(I_{XX} \times I_{YY}) + \frac{1}{2}(I_{XX} - I_{YY})\cos 90° - I_{XY}\sin 90°$

$$= I_{XX} + 0 - I_{XY}$$

$$= 30166.67 + 0 - (-18000)$$

$$= 84166.67 \text{ mm}^4$$

Also,

$$I_{ZZ} = I_{UU} + I_{VV}$$

$$= I_{XX} \times I_{YY}$$

or $\quad I_{VV} = I_{XX} + I_{YY} - I_{UU}$

$$= 2I_{XX} - I_{UU} \qquad (\because I_{XX} = I_{YY})$$

$$= 2(30166.6) - 48166.67$$

$= 12166.67 \text{ mm}^4$

5.17 : *Find the centroidal principal moment of inertia of the section shown in Fig. 5.56.*

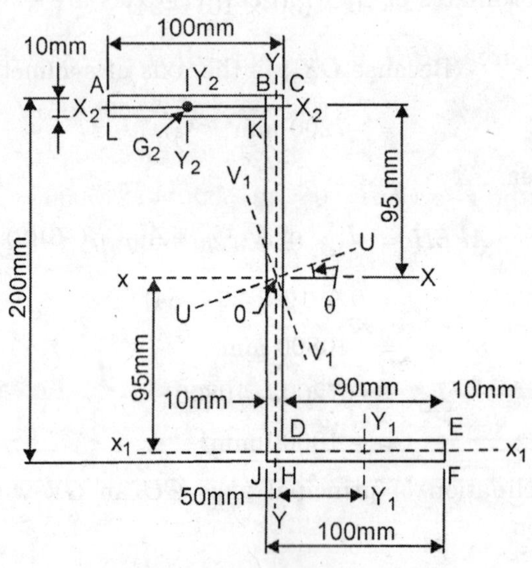

Fig. 5.54

Solution : Evidently the centroid G of the given Z-section is in the middle of web. Let the centroidal principal axes be GU and GV.

Further let G_1 and G_2 be the centroids of the rectangular areas $BCHJ$ and $ABKL$ respectively.

$I_{yy} = I_{yy}$ of area $BCHJ$ + I_{yy} of area $DEFH$ + I_{yy} of area $ABKL$

$$= \left(\frac{200 \times 10^3}{12}\right) + \left[\frac{10 \times 90^3}{12} + (90 \times 10)(50)^2\right] + \left[\frac{10 \times 90^3}{12}(90 \times 10)50^2\right]$$

$= 573.167 \times 10^4 \text{ mm}^4$

$I_{XX} = I_{XX}$ of area $BCHJ$ + I_{XX} of area $DEFH$ + I_{XX} of area $ABKL$

$$= \frac{10 \times 200^3}{12} + \left[\frac{90 \times 10^3}{12} + (90 \times 10)(95)^2\right] + \left[\frac{90 \times 10^3}{12} + (90 + 10)(95)^2\right]$$

$= 2292.6 \times 10^4 \text{ mm}^4$

I_{xy} of area $BCHJ$ = 0 since GX and GV are the axes of symmetry.

Now co-ordinates of G_1 wrt GX, GY axes are

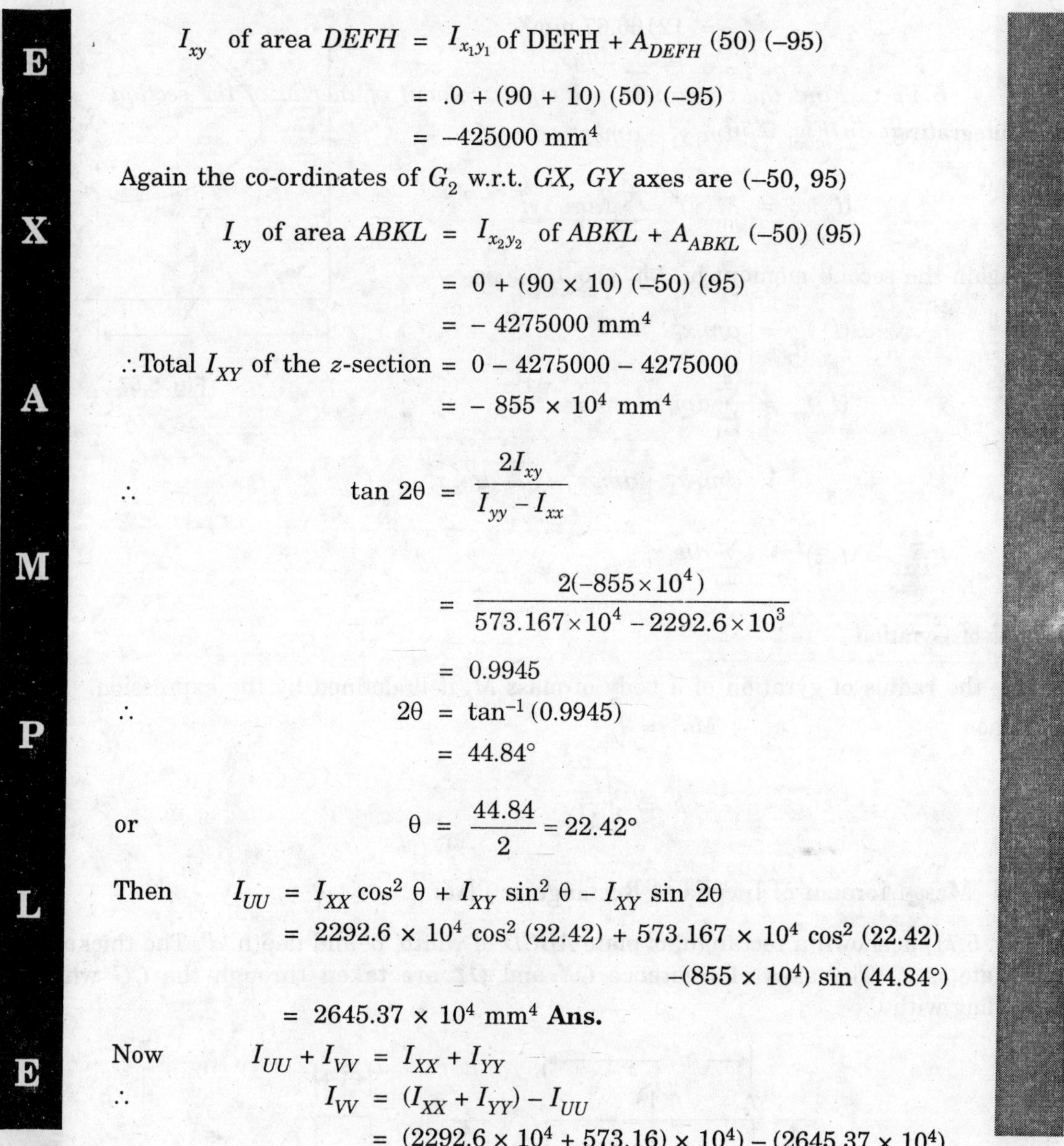

E X A M P L E

I_{xy} of area $DEFH$ $= I_{x_1y_1}$ of DEFH $+ A_{DEFH}$ (50) (–95)

$= .0 + (90 + 10)\ (50)\ (-95)$

$= -425000\ \text{mm}^4$

Again the co-ordinates of G_2 w.r.t. GX, GY axes are (–50, 95)

I_{xy} of area $ABKL$ $= I_{x_2y_2}$ of $ABKL + A_{ABKL}$ (–50) (95)

$= 0 + (90 \times 10)\ (-50)\ (95)$

$= -\ 4275000\ \text{mm}^4$

$\therefore$ Total I_{XY} of the z-section $= 0 - 4275000 - 4275000$

$= -\ 855 \times 10^4\ \text{mm}^4$

$$\therefore \quad \tan 2\theta = \frac{2I_{xy}}{I_{yy} - I_{xx}}$$

$$= \frac{2(-855 \times 10^4)}{573.167 \times 10^4 - 2292.6 \times 10^3}$$

$$= 0.9945$$

$$\therefore \quad 2\theta = \tan^{-1}(0.9945)$$

$$= 44.84°$$

or
$$\theta = \frac{44.84}{2} = 22.42°$$

Then $I_{UU} = I_{XX} \cos^2 \theta + I_{XY} \sin^2 \theta - I_{XY} \sin 2\theta$

$= 2292.6 \times 10^4 \cos^2 (22.42) + 573.167 \times 10^4 \cos^2 (22.42)$

$-(855 \times 10^4) \sin (44.84°)$

$= 2645.37 \times 10^4\ \text{mm}^4$ **Ans.**

Now $I_{UU} + I_{VV} = I_{XX} + I_{YY}$

$\therefore$ $I_{VV} = (I_{XX} + I_{YY}) - I_{UU}$

$= (2292.6 \times 10^4 + 573.16) \times 10^4) - (2645.37 \times 10^4)$

$= 220.397 \times 10^4\ \text{mm}^4$ **Ans.**

5.12 MASS MOMENT OF INERTIA AND RADIUS OF GYRATION

Fig. 5.57 shows a body of mass M with centre of gravity at $G(\bar{x}, \bar{y})$.

Let dm be an elemental mass having co-ordinates (x, y) with respect to XOY axis.

Then the second moment of mass dm about axis OX,

$$dI_{XX} = (dm)\ y^2$$

$$(I_m)_{xx} = \sum_{i=1}^{n} dm_i y_i^2$$

Integrating $\quad (I_m)_{xx} = dm_1 y_1^2 + dm_2 y_1^2 ... dm_n y_n^2$

$$(I_m)_{XX} = M(\bar{y})^2 = \sum_{i=1}^{n} (dm_1) y_i^2$$

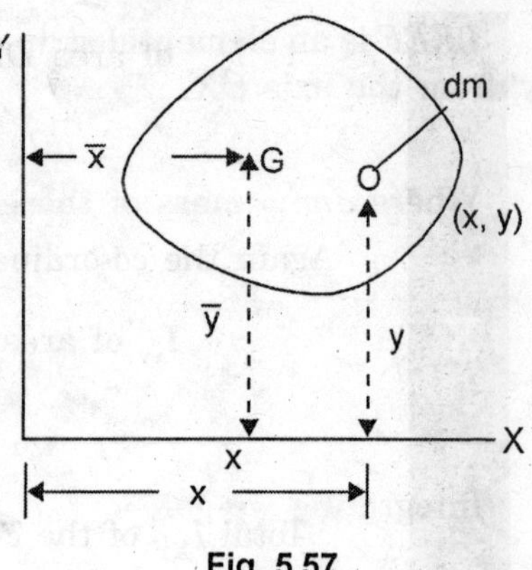

Fig. 5.57

Again the second moment of dm_i about Y-axis,

$$d\,(I_m)_{yy} = dm_i x_i^2$$

or $\quad (I_m)_{yy} = \sum_{i=1}^{n} dm_i\, n_i^2$

$$= dm_1 x_1^2 + dm_2 x_2^2 + ... + dm_n x_n^2$$

$$(I_m)_{yy} = M(\bar{x})^2 = \sum_{0=1}^{n} dm_i x_i^2 \qquad ...(5.103)$$

Radius of Gyration

It k is the radius of gyration of a body of mass M, it is defined by the expression.

Then $\quad Mk^2 = I_m$

$$k = \sqrt{\frac{I_m}{M}}$$

5.12.1 Mass Moment of Inertia of Rectangular Plate

In Fig. 5.71, is shown a rectangular plate $ABCD$ of width 'b' and depth 'd'. The thickness of the plate is 't'. The axes of reference OX and OY are taken through the CG which is coinciding with 0.

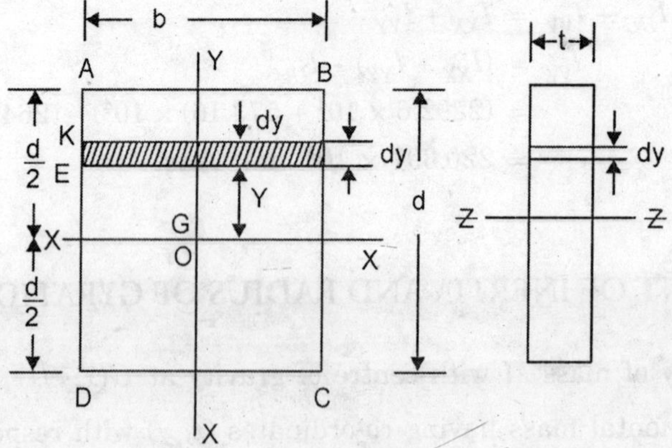

Fig. 5.58

LKEF is an elemental strip of width '*b*', depth *dy* and thickness *t*. It is situated at a height '*y*' from the axis *OX*.

$$\therefore \quad d(I_m)_{XX} = (dm)\, y^2$$

Where dm = mass of the elemental strip

$$= \text{density} \times \text{vol of strip}$$

$$= \rho\,(dV) = \rho\,(bt\ dy)$$

$$\therefore \quad d(I_m)_{XX} = \rho\ bt\ dy\ y^2$$

Integrating we get,
$$(I_m)_{XX} = \rho bt \int_{-d/2}^{d/2} y^2 dy = \rho bt \left[\frac{y^3}{3}\right]_{-\frac{d}{2}}^{d/2}$$

$$= \frac{\rho bt}{3}\left[\frac{d^3}{8}+\frac{d^3}{8}\right] = \frac{\rho bt\, d^3}{12}$$

$$= \rho t\left(\frac{bd^3}{12}\right) = \rho t\, I_{XX} \qquad \text{...(5.104)}$$

But I_{XX}, the second moment of area of rectangle about *XX* is given by,

$$I_{XX} = \frac{bd^3}{12}$$

$$\therefore \quad (I_m)_{XX} = \rho t\ I_{XX}$$

$$= \rho td \frac{bd^2}{12}$$

$$= \rho(btd)\cdot\frac{d^2}{12} = \rho V \cdot \frac{d^2}{12}$$

$$= M\frac{d^2}{12}$$

Where,
$$V = \text{Vol. of plate}$$
$$M = \text{Mass of plate} = \rho V$$

$$\therefore \quad (I_m)_{XX} = \frac{Md^2}{12} \qquad \text{...(5.105a)}$$

Similarly by taking the elemental strip parallel to *y*-axis we get,

$$(I_{mm})_{yy} = \frac{Mb^2}{12} \qquad \text{...(5.105b)}$$

5.12.1.1Mass Moment of Inertia of Rectangular Plate about base

From equation (5.105*a*),

$$(I_m)_{XX} = \frac{Md^2}{}$$

Now from theorem of parallel axis,

$$(I_m)_{DC} = (I_m)_{XX} + M\left(\frac{d}{2}\right)^2$$

$$= \frac{Md^2}{12} + \frac{Md^2}{4} = \frac{Md^2}{3} \quad ...(5.106)$$

5.12.2 Mass Moment of Inertia of Circular Disc

In Fig. 5.59 is shown a disc of Radius R and thickness t. Let ρ be density of the material.

Let there be an elemental area of ring shape whose radial thickness is dx and is in the shape of a circular ring of radius x.

Then dv = vol. of elemental ring of thickness t

$$= (2\pi\ x\ dx)\ t$$

$$(I_m)_{zz} = x^2 dm$$

$$= x^2(\rho\ dv)$$

$$= \rho x^2 (2\pi\ x\ dx\ t)$$

$$\therefore \quad (I_m)_{zz} = 2\pi\rho t \int_0^R x^3\, dx$$

Fig. 5.59

$$= 2\pi\rho t\left[\frac{x^4}{4}\right]_0^R$$

$$= \frac{2\pi\rho R^4 t}{2} = (\pi R^2 \rho t) \times \left(\frac{R^2}{2}\right)$$

$$= \frac{MR^2}{2} \quad ...(5.107)$$

Where, ρ = density of material

M = mass of disc

Now $(I_m)_{YY} = (I_m)_{XX}$

and by the theorem of perpendicular axes,

$$(I_m)_{ZZ} = (I_m)_{XX} + (I_m)_{YY}$$

$$= 2(I_m)_{XX}$$

$$\therefore \quad (I_m)_{XX} = (I_m)_{YY} = \frac{1}{2}(I_m)_{ZZ}$$

$$= \frac{1}{2}\frac{MR^2}{2} = \frac{MR^2}{4} \quad ...(5.108)$$

5.12.3 Mass Moment or Inertia of a Cylinder

Let the hollow cylinder be of outer radius R and inner radius r and length L (Fig. 5.60).

Let an elemental annular cylinder of radius x and radial thickness dx be taken. Then mass moment of inertia of an annular cylincler,

$$d(I_m)_{ZZ} = dm\ x^2$$
$$= (\rho\ dv)\ x^2$$
$$= \rho\ (2\pi\ x\ dx\ L)\ x^2$$
$$(I_m)_{ZZ} = 2\rho\pi L \int_r^R x^3\ dx$$
$$= 2\pi\rho L \left(\frac{x^4}{4}\right)_r^R$$
$$= \frac{\pi\rho L}{2}(R^4 - r^4)$$
$$= \frac{\pi\rho L}{2}(R^2 - r^2)(R^2 + r^2)$$
$$= \frac{\rho}{2}[\pi(R^2 - r^2)L](R^2 + r^2)$$
$$= \frac{1}{2}\rho(\text{vol of cylinder})\ (R^2 + r^2)$$

$\therefore$
$$= \frac{1}{2}\rho\, dv(R^2 + r^2)$$
$$= \frac{M}{2}(R^2 + r^2) \qquad ...(5.109)$$

Fig. 5.60

Now
$$(I_m)_{ZZ} = (I_m)_{XX} + (I_m)_{YY}$$
Also,
$$(I_m)_{XX} = (I_m)_{YY}$$
$\therefore$
$$(I_m)_{XX} = (I_m)_{XX} = \frac{1}{2}(I_m)_{ZZ}$$
$$= \frac{M}{4}(R^2 + r^2) \qquad ...(5.110)$$

5.12.4 Mass Moment of Inertia of a Ring of Uniform Cross-section

Let the mass of unit length of wire of given section be m *i.e.*, $m\ 2\pi R = M$ when M is total mass of ring and R the radius,

$$d_m = (R\ d\theta)\ m$$
$$d(I_m)_{ZZ} = (m\ R\ d\theta)\ R^2$$

Integrating $$(I_m)_{ZZ} = \int_{\theta=0}^{2\pi} m R^3\, d\theta = mR^3[\theta]_0^{2\pi}$$

$$= mR^3[\theta]_0^{2\pi}$$

$$= 2\pi\, mR^3 = (2\pi R\, m)\, R^2$$

$$= MR^2$$

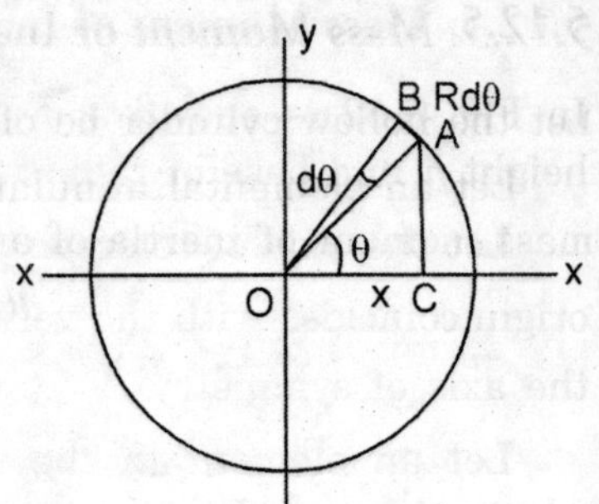

Fig. 5.61

Now from theorem of perpendicular are axes,

$$(I_m)_{ZZ} = (I_m)_{XX} + (I_m)_{YY}$$

$$= 2(I_m)_{XY}$$

$$(I_m)_{XX} = (I_m)_{YY}$$

$$= \frac{1}{2}(I_m)_{ZZ}$$

$$= \frac{1}{2}MR^2 \quad ...(5.111a)$$

Alternatively

From Fig. (5.74), $$d(I_m)_{XX} = dmy^2$$

$$= (m\, Rd\theta)\, y^2$$

$$= (m\, R\, d\theta)\, (R \sin\theta)^2$$

$$= mR^3 \int_0^{2\pi} \sin^2\theta\, d\theta$$

$$= \frac{mR^3}{2} \int_0^{2\pi} (2\sin^2\theta)\, d\theta$$

$$= \frac{mR^3}{2} \int_0^{2\pi} (1 - \cos 2\theta)\, d\theta$$

$$= \frac{mR^3}{2}\left[\theta - \frac{\sin 2\theta}{2}\right]_0^{2\pi}$$

$$= \frac{mR^3}{2}[(2\pi - 0) - (0)]$$

$$= \frac{2\pi mR^3}{2} = (2\pi Rm)\frac{R^2}{2}$$

or $$(I_m)_{xx} = \frac{MR^2}{2} \quad ...(5.111b)$$

5.12.5 Mass Moment of Inertia of a Right-circular cone about its axis of Symmetry

In Fig. 5.62 is shown a section of right circular cone of height h and base of radius r.

Let the axis of reference be taken as AX, AZ such that origin coincides with the vertex A of the cone. Then $\bar{Z}\bar{Z}$ is the axis of symmetry.

Let an element in the form of a disc EFH be taken whose radius is EF and thickness dz at distance of z from OX. Then from similar triangles ABK and AEF,

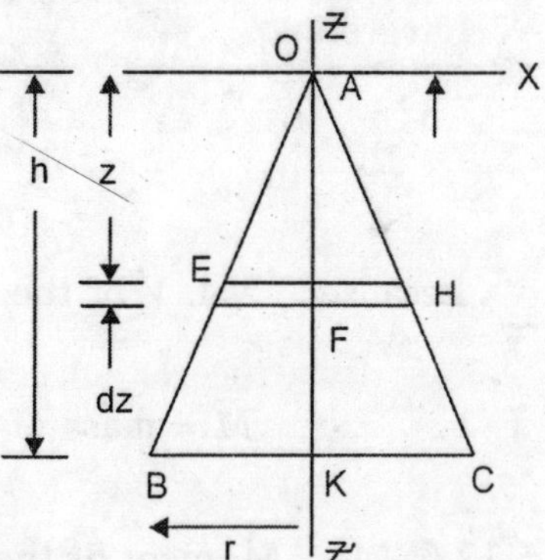

Fig. 5.62

$$\frac{EF}{AF} = \frac{BK}{AK}$$

or
$$EF = \frac{r}{h}\cdot z \qquad ...(5.112)$$

Then mass of the elemental disc of radius EF,

$$dM = \rho \times dV$$
$$= \rho(\pi\,(EF)^2\,dz)$$
$$= \rho\pi\left(\frac{r}{h}z\right)^2 dz$$

Now from equation (5.10) I_m of a disc about, normal to the section, centroidal axis is,

$$d(I_m)_{ZZ} = dm\cdot\frac{(EF)^2}{2}$$
$$= \rho\pi\left(\frac{r}{h}z\right)^2 dz\frac{(EF)^2}{2}$$
$$= \frac{\rho\pi}{2}\frac{r^2z^2}{h^2}dz\frac{r^2z^2}{h^2}$$
$$= \frac{\rho\pi}{2h^4}r^4\,z^4\,dz$$
$$= \frac{\pi\rho}{2}\frac{r^4}{h^4}\int_0^h z^4\,dz$$
$$= \frac{\pi}{4}\rho\frac{r^4}{h^4}\left[\frac{z^5}{5}\right]_0^h$$
$$= \frac{\pi}{2}\rho\frac{r^4}{h^4}\frac{h^5}{5}$$

$$= \frac{\pi}{2}\rho\frac{r^4 h}{5}$$

$$= \left(\frac{\pi r^2 h}{3}\rho\right)\frac{3r^2}{10} = \frac{3Mr^2}{10}$$

Because vol. V of the cone $= V = \dfrac{\pi r^2 h}{3}$ ...(5.113)

$\therefore$ M = mass of cone $= \rho V = \left(\rho\dfrac{\pi}{3}r^2 h\right)$

5.12.6 Mass Moment of Inertia of a sphere of Radius R

Let the axes OX, OY be taken though O the CG of sphere.

In Fig. 5.63 is shown an elemental disc of thickness dy at a right y from OX. The radius of the disc is BC.

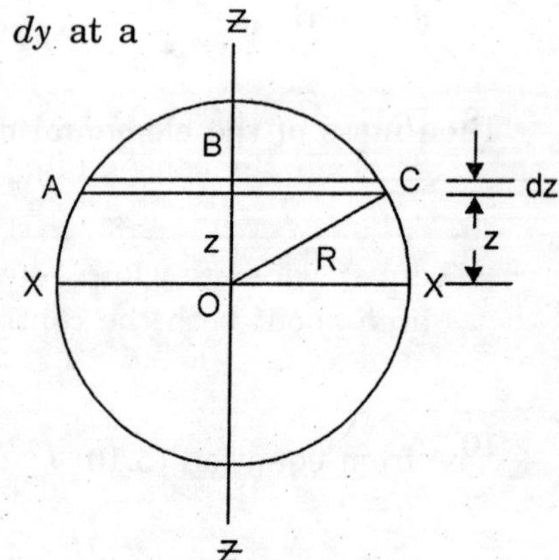

Fig. 5.63

$\therefore$ dM = Mm of elemental disc

$= \rho\pi(BC)^2 dy$

Now $BC^2 = (R^2 - z^2)$

$$\therefore \quad d(I_m)_{ZZ} = \frac{dM\,(BC)^2}{2}$$

$$= \frac{1}{2}\rho\pi(BC)^2 dz\,(BC)^2$$

$$(I_m)_{ZZ} = 2\int_0^R \frac{\rho\pi}{2}(x^2 - z^2)^2\,dz$$

$$= \int_{z=0}^{R} \rho\pi[R^4 + z^4 - 2R^2z^2]dz$$

$$= \rho\pi\left[R^4 z + \frac{z^5}{5} - \frac{2R^2z^3}{3}\right]_0^R$$

$$= \rho\pi\left[R^5 + \frac{R^5}{5} - \frac{2}{3}R^5\right]$$

$$= \frac{\rho\pi R^5}{15}[15 + 3 - 10] = \frac{8}{15}\rho\pi R^5$$

$$= \left(\frac{4}{3}\pi R^3\right)\rho\cdot\frac{2}{5}R^2$$

$$= \frac{2}{5}MR^2$$

PROBLEM FOR PRACTICE

1. Explain the difference between centroid and centre of gravity.
2. Define centre of gravity and centroid.
3. Define the terms "Area Moment of Inertia" and "Mass Moment of Inertia".
4. State and prove the theorems of parallel axes and perpendicular axes in the case of "Area Moment of Inertia".
5. Derive the expressions for second moment of area of a rectangle of width B and depth D about centroidal axis parallel to the depth.
6. Define the term "Radius of Gyration".
7. Show that the second moment of area of a circular section of diameter D about a diametric is $\frac{\pi D^4}{64}$.
8. Derive the expression for second moment of area of a triangle of height 'h' and base 'b' about the centroidal axis parallel to base.
9. A hollow rectangular section has wall thickness of 100 mm all around. The outside dimensions are width 1300 mm and depth 1200 mm. Calculate the second moment of area about both the centroidal axes XX and YY.

   (**Ans.** $I_{XX} = 955.3 \times 10^8$ mm^4, $I_{YY} = 1087.8 \times 10^8$ mm^4)

10. A cast iron channel section is shown in Fig. 5.64. Find the distance of centroid of the section from the base DF. Then calculate the value of second moment of area about centroidal axis parallel to DF.

    (**Ans.** $\bar{y} = 97.3$ mm, $I_{GX} = 2299.6971 \times 10^5$ mm^4)

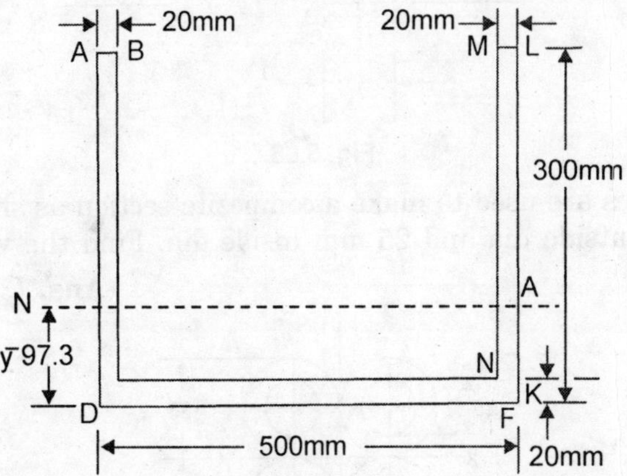

Fig. 5.64

11. Find the value of second moment of area of an I-section shown in Fig. 5.65 about XX.

    (**Ans.** 0.297×10^{10} mm^4)

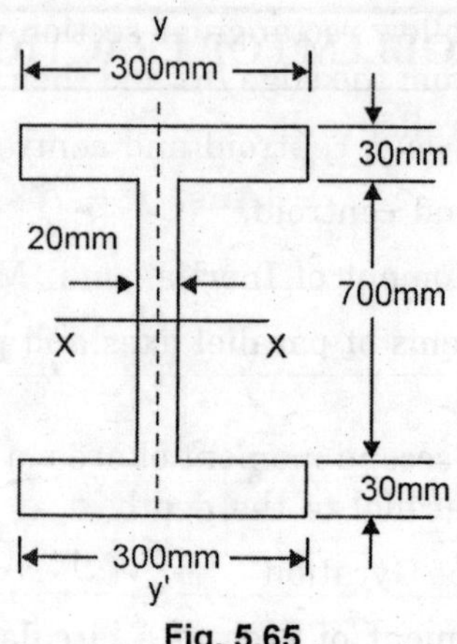

Fig. 5.65

12. Find the distance of centroid of the area of *T*-section shown in Fig. 5.66 from *AB* and then calculate Area Moment of Inertia about centroidal axis parallel to *AB*.

(Ans. $\bar{y}$ = 60 mm I_{XY} = 31.487 × 10^6 mm^4)

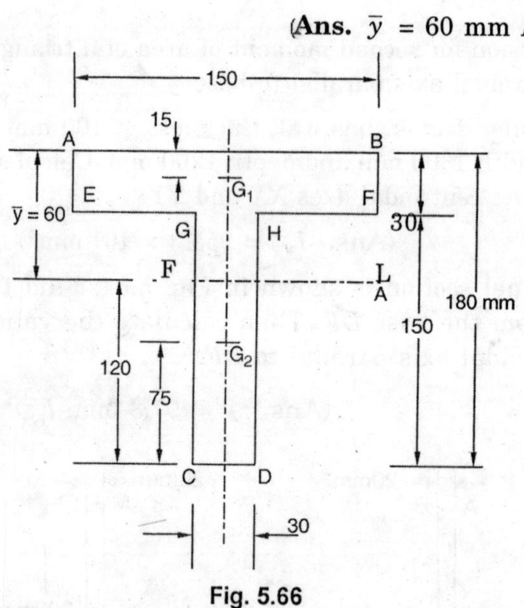

Fig. 5.66

13. Four similar tubes are used to make a composite section as shown in Fig. 5.67. Each tube is 35 mm outside dia and 25 mm inside dia. Find the value of I_{XX}.

(Ans. I_{XX} = 78.575 × 10^4 mm^4)

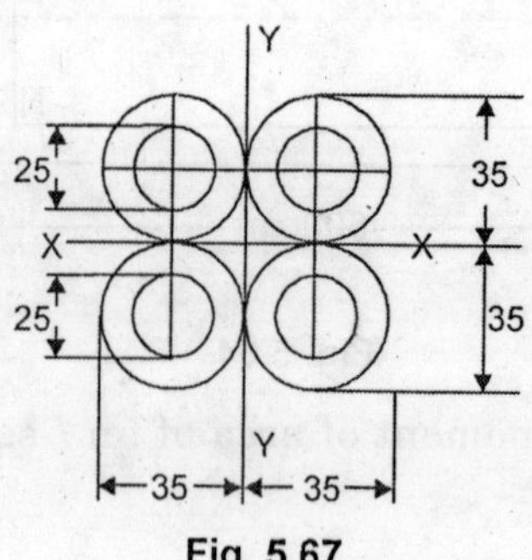

Fig. 5.67

14. In Fig. 5.68 is shown a hollow rectangular section which symmetrical about XX. Find the distance of centroid from the edge BC and then calculate I_{YY} about an axis passing through centroid and parallel to BC.

(Ans. $\bar{y}$ = 708 mm , $(I_G)_{YY}$ = 25.586 × 10^6 mm^4)

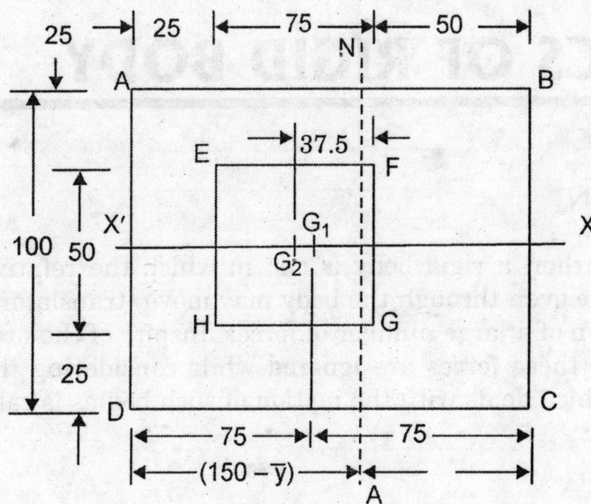

Fig. 5.68

15. A wooden beam rectangular section has a hole of 40 mm diameter as shown in Fig. 5.82. Find the distance of centroid from the base DC and then calculate $(I_{GG})_{XX}$.

(Ans. $(I_G)_{XX}$ = 24.95 × 10^6 mm^4, $\bar{y}$ = 75.64 mm)

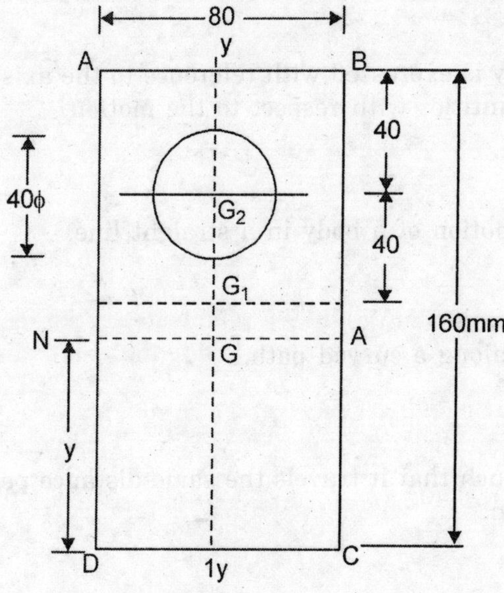

Fig. 5.69

□□□

CHAPTER 6

KINEMATICS OF RIGID BODY

6.1 INTRODUCTION

As already discussed earlier, a rigid body is one in which the relative position of any two particles does not change even through the body may move (translatory or rotational or both motions) under the action of a large number of forces. Inspite of the forces acting on the body and causing it to move, these forces are ignored while considering the motion of the body. The branch of science which deals with the motion of such bodies is called the KINEMATICS OF RIGID BODIES.

6.2 TYPES OF MOTION

6.2.1 Plane Motion

Plane motion of a body is the sum of the translation of the body with some point of the body plus rotation of the body about an axis passing through that point.

6.2.2 Absolute Motion

If the motion of a rigid body is expressed with reference to the axis on the earth, then these are called the absolute quantities with respect to the motion.

6.2.3 Rectilinear Motion

Rectilinear motion is the motion of a body in a straight line.

6.2.4 Curvilinear Motion

It is the motion of a body along a curved path.

6.2.5 Uniform Motion

If the motion of a body is such that it travels the same distance per unit time in the entire specified direction of motion.

6.2.6 Relative Motion

Relative motion is the motion in which quantities are measured with reference to moving axes.

6.3 RECTILINEAR VELOCITY

Linear or rectilinear velocity is defined as the rate of displacement per unit time along a straight line. It is a vector quantity and can be represented by an arrow head, the length of which is propotional to the magnitude of velocity and the arrowhead gives the direction of movement of the body.

Let a body be moving along a straight path. At any instant, if the displacement is ds in the duration dt their at that instant velocity v is expressed as,

$$v = \frac{ds}{dt} \quad \text{...(6.1)}$$

6.3.1 Acceleration

If the velocity is not uniform and it varies with time, the acceleration $\vec{a}$ is defined as the rate of change of velocity and mathematically it is represented by,

$$a = \frac{dv}{dt} = \frac{d^2 s}{dt^2} \quad \text{...(6.2)}$$

Further $\frac{dv}{dt}$ can also be written as,

$$\frac{dv}{dt} = \frac{dv}{ds} \times \frac{ds}{dt}$$

$$= v\frac{dv}{ds}$$

$$a = v \cdot \frac{dv}{ds} \quad \text{...(6.3)}$$

6.4 EQNATONS OF MOTION IN STRAIGHT LINE

For a body moving in a straight line, let the following be assumed,

u = initial velocity

v = final velocity after time t

a = acceleration of the body

s = distance travelled in time t

Then change of velocity in time t,

$$= v - u$$

$\therefore$ Acceleration

$$a = \frac{v-u}{t}$$

or

$$v = u + at \quad \text{...(6.4)}$$

It may be noted that in equation (6.4), u is the initial velocity at the start of motion *i.e.*, at $t = 0$ and a is the uniform acceleration which is not changing with time.

If ds is the distance covered in time dt at the instant when velocity is v at time t

Then $$ds = v\,dt$$

Total distance covered in time t,

$$\int_0^t ds = \int_0^t u\,dt = \int_0^t (u + at)\,dt$$

$$s = \int_0^t u\,dt + a\int_0^t t\,dt$$

$$s = ut + \frac{1}{2}at^2 \qquad \text{...(6.5)}$$

Substituting the value of t from equation (6.4) into equation (6.5) we get,

$$s = u\left(\frac{v-u}{a}\right) + \frac{1}{2}a\left(\frac{v-u}{a}\right)^2$$

$$= \left(\frac{v-u}{a}\right)\left[u + \frac{a}{2}\cdot\frac{v-u}{a}\right]$$

$$= \left(\frac{v-u}{a}\right)\left[u + \frac{v}{2} - \frac{u}{2}\right]$$

$$= \left(\frac{v-u}{a}\right)\left(\frac{v+u}{2}\right)$$

$$= \frac{v^2 - u^2}{2a}$$

i.e., $$v^2 - u^2 = 2as \qquad \text{...(6.6)}$$

6.5 BODY MOVING WITH UNIFROM VELOCITY ALONG A CURVED PATH

Let a body be moving with a uniform velocity v along a curvilinear path of radius 'r'. At any instant let it be at the position A of the path while the radius is r. After time dt it moves through an angle $d\theta$ to reach the position B. At the point A the velocity vector is $O'C$ which is directed at 90° to OA and at position B it is $O'D$ and directed at 90° to OB. Thus, as shown in Fig. (6.1 b) the change in the velocity is vector $\overline{CD}$ which is at right angles to $O'C$ *i.e.*, parallel to the this change in velocity has occurred in time dt

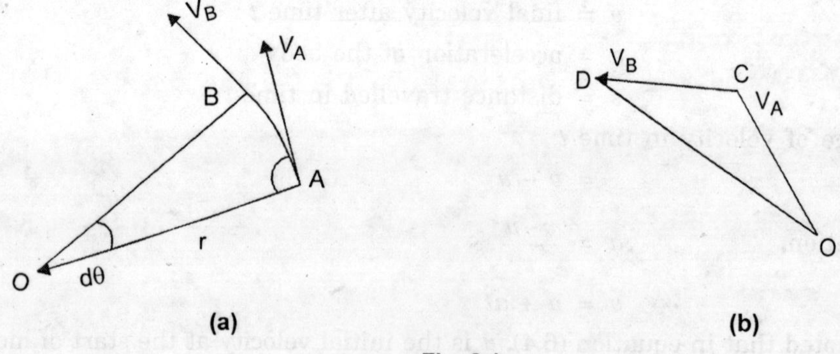

Fig. 6.1

Thus, the acceleration is $\frac{\overrightarrow{CD}}{dt}$

or $$a = \frac{\overrightarrow{CD}}{dt}$$

or $$a = v\frac{d\theta}{dt} \qquad ...(6.7)$$

But from Fig. 6.1*a* $$r\,d\theta = \overrightarrow{AB}$$

or $$r\frac{d\theta}{dt} = \frac{\overrightarrow{AB}}{dt}$$

$$= v$$

∴ $$\frac{d\theta}{dt} = \frac{v}{r} \qquad ...(6.8a)$$

or $$\frac{d\theta}{dt} = \omega \qquad ...(6.8b)$$

Where ω is the angular speed of the body about *O*. Therefore, from (6.7) and (6.8),

$$a = v\cdot\frac{v}{r} = \frac{v^2}{r} \qquad ...(6.9)$$

and it is directed towards *O*, this centre of rotation. It is called the centrepetal acceleration.

Also, from (6.8) and (6.9),

$$a = \frac{v^2}{r} = r\omega^2 \qquad ...(6.10)$$

As discussed in chapter 1, the equilibrium of the body can be considered for the application of equations of static equlibrium, if the inertia force is superimposed or the body (*D' Alembert* Principle).

Here, the inertia force

= (mass × acceleration) and is in opposite direction to the direction of acceleration

$$F_i = M\frac{v^2}{r} \qquad \text{(radially outward)}$$

$$= \frac{W}{g}\frac{v^2}{r} = \frac{W}{g}\omega^2 r \qquad ...(6.11)$$

This F_i is called centrifugal force.

EXAMPLE

6.1 : *The motion of a rigid body along a straight path is represented by equation,*

$$s = 16t + 4t^2 + 3t^3$$

where s is the distance moved in metres, t is the time in sec. Find (a) O′ the displacement, velocity and acceleration after 2 seconds (b). After what time the velocity will be zero and what will then be the distance covered and the value of acceleration (c). After what time the acceleration will be zero and displacement and velocity of the body.

Solution : Given data

$$s = 16t + 4t^2 - 3t^3 \quad ...(A)$$

$$\therefore \quad v = \frac{ds}{dt} = 16 + 8t - 9t^2 \quad ...(B)$$

$$a = \frac{dv}{dt} = 8 - 18t \quad ...(C)$$

(*a*) At = *t* = 2 sec

$$s = 16 \times 2 + 4 \times (2)^2 - 3(2)^3$$
$$= 32 + 16 - 24 = 24 \text{ m}$$
$$v = 16 + 8 \times 2 - 9 \times (2)^2$$
$$= 16 + 16 - 36 = -4 \text{ m/sec}$$
$$a = 8 - 18 \times 2 = -28 \text{ m/s}^2$$

(*b*) When zero, then from equation (*B*),

$$16 + 8t - 9t^2 = 0$$

or

$$9t^2 - 8t - 16 = 0$$

$$t = \frac{1}{18}\left[8 \pm \sqrt{64 + 64 \times 9}\right]$$

$$= \frac{1}{18}\left[8 \pm 8\sqrt{10}\right]$$

or

$$t = \frac{1}{18}[8 + 25.3] = 1.85 \text{ sec}$$

$$\therefore \quad s = 16 \times 1.85 + 4\,[1.85]^2 - 4\,(1.85)^3$$
$$= 29.6 + 13.69 - 25.32 = 17.97 \text{ m}$$

And

$$a = 8 - 18 \times 1.85 = -25.3 \text{ m/s}^2$$

(*c*) When

$$a = 0$$

or

$$8 - 18t = 0$$

$$t = \frac{8}{18} = 0.44 \text{ sec}$$

Then

$$s = 16 \times 0.44 + 4\,(.44)^2 - 3(.44)^3$$
$$= 7.04 + 0.7744 - 0.25555$$
$$= 2.5589 \text{ m}$$

and $v = 16 + 8\,(0.44) - 9\,(.44)^2$

$= 16 + 3.52 - 1.7424$

$= 17.777 \text{ m/s}^2$

6.2 : *The displacement 's' of a body moving along a straight line at time 't' is given by the equation,*

$$s = 20t + 3t^2 - 2t^3$$

Find (a) the velocity and acceleration at the start of motion (b) the time when the velocity will be maximum and the value of maximum velocity.

Solution : Given

(*a*) $s = 20t + 3t^2 - 2t^3$ ...(*A*)

$\therefore$ $v = \dfrac{ds}{dt} = 20 + 6t - 6t^2$ ...(*B*)

and $a = \dfrac{dv}{dt} = 6 - 12\,t$ ...(*C*)

$\therefore$ At $t = 0$ $v = 20 + 0 - 0 = 20$ m/s

and $a = 6 - 0 = 6 \text{ m/s}^2$ **Ans.**

(*b*) For max. velocity $\dfrac{dv}{dt} = 0$

$\therefore$ From equation (*c*) $\dfrac{dv}{dt} = 0 = 6 - 12t$

or $t = \dfrac{6}{12} = 0.5\,\text{sec}$

At $t = 0.5$ from (*B*) $v = 20 + 6 \times (.5) - 6\,(0.5)^2$

$= 20 + 3 - 1.5$

$= 21.5$ m/s **Ans.**

6.3 : *The distance travelled from starting point by a car is proportional to $(t)^3$. During the first 60 seconds it travels 216 m. Find the distance travelled during first 5 seconds after starting.*

Solution : Given data $= kt^3$ when k is a constant

$\dfrac{ds}{dt} = 3kt^2$

Now it is given that $s = 216$ m when $t = 80$ s

$\therefore$ $216 = k \cdot (60)^3$

$= 216000\,k$

$k = \dfrac{216}{216000} = \dfrac{1}{1000}$

$\therefore$ $s = \dfrac{t^3}{1000}$

EXAMPLE

EXAMPLE

$$v = \frac{ds}{dt} = \frac{3t^2}{1000}$$

and $$a = \frac{dv}{dt} = \frac{6t}{1000}$$

When $t = 5$ $$s = \frac{(5)^3}{1000} = 0.125 \text{ m/s}$$

$$v = \frac{ds}{dt} = \frac{3\times(5)^3}{1000} = \frac{75}{1000} = 0.075 \text{ m/s}$$

$$a = \frac{6\times 5}{1000} = 0.003 \text{ m/s}^2$$

6.4 : *On a national highway, a car is moving at a velocity of 120 km/hr. The car is brought to rest by applying brakes in 5 sec. Find the distance covered by the car before coming to rest and the value of the retardation.*

Solution : Given data $u = 120$ km/hr

$$= \frac{120\times 1000}{3600} = \frac{1000}{30} \text{ m/s} = \frac{100}{3} m/s$$

$$v = 0, \quad t = 5 \text{ secs}$$

(*a*) we have, $$v = u + at$$

$$0 = \frac{1000}{30} + a\times 5$$

or $$a = -\frac{1000}{150} = -\frac{20}{3} = -6.667 \text{ m/s}^2$$

(*b*) Again we know,

$$v^2 - u^2 = 2as$$

$$0 - \left(\frac{100}{3}\right)^2 = -2\times\frac{20}{3} \text{ s}$$

or $$s = \left(\frac{100}{3}\right)^2 \times \frac{3}{40} = \frac{100\times 100}{3\times 40} = 83.33 \text{ m}$$ **Ans.**

6.5 : *A bullet moving at a velocity of 200 m/s is fired into a block of wood and penetrates into it 300 mm and comes to rest. Again a bullet is fired with the same velocity into a similar wooden board 200 mm thick ,find the exist velocity if the bullet is in this case. Assume that the resistance offered to be uniform in both the cases.*

Solution : First case : Given data :

$$u = 200 \text{ m/s}$$
$$v = 0$$
$$s = 300 \text{ mm}$$
$$= 0.3 \text{ m}$$

We have, $v^2 - u^2 = 2as$

$$0 - (200)^2 = 2a \times 0.3$$

$$a = -\frac{(200)^2}{0.6} = -6.667 \times 10^4 \text{ m/s}^2$$

In the **Second case :** $s = 200 \text{ mm} = 0.2 \text{ m}$

$$u = 200 \text{ m/s}, a = -6.667 \times 10^4$$
$$v = ?$$
$$v^2 - u^2 = 2as$$

or $v^2 = 2(-6.667 \times 10^4)(0.2) + (200)^2$

or $v^2 = 2.667 \times 10^4 + 4 \times 10^4$

$$= 1.33 \times 10^4$$
$$v = 1.1545 \times 4^2 = 115.45 \text{ m/s}$$

E X A M P L E

6.6 : *After snatching a chain from a lady, snatchers ran away on a motorcycle with an acceleration of 1.5 m/s². The police party arrived the place of accident after 15 seconds and followed the snatchers on their jeep moving at a uniform velocity of 60 m/s. Find the time taken by the police party to catch the snatchers.*

Solution : Given data acceleration of motorcycle

$$a = 1.5 \text{ m/s}^2$$

and $u = 0$

Uniform velocity of Jeep = 60 m/s

Let the police party catch the snatchers after t secs.

Distance covered by the snatches in $(t + 15)$ sec

$$s = ut + \frac{1}{2}at^2$$
$$= 0 + \frac{1}{2} \times (1.5)(t + 15)^2$$
$$= 0.75 (t + 15)^2 \quad \text{...(A)}$$

This distance is covered by the police Jeep in t secs, at uniform velocity of 60 m/s.

$$s = ut$$
$$= 60\,t \quad \text{...(B)}$$

$\therefore$ From (A) and (B), $60t = 0.75 (t - 15)^2$

$$\frac{60 \times 4t}{3} = (t + 15)^2$$

$$80t = t^2 + 225 + 30t$$

or $$t^2 - 50t + 225 = 0$$

i.e., $$t = \frac{50 \pm \sqrt{(50)^2 - 4 \times 225}}{2}$$

$$= 25 \pm \frac{1}{2}\sqrt{2500 - 900}$$

$$= 25 \pm 20$$

$$ut = 5 \text{ or } 45 \text{ secs}$$

The required answer will be smaller of the two values *i.e.,* 5 sec.

6.7 : *A smugglars truck moving at a speed of 12 m/s crosses the police barrier without stopping. After 5 seconds the police party starts chasing the truck in a Jeep which started from rest with on acceleration of 1.2 m/s². Find the time elapsed where the police party will catch the truck.*

Solution : Let the time taken by the police jeep to catch the truck be t sec. Then distance travelled by truck in $(t + 5)$ see when moving at uniform speed of 12 m/s,

$$s = 12\ (t + 5) \qquad \text{...}(A)$$

Again the same distance is covered by jeep in t secs.

i.e., $$s = ut + \frac{1}{2}at^2$$

$$= 0 + \frac{1}{2} \times (1.2)t^2$$

$$= 0.6\ t^2 \qquad \text{...}(B)$$

Equation (A) and (B),

$$12(t + 5) = 0.6t^2$$

$$12t + 60 = 0.6t^2$$

or $$0.6t^2 - 12t - 60 = 0$$

$$t = \frac{12 \pm \sqrt{(12)^2 + 4 \times 60 \times 0.6}}{2 \times (.6)}$$

$$= \frac{1}{1.2}[12 \pm \sqrt{144 + 144}]$$

$$= \frac{1}{1.2}[12 \pm \sqrt{288}]$$

$$= \frac{1}{1.2} \pm [12 + 16.97]$$

$$\therefore \quad t = 24.14 \text{ secs } \textbf{Ans.}$$

6.6 DISTANCE MOVED IN n^{th} SECOND

Suppose a body starts moving with a velocity u and acceleration a. It is required to find the distance covered during the nth second.

The distance moved during nth second = Distance covered in n seconds

– distance moved in $(n - 1)$ second

Now the distance moved in n secs,

$$s_n = un + \frac{1}{2}an^2 \quad \text{...(6.12)}$$

Distance covered in $(n - 1)$ sec,

$$s_{n-1} = u(n-1) + \frac{1}{2}a(n-1)^2 \quad \text{...(6.13)}$$

∴ Subtracting (6.13) from (6.12) we get,

$$s_{nth\ sec} = s_n - s_{n-1}$$

$$= \left(un + \frac{1}{2}an^2\right) - \left[u(n-1) + \frac{a}{2}(n-1)^2\right]$$

$$= u[n-(n-1)] + \frac{a}{2}[n^2 - (n-1)^2]$$

$$= u + \frac{a}{2}(2n-1) \quad \text{...(6.14)}$$

EXAMPLE

6.8 : *A body starts moving with a velocity of 14 m/s and uniform acceleration of 5 m/s². Find the distance covered during the 10th sec.*

Solution : Let S_{10} be the distance travelled in 10 secs. and S_9 during the period of 9 secs.

Then
$$S_{10} = 10u + \frac{1}{2}a(10)^2$$

$$= 10 \times 14 \times \frac{1}{2} \times 5 \times 100$$

$$= 140 + 250 = 390$$

And
$$S_9 = 9u + \frac{1}{2}a(9)^2$$

$$= 9 \times 14 + \frac{1}{2} \times 5 \times 81$$

$$= 126 + 202.5$$

$$= 328.5 \text{ m}$$

∴ Distance covered during 10th sec

$$= 390 - 328.5$$

$$= 61.5 \text{ m } \textbf{Ans.}$$

EXAMPLE

6.9 : *A car starts with a velocity u and travelling at uniform acceleration covers 36.5 m during 5th sec and 61.5 m during 10th sec. Determine the starting velocity and the value of uniform acceleration.*

Solution : Let starting velocity $= u$

and acceleration $= a$

The $$S_{10} = 10u + \frac{1}{2}a\,(10)^2$$

$$S_9 = 9u + \frac{1}{2}a\,(9)^2$$

$\therefore$ Distance moved during 10th sec,

$$61.5 = S_{10} - S_9$$

$$= \left[10u + \frac{1}{2}a(10)^2\right] - \left[9u + \frac{1}{2}a\,(9)^2\right]$$

$$= u + \frac{a}{2}(100 - 81)$$

or $$61.5 = u + 9.5a \qquad ...(A)$$

Similarly, $$S_5 = 5u + \frac{1}{2}\,a(5)^2$$

$$S_4 = 4u + \frac{1}{2}\,a(4)^2$$

or $$36.5 = S_5 - S_4$$

$$= \left[5u + \frac{1}{2}a(5)^2\right] - \left[4u \times \frac{1}{2}a(4)^2\right]$$

$$= u + \frac{a}{2}(25 - 16)$$

or $$36.5 = u + 4.5a \qquad (B)$$

Subtracting (*B*) from (*A*),

$$61.5 - 36.5 = (9.5 - 4.5)\,a$$

$$25 = 5a$$

$$a = 5 \text{ m/sec}^2$$ **Ans.**

Then from (*B*), $$36.5 = u + 4.5 \times 5$$

$$= u + 22.5$$

$$u = 36.5 - 22.5$$

$$= 14 \text{ m/s}$$ **Ans.**

6.7 EQUATIONS OF MOTION DUE TO GRAVITY

Whenever a body is moving in the vertical direction either down towards the centre of earth or upwards away from the centre of earth its motion is affected due to te attraction of earth. The acceleration due to gravity, g will accelerate the body while it is moving downwords and the retardation equal to g will be there, when it is moving upwards.

i.e., for downwords movement acceleration $a = +g$

For upwards motion $a = -g$

Thus, if a body is projected upwards with initial velocity u after sometimes its velocity will become zero due to uniform retardation of value g and then it will start moving down from the highest point where its velocity had reached the value zero. It will be equvalent to starting from rest with acceleration of g, when it has started moving down. Thus, if it is projected upwards with a velocity u and 'h' is the height of the highest point when the velocity will be zero, we have,

$$v^2 - u^2 = 2as$$

Here, $v = 0$ $\quad a = -g$ $\quad$ and $\quad s = h$

$$0 - u^2 = -2 \times gh$$

$$h = \frac{u^2}{2g} \qquad \text{...(6.15)}$$

EXAMPLE

6.10 : *A stone is dropped from the top of a cliff 100 m high. Find the time taken by the stone to reach the ground. Take g = 9.81 m/s².*

Solution : $u = 0,\ h = 100 \text{ m},\ a = g = 9.81 \text{ m/s}^2$

$$s = ut + \frac{1}{2}gt^2$$

$$100 = 0 + \frac{1}{2} \times 9.81 \times t^2$$

$$\therefore \quad t^2 = \frac{2 \times 100}{9.81} = 20.381$$

$$\therefore \quad t = 4.515 \text{ sec.}$$ **Ans.**

6.11 : *A stone is thrown up with a velocity of 30 m/s. Find the time when the stone will be at a height of 30 m. Take g = 9.81 m/s².*

Solution : Given $u = 30 \text{ m/s}$

$s = 30$ m (above ground)

$a = -g = -9.81 \text{ m/s}^2$

$$s = ut + \frac{1}{2}at^2$$

$$30 = 30 \times t - \frac{1}{2} \times 9.81t^2$$

EXAMPLE

$$\text{or} \quad 60 = 60t - 9.81t^2$$

$$\text{or} \quad 9.81t^2 - 60t + 60 = 0$$

$$t = \left[\frac{60 \pm \sqrt{(60)^2 - 4 \times 60 \times 9.81}}{2 \times 9.81}\right]$$

$$= \frac{1}{2 \times 9.81}\left[60 \pm \sqrt{1245.6}\right]$$

$$= \frac{60 \pm 35.293}{2 \times 9.81}$$

i.e., $t = t_1 = 1.259$ sec

or $t = t_2$

$= 4.856$ sec **Ans.**

When $t = t_1$ the body will be moving up and when $t = t_2$ it will be moving down which is shown below,

At $t = t_1$, $v = u - gt_1$

$= 30 - 9.81 \times 1.259$

$= 17.64$ m/s

At $t = t_2$, $v = 30 - 9.81 \times 4.586$

$= 30 - 47.64$

$= -17.64$ m/s **Ans.**

6.12 : *A stone is dropped into a well. The sound of impact with water was heard after 5 secs. If the velocity of sound in air is 350 m/s, what is the depth of water in the well? Take g = 9.81.*

Solution : Let the depth of water in well be = 'h' meter and let t_1 is the time taken for the stone to reach water level. Then,

$$S = ut_1 + \frac{1}{2}gt_1^{\,2}$$

$$h = ut_1 + \frac{1}{2}gt_1^{\,2}$$

$$\text{or} \quad h = 0 + \frac{1}{2} \times 9.81 \times t_1^{\,2}$$

$$t_1^{\,2} = \frac{2.h}{9.81} \quad \text{or} \quad t_1 = \sqrt{\frac{2h}{9.81}}$$

If t_2 is the time taken by sound to cover a distance equal to 'h', then,

$$h = t_2\, u_s = t_2 \times 350$$

$$t_2 = \frac{h}{350}\text{ sec}$$

Now it is given $t_1 + t_2 = 5$

$$\therefore \quad \sqrt{\frac{2h}{9.81}} + \frac{h}{350} = 5$$

$$\sqrt{\frac{2h}{9.81}} = 5 - \frac{h}{350}$$

$$= \frac{1750 - h}{350}$$

or
$$\frac{2h}{9.81} \times (350)^2 = (1750 - h)^2$$

$$24974.5h = 30625 + h^2 - 3500h$$

or
$$h^2 - 28474.5h + 3062500 = 0$$

$$h = \frac{1}{2}\left[28474.5 \pm \sqrt{(28474.5)^2 - 4 \times 306300}\right]$$

$$= \frac{1}{2}\left[28474.5 \pm \sqrt{8.108 \times 10^8 - .1225 \times 10^8}\right]$$

$$= \frac{1}{2}[2.8474 \times 10^4 \pm \sqrt{7.9354 \times 10^8}]$$

$$= \frac{1}{2}[2.8474 \times 10^4 \pm 2.817 \times 10^4]$$

$$= \frac{1}{2} \times 5.6575 \times 10^4 \text{ or } \frac{1}{2} \times 300$$

$$= 2.83 \times 10^4 \text{ m or } 150 \text{ m}$$

The answer will be smaller of the two *i.e.*,

$$h = 150 \text{ m}$$

6.13 : *A stone is dropped from the top of a tower. It passes two points A and B in 0.2 sec. If the point B is 10 m vertically below A, Find the distance between the top of tower and point A.*

Take $g = 9.8\ m/s^2$

Solution : Let the time taken to cover the distance h from the top of tower to reach the point A *be* t.

$$h = \frac{1}{2}t^2 \qquad ...(A)$$

Then $(t + 0.2)$ sec will be the time taken by stone to cover a distance $(h + 10)$ m

EXAMPLE

EXAMPLE

i.e., $$(h + 10) = \frac{1}{2}g(t+0.2)^2 \quad ...(B)$$

Subtract (*A*) from (*B*),

$$(h + 10) - h = \frac{g}{2}[(t+0.2)^2 - t^2]$$

or $$10 = \frac{g}{2}[0.2(2t+0.2)]$$

$$\frac{10 \times 2}{9.8 \times 0.2} = (0.2 + 2t)$$

$$10.2 = 0.2 + 2t$$

or $$2t = 10$$

i.e., $$t = 5 \text{ sec}$$

$\therefore$ $$h = 0 + \frac{1}{2}gt^2$$

$$= \frac{1}{2} \times 9.8 \times (5)^2$$

$$= 122.5 \textbf{ Ans.}$$

6.8 RIGID BODY ANGULAR MOTION

Let a body B be moving in a circular path of radius r. Let the reference line be OA from which it starts rotating. The line OA is called the initial line. The angle θ through which the body rotates about the centre 0 is called the angular displacement. Let the body move further through an angle $d\theta$ in time dt, then $\frac{d\theta}{dt}$ is called the angular speed of rotation about the take origin 'O' and represented by ω.

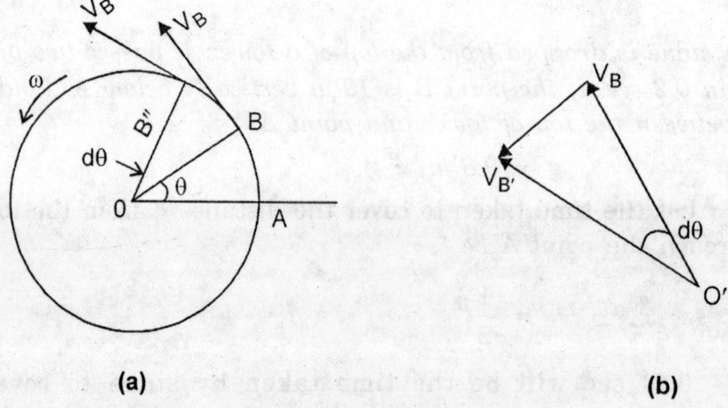

Fig. 6.2

Then $$\omega = \frac{d\theta}{dt} \quad \text{...(6.15)}$$

Further if v is the perpheral velocity of B, we notice that,

$$v = \frac{arc\ BB'}{dt}$$

$$= \frac{rd\theta}{dt} = r\omega \quad \text{...(6.16a)}$$

Further centripetal acceleration in the radial direction,

$$a = \frac{dv}{dt}$$

$$= \frac{d}{dt}\left[r\frac{d\theta}{dt}\right] \quad \text{for (6.16a)}$$

$$= r\frac{d^2\theta}{dt^2} = r\alpha \quad \text{...(6.16b)}$$

Also, for Fig. (6.2b), $$a = \frac{dv}{dt} = \frac{vd\theta}{dt}$$

$$= v[\omega]$$

$$= r\omega(\omega) = r\omega^2 \quad \text{...(6.17a)}$$

$$= r\left(\frac{v}{r}\right)^2 = \frac{v^2}{r}$$

Angular acceleration, $$\alpha = \frac{d\omega}{dt}$$

$$= \frac{d}{dt}\left(\frac{d\theta}{dt}\right)$$

$$= \frac{d^2\theta}{dt^2} \quad \text{...(6.17b)}$$

6.9 UNIFORMLY ACCELERATED ANGULAR MOTION

$$\alpha = \frac{d\omega}{dt}$$

or $$d\omega = \alpha\ dt$$

Integrating $\omega = \alpha t + c$ (where c is constant of integration)

If the initial velocity at time $t = 0$ & $w = w_0$

then $= 0 + c$

or $c = \omega_0$

i.e., $\omega = \omega_0 + \alpha t$ ...(6.18)

or $$\frac{d\theta}{dt} = \omega_0 + \alpha t$$

or $$d\theta = \omega_0\, dt + \alpha\, tdt$$

Integrating $$\theta = \omega_0 t + \frac{1}{2}\alpha t^2 + c_2 \quad ...(6.19)$$

But $\theta = 0$

at $t = 0$

$\therefore$ $0 = 0 + 0 + e_2$

or $e_2 = 0$

$\therefore$ $$\theta = \omega_0 t + \frac{1}{2}\alpha t^2 \quad ...(6.20)$$

Then equation (6.18) and (6.20),

$$t = \frac{\omega - \omega_0}{\alpha}$$

i.e., $$\theta = \omega_0\left(\frac{\omega - \omega_0}{\alpha}\right) + \frac{1}{2}\alpha\left(\frac{\omega - \omega_0}{\alpha}\right)^2$$

$$= \left(\frac{\omega - \omega_0}{\alpha}\right)\left(\omega_0 + \frac{1}{2}(\omega - \omega_0)\right)$$

or $$\theta = \left(\frac{\omega - \omega_0}{\alpha}\right)\left(\frac{\omega + \omega_0}{2}\right)$$

$$= \frac{1}{2\alpha}[\omega^2 - \omega_0^2]$$

or $$\omega^2 - \omega_0^2 = 2\alpha\, \theta \quad ...(6.21)$$

Note : Units of θ are radians and that of ω_1 the angular speed are rad/sec. In most of the industrial applications, the speed is given in revolution per minutes in short rpm the relationship between ω and, rpm is given below.

Now 1 revolution $= 2\pi$ radians

$\therefore$ N rpm $= 2\pi N$ rad/min

or $$\omega = \frac{2\pi N}{60} \text{ rad/sec} \quad ...(6.22)$$

6.14 : *A wheel rotating about a fixed axis at 20 r.p.m. is uniformly accelerated for 70 seconds during which time it finds 50 revolutions find :*

(a) Angular velocity of the end of the interval.

(b) Time required for speed to reach 100 rpm.(UP Tech Univ. 2000-2001)

Solution : Given data

$$\omega_0 = 20 \text{ rpm} = \frac{20 \times 2\pi}{60} \text{ rad/sec}$$

$$= \frac{2}{3}\pi \text{ rad/sec} = 2.09$$

$$t = 70 \text{ sec}$$

$$\theta = 50 \text{ revolution}$$

$$= 50 \times 2\pi = 100\pi \text{ rad}$$

$$\theta = \omega_0 t + \frac{1}{2}\alpha t^2$$

$$100\pi = \frac{2\pi}{3}(70) + \frac{1}{2}\alpha(70)^2$$

$$314.16 = 146.61 + 2450\ \alpha$$

$$\therefore \quad \alpha = \frac{167.55}{2450} = 0.068 \text{ rad/s}^2$$

$$\omega_1 = \omega_0 + \alpha t$$

$$= \frac{2\pi}{3} + 0.068 \times 70 = 2.094 + 4.76$$

$$= 6.856 \text{ rad/sec}$$

$$\omega_2 = 100 \text{ rpm} = \frac{2\pi}{60} \times 100 = 10.47 \text{ rad/sec}$$

$$\omega_2 = \omega_0 + \alpha t$$

$$10.47 = 2.09 + 0.068 \times t$$

$$t = \frac{8.38}{0.068} = 123 \text{ sec} \quad \textbf{Ans.}$$

6.15 : *A train starts from rest and moves along a curved track of 800 m radius with uniform acceleration until it attains a velocity of 72 km/m at the end of 3 minutes. Determine the tangential, normal and total acceleration (in m/s^{2}) of the train at the end of second minute.*

(UP Tech Univ 2000-01)

Solution : Given data $\omega_0 = 0$

EXAMPLE

E X A M P L E

$$r = 800 \text{ m}$$

After 180 sec, $v = 72$ km/hr

$$= \frac{72 \times 1000}{3600} \text{ m/sec}$$

$$= 20 \text{ m/s}$$

$$\omega = \frac{v}{r} = \frac{20}{800} = \frac{1}{40} \text{ rad/sec}$$

Let the tangential acceleration be α after 3 minutes,

$$\omega = \omega_0 + \alpha t$$

$$\frac{1}{40} = 0 + \alpha \times 180$$

$$\alpha = \frac{1}{40 \times 180} \text{ rad/s}^2$$

After two moments *i.e.*, 120 sec,

$$\omega_{120} = \omega_0 + \alpha \times 120$$

$$= 0 + \frac{1}{40 \times 180} \times 120 = \frac{1}{60} \text{ rad/s}$$

Now after 2 minutes,

Centripetal acceleration or radial accderation,

$$a_{rc} = \omega^2 r \qquad \text{(for equation 6.17}a\text{)}$$

$$= \left(\frac{1}{60}\right)^2 \times 800$$

$$= 0.222 \text{ m/s}^2 \qquad \text{(for equation 6.16}b\text{)}$$

Tangential acceleration $= r\alpha$

or $$a_t = 800 \times \frac{1}{40 \times 180}$$

$$= 0.111 \text{ m/s}^2$$

Resultant acceleration $= \sqrt{a_t^{\,2} + a_r^{\,2}}$

$$= \sqrt{(0.111)^2 + (0.222)^2}$$

$$= 0.246 \text{ m/s}^2$$

6.16 : *The rotor of an electric motor uniformly accelerates at a speed of 1800 rpm from rest in 5 seconds and then the rotor decelerate to stop. If the total time elapsed from start to stop is 12.3 seconds, Determine the number of revolutions made during acceleration and deceleration. What would be the reason for larger time taken during stopping?*

(U.P. Tech Univ. 2002-03)

(U.P. Tech Univ. 2002-03)

Solution : Given data :

$$\omega_0 = 0$$

$$N = 1800 \text{ rpm}$$

or $$\omega = \frac{1800 \times 2\pi}{60}$$

$$= 60\pi \text{ rad/s}$$

$$t = 5 \text{ sec}$$

We know $$\omega = \omega_0 + \alpha t$$

$$60\pi = 0 + \alpha \times 5$$

$\therefore$ $$\alpha = \frac{60\pi}{5}$$

$$= 37.699 \text{ rad/sec}$$

Angle covered in 5 sec,

$$\theta = \omega_0 t + \frac{1}{2}\alpha t^2$$

$$= 0 + \frac{1}{2} \times 12\pi \times (5)^2$$

$$= 150\pi \text{ rad}$$

Corresponding number. of revolution

$$= \frac{150\pi}{2\pi} = 75 \text{ rev.}$$

Deceleration time $= 12.3 - 5$

$$= 7.3 \text{ sec}$$

Here, $$\omega_0 = 60\pi$$

and $$\omega_2 = 0$$

$$\omega_2 = \omega_0 + \alpha_2 t$$

$$0 = 60\pi + \alpha \times 7.3$$

$$\alpha_2 = -\frac{60\pi}{7.3} = -25.82$$

Now $$\omega_2^2 - \omega_0^2 = -2\alpha_2 \theta_2$$

$$0 - (60\pi)^2 = -2 \times 25.82\, \theta_2$$

$$\theta_2 = \frac{(60\pi)^2}{2 \times 25.82} = 688 \text{ rad}$$

$$= \frac{688}{2\pi}$$

EXAMPLE

E X A M P L E

$$= 109.5 \text{ revolution}$$

In this second case the starting velocity is more than in the first case, therefore, θ_2 is bigger. Moreover due to inertia of the wheel, continues to rotate during stopping cycle. Therefore,

$$\theta_2 > \theta_1$$

6.17 : *The motor of a belt conveyer running of 720 rpm when the electric supply to the motor was cut off. The motor came to rest after making 250 revolutions. Determine the angular retardation and the time taken to come to rest.*

Solution : Given data $N = 720$ rpm

$$\omega_0 = \frac{2\pi N}{60}$$

$$= \frac{2\pi \times 720}{60}$$

$$= 24\pi \text{ rad/s}$$

$$\theta = 250 \times 2\pi \text{ radius,}$$

when $\omega = 0$

$$\omega_0^2 - \omega^2 = -2\alpha\,\theta$$

$$(24\pi)^2 = (0)^2$$

$$= -\,2\alpha \times 250\pi$$

$$\therefore \quad \alpha = -\frac{(24\pi)^2}{500\pi}$$

$$= -3.619 \text{ rad/sec}^2$$

If t is the time to come to rest, then,

$$\omega = \omega_0 + \alpha t$$

$$0 = 24\pi - 3.619 \times t$$

$$t = \frac{24\pi}{3.619}$$

$$= 20.83 \textbf{ Ans.}$$

6.18 : *The equation of motion of a particle moving in a straight line is given by*

$$s = 18t + 3t^2 - 2t^3$$

where s is the distance covered from the starting point in metres at the end of t seconds find:

(a) The velocity and acceleration at start.

(b) The maximum velocity of the particle. *(U.P. Tech. Univ. 2002-03)*

EXAMPLE

Solution : Given $s = 18t + 3t^2 - 2t^3$

$\therefore$ Differentiating $\dfrac{ds}{dt} = v$

$$= 18 + 6t - 6t^2$$

and $a = \dfrac{dv}{dt}$

$$= 0 + 6 - 12t = 6 - 12t$$

Therefore, at $t = 0$

$$v = \left(\frac{ds}{dt}\right)_{t=0} = 18 \text{ m/s}$$

$$a = \frac{dv}{dt}$$

$$= 6 - 12 \times 0 = 6 \text{ m/s}^2$$

For maximum velocity,

$$\frac{dv}{dt} = 0$$

i.e., $6 - 12t = 0$

or $t = \dfrac{t}{12} = 0.5 \text{ secs}$

$\therefore$ At $t = 0.5,$

$$v = 18 + 6t - 6t^2$$

$$= 18 \times (6 \times 0.5) - 6\,(0.5)^2$$

$$= 18 + 3 - 1.5$$

$$= 19.5 \text{ m/s } \textbf{Ans.}$$

6.10 RELATIVE VELOCITY

So far we have been discussing about the velocity of a body *'A'* with respect to earth which has been called as the absolute velocity. However, when the velocity of a body is observed while the observer is sitting on a platform *'B'* which itself is moving, then the observed velocity of the body will be relative velocity of *A* with respect to *B*. As an example, if both the body *A* and body *B* are moving parallel to each other with the same velocity, then the observer will feel that body *A* is at rest or not moving. However, when *A* is moving in opposite direction to the velocity of *B*, then the observer will feel that *B* is moving very fast (actually at double the speed) away from *B*.

Thus, the relative velocity of *B* with respect to *A* is defined as

$$\overrightarrow{V_{BA}} = \overrightarrow{V_B} - \overrightarrow{V_A} \qquad \text{...(6.23)}$$

Where $\overrightarrow{V_{BA}}$ = **Relative velocity of *B* with respect to *A***

$\overrightarrow{V_B}$ = **Velocity vector of *B***

$\overrightarrow{V_A}$ = **velocity vector of *A***

Let $\overrightarrow{V_A}$ and $\overrightarrow{V_B}$ be the velocity vectors of *A* and *B* respectively. Then find the resultant of V_B and $(-V_A)$ as shown in Fig. (6.3*b*) is the relative velocity of *B* with respect to *A*. Analytically we find the components of V_A and V_B along any two chosen perpendicular axis (say *XX* and *YY* axis). Then take difference of the components along both *X* and *Y* directions.

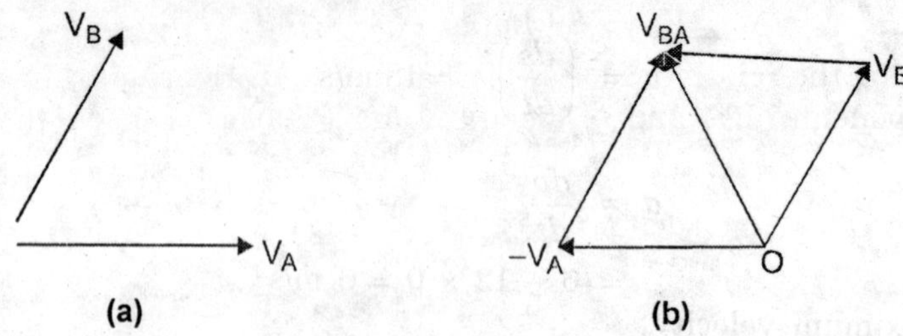

Fig. 6.3

i.e., $$V_{r_x} = \left(\vec{V}_{B_x} - \vec{V}_{A_x}\right) \quad ...(6.24)$$

and $$V_{r_y} = \left(\vec{V}_{B_y} - V_{A_y}\right) \quad ...(6.25)$$

Then resultant relative velocity v_r of *B* with respect to *A* will be,

$$v_r = \sqrt{V_{r_x}^2 + V_{r_y}^2} \quad ...(6.26)$$

The resultant will not be inclined to v_{r_x} at 0 given by,

$$\tan\theta = \left(\frac{V_{r_y}}{V_{r_x}}\right)$$

$$\theta = \tan^{-1}\left(\frac{V_{r_y}}{V_{r_x}}\right)$$

6.10.1 Relative Distance

Let V_A and V_B be the velocities of two bodies *A* and *B* respectively in a plane. At some instant of time, let this respective positions be at points *C* and *D* (Fig. 6.5). Suppose we want to find the distance between these bodies after time *t*. We first calculate the relative velocity of Body *B* with respect to that of body *A*. For this we superimpose $-V_A$ or V_B at the point *D*. Then vector $\overline{DE}$ represents the V_{BA}. This distance moved corresponding to the relative velocity V_{BA} in time interval *t* is say *EF*.

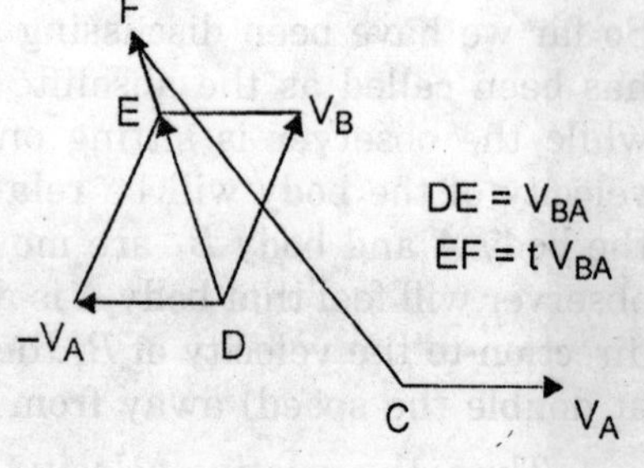

Fig. 6.5

Where, $$EF = t \times V_{BA} \quad ...(6.28)$$

EXAMPLE

6.19 : *Two trains (one express and other Shatabdi) leave simutaneously Delhi junction railway station. Shatabdi has a velocity of 120 km/hr and moves in the direction N 30° W and the express train has a velocity 60 km/hr in the direction S 60° E. Find the relative velocity of Shatabdi with respect to express train and the distance between them after 30 minutes.*

Solution : Fig. 6.6 shows the velocity vectors V_S and V_E, the velocities of Shatabdi and express trains respectively at the start of the trains,

$$V_S = 120 \text{ km/hr}$$
$$V_E = 60 \text{ km/hr}$$

For finding the relative velocity of Shatabdi w.r.t. Express train, we take the components of V_S and V_E along OX and OY axis (as shown in Fig. 6.6)

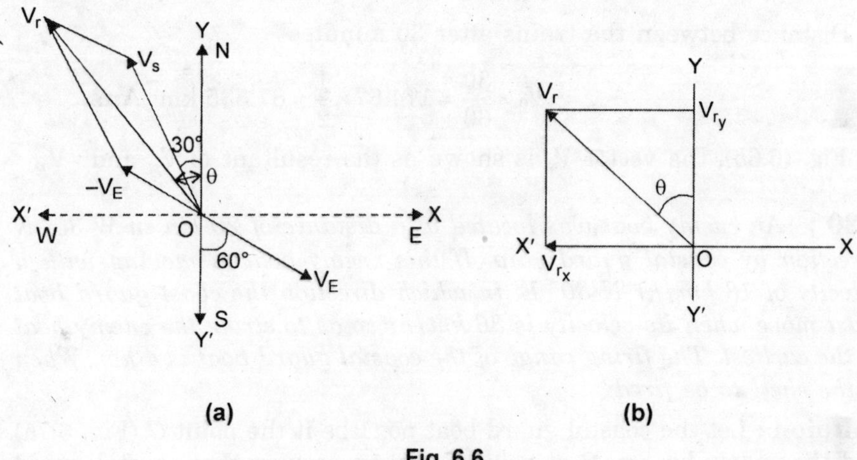

Fig. 6.6

$$V_{S_x} = V_S \sin 30$$

$$= -120 \times \frac{1}{2} = -60 \text{ km/hr}$$

$$V_{S_y} = V_S \cos 30$$

$$= 120 \times .866 = 103.92$$

$$-V_{E_x} = -V_E \sin 60$$

$$= -60 \times 0.866$$

$$= -51.96 \text{ km/hr}$$

$$\therefore \quad -V_{E_y} = -[-V_E \cos 60] + 60 \times \frac{1}{2} = 30 \text{ km/sec}$$

$$\therefore \quad V_{r_x} = V_{S_x}[-V_{E_x}]$$

$$= -60 - 51.96 = 111.96 \text{ km/sec}$$

$$V_{r_y} = V_{S_y} + (-V_{E_y}) = 103.92 + 30 = 133.52$$

$$\therefore \quad V_r = \sqrt{(V_{r_x})^2 + (V_{r_y})^2}$$

$$= \sqrt{(-111.96)^2 + (133.92)^2}$$

$$= \sqrt{12535.04 + 17924.56}$$

$$= \sqrt{30509.6} = 174.67 \text{ km/hr}$$

V_r is in the direction θ° west of N as shown in Fig. 6.6(*b*) where,

$$\tan\theta = \frac{|V_{r_x}|}{|V_{r_y}|} = \frac{111.96}{133.92} = 0.8360$$

$$\theta = 39.896°$$

∴ Distance between the trains after 30 minutes

$$= V_2 \times \frac{30}{60} = 174.67 \times \frac{1}{2} = 87.335 \text{ km} \quad \textbf{Ans.}$$

In Fig. (6.6*b*), the vector V_r is shown as the resultant of V_S and $-V_E$

6.20 : *An enemy boat was located at a distance of 20 km in W 30° N direction by coastal guard ship. If this enemy boat is moving with a velocity of 18 km/hr N 30° E, in which direction the coast guard boat must move when its velocity is 36 km/hr so as to strike the enemy boat at the earliest. The firing range of the coastal guard boat is 4 km. When is the shell to be fired?*

Soltuion : Let the coastal guard boat no. 1 be it the point C (Fig. 6.7a) and the enemy boat no.2 at point E. Let us assume that coastal guard boat having velocity V_1 = 36 km/hr be set to move in N θ° W direction.

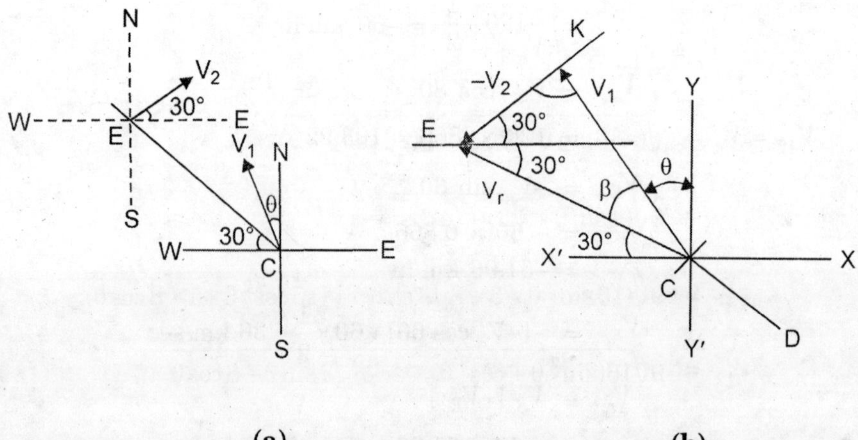

Fig. 6.7

Also, it is given that,

$$\vec{V}_2 = 18 \text{ km/hr}$$

We proceed to find the relative velocity, of coastal guard boat no. 1 w.r.t to boat no. 2. For this we take origin at C and mark the X and Y axis as shown in Fig. 6.7

Then $$V_{1_x} = -V_1 \sin\theta$$

$$V_{1_y} = V_1 \cos\theta$$

Also, $$-V_{2_x} = -V_2 \cos 30$$

$$-V_{2_y} = -V_2 \sin 30$$

If V_r is the relative velocity of boat no. 1 w.r.t boat no.2,

$$V_{r_x} = V_{1_x} + (-V_{2_x})$$

$$= -V_1 \sin\theta + (-V_2 \cos 30)$$

$$= -36\sin\theta - 18\times\frac{\sqrt{3}}{2}$$

$$= -36\sin\theta - 9\sqrt{3}$$

$$V_{r_y} = V_{1_y} + (-V_2)_y$$

$$= V_1 \cos\theta + (-V_2 \sin 30°)$$

$$= 36\cos\theta - 18\times\frac{1}{2}$$

$$= 36\cos\theta - 9$$

$$\therefore \quad V_r = V_{12} = \sqrt{(V_{r_x})^2 + (V_{r_y})^2}$$

$$\therefore \quad V_{12} = V_r = \sqrt{(-36\sin\theta - 9\sqrt{3})^2 + (36\cos\theta - 9)^2}$$

$$= 9\sqrt{(4\sin\theta + \sqrt{3}) + (4\cos\theta - 1)^2}$$

$$= 9\sqrt{(16\sin^2\theta + 3 + 8\sqrt{3}\sin\theta) + (16\cos^2\theta + 1 - 8\cos\theta)}$$

$$V_{12} = 9\sqrt{16(\sin^2\theta + \cos^2\theta) + 4 + 8\sqrt{3}\sin\theta - 8\cos\theta}$$

$$= 9(20 + 8\sqrt{3}\sin\theta - 8\cos\theta)^{\frac{1}{2}} \qquad ...(A)$$

If α is the inclination of resultant relative velocity with X-axis.

EXAMPLE

E X A M P L E

$$\tan\alpha = \frac{V_{r_y}}{V_{r_x}} = \frac{36\cos\theta - 9}{-36\sin\theta - 9\sqrt{3}}$$

But here V_r is to be directed along *CE*

i.e., $\alpha = 150°$

$$\tan 150 = -\frac{(4\cos\theta - 1)}{(4\sin\theta + \sqrt{3})}$$

$$-\frac{1}{\sqrt{3}} = -\frac{(4\cos\theta - 1)}{(4\sin\theta + \sqrt{3})}$$

or $$(4\cos\theta - 1)\sqrt{3} = 4\sin\theta + \sqrt{3}$$

$$4\sqrt{3}\cos\theta - \sqrt{3} = 4\sin\theta + \sqrt{3}$$

$$4\sqrt{3}\cos\theta = 4\sin\theta + 2\sqrt{3}$$

or $$\cos\theta = \frac{\sin\theta}{\sqrt{3}} + \frac{1}{2} \quad ...(B)$$

Let $\sin\theta = x$ Then $\cos\theta = \sqrt{1 - x^2}$

Equation *B* reduces to,

$$\sqrt{1 - x^2} = \frac{x}{\sqrt{3}} + \frac{1}{2}$$

Squaring both side,

or $$(1 - x^2) = \frac{x^2}{3} + \frac{1}{4} + \frac{x}{\sqrt{3}}$$

$$12\sqrt{3}(1 - x^2) = 12\sqrt{3}\left[\frac{x^2}{3} + \frac{1}{4} + \frac{x}{\sqrt{3}}\right]$$

or $$12\sqrt{3} - 12\sqrt{3}x^2 = 4\sqrt{3}x^2 + 3\sqrt{3} + 12x$$

$$9\sqrt{3} = 16\sqrt{3}x^2 + 12x$$

or $$16x^2 + \frac{12}{\sqrt{3}}x - 9 = 0$$

or $$16x^2 + 4\sqrt{3}x - 9 = 0$$

or $$x = \frac{-4\sqrt{3} \pm \sqrt{(4\sqrt{3})^2 + 64 \times 9}}{32}$$

$$= \frac{-4\sqrt{3} \pm \sqrt{48 + 64 \times 9}}{32}$$

$$= \frac{-4\sqrt{3} \pm 4\sqrt{3 + 36}}{32}$$

$$= \frac{1}{8}[-\sqrt{3} \pm \sqrt{39}]$$

$$= \frac{1}{8}\left[-\sqrt{3} \pm \sqrt{3}\,\sqrt{13}\right]$$

$$= \frac{\sqrt{3}}{8}[-1 \pm \sqrt{13}]$$

$$= 0.2165\ [-1 \pm 3.605]$$

$$= (0.2165 \times 2.605) \text{ or } 0.2165 \times 4.605$$

i.e., either $x = 0.564$ or -0.991

Taking 185 value as the other value is trivial

$$x = 0.564 = \sin\theta$$

$$\theta = 34.33°$$

For $\theta = 34.33°$ we get the value of V_{12} from equation (*A*)

$$V_{12} = V_r = 9[20 + (8\sqrt{3} \times .564) - (8 \times .8258)]^{\frac{1}{2}}$$

$$= 9[20 + 7.815 - 6.606]^{\frac{1}{2}}$$

$$= 9\sqrt{21.209} = 41.445/\text{hr}$$

Now range of gun = 4 km

Let T hrs be the time taken to bring the boat B within the firing range of gun at boat A

$T \cdot V_r = (20 - 4)$ (as the distance to be covered, by the shell is 16 km)

$$\therefore \quad T = \frac{16}{41.445} \text{ hrs}$$

$$= 0.386 \text{ hrs.}$$

$$= \textbf{23.16 minutes Ans.}$$

E X A M P L E

Alternate Solution

A simpler way to solve this problem will be to draw the vector diagram of the velocities which is shown in Fig. (6.7*b*). Draw the vector $-V_2$ to same scale at 30° to horizontal. The length *KE* will be proportional to 18 km/hr. From point *E* draw a line *EC* at 30° to horizontal as V_r is along *W* 30° *N*. Then with *K* as centre draw an arc KC of radius 36 km/hr. Then mark the *X* and *Y* axis taking origin at *C*. Then CK will represent the vector $\vec{V_1}$ in direction and magnitude and *CE* will be $\vec{V_r}$.

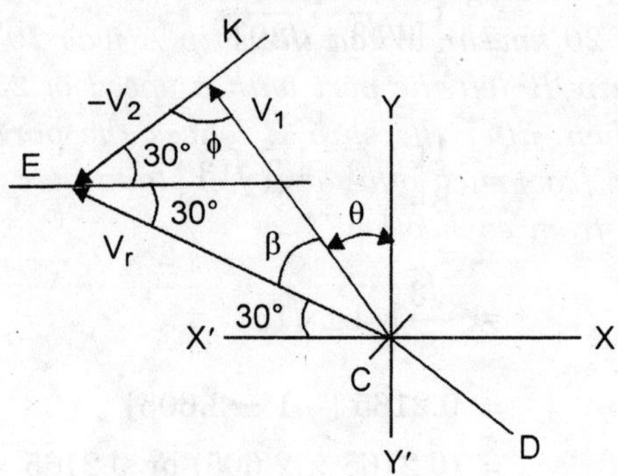

Fig. 6.7 (c)

The angle *KEC* of the velocity vector triangle is 60°. Let angle *ECK* be β and *EKC* be ϕ. Then from the sine rule applied to triangle *ECK*,

$$\frac{CK}{\sin 60} = \frac{KE}{\sin\beta} = \frac{EC}{\sin\phi}$$

i.e.,
$$\frac{CK}{\sin 60} = \frac{KE}{\sin\beta}$$

or
$$\sin\beta = \frac{KE\sin 60}{CK} = \frac{18}{36}\times\frac{\sqrt{3}}{2} = \frac{\sqrt{3}}{4} = 0.433$$

or
$$\beta = \sin^{-1}(0.433) = 25.66°$$

$$\therefore \quad \phi = 180 - (60 + 25.66)$$
$$= 94.34°$$

$$\therefore \quad EC = \frac{CK}{\sin 60}\times\sin 94.34°$$

$$= \frac{36\times 2}{\sqrt{3}}\times 0.9971 = 41.45 \text{ km/hr}$$

$$= V_r$$

Firing range = 4 km

∴ Distance to be covered with velocity,

$$V_r = 20 - 4 = 16 \text{ km}$$

$$T = \frac{16}{41.43} = 0.386 \text{ hr}$$

$$= \textbf{23.16} \text{ minutes}$$

Also, angle
$$\theta = 90 - (30 + 1)$$
$$= 90 - (30 + 25.66) = \mathbf{34.34°}$$

θ is the angle which gives the direction of the velocity vector V_1.

6.21 : *Ship $\overline{A}$ moving in the direction N 60° E approaches the port with a speed of 20 km/hr. When the ship A was 70 km away from the port, another ship 'B' left the port with a speed of 25 km/hr towards S 30° W. Find when, after the ship B leaves the port, they will be at a distance of 3 km from each other and how long they will remain with in 30 km distance from each other?*

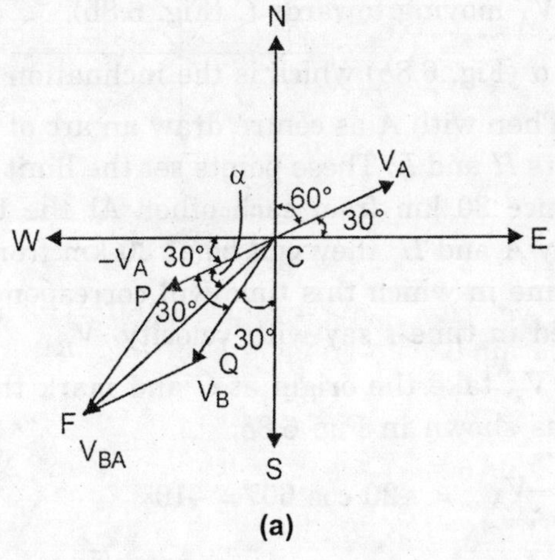

(a)

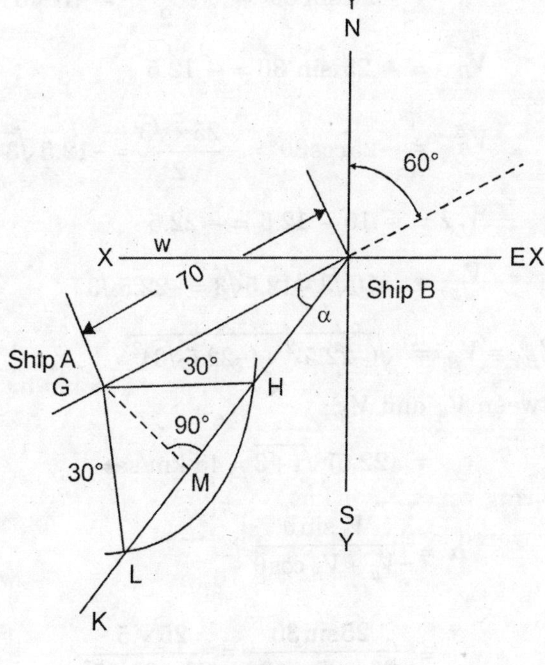

(b)

Fig. 6.8

Solution : Let V_A and V_B be the velocity of the two ships respectively (Fig. 6.8*a*).

EXAMPLE

E X A M P L E

Then the relative velocity of V_{BA} will be the resultant of $\vec{V}_B$ and $-\vec{V}_A$

$$V_A = 20 \text{ km/hr}$$

and $$V_B = 25 \text{ km/hr}$$

V_{BA} is vector $\overline{CF}$ in Fig. (6.8a) when the ship B was to leave the port *i.e.*, point C the ship A was at the point G, 70 km from C along the line parallel to vector V_A moving towards C (Fig. 6.8b).

BK is at an angle α (Fig. 6.8*b*) which is the inclination of the $\vec{V}_{BA}$ with V_A (in Fig. 6.8*b*). Then with A as centre draw an arc of radius 30 km to cut the BK of points H and L. These points set the limit of the time they will be at a distance 30 km from each other. At the time when their location is given by A and H, they will be at 30 km from each other for the first an last time in which this time will correspond to the relative distance BH moved in time t say with velocity V_{BA}.

For finding $V_{BA} = V_r$, take the origin at C and mark the X and Y axes along CE an CN as shown in Fig. 6.8*b*.

$$-V_{A_x} = -20 \cos 60° = -10$$

$$-V_{A_y} = -20 \sin 60° = -\frac{20 \times \sqrt{3}}{2} = -10\sqrt{3}$$

$$V_{B_x} = -25 \sin 30 = -12.5$$

$$V_{B_y} = -25 \cos 30° = -\frac{25 \times \sqrt{3}}{2} = -12.5\sqrt{3}$$

$$V_{r_x} = -10 - 12.5 = -22.5$$

$$V_{r_y} = -10\sqrt{3} - 12.5\sqrt{3} = -22.5\sqrt{3}$$

$$V_{BA} = V_R = \sqrt{(-22.5)^2 + (-22.5\sqrt{3})^2}$$

It θ is angle between V_A and V_B,

$$= (22.5)\sqrt{1+3} = 45 \text{ km/sec}$$

Then the $$\alpha = \frac{V_1 \sin\theta}{V_a + V_B \cos\theta}$$

$$= \frac{25 \sin 30}{20 + 25 \cos 30} = \frac{25 \times .5}{20 + 21.65}$$

$$= \frac{12.5}{41.65} = 0.2881$$

$$\therefore \quad \alpha = \tan^{-1}(.288) = 16.07°$$

In $\Delta\, CAM$

$\therefore$ Form Fig. in (6.8 *b*),

$$AM = 70 \sin \alpha$$
$$= 70 \times \sin 16.07$$
$$= 19.37$$
$$CM = 70 \cos \alpha = 70 \times \cos 16.07^\circ = 67.26 \text{ km}$$

$\therefore$ From ΔAMH,
$$MH = \sqrt{(30)^2 - (19.37)^2}$$
$$= \sqrt{900 - 325.19}$$
$$= 29.37 \text{ km}$$
$$= ML$$

$\therefore$ Distance
$$CL = CM - ML$$
$$= 67.26 = 29.37$$
$$= 37.89 \text{ km}$$

Time required to cover 37.89 km with velocity equal to V_r

$$t_1 = \frac{37.89}{45} = 0.842 \text{hrs}$$
$$= .842 \times 60 = 50.52 \text{ minute}$$

Time to cover distance,
$$CL = \text{CM} + ML$$
$$= 67.26 + 29.37$$
$$= 96.63$$
$$t_2 = \frac{96.63}{V_r} = \frac{96.63}{45}$$
$$= 2.147 + 60$$
$$= 128 \text{ m/s}$$

The distance apart between A and B will be less than 30 km from t_1 & t_2 m/s *i.e.* from 50.52 minutes upto 128 minutes

Length of duration $= 128 - 50.52$
$= 78.32$ **Ans.**

EXAMPLE

6.22 : *Two railway stations A and B are 60 km apart. An express non stop train passes station A_1 at some instant of time with a uniform speed of 120 km/hr toward station B. After 10 minutes Shatabdi train passes the station B with a uniform speed of 130 km/hr towards station A. They are moving on parallel tracks. The length of Express and Shatabdi trains are 150 m and 175 m respectively. Find the time lapse from the instant Shatabdi train passed the station B, when the engines of these trains will be crossing each other and the time taken to cross each other's last bogie.*

E X A M P L E

Solution : Let V_E = Velocity of express train

$= 120$ km/hr

V_s = Velocity of Shatabdi train

$= 130$ km/hr

d = Distance covered in 10 mts by the express train

$$= \frac{10}{60} \times 120 = 20 \text{ km.}$$

Let t be the time taken for the engines to cross each other.

Then d_E = Distance move by the express train in minutes

$$= t \times \frac{120}{60} = 2t \text{ minutes}$$

and d_S = Distance moved by the Shatabdi train

$$= t \times \frac{130}{60} \text{ km}$$

Total distance between the station = 60 km

Then $20 + d_E + d_S = 60$

or $$20 + t \times \frac{120}{60} + t \times \frac{130}{60} = 60$$

$$\frac{250}{60} t = 40$$

$$t = \frac{40 \times 60}{250} = \frac{240}{25}$$

$$= 96 \text{ minutes.}$$

Thus, the two engines will be crossing each other 96 minutes after the start of Shatabdi train **Ans.**

(*b*) Relative velocity of V_R Shatabdi train w.r.t. passenger train will be

$$V_R = 130 + 120 = 250 \text{ km/m}$$

Total length of the two trains = 150 + 175

= 325 m = 0.325 km

Let t, be the time taken to cross each other's last bogie,

$$\therefore \quad t_1 = \frac{0.325}{250} = \text{hr}$$

$= 0.0013$ hrs

$= 0.0013 \times 3600$

$= 4.68$ secs **Ans.**

6.11 GENERAL PLANE MOTION

A rigid body is said to have general plane motion if it has both the translation and rotory motions simultaneously. Railway wheel, for example, when rolls on the railway track without slipping, it has evidently rotary motion as well as its axile moves forward linearly.

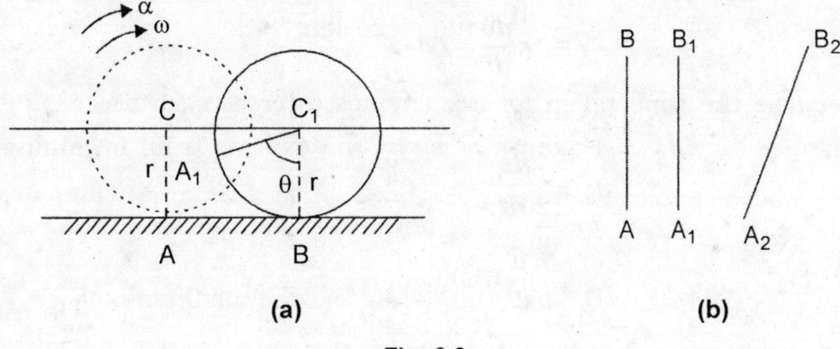

Fig. 6.9

Let a wheel of radius 'r' be rolling on a plane surface AB without slipping. Let the initial position of the rollar be represented by the left circle in Fig. 6.3 and let it be rotating at angular speed of ω and angular acceleration α (both clockwise taken as positive) so as to occupy position on the right after an interval of time, when it was rotated through an angle θ about the central axis. Thus, the wheel has translateral as well as rotary motion CC_1 is linear displacement and it has rotary motion as well which is the rotation of A through an angle BCA. Similarly, if a link or a member of a mechanism bodily moves so that all the points on it move the same distance and all move in the same direction, it will have translatory motion as shown in log 6.10(b). In Fig. 6.9 (a), the link AB has both translating and rotational motion.

6.11.1 Motion of a Wheel Rolling on Plane Surface without Slipping

Let a wheel of radius r roll without slipping on a plane surface. The rotational speed of wheel is w rad/s and its angular acceleration α rad/s^2 both clockwise taken as positive. Let the original position be as shown dotted and the solid line shows the position when the wheel has rotated through an angle θ. The point A has moved to the position A_1 and the position of point of contact has now been occupied by point B. So the linear motion CC_1 of the centre of wheel is equal to AB which also is equal to length of arc BA_1.

i.e., $$s_c = AB = \overrightarrow{BA}$$

$$= r\theta$$

or $$s_c = r\theta \qquad ...(6.29)$$

Where, r = Radius of wheel

θ = Rotation of wheel in time t with angular velocity w rad/sec

s_c = Linear distance moved by the centre of wheel in time t

v_c = Linear velocity of c

a_c = Linear acceleration of c

i.e., α = Angular acceleration of wheel

Therefore, differentiating equation (6.29) we get,

$$v_c = \frac{ds}{dt} = \frac{d}{dt}(r\theta)$$

$$= r\frac{d\theta}{dt} = r\omega$$

or $$v_c = r\omega \quad ...(6.30)$$

And $$a_c = \frac{dv_c}{dt}$$

$$= \frac{d}{dt}(r\omega)$$

$$= r\frac{d\omega}{dt}$$

$$= r\frac{d^2\theta}{dt^2}$$

or $$a_c = r\alpha \quad ...(6.31)$$

Important Note :

From equations (6.29), (6.30) and (6.31), we conclude that when the wheel of radius r rotates without slipping the linear displacement, velocity and acceleration of the centre of rotation are individualy equal to r times (where r is the radial distance of any point circumference of wheel), therefore the angular displacement of the point B at a distance r_B form centre of rotation, angular velocity angular acceleration are ω_{BA} and α_{BA} of the point B respectively. We re-write these results below,

$$s_c = r_{BC}\,\theta_{BC} \quad ...(6.29)$$

$$v_c = r_{BC}\frac{d\theta_{BC}}{dt}$$

$$= r_{BC}\omega_{BC} \quad ...(6.30)$$

$$a_c = r_{BC}\,\alpha_{BC} \quad ...(6.31)$$

where, r_{BC} = Radius of rotation of B w.r.t. centre C

θ_{BC} = Angular displacement of B with respect to the centre of rotation C

ω_{BC} = Relation angular velocity of B w.r.t. the into of rotation C

a_c = Linear acceleration of B and its direction is tangential to circular path of rotation of B *i.e.*, normal to CB. It is also called the tangential acceleration of B and is normal to CB.

6.23 : *A wheel of radius 1.5 m rolls without slipping in the vertical plane with rotational speed of 6.5 rad/sec about the centre C and angular acceleration of 4.5 rad/s² both, these are clockwise. On the wheel three points A, B and D are marked as shown in Fig. (6.10a). Determine the velocity and acceleration of these points.*

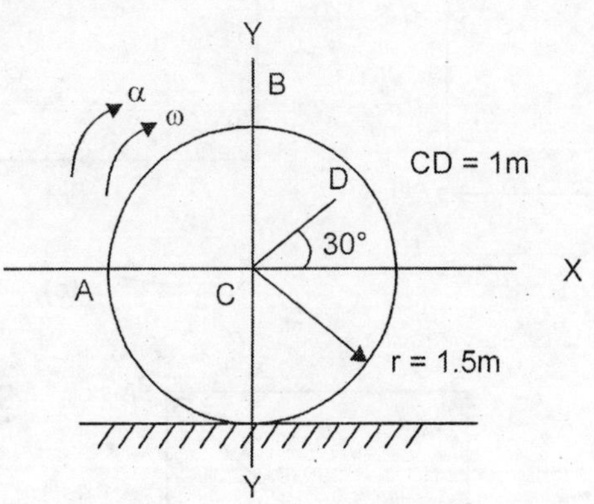

Fig. 6.10 (a)

Solution : Given data

Radius of A w.r.t C

$$r_{AC} = r_{BC} = \text{radius of } B \text{ w.r.t. } C$$
$$= 1.5 \text{ m}$$
$$CD = 1 \text{ m}$$
$$\omega \text{ of wheel} = \text{Angular velocity of wheel}$$
$$= 6.5 \text{ rad/sec}$$
$$\alpha = \text{Angular acceleration of wheel about } C$$
$$= 4.5 \text{ rad/s}^2$$
$$s_c = r\theta = 1.5 \times \theta$$

Therefore, $v_c = r\omega = 1.5 \times 6.5 = 9.75$ m/s along CX

$a_c = r\alpha = 1.5 \times 4.5 = 6.75$ m/s² along CX

(*b*) For the Point *A* : Velocity of A with respect to C

$$v_{AC} = r_{AC}\,\omega$$
$$= 1.5 \times 6.5$$
$$= 9.75 \text{ m/s} \quad (\text{Normal to } CA)$$

Acceleration of A w.r.t. C along tangential direction

$$= (a_{AC})_T$$
$$= r_{AC} \cdot \alpha = 1.5 \times 4.5 = 6.75 \text{ m/s}^2 \text{ along tangential direction}$$

Acceleration of A w.r.t. C along radially inward direction AC

$$(a_{AC})_R = \omega^2 \times r_{AC} = (6.5)^2 \times 1.5$$
$$= 63.375 \text{ m/s}^2$$

E X A M P L E

Therefore, the vector diagram for velocities of the point A, is as shown in Fig. (6.12) and 'a' for acceleration in Fig. (6.12c)

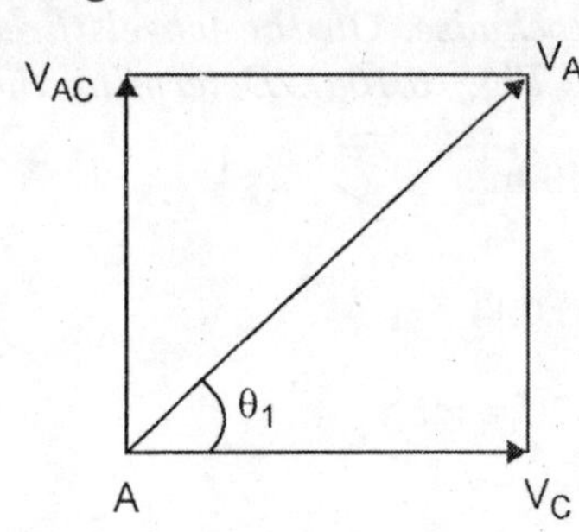

(b)

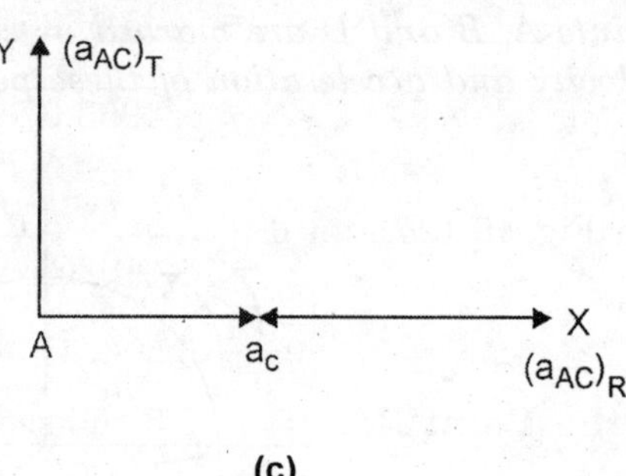

(c)

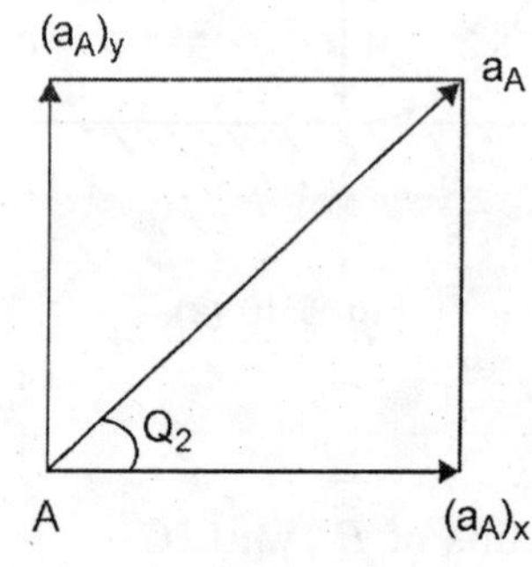

(d)

Fig. 6.10

$$v_A = \sqrt{v_C^2 + (v_{AC})^2}$$

$$= \sqrt{(9.75)^2 + (9.75)^2}$$

From Fig. (6.12b), $= (9.75)\sqrt{2} = 13.79 \text{ m/s}$

and $\tan \theta_1 = \dfrac{v_{AC}}{v_c} = \dfrac{9.75}{9.75} = 1$ or $\theta_1 = 45°$

$(a_A)_Y$ = Algebric sum of the components along CY direction

$\therefore$ $(a_A)_Y = [(a_{AC})_T]_y + [a_c]_Y + [(a_{AC})_R]_Y$

$= (a_{AC})_T + 0 + 0$

$= 6.75 \text{ m/s}^2$

$(a_A)_X = [(a_{AC})_T]_X \times [a_c]_X + [(a_{AC})_R]_X$

= Algebric sum of components along CX direction

$= 0 + 6.75 + 63.375$

$= 70.125 \text{ m/s}^2$

$\therefore$ $a_A = \sqrt{(a_A)_Y^2 + (a_A)_X^2}$

$$= \sqrt{(6.76)^2 + (70.125)^2}$$

$$= \sqrt{45.5625 + 4917.5156}$$

$$= \sqrt{4963.078} = 70.449 \text{ m/s}^2$$

From Fig. (6.12*d*), $\tan \theta_2 = \dfrac{(a_A)_Y}{(a_A)_X} = \dfrac{6.75}{70.125} = 0.096$

$\therefore \quad \theta_2 = 5.5°$ with CX

For the Point *B*: $\omega = 6.5$ rad/sec

$v_c = r\omega$

$= 1.5 \times 6.5 = 9.75$ m/s (along CX)

v_{BC} = velocity of B w.r.t C $= r_{BC}\,\omega$

$= 1.5 \times 6.5 = 9.75$ m/s

Along the tangent at B parallel to CX axis

Also, $\alpha = 4.5$ m/s^2

$\therefore \quad a_c$ = acceleration of

$C = r\alpha = 1.5 \times 4.5$

$= 6.75$ m/s^2 (along CX)

$(a_{BC})_T$ = Acceleration of B w.r.t. C along tangential direction *i.e.*, parallel to CX

$= r_{BC} \times \alpha$

$= 1.5 \times 4.5 = 6.75$ m/s^2

$(a_{BC})_R$ = centerpetal acceleration along BC *i.e.*, radially inward direction

$= r_{BC}\,\omega^2 = 1.5 \times (6.5)^2 = 63.375$ m/s^2

Therefore, if $(v_{BC})_X$ and $(v_{BC})_Y$ are the algebric sum of the components of velocity along X and Y directions.

We have, $(v_C)_X = v_C = 9.75$ m/s

$(v_C)_Y = (v_C \cos 90) = 0$

and $(v_{BC})_X = [(v_{BC})_T]_X + [(v_{BC})]_X$

$= (v_{BC})_T + 0$

$= 9.75$ m/s

Since $(v_{BC})_Y = (v_{BC})_T \cos 90 = 0$

$\therefore \quad (v_{BC})_X = 9.75 + 9.75$

$= 19.5$ m/s along tagent at B

and $(v_{BC})_Y = 0$

EXAMPLE

E X A M P L E

$$\therefore \quad v_B = \sqrt{(v_{BC})_X^2 + (v_{BC}^2)_X^2}$$

$$= (v_{BC})_X = 19.5 \text{ m/s}$$

(along the direction parallel CX)

For Acceleration of *B*

There are four different terms to be added vectorialy.

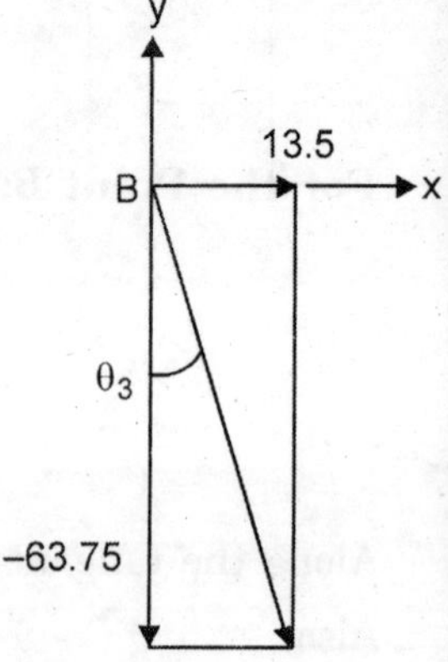

Fig. 6.10 (e)

$(a_C)_X$ = 6.75 along CX

$(a_C)_Y$ = zero

$[(a_{BC})_T]_X$ = 6.75 m/s² along tangent at B parallel to CH axis

$[(a_{BC})_T]_Y$ = zero as it is perpendicular to CX

$[(a_{BC})_R]_X$ = zero as it is perpendicular to CX axis

$[(a_{BC})_R]_Y$ = component perpendicular to CH directed for B & C

$= -63.375 \text{ m/s}^2$

$$[a_{BC}]_X = [a_C]_X + [(a_{BC})_T]_X + [(a_{BC})_R]_X$$
$$= 6.75 + 6.75 + 0$$
$$= 13.5 \text{ m/s}^2$$

$$[a_{BC}]_X = [ac]_Y + [(a_{BC})_T]_Y + [(a_{BC})_R]_Y$$
$$= 0 + 0 + [-63.75]$$

$$(a_B) = \sqrt{(a_{BC})_X^2 + (a_B)_Y^2}$$
$$= \sqrt{(13.5)^2 + (-63.75)^2}$$
$$= \sqrt{182.25 + 4016.39}$$
$$= \sqrt{4188.64} = 64.797 \text{ m/s}^2$$

Then Fig. (6.10*e*) $\tan\theta_3 = \dfrac{|(a_B)_X|}{|(a_B)_Y|} = \dfrac{13.5}{+63.75} = 0.21176$

$$\theta_3 = 11.956°$$

For the Point *D*, $r_{DC} = 1.0 \text{ m}$

$$v_C = r\,\omega$$
$$= 1.5 \times 6.5$$
$$= 9.75 \text{ along } CX$$
$$a_C = r\alpha$$
$$= (1.5)\,4.5$$
$$= 6.75 \text{ m/s}^2 \text{ along } CX\ (v_D)_X$$

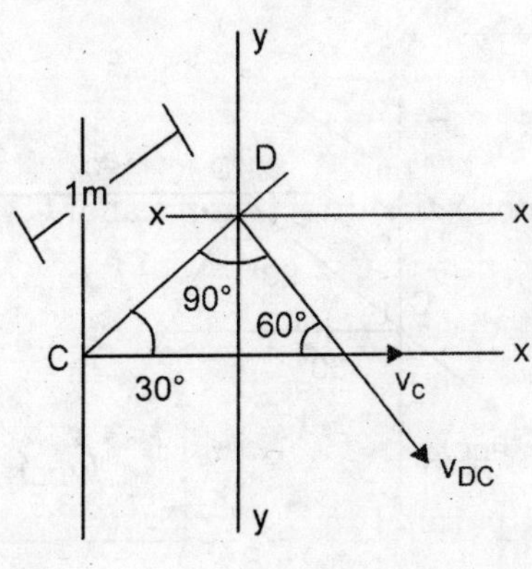

(f) (g)

Fig. 6.10

$$v_{DC} = r_{DC}\,\omega$$

$= 1 \times 6.5 = 6.5$ m/s along tangential direction to radius CD

i.e., along perpendicular to CD *i.e.*, at 60° downward to CX

$$\therefore \qquad \vec{V}_D = \vec{V}_C + \vec{V}_{DC}$$

or

$$(V_D)_X = (V_C)_X + (V_{DC})_X$$

$$= 9.75 + V_{DE} \cos 60$$

$$= 9.75 + 6.5 \times \frac{1}{2}$$

$$= 9.75 + 3.25$$

$$= 13.0 \text{ m/s}$$

and

$$(V_D)_Y = (V_C)_Y + (V_{DC})_Y$$

$$= 0 - 6.5 \sin 60 = \frac{6.5 \times \sqrt{3}}{2}$$

$$= -5.629 \text{ m/s}$$

From Fig. (6.10*g*),

$$V_D = \sqrt{(V_D)_X^2 + (V_D)_Y^2} = \sqrt{13^2 + (-5.629)^3}$$

$$= \sqrt{169 + 31.6856}$$

$$= \sqrt{200.6856} = 14.1663 \text{ m/sec}$$

$$\tan \theta_4 = \frac{|(V_D)_Y|}{(V_B)_X} = \frac{5.629}{13} = 0.433$$

From Fig. (6.10*g*),

$$\theta_4 = \tan^{-1}(0.433)$$

$$= 23.47°\text{C} \textbf{ Ans.}$$

Acceleration of *D*

$$a_C = 1.5 \times 4.5$$
$$= 6.75 \text{ m/s}^2 \text{ along } DX$$
$$(a_{DC})_T = r_{DC}\,\alpha = 1 \times 4.5$$
$$= 4.5 \text{ m/s}^2 \text{ perpendicular to } CD$$
i.e., at D

$$(a_{DC})_A = \text{centripetal acceleration of } D \text{ w.r.t. } C$$
$$= r_{DC}\,\omega^2$$
$$= 1 \times (6.5)^2$$
$$= 42.25 \text{ m/s}^2 \text{ along } DC \text{ is radially inward direction}$$

Fig. 6.10 (h)

$$(a_D)_X = \text{Algebric sum of the components along direction } DX$$
$$= (a_C)_X + (a_{DC})_T \cos 60 - (a_{DC})_R \cos 30$$
$$= 6.75 + 4.5 \times \frac{1}{2} - (42.25)\frac{\sqrt{3}}{2}$$
$$= 6.75 \times 2.25 - 36.546$$
$$= -27.546 \text{ m/s}^2$$

and $$(a_D)_Y = (a_C)_X(0) - (a_{BC})_T \sin 60 - (a_{BC}) \sin 30$$
$$= 0 - \frac{4.5 \times \sqrt{3}}{2} - (42.25) \times \frac{1}{2}$$
$$= -3.8971 - 21.125$$
$$= -25.0.221 \text{ m/s}^2$$

$$\therefore \quad a_D = \sqrt{(a_{DC})_X^2 + (a_{DC})_Y^2}$$
$$= \sqrt{(-27.546)^2 + (-25.022)^2}$$
$$= \sqrt{758.782 + 626.050}$$
$$= \sqrt{1384.83} = 37.21 \text{ m/s}^2$$

$$\therefore \quad a_D = 37.21 \text{ m/s}^2$$

$$\tan \theta_6 = \frac{|(a_{DC})_X|}{|(a_{DC})_Y|} = \frac{-27.564}{-25.0221}$$
$$= 1.1015$$
$$\theta_6 = \tan^{-1}(1.1015)$$
$$\theta_6 = 47.765 \text{ degrees}$$

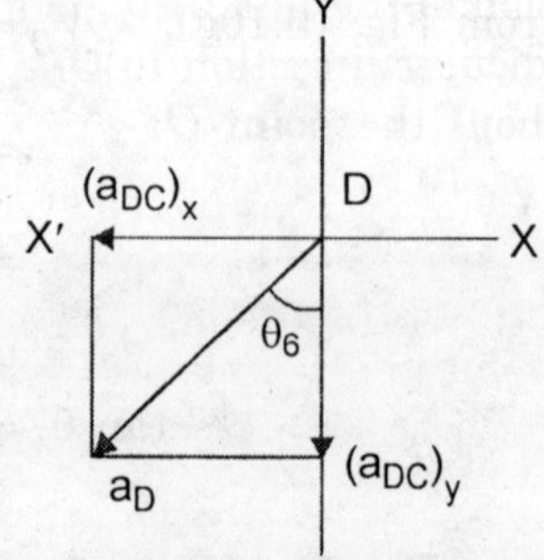

Fig. 6.10 (i)

6.12 COMBINED MOTION OF TRANSLATION AND ROTATION

Let there be a link or member of some mechanism A_0B_0 which after a short interval of time goes to the position A_2B_2 (Fig. 6.13). Evidently, the link has rotated a bit as well as there is linear motion of the link. This motion can be split up into two motions. Firstly from A_0 B_0 to A_2B_1 where $A_0B_0B_1A_2$ is a parallelogram. This will amount to linear motion of the link moving from A_0 B_0 to A_2B_1 such that each and every particle of the link has moved linearly through the same distance and along the same direction, thus, it constitutes the motion of translation. During the second part of motion it can be considered that the link has rotated from the position A_2B_1 to A_2B_2, that has rotated through an angle α.

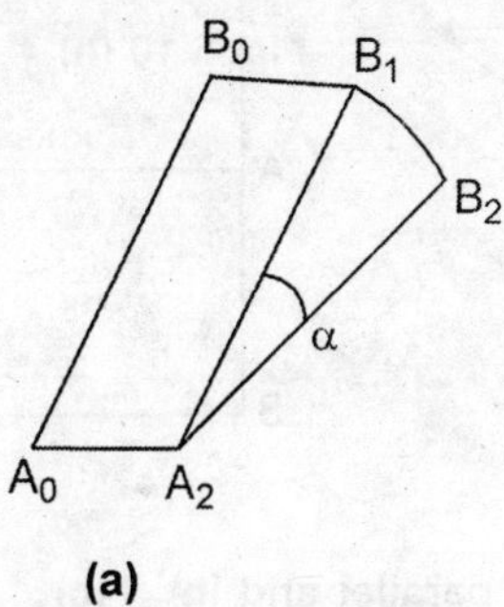

(a)

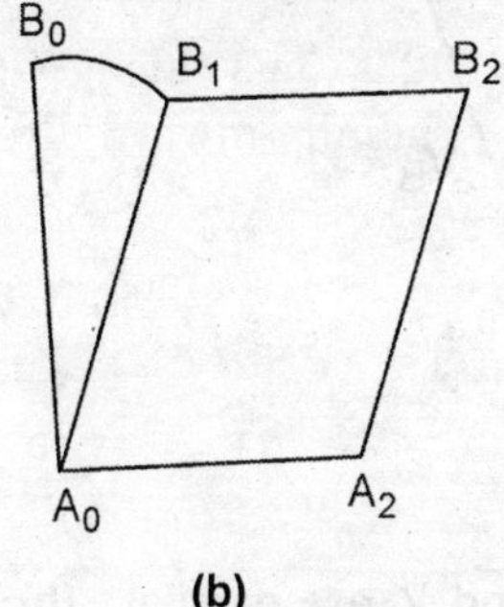

(b)

Fig. 6.11

Above motion can also be split in to two motions, Rotation through angle α first and then the translation as in Fig. 6.12(*d*), when $A_0B_1B_2A_2$ is a parallelogram.

Still further since the rotational motion and translation occur simultaneously, it can be assumed that the motion is entirely of rotation about a certain point, such that the point is stationary, called the centre of rotation at that instant and the entire link is rotated about its centre of rotation. Next moment the centre of rotation will change, that is why this is called the intantaneous centre of rotation. And in the short interval of time *dt* the whole body rotates at *C* with an angular speed of ω. Thus, the location of the centre of rotation of the link will go on changing. Thus, at any particular instance, *C* will be called the instantaneous centre of rotation of the link. Since the time interval *dt* is small. The right-bisectors of the chords A_1A_2 and B_1B_2 are drawn to intersect at point *O* which is called the instantaneous centre of rotation of the link. Any point *P* on the link will be moving in the perpendicular direction to *OP*. If ω is the angular velocity of the link *AB* about the point *O*.

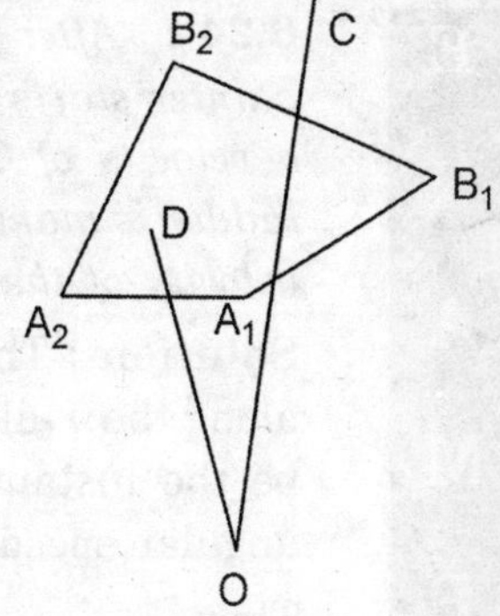

Fig. 6.12

Thus,
$$v_P = \omega \times (OP)$$
$$v_A = \omega\,(OA)$$
$$v_B = \omega\,(OB)$$

$$\therefore \quad \frac{v_A}{v_B} = \frac{\omega(OA)}{\omega(OB)}$$

$$\omega \frac{v_A}{v_B} = \frac{OA}{OB} \qquad \text{...(6.29)}$$

6.12.1 Important Note

(*a*) If the directions of the velocities of the ends of link *i.e.*, A and B, then the instantaneous centre of rotation of the link AB is obtained by drawing perpendicular lines from A and B to the respective directions of the velocities of A and B.

(*b*) If V_A and V_B are parallel but unequal, then the instantaneous centre of rotation of link AB obtained by locating the point of intersection of the centre line of the link AB and the line joining the extremities of the vectors V_A and V_B.

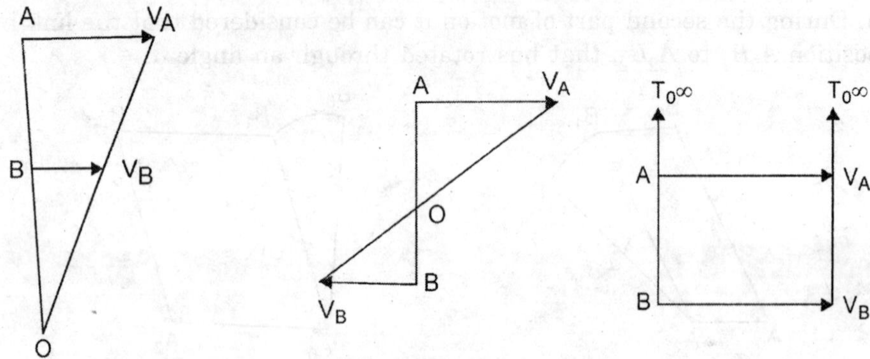

(*a*) Vector V_A and V_B are parallel and having same direction. **(*b*) V_A and V_B are parallel and in opposite direction.** **(c)**

Fig. 6.13

(*c*) If V_A and V_B are parallel and equal in magnitude, then the instantaneous centre of rotation will be at infinity.

(*d*) The instantaneous center of rotation may or may not lie an the body. If it lies on the body, its velocity will be zero at this instant. If it is outside the body, even then its velocity will be zero. As a matter of fact the point is used to find the tangential velocity of a point on the body perpendicular to OP.

EXAMPLE 6.24 : *After finishing the white washing of the exterior of a house, the painter starts pulling the lower end A of the 10 m long ladder with the a velocity of 0.5 m/sec, for the purpose of removing the ladder. If the ladder is making an angle of 60° with the horizontal, find the downward velocity of the upper end B of the ladder.*

Solution : The V_A is along the horizontal direction CA and V_B will be along the Y-direction BC. The normals at A and B meet at 0 which will be the instantaneous centre of rotation of the ladder AB. If ω is the angular speed of rotation of the ladder about O,

Then
$$V_A = \omega \times (OA)$$
$$= \omega \times (AB \sin 60)$$
$$= \omega \times \frac{10\sqrt{3}}{2} \qquad ...(A)$$

and $\quad v_B = \omega \times (OB)$

$$= \omega \times (AB \cos 60)$$

$$= \omega \times 10 \times \frac{1}{2} \quad ...(B)$$

$\therefore$ Dividing (B) by (A),

$$\frac{V_B}{V_A} = \frac{1}{2} \times \frac{2}{\sqrt{3}} = \frac{1}{\sqrt{3}}$$

$$v_B = v_A \times \frac{1}{\sqrt{3}}$$

$$= 0.5 \times \frac{1}{\sqrt{3}}$$

$$= 0.2887 \text{ m/s } \textbf{Ans.}$$

Fig. 6.14

6.25 : *A cylindrical roller 500 mm diameter is in contact with two conveyer belts at its top and bottom as shown in Fig. 6.15. If the belts A and B run at uniform speeds of 5 m/s and 3m/s, find the linear velocity and the angular velocity of the roller.*

(U.P. Tech Univ. 2000-01)

Solution : Given data : Roller diameter = 500 mm

$$r = \frac{500}{2} = 250 \text{ mm}$$

$$V_A = 5 \text{ m/s}$$

$$V_B = 3 \text{ m/s}$$

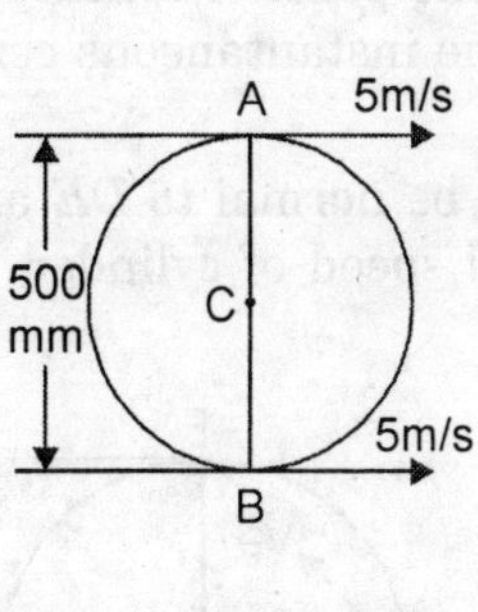

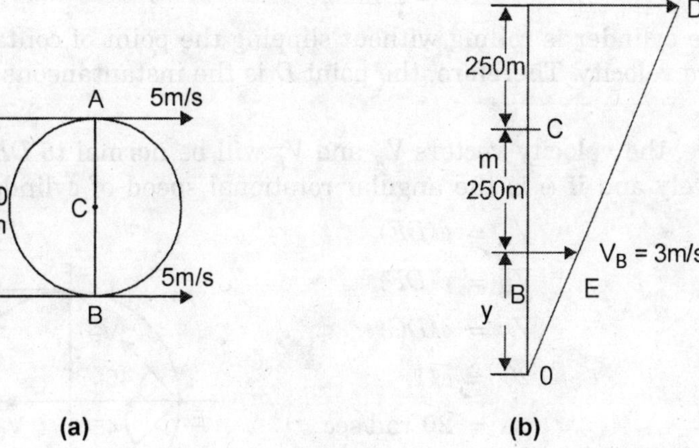

Fig. 6.15

It is given that the points A and B have velocities V_A = 5m/s and V_B = 3m/s. That the velocities are parallel and unequal. Therefore, the instantaneous centre of rotation will be found by joining the extremities of the vectors

V_A and V_B and then taking the intersection of the line joining the extermities of these vectors with the line *AB*.

In Fig. 6.15 (*b*) vectors V_A and V_B are drawn as *AD* and *BE* respectively joining the points *D* and *E* + and take the intersection of *AB* with *DE*. Thus, *'O'* is the instantaneous centre of rotation.

There if ω is the angular speed of roller,

$$\frac{V_A}{V_B} = \frac{\omega(AO)}{\omega(BO)} = \frac{\omega(0.50+y)}{\omega(y)} \text{ when } BO = y$$

i.e., $$\frac{5}{3} = \frac{0.5+y}{y}$$

or $$5y = 1.5 + 3y$$

or $$2y = 1.5$$

or $$y = 0.75 \text{ m}$$

Now $$V_B = \omega y = \omega \times .75$$

$$3 = \omega \times 0.75$$

or $$\omega = \frac{3}{0.75} = 4 \text{ rad/s} \quad \textbf{Ans.}$$

∴ $$V_C = (OC)\,\omega = (y + 0.25)\,\omega$$

$$V_C = (0.75 + 0.25)\,\omega = 4 \text{ m/s} \quad \textbf{Ans.}$$

EXAMPLE **6.26 :** *A cylinder of radius 1 m rolls without slipping along a horizontal plane AB. Its centre has uniform velocity of 20 m/s. Find the velocity of points E and F on the circumference of the cylinder as shown in Fig. 6.16.*

(U.P. Tech Univ. 2001-02)

Solution : Given data $V_C = 20$ m/s

$DF = 1$ m

Since the cylinder is rolling without slipping the point of contact *D* will have zero velocity. Therefore, the point *D* is the instantaneous centre of rotation.

Therefore, the velocity vectors V_E and V_F will be normal to *DE* and *DF* respectively and if ω is the angular rotational speed of cylinder,

$$V_E = \omega(DE)$$

$$V_F = \omega(DF)$$

$$V_C = \omega(DC)$$

$$20 = \omega(1)$$

i.e., $$\omega = 20 \text{ rad/sec}$$

From $$\Delta\, ECD,\ DE^2 = EC^2 + CD^2$$

or $$DE = \sqrt{(1)^2 + (1)^2}$$

$$DE = \sqrt{2}$$

∴ $$V_E = \omega \times (DE)$$

Fig. 6.16

$$= \sqrt{2}\,\omega = 20\sqrt{2}$$
$$= 28.28 \text{ m/s}$$

and it is directed at 45° to horizontal

Also, $$V_F = \omega \times DF$$
$$= 20 \times 2 = 40 \text{ m/s}$$

It is directed along horizontal direction parallel to *CX*.

6.27 : *The crank radius of a reciprocating engine is 250 mm and the length of the connecting rod is 1125 mm, The crank is rotating at 420 rpm. Find the speed of the piston when the crank is at 40° from the inner dead centre.*

Solution : Given data :

$$AB = 1125 \text{ mm}$$
$$CB = r = 250 \text{ mm}$$
$$N = 420 \text{ rpm}$$
$$\therefore \quad \omega = \frac{2\pi \times 420}{60}$$
$$= 14\pi \text{ rad/sec}$$

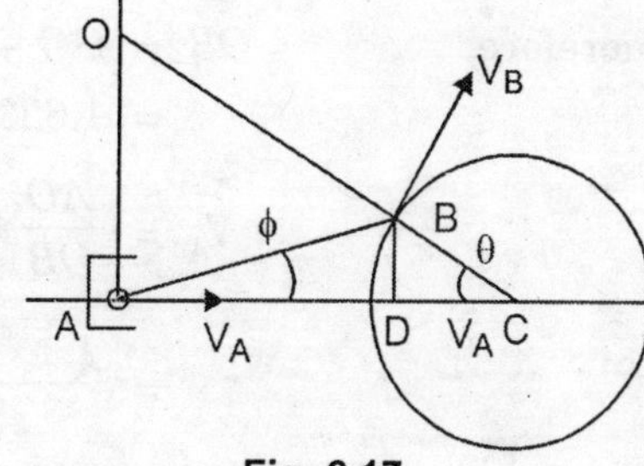

Fig. 6.17

The piston is moving along the straight line *AC*. Therefore, the instantaneous centre will be on perpendicular line to *AO*. Also, velocity of *B*, V_B is moment to *BC* therefore, the instantaneous centre of rotation will lie on be *BC*.

Thus, *O* is the instantaneous centre of rotation,

$$\therefore \quad \frac{V_A}{V_B} = \frac{AO}{OB} \qquad ...(A)$$

If *BD* is moment to *AC* and $\angle BAD = \phi$

Then $$BD = BC \sin\theta = 250 \sin 40$$

Also, from ΔBAD $$BD = AB \sin\phi$$

or $$250 \sin 40 = 1125 \sin\phi$$

or $$\sin\phi = \frac{250 \times 0.6428}{1125} = 0.1428$$

$$\therefore \quad \phi = 8.21°$$

$$\therefore \quad AC = AD + DC$$
$$= AB \cos\phi + BC \cos\theta$$
$$= 1125 \cos(8.21°) + 250 \cos 40$$
$$= 1125 \times 0.9897 + 250 \times 8.7660$$
$$= 1121.62 + 191.57$$
$$= 1313.10 \text{ mm} = 1.313 \text{ m}$$

EXAMPLE

From ΔAOC $\quad \dfrac{AO}{AC} = \tan 40$

$$AO = AC \tan 40$$
$$= 1.313.19 \times .8391$$
$$= 1.1 \text{ m}$$

And $\quad V_B = r\omega$
$$= 0.25 \times 14\pi$$
$$= 10.9955 \text{ m/s}$$

Now $\quad OC = \dfrac{AC}{\cos 40} = \dfrac{1.313}{0.809} = 1.623$

Therefore, $\quad OB = OC - BC$
$$= 1.623 - 0.25 = 1.37$$

$$V_A = \frac{AO}{OB} \times V_B$$

$$= \frac{1.1 \times 10.9955}{OB} - \frac{1.1 \times 10.9955}{1.37}$$

$$= 8.829 \text{ m/s}$$

6.13 MOTION WITH VARIABLE ACCELERATION

EXAMPLE

6.28 : *A car moving on a straight road has an equation of motion given by ,*

$$s = 2t^3 - 45t + 10$$

where, $\quad s$ = *distance travelled (in m)*

t = *time in seconds*

Find (a) the time required for the car to reach the velocity of 68 m/s starting from rest (b) the acceleration of the car when the velocity is 32 m/s (c) The distance travelled by the car during the first 4 secs.

Solution : Gives $\quad s = 2t^3 - 4t + 10 \qquad ...(A)$

Differentiating w.r.t. 't',

$$v = \frac{ds}{dt} = 6t^2 - 4 \qquad ...(B)$$

Now we have, $\quad v = 6t^2 - 4$

When $\quad v = 68 \text{ m/s} = 6t^2 - 4$

or $\quad 6t^2 = 68 + 4 = 72$

or $\quad t^2 = 12$

or $\quad t = \sqrt{12} = 3.46$ sec **Ans.**

(*b*) When $v = 32$, then we have from (*B*)

$$32 = 6t^2 - 4$$

i.e., $6t^2 = 36$ or $t^2 = 6$

$$t = \sqrt{6} = 2.45$$

Then from equation *A* by differentiating,

$$a = \frac{dv}{dt} = 12t$$

$$= 12 \times 2.45$$

$$= 29.40 \text{ sec } \textbf{Ans.}$$

(*c*) $s = 2t^3 - 4t + 10$

at $t = 4$ $(s)_4 = 2 \times 64 - 4 \times 4 + 10$

$$= 128 - 16 + 10 = 122 \text{ m}$$

Again then equation (*A*),

At $t = 0$ $(s)_0 = 0 - 0 + 10 = 10$ m

Thus, the time is set of zero when the displacement is 10 m then

$$s_4 - s_0 = 122 - 10$$

$$= 112 \text{ m } \textbf{Ans.}$$

6.29 : *Distance travelled by a car on a highway is directly proportional to t^3 where t is the time taken with respect to rest position. The car covered 216 m during first 60 sec. Find displacement, velocity and acceleration at the start when t = 10 sec.*

Solution : Given $s = ct^3$

where C is a constant

where $s = 216$ m $t = 60$s

$$216 = C \times (60)^3$$

or $$C = \frac{216}{60 \times 60 \times 60} = 0.001$$

$\therefore$ $s = 0.001\, t^3$

Differentiating $$\frac{ds}{dt} = v = 0.003t^2$$

and $$a = \frac{dv}{dt} = \frac{d^2s}{dt^2} = 0.006t$$

(*a*) At start $t = 0$

$\therefore$ $v = 0$ that is it starts from rest

A 10 secs *i.e.*, when $t = 10$

$$s = 0.001 \times 10^3$$

$$= 1 \text{ m } \textbf{Ans.}$$

E X A M P L E

$$v = 0.003t^2$$
$$= 0.003 \times 10^2 = 0.3 \text{ m/s}$$

and $$a = \frac{dv}{dt} = 0.006t$$
$$= 0.006 \times 10 - 0.06 \text{ m/s}^2 \textbf{ Ans.}$$

6.30 : *A body starts from rest with an acceleration a given by the following equation*

$$a = 4 - \frac{1}{6}t^3$$

Find (i) the velocity and distance travelled in 5 secs (ii) when will the body come to rest and what will then be the distance covered.

Solution : Given $$a = 4 - \frac{1}{6}t^3 \qquad ...(B)$$

or $$\frac{dv}{dt} = 4 - \frac{1}{6}t^3$$

or $$dv = \left(4 - \frac{1}{6}t^3\right) dt$$

Integrating $$v = 4t - \frac{1}{6}\left[\frac{t^4}{4}\right] + C_1$$

where C is a constant of integration. We have the end condition that $v = 0$ at $t = 0$ as the body starts from rest

$$\therefore \quad 0 = 0 - 0 + C_1$$

or $$C_1 = 0$$

or $$v = 4t - \frac{t^4}{24} \qquad ...(B)$$

or $$\frac{ds}{dt} = 4t - \frac{t^4}{24}$$

or $$ds = \left(4t - \frac{t^4}{24}\right) dt$$

Again integrating $$s = \frac{4t^2}{2} - \frac{1}{24}\left(\frac{t^5}{5}\right) + C_2$$

Where C_2 = Another constant of integration

Again $$s = 0 \text{ at } t = 0$$
$$0 = 0 + C_2 \quad \text{or} \quad C_2 = 0$$

$$\therefore \qquad s = 2t^2 - \frac{t^4}{72} \qquad ...(C)$$

When $t = 5$ we have from equation (B)

$$v = 4 \times 5 - \frac{(5)^3}{18}$$

$$= 20 - 6.944$$

$$= 19.056 \text{ m/s}$$

And from equation (C), the distance after $t = 5s$

$$s = 2t^2 - \frac{t^4}{72}$$

$$= 2 \times 5^2 - \frac{(5)^4}{72}$$

$$= 50 - 3556$$

$$= 46.444 \text{ m}$$

(*ii*) The body will come to rest again when $v = 0$

$\therefore$ From equation (B) when $v = 0$

$$0 = v = 4t - \frac{t^3}{18}$$

or $$4t = \frac{t^3}{18}$$

or When $t = 0$ or $t^2 = 72$

$t = 0$ refers to the starting time

Then it will come to rest

When $t = \sqrt{72} = 8.485$ sec

At $t = 8.485$ s

We have from equation (C) or $t^2 = 72$

$$s = 2t^2 - \frac{t^4}{72}$$

$$= t^2\left(2 - \frac{t^2}{72}\right)$$

$$= 72\left(2 - \frac{72}{72}\right) = 72 \text{ m} \quad \textbf{Ans.}$$

6.31 : *A body starts from rest with an acceleration given by*

$$a = 10 - 0.006\, s^2$$

EXAMPLE

E X A M P L E

Find (i) the velocity of the body when it has travelled 20 m (ii) The distance covered by the body when it comes to rest.

Solution : Given $a = 10 - 0.006\ s^2$ ...(*A*)

Now $a = \dfrac{dv}{dt} = \dfrac{dv}{ds} \cdot \dfrac{ds}{dt} = v \cdot \dfrac{dv}{ds}$

∴ Equation (*A*) reduces to

$$v \cdot \frac{dv}{ds} = 10 - 0.006\ s^2$$

or $v\ dv = (10 - 0.006\ s^2)\ ds$

Integrating we get $\dfrac{v^2}{2} = 10s - 0.006\left(\dfrac{s^3}{3}\right) + C_1$

Where C_1 is a constant of integration

Now at $s = 0,\ v = 0$

$$0 = 0 - 0 + C_1 \text{ or } C_1 = 0$$

∴ $\dfrac{v^2}{2} = 10s - 0.002s^3$

or $v^2 = 20s - 0.004s^3$ ...(*B*)

When $s = 20$ m, we have from (*B*)

$$(v)^2 = 20 \times 20 - 0.004\ (20)^3$$

$$v^2 = 400 - 0.004 \times 8000$$

$$= 400 - 32 = 368$$

∴ $72 = \sqrt{368} = 19.2$ m/s **Ans.**

(*ii*) It is required to find *s* when,

$$v = 0$$

From equation (*B*) by putting,

$$v = 0$$

$$0 = 20s - 0.004s^3$$

$$= s(20 - 0.004s^2)$$

i.e., either $s = 0$ or $s^2 = \dfrac{20}{0.004}$

How $s = 0$ sufers to starting position the other required solution

$$s^2 = \frac{20}{0.004}$$

or $s^2 = 500$

or $s = \sqrt{5000} = 70.7$ m **Ans.**

PROBLEM FOR PRACTICE

1. Define "Rigid Body".
2. Define the term 'Uniform Motion' 'Velocity' and 'Acceleration'.
3. Explain giving examples, the difference between uniform motion and uniformaly accelerated motion.
4. Derive the following equations of motion

$$v = u + at$$

$$s = ut + \frac{1}{2}at^2$$

$$v^2 - u^2 = 2as$$

5. A body is projected vertically upwards with a velocity u. What he its velocity at the highest point and the distance of the point from earth. Also, find the velocity when it touches back the earth.
$\left(\textbf{Ans. } 0, \frac{u^2}{2g}, u\right)$
6. An electric train starts from rest and within 10 secs it attains the speed of 60 km/hr, find the acceleration (in m/s^2). (**Ans.** 1.667 m/s^2)
7. A car is moving at a velocity of 80 km/hr, when the brakes are applied due to some emergency. The car comes to rest within 5 secs. Find the value of retardation and the distance moved before coming to rest. (**Ans.** 4.44 m/s^2; 55.55 m)
8. A car is travelling at a speed of 70 km/hr when the driver suddenly applied brakes and the car came to rest after skidding a distance of 50 m. Find the time taken by car to come to rest and the retardation after the brakes were applied. (**Ans.** t = 5.14 sec, 3.78 m/s)
9. A stone A is dropped from the top of a 100 m high tower. At the same time a stone B is thrown upwards, from the foot of the tower to meet the stone A , at a height of 40 m. Find the velocity of the stone B with which it was thrown up. (**Ans.** 28.6 m/s)
10. A wheel was rotating at an angular speed of 30 r.p.m. when it started accelerating at a uniform rate. After 50 seconds it had rotated 40 revolutions under the action of acceleration. Determine the angular velocity at the end of 50 sec. Also, find the time required for the speed to reach 80 rpm value. (**Ans.** 6.91 rad/s; 69.44s)
11. Rotor of a motor starts rotating from rest with uniform acceleration of 1.5 rad/s^2. What will be its angular velocity after 60s.
(**Ans.** 90 rad/s)
12. The angle of rotation of flywheel is governed by the relation.

$$\theta = t^3 - 4t^3 + 10t + 5$$

Where θ is the angular displacement in radians and t is the time in seconds from rest. Calculate the angular velocity and angular acceleration, when t = 0 and when t = 5 sec.
(**Ans.** (*i*) 10 rad/s, 4.5 rad/s, (*ii*) 6 rad/s, 22 rad/s)
13. A 11 m long ladder is to be placed with its top on the wall and the lower end on ground. It is being pushed at the lower end towards the wall with a velocity of 5 m/s. Find the velocity of the upper end where the inclination of the ladder with the horizontal is 60°. [**Ans.** 2.88 rad/sec]
14. An army vehicle is moving at a speed of 45 Km/s. Its wheel radius is 500 mm. Find the velocity of the top of the wheel relative to a person:

(*i*) Sitting on the vehicle

(*ii*) Sitting on the ground (**Ans.** (*i*) 12.5 m/s, (*ii*) 25 m/s))

❑❑❑

CHAPTER 7

KINETICS OF RIGID BODIES

7.1 INTRODUCTION

While dealing the kinematics of rigid bodies, in Chapter 6, the subject matter relating to displacement, relating and acceleration of rigid bodies were studied in detail without taking into account the forces which produce their motions. Kinetics is the branch of mechanics which takes into account the forces which produce the acceleration which in turn results in the motion of rigid bodies.

7.2 PLANE MOTION

If the motion of a rigid body is such that all the particles which constitute the body remain individually at a constant distance with reference to a fixed plane, it is known as plane motion. In case of rotation and translatory motion of the body, the plane motion will be the sum of the translation of the body with respect to some point of the body plus rotation of the body about an axis passing through that point.

7.2.1 Force, mass, acceleration and D'Alembert's principle

7.2.2 Mass

Mass is the quality of matter constant in a body. The unit of mass in SI units is 1 kilogram.

7.2.3 Force

As per Newton's Law of continue a body will continue to be in the state of rest or of uniform motion in a straight line unless it is compelled to change that state by some external agency acting on it. This leads to the existance of and definition of force which changes or tends to change the condition of rest or uniform motion in straight line of the body.

Further Newton's Second Law of motion states that the rate of change of momentum of a body is directly proportional to the impressed force and it takes place in the direction of the force applied. Mathematically it is expressed as,

Force ∝ Rate of change of momentum

∝ Rate of change of (Mass × velocity)

or $$F \propto \frac{d}{dt}(m.v.) \qquad \text{....(7.1)}$$

where, F = Force

m = mass of the body

v = velocity of the body

Since mass does not change, we get from equation (7.1)

$$F = km \cdot a \quad ...(7.2)$$

where, a = acceleration of the body

k = constant

Therefore if, $m = 1\ kg$

and $a = 1\ m/s^2$

Then from (7.2), $F = k \times 1 \times 1$

or $1 = k \times 1$

i.e., $k = 1$

Now if $m = 1$ kg $\quad a = 1$ m/s^2

Then

$$F = m \cdot a \quad ...(7.3)$$

Thus, if a unit force (*i.e.*, 1 Newton) and (one Newton) is defined as the force which will generate an acceleration of 1 m/s^2 in the body of mass 1 kg, then the eqation (7.3) is universally true.

In case instead of one force the number of forces acting on the body is large, then these forces can be replaced by the force which will be resultant of all the forces applied.

$$R = \Sigma F = ma \quad ...(7.4a)$$

where, m = mass of the ryth

a = acceleration

$$R - ma = 0 \quad ...(7.4b)$$

French Mathematician D' Alembert argued, on the basis of equation (7.4*b*), that the term ($m \times a$) be booked upon as a force of magnitude ($m \times a$) applied in the opposite direction of motion and termed it as Inertia Force in the opposite direction to that of acceleration. He then concluded that then the equation 7.4 as an equation of equilibrium and gave the following statement which is known as D'Alemberts Principle.

"The system of forces acting on a body in motion is in dynamic equilibrium with inertia force of the body."

7.3 KINETICS OF BODIES IN TRANSLATION

Let a body be acted upon by a large number of forces F_1, F_2, F_3... F_n whose resultant is F_R (Fig. 7.1) due to which let the acceleration be 'a', which will be along the direction of F_R will start moving along this direction with acceleration 'a'. However, if a force $F_i = -m \times a$ on the opposite direction of F_R is superimposed this body will be at rest and the equation of static equilibrium can then be applied. Thus, force F_i is called the inertia force. According to Newton's first law of motion, a body will continue to be in the state of rest or uniform motion along a straight line unless acted upon by an external

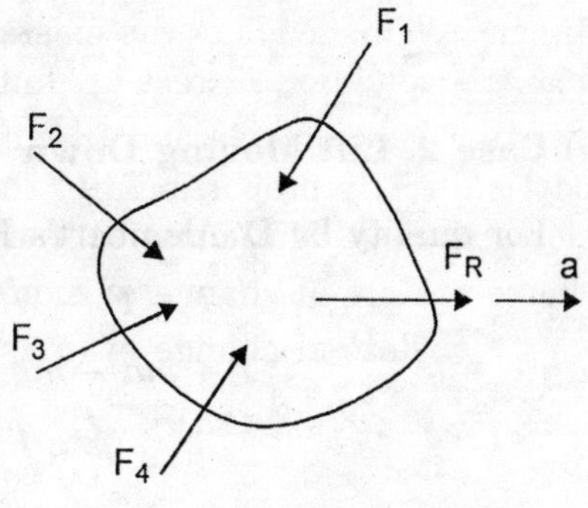

Fig. 7.1

force. This means that a body has a tendency to continue to remain at rest or of uniform motion. This tendency is called the inertia and that is why $F_i = -ma$ is called the inertia force.

7.3.1 Motion of A Lift (Elevator)

(*a*) (Case 1 Lift Moving up)

Let m = mass carried by the lift including its own

W = mg (weight carried by lift including its own)

a = uniform acceleration of the lift

T = Tension in the wire rope to supporting the cage

Then by D'Alembert's principle, Inertia force *(ma)* will be acting downward (Fig. 7.2*b*).

Then for equilibrium from Fig. 7.2*b*,

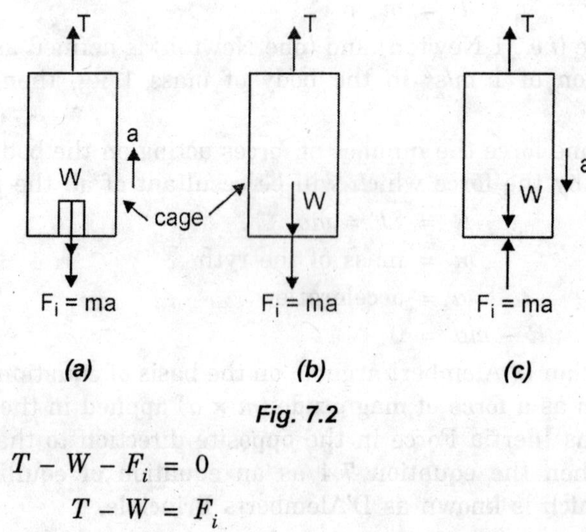

Fig. 7.2

$$T - W - F_i = 0$$

$$T - W = F_i$$

$$= ma$$

$$T - W = ma$$

$$T - mg = ma$$

$$T = ma + mg$$

$$= m(a + g)$$

(*b*) Case 2. Lift Moving Down

For quality by D' Alembert's Principle

$$T + F_i - W = 0$$

$$T + ma - mg = 0$$

$$T = mg - ma$$

$$= m(g - a)$$

7.1 : *A man having weight 700 N is going up and down moving in a lift. The weight of the lift chamber is 10 kN. If the lift is moving with an acceleration of 1.5 m/s², Find in the following cases in reaction of the man on the floor of the left chamber.*

(i) *When the lift is going up*

(ii) *When the lift is going down*

Solution : (*a*) Lift acceleration = 1.5 m/s²

$$\text{Weight of man} = 700 \text{ N} = W_2 \qquad \therefore \quad m_2 = \frac{W_2}{g}$$

$$\text{Weight of lift chamber} = 10 \times 1000$$

$$= 10000 \text{ N} = W$$

Fig. 7.3*a* shows is the free body diagram of the man whose weight is 700 N. The inertia force (700 x *a*) will be downwards as the acceleration is upwards. Therefore, from D'Alembert's principle for equilibrium.

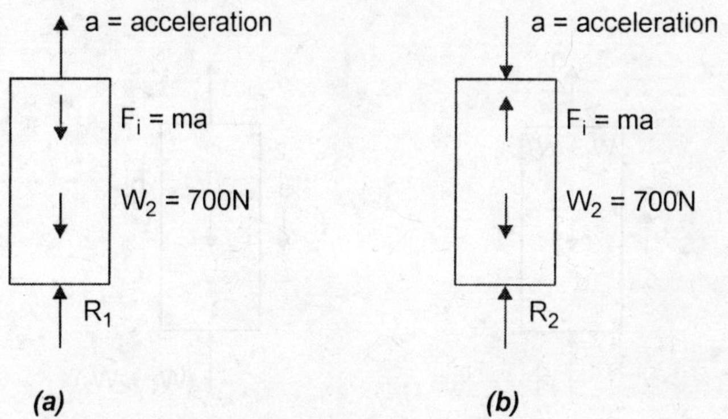

Fig. 7.3

$$F_i + W_2 = R_1$$

$$\frac{700 \times 1.5}{g} + 700 = R_1$$

$$700\left(\frac{1.5+1}{g}\right) = R_1$$

$$700\left(\frac{1.5}{9.81}+1\right) = R_1$$

or $$R_1 = 807.03 \text{ N}$$

(*b*) Case & when the lift is moving down

The free body diagram of the mass is shown in Fig. *b*.

The acceleration acting downwards to the body, therefore the F_i will be upwards. Thus, for equilibrium.

$$F_i = w_2 + R_2$$

EXAMPLE

E X A M P L E

$$\left(\frac{W_2}{g}\cdot a\right) = W_2 - R_2$$

or $$\frac{700}{9.81}\times 1.5 = 700 - R_2$$

or $$R_2 = 700\left(-\frac{1.5}{9.81}\times 1\right)$$

$$= 592.96 \text{ N}$$

7.2 : *In Example 7.1, find the tension in the rope in both the cases.*

Solution : Case I : Lift moving up F_i is downward.

$$= \left(\frac{W_1 + W_2}{g}\right) a$$

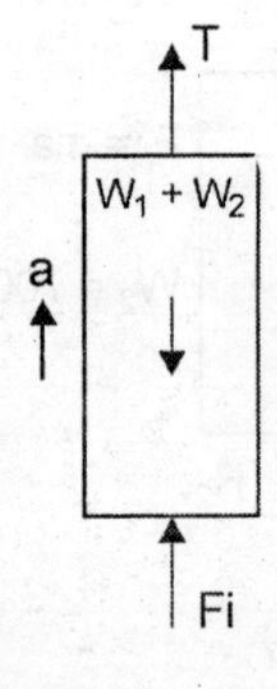

(a)

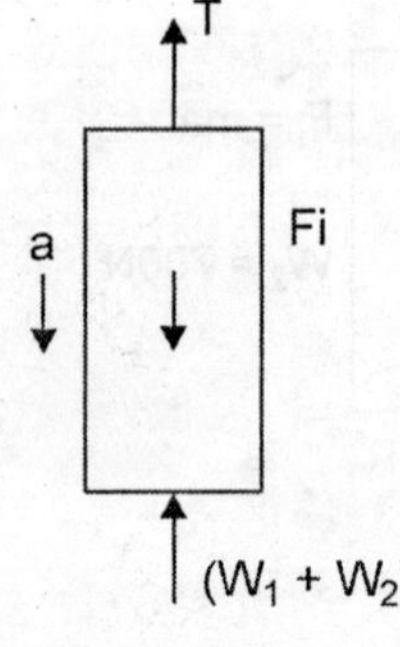

(b)

Fig. 7.4

For equilibrium with D' Alembert's principle,

$$T = F_i + (W_1 + W_2)$$

$$= \left(\frac{W_1 + W_2}{g}\right) a + (W_1 + W_2)$$

$$= (W_1 + W_2)\left(\frac{a}{g} + 1\right)$$

$$= (10000 + 700)\left(\frac{1.5}{9.81} + 1\right)$$

$$= 10700\ (0.1529 + 1)$$

$$= 12336 \text{ N}$$

$$= 12.36 \text{ kN}$$

Case II : Lift moving down. Therefore, F_i number upwards. Therefore, considering equilibrium under D. Alembert's principle.

$$T + F_i = (W_1 + W_2)$$

$$T = (W_1 + W_2) - \left(\frac{W_1 + W_2}{g}\right) a$$

$$= 10700 - \frac{10700 \times 1.5}{9.81}$$

$$= 10700\left(1 - \frac{1.5}{9.81}\right)$$

$$= 9063.97 \text{ N}$$

$$= 9.064 \text{ kN Ans}$$

EXAMPLE

7.4 ACCELERATING MOVEMENT OF BODIES ON HORIZONTAL AND INCLINED ROUGH PLANES

7.4.1

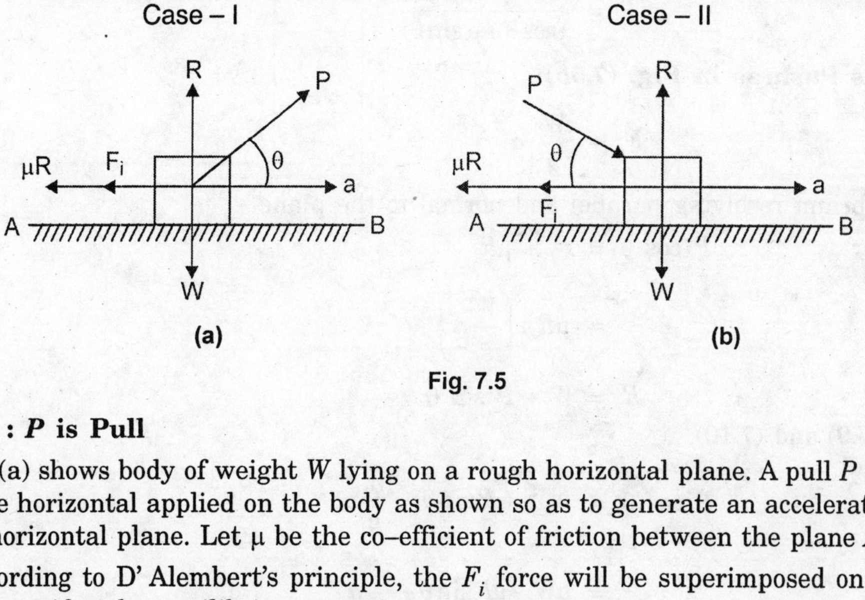

Fig. 7.5

Case I : *P* is Pull

Fig. 7.5(a) shows body of weight W lying on a rough horizontal plane. A pull P inclined at θ° with the horizontal applied on the body as shown so as to generate an acceleration *'a'* on the rough horizontal plane. Let μ be the co–efficient of friction between the plane AB and body.

According to D' Alembert's principle, the F_i force will be superimposed on the leftward side to consider the equilibrium equation,

$$F_i = \frac{W}{g} \cdot a$$

$$R = \text{Reaction of table on body}$$

$$\mu R = \text{Frictional force}$$

Resolving the forces along and normal to the plane,

$$P \cos \theta = \mu R + F_i$$

$$= \mu R + \left(\frac{W}{g}a\right) \quad ...(7.6)$$

and

$$P \sin \theta + R = W$$

$$R = W - P \sin \theta \quad ...(7.7)$$

Substituting the value of R from (7.7) into (7.6),

$$P \cos \theta = \mu[W - P\sin\theta] + \frac{W}{g}a$$

$$= \mu W - \mu P \sin\theta + \frac{W}{g}a$$

$$P \cos \theta + \mu P \sin \theta = \mu W + \frac{W}{g}a$$

$$P(\cos \theta + \mu \sin \theta) = W\left(\mu + \frac{a}{g}\right)$$

$$\therefore \quad P = \frac{W\left(\mu + \dfrac{a}{g}\right)}{(\cos\theta + \mu\sin\theta)} \quad ...(7.8)$$

Case II : *P* is Push as in Fig. (7.5*b*)

$$F_i = \frac{W}{g} \cdot a$$

For equilibrium resolving parallel and normal to the plane

$$P \cos \theta = F_i + \mu R$$

$$= \mu R + \left(\frac{W}{g}a\right) \quad ...(7.9)$$

and

$$R = W + P \sin \theta \quad ...(7.10)$$

$\therefore$ From (7.9) and (7.10),

$$P \cos \theta = \mu[W + P\sin\theta] + \frac{W}{g}a$$

$$= \mu W + \mu P \sin\theta + \frac{W}{g}a$$

$$P\cos \theta - \mu P \sin \theta = W\left(\mu + \frac{a}{g}\right)$$

$$\therefore \quad P = \frac{W\left(\mu + \dfrac{a}{g}\right)}{(\cos\theta - \mu\sin\theta)} \quad ...(7.11)$$

7.5 ACCELERATED MOTION UPON INCLINED PLANE

Let a force P(PULL) is applied parallel to the plane AB on which the body of weight to and gets accelerated upwards. Then F_i, the inertia force will be acting in downward direction.

For the equilirbium, resolving the forces along and normal to the plane AB,

$$F_i = \frac{W}{g}a$$

$$P = \mu R + F_i + W \sin \theta$$

$$= \mu R + \frac{W}{g}a + W \sin \theta \quad ...(7.12)$$

Fig. 7.6

and $$R = W \cos \theta \quad ...(7.13)$$

From equations (7.12) and (7.13),

$$P = \mu W \cos \theta + \frac{W}{g}a + W \sin \theta$$

$$P = W\left[\mu \cos \theta + \sin \theta + \frac{a}{g}\right] \quad ...(7.14)$$

EXAMPLE

7.3 : *Blocks A and B weighing 100 N and 40 N respectively are connected by light string. The blocks A and B are placed on inclined planes as shown in Fig. 7.7. The string passes over a frictionless pulley. The co-efficient of friction is 0.2 on the rough planes. Find the tension in the string and the velocity of the block A after 2 seconds.*

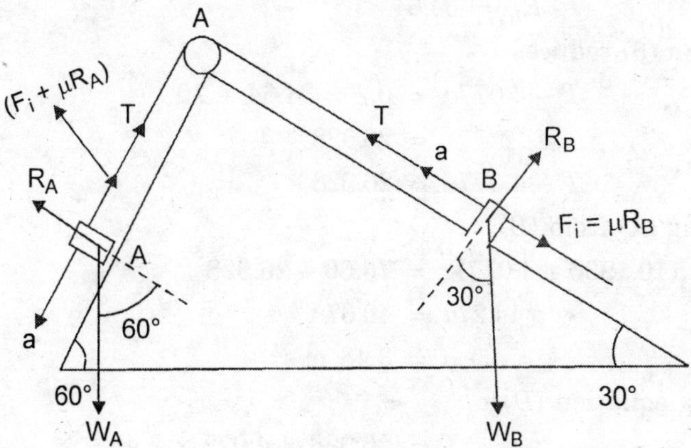

Fig. 7.7

E X A M P L E

Solution : Since

$$W_A = 100 \text{ N}$$
$$W_B = 40 \text{ N}$$

The set will be accelerating with block A moving downwards therefore, F_{i_A} will be acting upwards and on block B, F_{i_B} is acting downwards. Let the tension in string be T.

For the equilibrium of A,

$$T + F_{i_A} + \mu R_A = W_A \sin 60$$

or $$T + \left(\frac{100}{9.81} \times a\right) + 0.2 R_A = 100 \times \frac{\sqrt{3}}{2} \quad ...(A)$$

and $$R_A = W_a \cos 60 = \frac{1}{2} \times 100 = 50$$

Equation (A) reduces to,

$$T + \left(\frac{100}{9.81} \cdot a\right) + 0.2 \times 50 = 50\sqrt{3}$$

or $$T + 10.1936a = 76.60 \quad ...(B)$$

For equilibrium of (B) resolving forces parallel and normal to 30° plane,

$$T = \left(\frac{W_b}{g} a\right) + \mu R_B + W_B \sin 30$$

$$= \left(\frac{40}{9.81} \times a\right) + 0.2 \times R_B + 40 \times \frac{1}{2}$$

or $$T = (4.077a) + .2 R_B + 20$$

or $$T - 4.077a = .2 R_B + 20 \quad ...(C)$$

Also, $$R_B = W_B \cos 30$$
$$= 40 \times 0.866$$
$$R_B = 34.64$$

∴ Equation (6) reduces

$$T - 4.077a = 0.2 \times 34.64 + 20$$
$$= 26.928$$
$$T - 4.077a = 26.928 \quad ...(D)$$

Subtracting (A) from (B),

$$(10.1936 + 4.077)a = 76.60 - 26.928$$

or $$14.27a = 49.67$$
$$a = 3.48 \text{ m/s}^2 \quad (E)$$

Then from equation (D),

∴ $$T = 26.928 \times 4.077 \times 3.48$$
$$= 41.116 \text{ N} \quad ...(F)$$

Knowing the value 'a' from equation (E), the velocity of block A after 2 seconds is obtained by using the relationship.

$$v = u + at$$
$$= 0 + 3.48 \times 2$$
$$= 6.96 \text{ m/s Ans.}$$

7.4 : *Two blocks A and B weighing 60 N and 40 N respectively are hung at each end of a thin rope over a frictionless pulley. Find the acceleration with which the block A will be coming down. Also, find the tension in the rope.*

Solution : Let the acceleration be $= a$

Since Block A moves down with acceleration 'a', the inertia force $(F_i)_A$ will be upward, however on Block B $(F_i)_B$ will be downward as per D' Alembert's principle. For the equilibrium of Block A,

$$\Sigma Y = 0$$

or $$T + (F_i)_A - W_A = 0$$

$$T + \left(\frac{W_A}{g}a - W_A\right) = 0$$

or $$T + \frac{60}{9.81}a = 60$$

or $$T + 6.116a = 60 \qquad ...(A)$$

For the equilibrium of Block B,

$$T - (F_i)_B = W_B$$
$$= 40$$

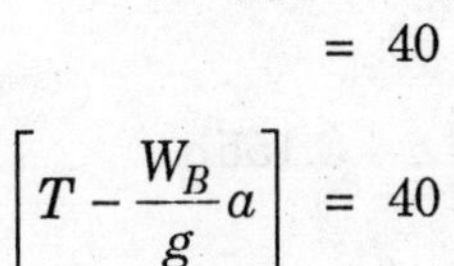

$$\left[T - \frac{W_B}{g}a\right] = 40$$

$$T - \frac{40}{9.81}a = 40$$

$$T - 4.3956a = 40 \qquad ...(B)$$

Fig. 7.8

Subtract (B) from (A),

$$a(6.116 + 4.3956) = 20$$
$$11.032a = 20$$

$$a = \frac{20}{11.032}$$
$$= 1.813 \text{ m/s}^2$$

Then from (B), $$T = 40 + 1.813 \times 4.3956$$
$$= \mathbf{57.965\ Newton\ (N)\ Ans.}$$

EXAMPLE

7.5 : *A block 'A' weighing 80 N rests on a rough plane inclined to the horizontal at 30°. The block A is pulled up the plane by a string whose other end after passing over a smooth pulley hangs vertically down and to this end is attached a weight 600 N as shown in Fig. 7.9. If co–efficient of friction is 0.2, find:*

(a) Tension in string

(b) The acceleration

(c) The distance moved by body A in 1 sec

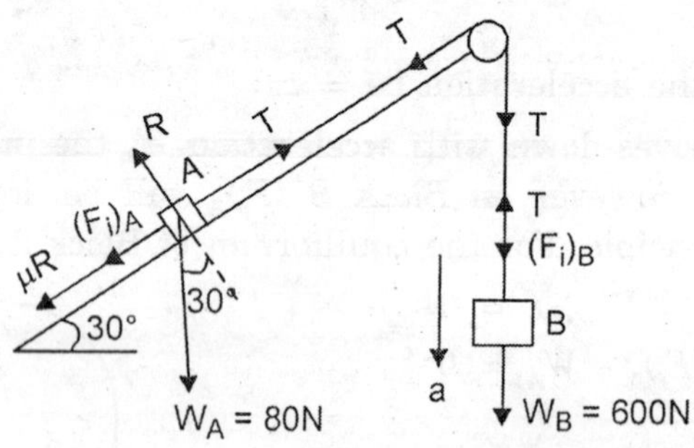

Fig. 7.9

Solution : Given

$$W_A = 80 \text{ N}$$
$$W_B = 600 \text{ N}$$
$$\mu = 0.2$$
$$\theta = 30°$$

Let the acceleration of blocks = a

The
$$(F_i)_A = M_A \cdot a$$
$$= \left(\frac{80}{9.81}\right) a = 8.155a$$

For Body B
$$W_B = 600 \text{ N}$$
$$(F_i)_B = \frac{600}{9.81} a$$

For equilibrium of Block B we have,

$$T + (F_i)_B = W_B$$
$$T + \frac{600}{9.81} a = 600$$

or
$$T + 61.16a = 600 \quad (A)$$

For Equilibrium of Block A

Resolving forces normal and along the plane.

Solution $\quad R = W_A \cos 30$

and $\quad \mu R + (F_i)_A + W_A \sin 30 = T$

$$0.2\left(W_A \cdot \frac{\sqrt{3}}{2}\right) + \frac{W_A}{g}a + W_A\frac{1}{2} = T$$

or $$0.2 \times 80 \times \frac{\sqrt{3}}{2} + \frac{80}{9.81} \times a + \frac{80}{2} = T$$

or $$T = 13.85 + 8.155a + 40$$

$$T = 8.155a + 53.85 \quad ...(B)$$

From equation (*A*) and (*B*),

$$(8.155a + 53.85) + 61.16a = 600$$

$$69.315a = 546.75$$

or $$a = 7.879 \text{ m/s}^2 \textbf{ Ans.}$$

∴ From (B), $$T = 8.155 \times 7.879 + 53.85$$

$$= 63.764 + 53.85$$

$$= 117.6 \text{ N} \quad \textbf{Ans.}$$

Distance travelled by block *A* in sec,

$$s = ut + \frac{1}{2}at^2$$

$$= 0 + \frac{1}{2} \times 7.879 \times (1)^2$$

$$= 3.9395 \text{ m} \quad \textbf{Ans.}$$

EXAMPLE

7.6 CIRCULAR MOTION (ABOUT FIXED AXIS)

If a body is moving along a circular path it is called circular motion about an axis passing through the centre of circular path. Connected with this motion, the angular displacement is defined as the angle in radians the radial line connecting the centre to the body, moves from a certain reference line is called the angular displacement. In fig. 7.10, *OX* is the inertial reference line and radial line *OA* has moved through angle θ to occupy the position *OB*. Thus, the displacement corresponding to the position *OB* is angle θ and is expressed in radians.

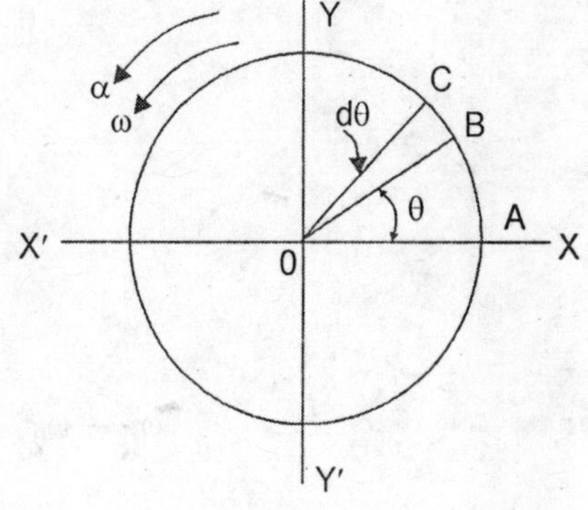

Fig. 7.10

Again in time, *dt* let the radial line move further to position *OC* an additional angle *dθ* in time *dt*, then the angular velocity ω is given by,

$$\omega = \frac{d\theta}{dt} \text{ rad/s} \quad ...(7.14)$$

Similarly, if the velocity does not remain constant and it increases by *dω* in time *dt*, then angular acceleration α is expressed as,

$$\alpha = \frac{d\omega}{dt} = \frac{d^2\theta}{dt^2} \qquad \text{...(7.15)}$$

7.6.1 Equations Controlling the Angular Motion

Let ω_0 be the initial velocity and ω the velocity after t second with angular acceleration α.

Then $$W_1 = \omega_0 + \alpha t \qquad \text{...(7.16)}$$

Also, if θ is the angle traversed during the time t secs.

Then average velocity $$= \left(\frac{\omega_0 + \omega}{2}\right) = \omega_{av}$$

$\therefore$
$$\theta = W_{av} \times t$$
$$= \left(\frac{\omega_0 + \omega}{2}\right)t$$
$$= \left(\frac{\omega_0}{2} + \frac{\omega}{2}\right)t$$
$$= \left[\frac{\omega_0}{2} + \frac{\omega_0 + \alpha t}{2}\right]t$$

From equation (7.16), $$= \left[\omega_0 + \frac{1}{2}\alpha t\right]t$$

or $$\theta = \omega_0 t + \frac{1}{2}\alpha t^2 \qquad \text{...(7.17)}$$

From equation (7.16) and (7.17) we have,

$$\theta = \omega_0\left(\frac{\omega - \omega_0}{\alpha}\right) + \frac{1}{2}\alpha\left(\frac{\omega - \omega_0}{\alpha}\right)^2$$
$$= \left(\frac{\omega - \omega_0}{\alpha}\right)\left[\omega_0 + \left(\frac{\omega - \omega_0}{2}\right)\right]$$
$$= \left(\frac{\omega - \omega_0}{2}\right)\left[\frac{\omega + \omega_0}{2}\right]$$

or $$\omega^2 - \omega_0^2 = 2\alpha\theta \qquad \text{...(7.18)}$$

7.7 MOMENTUM

7.7.1 Momentum of Translation

Let a body of mass m kg be moving with a velocity v m/s, then the product $(m \times v)$ is called the momentum of the body. If momentum is M, then,

$$M = mv \qquad \text{...(7.19)}$$

7.7.2 Angular Momentum (Moment of Momentum)

Angular momentum is the product of mass moment of Inertia of the body about the axis with respect to which the body is rotating with angular velocity ω about the same axis.

i.e., Angular momentum $= I \times \omega$ ...(7.20)

Where, I = mass moment of inertia of the body about the due is of rotating

$$= \int (dm)\, r^2$$

Fig. (7.11) show an elemental mass dm at a distance of r from the axis of rotation of the body.

Then momentum of elemental mass dm

$$= dm\,(v)$$

$$= dm\,(r\omega)$$

∴ Moment of momentum of the body

$$= \oint dm(r\omega) \times r$$

$$= \oint dm(r^2) \times \omega$$

or $$= \omega \oint (dm)\, r^2 = I\omega \quad ...(7.21)$$

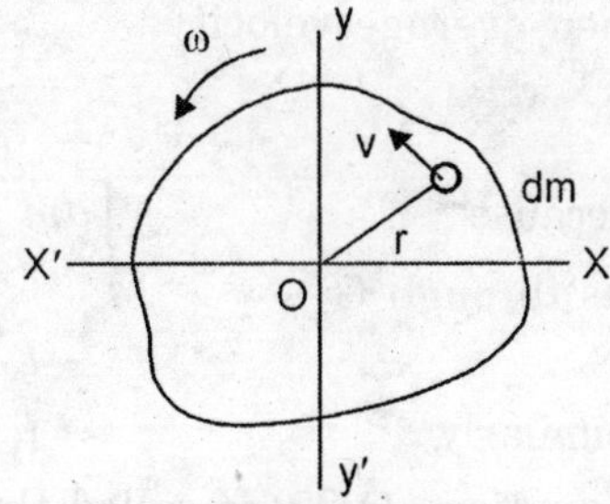

Fig. 7.11

7.8 MASS MOMENT OF INERTIA OF VARIOUS BODIES

Before we proceed to find the mass moment of Inertia (MOI) or simply I_m, we will discuss the theorems of parallel and perpendicular axes governing the MOI.

(a) Theorem of Parallel Axes

Let G be the centre of gravity of the body and take the origin O at the point G. The axes XOX', YOY' pass through G. An elemental mass dm be at a point whose co–ordinates are (x, y) and its distance from O is r.

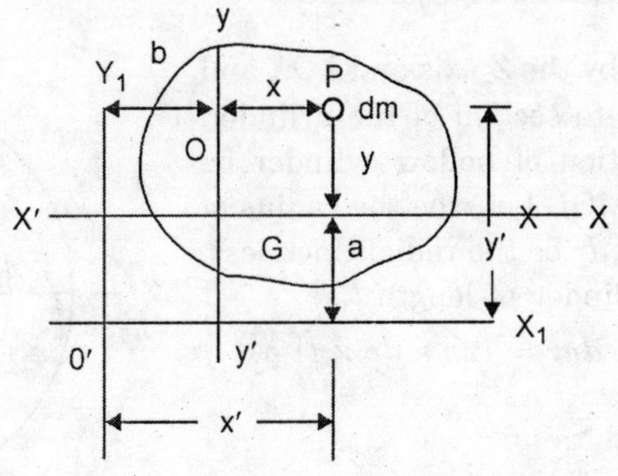

Fig. 7.12

Then $$I_{XX} = \oint dm\, y^2$$

$$I_{YY} = \oint (dm)x^2$$

Let O' be an other point where ones $X'X'$ and $Y'Y'$ are taken. With respect to new axes X' O X an $YO'Y'$, co–ordinates of P are. $(x + b)$, $(y + a)$. In other words, the distance between XX and $X'X'$ equal to 'a' and that between YX an $Y'Y'$ is 'b'.

Then $$I_{X'X'} = \int (dm)\,(y+a)^2$$

$$= \int (y^2 + a^2 + 2ay)\,dm$$

$$= \int y^2 dm + a^2 \int dm + 2a \int dm\, y$$

$$= I_{XX} + a^2M + \text{zero}$$

because $\int dm\, y$ = First moment of mass about y-axis is zero as y-axis passes through G

$$\therefore \quad I_{X'X'} = (I_G)_{XX} + Ma^2$$

Similarly, $$I_{Y'Y'} = (I_G)_{YY} + Mb^2 \qquad \text{...(7.22)}$$

Equation (11.22) is called the theorem of parallel axes.

Now $$r^2 = x^2 + y^2$$

$$\therefore \quad I_{ZZ} = \oint dm(r^2)$$

$$= \int dm\,(x^2 + y^2)$$

$$= \int dm\, y^2 + \int dm\, x^2$$

$$I_{ZZ} = I_{XX} + I_{YY} \qquad \text{...(7.23)}$$

Equation (7.23) is called the theorem of perpendicular axes.

7.8.2 MASS MOMENT OF INERTIA OF DIFFERENT BODIES

7.8.2.1 Mass Moment of Inertia of Cylinder

Let the axis of the cylinder by the Z axis and XOX and XOY are axis with respect to section of the cylinder. Let an elemental ring section of hollow cylinder be considered for finding the MOI. Let r be the radius of this elemental cylinder and dr be the radial thickness, then mass of elemental cylinder of length l.

$$dm = (2\pi\, r\, dr \times l)\, \rho$$

Where ρ is the density

Then $$I_{ZZ} = \int_{r=0}^{R} [\rho\, 2\pi r\, dr\, l] r^2$$

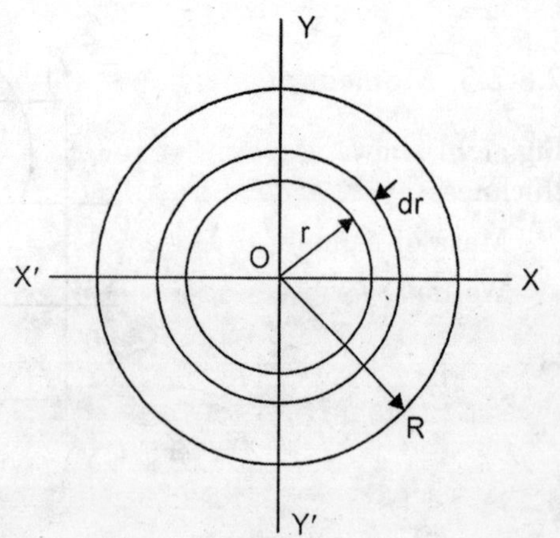

Fig. 7.12

$$= \rho 2\pi l \int_0^R r^2 dr$$

$$= 2\pi l \rho \left[\frac{r^4}{4}\right]_0^R$$

$$= \frac{\pi}{2} l \rho R^4 = (\pi R^2 l \rho)\frac{R^2}{2}$$

$$I_{ZZ} = \frac{MR^2}{2} \qquad \text{...(7.24}a\text{)}$$

where M = mass of cylinder

Then from equation (11.23),

$$I_{XX} = I_{YY} = \frac{1}{2} I_{ZZ} \qquad \text{...(7.24}b\text{)}$$

$$= \frac{MR^2}{4} \qquad \text{...(7.25)}$$

7.8.2.2 Moment of Inertia of a Disc

In equation (7.24), the MOI of a solid cylinder of length l has been found about I_{ZZ}.

That is $\quad I_{ZZ} = \dfrac{MR^2}{2}$

where $\quad M = \pi r^2 l \rho$

If l is replaced by t, the thickness of the disc of radius R,

Then for disc $\quad I_{ZZ} = \dfrac{M_d R^2}{2} \qquad$ ...(7.25)

where $\quad M_D$ = mass of disc

$= (\pi R^2 \; t\rho)$

7.8.2.3 Moment of Inertia of a Sphere

Fig. 7.14 shows a sphere of radius R. An elemental disc of thickness dy at a height y,

Mass of elemental disc $\quad dM_d = \pi r^2 \rho \; dy$

We have form equation (7.25),

$$dI_{yy} = \frac{(dM_d)r^2}{2}$$

$$I_{yy} = 2\int_0^R \left(\frac{1}{2}\pi r^2 \rho dy) r^2\right) \qquad \text{...(7.26)}$$

Let $\quad \angle COX = \theta$

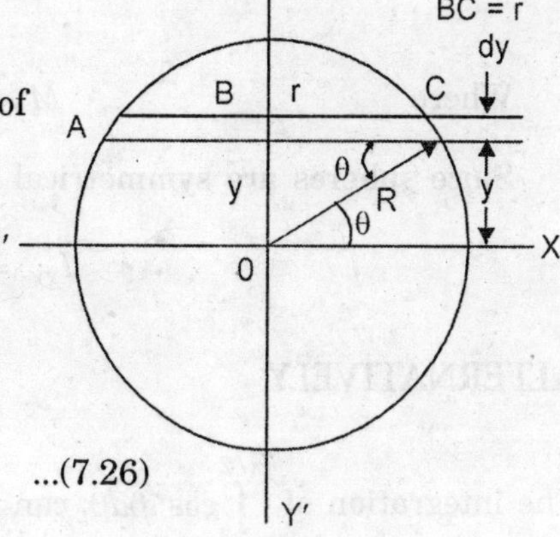

Fig. 7.14

Then
$$y = R \sin \theta$$
$$dy = R \cos \theta \, d\theta$$

Then eqn. (7.26) reduces to

$$I_{yy} = \rho\pi \int_0^R r^4 dy = \pi\rho \int_{\theta=0}^{\pi/2} (R\cos\theta)^4 (R\cos\theta) d\theta$$

$$= \pi p R^5 \int_0^{\pi/2} \cos^5 \theta d\theta$$

$$= \pi\rho R^5 \int_0^{\pi/2} \cos\theta[1 - \sin^2\theta]^2 d\theta$$

$$= \pi\rho R^5 \int_0^{\pi/2} \cos\theta(1 + \sin^4\theta - 2\sin^2\theta) d\theta$$

$$= \pi\rho R^5 \int_0^{\pi/2} \cos\theta d\theta + \int_0^{\pi/2} \sin^4\theta\cos\theta d\theta - 2\int_0^{\pi/2} \sin^2\theta\cos\theta d\theta$$

or
$$I_{yy} = \rho\pi R^5 \left[\sin\theta + \frac{\sin^5\theta}{5} - 2\frac{\sin^3\theta}{3}\right]_0^{\pi/2}$$

$$= \rho\pi R^5 \left[1 + \frac{1}{5} - \frac{2}{3}\right]$$

$$= \frac{8}{15}\pi\rho R^5 = \left(\frac{4}{3}\pi R^3 \rho\right) R^2 \left(\frac{2}{5}\right)$$

$$\boxed{\rho = \frac{\text{Mass}}{\text{Volume}}}$$

$$I_{YY} = \frac{2MR^2}{5}$$

Where
$$M = \frac{4}{3}\pi R^3 \rho = \text{vol} \times \rho = \text{mass of sphere}$$

Since spheres are symmetrical about all the three axes,

$$I_{XX} = I_{yy} = I_{ZZ} = \frac{2}{5}MR^2$$

ALTERNATIVELY

The integration of $\int_0^{\pi/2} \cos^5 \theta d\theta$ can be done with the help of Gama functions

$$\int_0^{\pi/2} \cos^5 \theta d\theta = \int_0^{\pi/2} \cos^5 \theta \sin^0 \theta d\theta$$

We know $\int_0^{\pi/2} \cos^m \theta \sin^n \theta d\theta$

$$= \frac{\left|\frac{m+1}{2}\right. \left|\frac{n+1}{2}\right.}{2\left|\frac{m+n+2}{2}\right.}$$

and $\quad \left|n+1\right. = n\left|n\right.$

and $\quad \left|1\right. = 1$

$$\left|\frac{1}{2}\right. = \sqrt{\pi}$$

Here, put $m = 5$ and $n = 0$

$$I_{YY} = \frac{\left|\frac{5+1}{2}\right. \left|\frac{0+1}{2}\right.}{2\left|\frac{5+0+2}{2}\right.}$$

$$= \frac{\left|3\right. \left|\frac{1}{2}\right.}{2\left|\frac{7}{2}\right.} = \frac{(2\times 1)\left(\sqrt{\pi}\right)}{2\left(\frac{5}{2}\times\frac{3}{2}\times\frac{1}{2}\sqrt{\pi}\right)}$$

$$= \frac{8}{15}$$

$$\therefore \quad I_{yy} = \pi\rho R^5 \times \frac{8}{15} = \left(\frac{4}{3}\pi\rho R^3\right)\frac{2}{5}R^2$$

$$= \frac{2}{5}MR^2 \qquad \text{...2.27}$$

7.8.2.4 Moment of inertia of a rod about YY at midpoint

Part I

Moment of inertia of rod about Y Y at midpoint Fig. 7.15 show a rod of length L. Let m be the mass of unit length of the rod. It is desired to find the MOI about yy.

Let an elemental mass dm be situated at a distance x and its length is dx

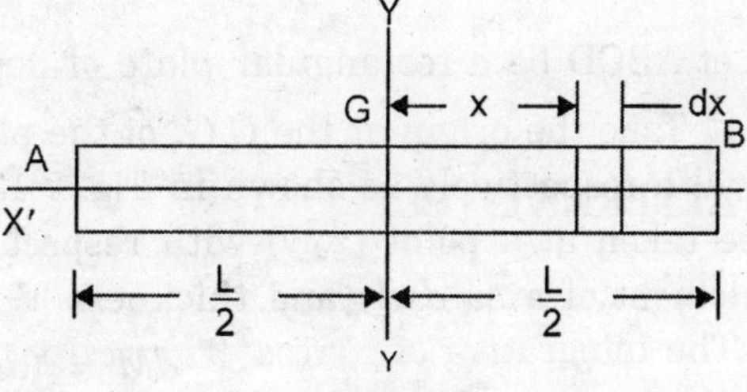

Fig. 7.15

Then $$I_{yy} = \int_{-L/2}^{L/2} dMx^2 = \int_{-L/2}^{L/2} (mdx)(x^2) = \int_{-L/2}^{L/2} mx^2\,dx \quad [dM = mdx]$$

$$= m\left[\frac{x^3}{3}\right]_{-\frac{L}{2}}^{L/2} = \frac{m}{3}\left[\frac{L^3}{8}+\frac{L^3}{8}\right]$$

$$I_{yy} = \frac{mL^3}{12} = \frac{ML^2}{12} \qquad [M = mL] \quad ...7.28$$

Part II: Moment of inertie of rod about YY at one end

In case the axis of rotation is Y_1Y_1 at one end and orgin is at A (Fig. 7.16)

Then $$I_{y_1y_1} = \int_0^L dMx^2$$

$$= \int_0^L (mdx)x^2$$

$$= m\left[\frac{x^3}{3}\right]_0^L$$

Fig. 7.16

$$I_{y_1y_1} = \frac{mL^3}{3} = \frac{ML^2}{3} \qquad [\because M = mL] \quad ...(7.29\ a)$$

Atternatively applying theorem of parallel axes,

$$I_{y_1y_1} = I_G + M\,(AG)^2$$

$$= I_{yy} + M\left(\frac{L}{2}\right)^2$$

$$= \frac{ML^2}{12} + \frac{ML^2}{4} = \frac{ML^2}{3} \qquad ...(7.29b)$$

7.8.2.5 Mass Moment of Ineratia of a Rectangular Plate

Let ABCD be a rectangular plate of depth d, width b and theorem t.

Take the origin at the $C.G.$ of the plate and make the XX and YY axes along length and width respectively as shown in Fig. 7.17. Let an elemental area of length dx, and width dy be taken at a point (x, y) with respect to XX and YY axes. Let dM be elemental mass of element of area $dxdy$ and thickness 't' and density = ρ

$$dM = (dx.\ dy)t\rho$$

$$I_{xx} = \int_{-\frac{b}{2}}^{b/2}\int_{\frac{-d}{2}}^{d/2} dmy^2 = \iint(\rho.t.dxdy)(y^2) = \rho\int_{n,\frac{-b}{2}}^{b/2} t\left[\int_{\frac{-d}{2}}^{d/2} dyy^2\right]dx$$

$$= \rho tx\int_{x=-\frac{b}{2}}^{b/2}\left[\frac{y^3}{3}\right]_{-d/0}^{d/2} dx$$

$$= \frac{\rho t}{3}\int_{x=-\frac{b}{2}}^{b/2}\left(\frac{d^3}{8}+\frac{d^3}{8}\right)dx$$

$$= \frac{\rho td^3}{3\times 4}[x]_{-\frac{b}{2}}^{+b/2}$$

$$= \frac{\rho td^3}{12}\left[\frac{b}{2}+\frac{b}{2}\right]$$

$$= \frac{\rho tbd^3}{12} = \frac{Md^2}{12} \quad \text{...(7.30)}$$

Fig. 7.17

For

$$I_{yy} = \int_{y=-\frac{d}{2}}^{d/2}\int_{y=\frac{-b}{2}}^{b/2} dMx^2$$

$$= \int_{\frac{-d}{2}}^{d/2}\int_{\frac{-b}{2}}^{b/2}(\rho.t.dxdy)(x^2)$$

$$= \rho t\int_{y=-\frac{d}{2}}^{+d/2}\left[\int_{-\frac{b}{2}}^{b/2} x^2dx\right]dy$$

or

$$R_{yy} = \rho t\int_{y=-\frac{d}{2}}^{d/2}\frac{1}{3}\left[\frac{b^3}{8}+\frac{b^3}{8}\right]dy$$

$$= \frac{\rho t}{3}\int_{-\frac{d}{2}}^{d/2}\left(\frac{b^3}{4}\right)dy = \frac{\rho t}{12}b^3[y]_{-\frac{d}{2}}^{\frac{d}{2}}$$

$$= \frac{\rho tb^3}{12}\times d = \left(\frac{\rho tbd}{12}\right)b^2$$

$$R_{yy} = \frac{Mb^2}{12} \quad \text{....(7.31)}$$

By using theorem of perpudcular axis we get

$$I_{zz} = R_{xx} + R_{yy}$$

$$= \frac{Md^2}{12} + \frac{Mb^2}{12}$$

$$= \frac{M}{12}\left(b^2 + d^2\right) \quad ..(7.32)$$

7.8.2.6 Mass Moment of Inertia of a Thin Rim of Uniform Section

Let R be the mean radius of the rim of cross–sectional Area A and ρ its density, then mass per unit length m is

$$m = (A \times 1)\rho = A\rho$$

Let an element of length $R\, d\theta$

Then $$I_{XX} = \int_{\theta=0}^{2\pi} (dM)(R\sin\theta)^2$$

Here, $$dm = (Rd\theta)A\rho$$

$\therefore$ $$I_{XX} = \int_{\theta=0}^{2\pi} RA\rho d\theta (R\sin\theta)^2$$

$$= R^3 A\rho \int_0^{2\pi} \sin^2\theta d\theta$$

$$= \frac{R^3 A\rho}{2}\int_0^{2\pi} 2\sin^2 0\, d\theta \Rightarrow \frac{R^3 A\rho}{2}\int_0^{2\pi} 1 - \cos 2\theta\, d\theta$$

$$= \frac{R^3 A\rho}{2}\left[\theta - \frac{\sin 2\theta}{2}\right]_0^{2\pi}$$

$$= \frac{R^3 A\rho}{2}(2\pi)$$

$$= \pi R^3 A\rho$$

$$= (2\pi \mathrm{R}\, A\rho)\frac{R^2}{2}$$

Fig. 7.18

or $$I_{XX} = \frac{MR^2}{2} \quad ...(7.33)$$

Also, $$I_{YY} = I_{XX} = \frac{1}{2} I_{ZZ}$$

$\therefore$ $$I_{ZZ} = 2I_{XX} = MR^2 \quad ...(7.34)$$

7.8.2.7 Mass Moment of Inertia of a Right Circular Cone About its Axis of Rotation

Let the height of the right circular cone be 'h' and radius of the base be R.

Let an elemental disc of thickness d_z be taken at a depth of Z from Apex. Then for the elemental disc we have from, equation 7.25,

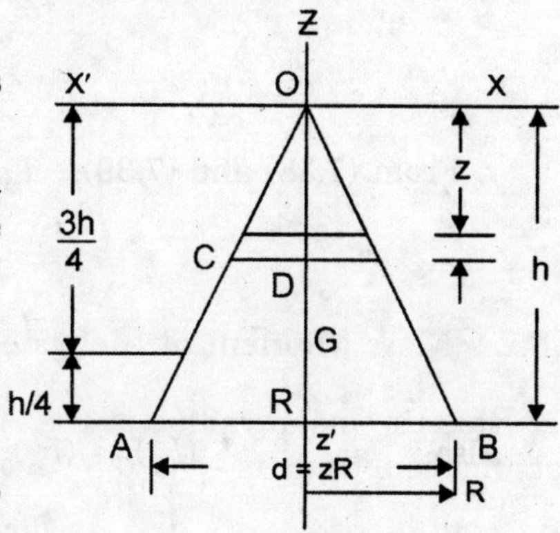

Fig. 7.19

$$DI_{zz} = \frac{1}{2}\text{(mass of disc) (radius)}^2$$

$$= \frac{1}{2}(dM)(CD)^2 \quad \text{...(7.35)}$$

Now from similar triangles OAF and OCD, radius of disc is given by,

$$CD = \frac{AF}{h}z$$

$$= \frac{R}{h}z \quad \text{...(7.36)}$$

$$\text{Mass of disc} = dM = \text{Vol} \times \text{density}$$

$$= \rho(CD)^2 dz \times \pi$$

$$= \pi\rho\frac{R^2}{h^2}z^2 dz \quad \text{...(7.37)}$$

Equation (7.35), (7.36) and (7.37),

$$\therefore \quad I_{zz} = \int dI_{zz} = \int_{z=0}^{h}\frac{1}{2}\left[\rho\pi\frac{R^2z^2}{h^2}dz\right]\frac{R^2}{h^2}z^2$$

$$I_{zz} = \frac{\pi\rho R^4}{2h^4}\int_{z=0}^{h} z^4 dz$$

$$= \frac{\rho\pi}{2}\frac{R^4}{h^4}\left[\frac{z^5}{5}\right]_0^h$$

$$= \frac{\pi}{10}\rho R^4 h \quad \text{...(7.38)}$$

Now, mass of cone $\quad M = \int_{z=0}^{h}\pi\{(CD)^2\}dz\rho$

$$= \int_0^h \pi\rho\frac{R^2}{h^2}z^2 dz$$

$$= \frac{\pi R^2 \rho}{h^2}\left[\frac{z^3}{3}\right]_0^h = \frac{\pi \rho R^2 h}{3}$$

$\therefore$ From (7.38) and (7.39),
$$I_{zz} = \left(\pi \frac{\rho R^2 h}{3}\right)\frac{3}{10}R^2 \qquad ...(7.40)$$

$$I_{zz} = \frac{3M}{10}R^2$$

Also, $(I_{zz})_G - (I_{zz})_0 = \frac{3M}{10}R^2$ (Where axis are having GM)

N/W applying the theorem of perpendicular axes

We get,
$$(I_x)_G = (I_y)_G = \frac{1}{2}I_{zz}$$

$$\therefore \quad (I_{xx})_G = (I_{yy})_G = \frac{3}{20}MR^2$$

Note: The X, Y and Z axes have origin at O, the vertical of the cone and Z axis is shell theorem as G lies on it.

Therefore, now using theorem of parallel axis by trnsferring from $(I_{xx})_G$ to $(I_{xx})_0$

i.e.,
$$(I_{xx})_0 = (I_{xx})_G + M(G_0)^2$$

$$= \frac{3}{20}MR^2 + M\left(\frac{3}{4}h\right)^2$$

$$= \frac{3M}{20}R^2 + \frac{9}{16}Mh^2$$

$$\therefore \quad (I_{yy})_0 = (I_{xx})_0 = \frac{3M}{4}\left(\frac{R^2}{5} + \frac{3}{4}h^2\right) \qquad ..7.41$$

7.9 RADIUS OF GYRATION

Radius of Gyration is that distance from the axis, square of which when multiplied with the mass of the body gives the mass moment of inertia of body about that axis.

Then if K = Radius of gyration

Then $I = MK^2$

or
$$K = \sqrt{\frac{I}{M}} \qquad (7.42)$$

7.9.1

Radius of Gyration of circular disc about centroidal axis normal to the plane of the disc of radius R and mass M. Using equation (7.24), we get

$$I_{zz} = \frac{MR^2}{2} = \text{MK}^2$$

$$\therefore \quad K = \frac{R}{\sqrt{2}} \qquad \text{...(7.48)}$$

7.9.2

Radius of Gyration of circular cylinder of radius R and mass M. Here, again we have,

$$I_{zz} = \frac{MR^2}{2} = \text{MK}^2$$

$$\therefore \quad K = \frac{R}{\sqrt{2}} \qquad \text{...(7.44)}$$

7.8.3 Radius of Gyration of solid sphere of radius R and mass M

We have from equation (7.27),

$$I_{zz} = \frac{2}{5}MR^2 = MK^2$$

$$\therefore \quad K = \sqrt{\frac{2}{5}}R \qquad \text{...(7.45)}$$

7.9.4 Radius of Gyration of Cone About Axis of Symmetry

From equation (7.40), $I_{zz} = \frac{3}{10}MR^2 = MK^2$

$$K = \sqrt{\frac{3}{10}}R \qquad \text{...(7.46)}$$

7.9.5 Radius of Gyration or Rectangular Disc About ZZ.

We have from equation (7.32),

$$I_{zz} = \frac{M}{12}(b^2 + d^2) = MK^2$$

$$\therefore \quad K = \sqrt{\frac{b^2 + d^2}{12}} \qquad \text{...(7.47)}$$

7.9.6 Radius of Gyration of rim of radius R

From equation (7.34),

$$I_{zz} = MR^2 = MK^2$$

i.e., $K = R$ ...(7.48)

7.10 LAWS OF ROTARY MOTION

7.10.1 Newton's First Law

It states that a body continues to be in the state of rotation about an axis with constant angular velocity until compelled by an external torque to change the state.

However, in actual practice we notice that when a body is rotating at certain rotational speed is left on its own, it will slowly come to rest even without the application of external torque. Evidently the air resistance and bearing friction act to provide the opposing torque to reduce the speed and bringing it to rest.

7.10.2 Newton's Second Law for Rotary Motion

It states that rate of change angular momentum of the relating body is proportional to the torque applied on the body and takes place in the direction of torque applied.

Consider a body of mass M and moment of inertia I with K as radius of gyration starging of having an initial angular speed of ω_0 is acted upon by a torque. T to raise the speed to the level of ω in time t sec.

Then, Initial angular momentous

$$= \text{(Moment of Inertia)} \times \text{angular velocity.}$$

$$= I\omega_0$$

Final angular momentum $= I\omega$

Therefore, change of angular momentum

$$= I\omega - I\omega_0$$

$$= I(\omega - \omega_0)$$

Then rate of change of angular momentum.

$$= \frac{I(\omega - \omega_0)\omega_0}{t}$$

$$= I\left[\frac{\omega - \omega_0}{t}\right] = \text{angular acceleration}$$

$$= I\alpha \qquad \text{...(7.48)}$$

But as per the statement of second law, it is proportional to torque applied.

$$\therefore \qquad T \propto I\alpha$$

or $$T = CI\alpha \qquad \text{...(7.50)}$$

Where, C = Constant of proportionality

However, we define unit torque as that which produces unit angular acceleration in a body having unit moment of inertia. Then from equation (7.50),

$$1 = C \times 1 \times 1$$

or $$C = 1$$

Then equation (7.50) redus to $T = I\alpha$ ...(7.51)

7.11 WORK

When a force acts on a body as a result of which the body moves through a certain distance, work is then said to be done. The work done by the force is defined as the product of the force and the distance moved by the body as result of the force.

$$\text{Work done} = \text{Force} \times \text{distance moved}$$

It P = the force applied.

s = distance moved by the body and it force P

Then $$\text{Work done} = P \times s \quad ...(7.54)$$

In case the body does not move along the direction of the force applied, then the work done is calculated by multiplying the component of force along the direction of movement of the body and the distance moved by the body.

That $$W.D. = (P \cos \theta) \times s \quad ...(7.53)$$

Where θ is the inclination of the movement direction with the direction of force applied.

From the definition of the work, evidently the unit of work will be obtained by multiplying the unit of force by the unit of distance moved.

i.e., unit $$\textit{Work done} = \text{(Unit force) unit)(distance)}$$

$$= 1 \text{ Newton} \times 1 \text{ metre}$$

i.e., $$= 1\ Nm$$

Unit work *i.e.,* 1 Nm is called JOULE. Since One N-m is a very small unit, more common unit in work is 1 kJ = 1000 J (kilo joule).

i.e., $$1kJ = 1000J$$

$$= 1000\ N\text{-}m$$

7.11.1 Work Done by Variable Force

In actual practice in most of the cases, the force may not be constant but may be changing.

Let P be the force at any instant the distance moved by ds, the work done is given by

$$\textit{Work done} = \int P ds \quad ...(7.53)$$

Substituting the value of P as a function of s, we get from eqn. (7.53),

$$\textit{Work done} = \int_0^s P(ds)$$

Where P is variable and a function of s.

7.12 ENERGY

Energy is defined on capacity to do work. Energy exists in numerous forms, namely mechanical, kinetic, potential, internal, magnetic, chemical, nuclear and electrical. However, in this course we will concentrate on mechanical form which consists of kinetic energy and potential energy.

7.12.1 Kinetic Energy

Kinetic energy is the capacity to do work due to the motion at the velocity v while potential energy is the capacity to do work due to the position of the body. If a weight of W is allowed to fall from a height of h from the ground level potential energy with respect to the ground level will be given by Wh Joules.

Kinetic energy is this capacity to do work due to the motion of the body. Consider a vehicle moving at a velocity of v *m/s*. and the engine of the vehicle is stopped. Still the vehicle will not come to rest immediately but will move a cetian distance before coming to rest. We know from the equation of motion the distance moved will then be obtained from following equation, where 'a' is retardation

$$2as = 0 - v^2$$

or
$$a = -\frac{v^2}{2s} \qquad ...(7.59)$$

That is the retardation which is in opposite direction of the original velocity 'v'. Accordingly the force, as per D' Alembert's principle will be in the direction of v itself and the vehicle has moved a distance in the direction of inertia force F_i.

$$\therefore \quad \text{Work done} = F_i \times s$$

$$= \left(\frac{W}{g}a\right)$$

But
$$a = \frac{v^2}{2s} \qquad \text{(From equation 7.59)}$$

Note: It may be noted that equation 7.59 'a' has negative value. That is it is retardation and gives the direction of retardation which is opposite to that of v. This will mean that F_i is in the direction of v.

$$\therefore \quad \text{Work done} = \text{Force} \times \text{distance moved}$$

$$= \left(\frac{W}{g}\frac{v^2}{2s}\right)s$$

$$= \frac{1}{2}\frac{W}{g}v^2 \text{ Joule}$$

or
$$W.D = \frac{1}{2}mv^2 \qquad ...(7.60)$$

This work which was stored in the body by virtue of velocity is called the kinetic energy.

7.13 POTENTIAL ENERGY

The energy that a system possesses due to elevation in the gravitational field is called the potential energy and is given by,

$$PE = mgh \qquad ...(7.61)$$

Where m is the mass of the body, g the acceleration due to gravity, and h the elevation of the centre of gravity of the body with respect to certain chosen datum or reference line.

7.14 POWER

Power is defined as the rate of doing work. Its unit is J/s or Watt.

7.15 WORK EQUATION FOR TRANSLATION

Consider a body weighing W is under the action of large number of forces. At some point of time let its position be at 'A', when its velocity is u and acceleration 'a'. (Fig. 7.20) Let R he the resultant of the forces.

Then mark the X-axis along the direction of resultant R and then acceleration will be given by the relation.

$$R = \frac{W}{g}a$$

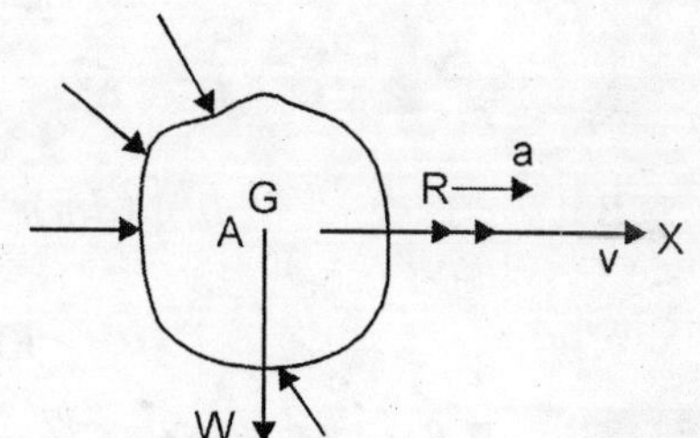

Fig. 7.20

or
$$a = \frac{Rg}{W} \text{ along } AX.$$

work done if it moves a distance ds in time dt,

$$R.ds = \frac{W}{g}ads = \frac{W}{g}\frac{dv}{dt}ds$$

$$= \frac{W}{g}\frac{dv}{ds}\frac{ds}{dt}ds$$

$$= \frac{W}{g}v\frac{dv}{ds}ds$$

$$Rds = \frac{W}{g}vdv$$

Total *work done* as it moves from A to B when its velocity increases from u to v,

$$\int_A^B Rds = \int_u^v \frac{W}{g}vdv$$

$$W.D._A^B = \frac{1}{2}\frac{W}{g}\left[v^2 - u^2\right]$$

$$W.D._A^B = RS = \frac{1}{2}m(v^2 - u^2) \qquad ...(7.61)$$

$$W.D \text{ from } A \text{ to } B = \text{change in kinetic energy}$$

$$= \frac{1}{2}mv^2 - \frac{1}{2}mu^2$$

7.16 KINETIC ENERGY OF ROTATION ABOUT FIXED AXIS

Consider a body rotating with angular velocity of ω rad/s about fixed axis at O. Let an elemental mass dm be located at a radius r.

Then velocity of the elemental mass

$$v = r\omega$$

Let ΔKE = Kinetic energy of this elemental mass

$$= \frac{1}{2}(dm)v^2$$

$$\Rightarrow \quad KE = \frac{1}{2}dm\omega^2 r^2 \qquad ...(7.62)$$

Integrating equation (7.62) for the whole body,

$$KE = \int \frac{1}{2}\omega^2 r^2 dm$$

$$= \frac{1}{2}\omega^2 \oint (dm) r^2$$

But $\oint (dm)r^2 = I_{ZZ}$

$$\therefore \quad KE = \frac{1}{2} I_{zz}\omega^2 \qquad ...(7.63)$$

Fig. 7.21

7.6 : *Moment of inertia of the rotor of a traction motor is 500 kgm². When the motor is started it attains the speed of 720 rpm from rest to in 10 sec. Find the accelerating torque and change in kinetic energy in this direction.*

Solution: Intial velocity $= \omega_0 = 0$ rad/see

Final velocity = 720 rpm

$$= 720 \times \frac{2\pi}{60} = 24\pi \text{ rad/s}$$

Now we know, $\omega = \omega_0 + \alpha t$

$$24\pi = 0 + \alpha \times 10$$

$$\therefore \quad \alpha = \frac{24\pi}{10} = 7.54 \text{ rad/s}^2$$

Now moment of inertia = 500.0 kg.m²

$\therefore$ Accelerating torque $= T = I\alpha$

$= 500 \times 7.54$

$= 3770$ *N-m*

Change in KE = $$\Delta K.E = \frac{1}{2}I\omega^2 - \frac{1}{2}\omega_0^2$$

$$= \frac{1}{2} \times 500 \times (7.54)^2 - 0.0$$

$= 14212.9$ *N-m* is *J*

$= 14.2125$ *kJ*

7.17 WORK ENERGY EQUATION FOR ROTATION

Consider a body rotating about an axis through O with angular acceleration α. Let dm be an elemental mass at a distance r from the axis. If δF is the resultant force on the elemental mass,

Then $\delta F = dm \times a$

Where a is the tangential acceleration of δm

But we know $a = r\alpha$

and $v = r\omega$

$\therefore \quad \delta F = r\,\alpha\, dm$

Fig. 7.22

∴ Rotational moment of dF about O

$$dM = (r\alpha\, dm)r \qquad ...(7.63)$$

Integrating equation (7.63, total moment),

$$M = \int \alpha r^2 dm = \alpha \int r^2 dm$$

$$M = I_{zz}\alpha \qquad ...(7.64)$$

$W.D$ by the elemental mass for rotation = θ

$$M_1\theta = (I_{zz}\,\alpha)\theta \qquad ...(7.65)$$

But $\omega^2 - \omega_0^{\,2} = 2a\theta$ ...(7.66)

Eqn (7.65) and (7.66), $M_1\theta = I_{zz}\,(d\theta)$

i.e., work done by rotational moment $= I_{zz}\left(\dfrac{\omega^2 - \omega_0^2}{2}\right)$

$$= \frac{1}{2}I_{zz}\omega^2 - \frac{1}{2}I_{zz}\omega_o^2$$

$= $ Change in $K.E$ ...(7.67)

EXAMPLE

7.7 : *Two blocks A and B weighing 240 N and 80 N respectively are supported at the ends of a rope (of negligible weight) , which passess on the rough surface of the pulley (Fig. 7.23) so that there is no slip between rope and rough surface of the pulley.* Assume pulley to be solid disc of radius 25mm. Bearing friction may be neglected. Find the torsion in the rope.

Solution: $W_1 = 240N$

$W_2 = 80\ N$

Let the acceleration of A in the downward direction be a and velocity of A is V_1. As there is no slip between the rope and pulley,

The peripheral velocity of Pulley

$= v_1$

$= r\omega$ (wheere ω is the angular speed at any instant)

$= v_2$

E X A M P L E

Also $a_1 = a_2$

$r_1 = r_2$

For the motion of A

Net downward force on

$$A = (W_1 - T)$$

$$(F_i)_A = \text{mass acceleration}$$

$$= \frac{w}{g}a\text{, (upward)}$$

$$\therefore \quad W_1 - T = \frac{W_1}{g}a, \quad \text{...(A)}$$

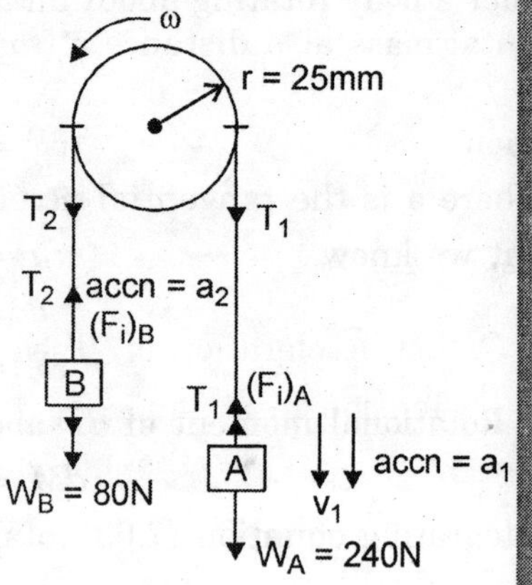

For the motion of block B,

$$\text{Now upward force} = T - W_2 = \frac{W_2}{g}a$$

Fig. 7.23

Acceleration 'a' is upwards.

$$\therefore \quad (F_i)_B = \frac{W_z}{g}.a \quad \text{...(B)}$$

Adding equations (A) and (B) we get,

$$W_1 - W_2 = \frac{W_1}{g}a + \frac{W_2}{g}a$$

$$= \frac{a}{g}(W_1 + W_2)$$

$$\therefore \quad a = \frac{g(W_1 - W_2)}{(W_1 + W_2)} = 9.81\frac{(240-80)}{(240-180)}$$

$$= \frac{9.81 \times 160}{320}$$

$$= 4.905\ m/s^2$$

$$\therefore \text{ equation } B, \quad T = W_2 + \frac{W_2}{g}a$$

$$= 80 + \frac{80}{9.81} \times 4.905$$

$$= 120\ N \textbf{ Ans.}$$

7.8 : *Find the tension in the string and acceleration of blocks A and B weighing 200 N and 50 N respectively connected by a string and frictionless and weightless pulleys as shown in Fig. (7.24).*

(U.P. Tech Univ. 2002-2001)

Solution : Given Data

$$W_A = 200 \text{ N}$$

$$m_A = \frac{200}{9.81} \text{ kg}$$

$$W_B = 50 \text{ N}$$

$$m_B = \frac{50}{9.81} \text{ kg}$$

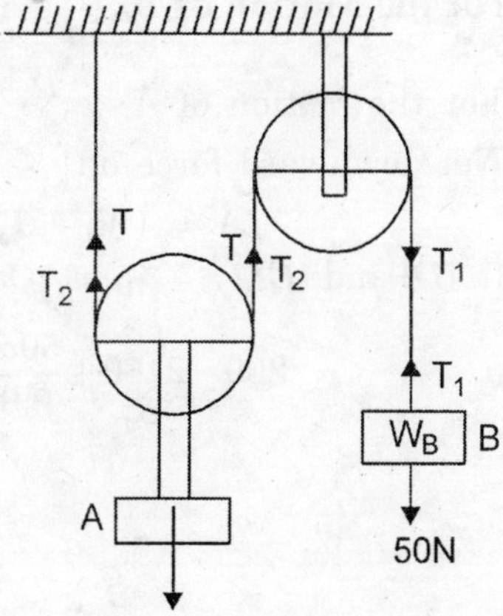

Fig. 7.24

Let acceleration of 'B' be a (downward)

Tension in string = T

s_A = Displacement of A upward

s_B = Displacement of B downward

Also $\quad s_B = 2s_A \quad$ ...(A)

Since these displacements will occur in the same time t,

$$\therefore \quad v_B = \frac{s_b}{t}$$

$$v_A = \frac{s_a}{t}$$

$$\frac{v_A}{v_B} = \frac{s_a}{t} \times \frac{t}{s_b} = \frac{s_a}{s_b} = \frac{1}{2}$$

$$\therefore \quad \frac{v_A}{v_B} = \frac{1}{2} \quad \text{...(B)}$$

Also, $\quad \frac{a_A}{a_B} = \frac{1}{2} \quad$ or $\quad a_A = \frac{1}{2} a_B \quad$ ...(C)

Now for the motion of (upward)

$$(T - W_B) = m_B \, a_B$$

$$= \frac{50}{9.81} \times a_B \quad (C)$$

For the motion of A (downward)

$$(W_A - 2T) = m_A \frac{a_B}{2} \quad \left(\text{Since } a_A = \frac{a_B}{2}\right)$$

$$(200 - 27) = \frac{m_A}{9.81} \frac{a_B}{2} \quad \text{...(D)}$$

$$= \frac{200}{9.81} \times \frac{a_B}{2}$$

E X A M P L E

For the Motion of *B*,

$$T - 50 = m_B\, a_B$$

$$T = 50 + \frac{50}{9.81} \cdot a_B \qquad \text{...}(E)$$

In (*D*) and (*E*),

$$200 - 2\left[80 + \frac{50 a_B}{9.81}\right] = \frac{100}{9.81} a_B$$

$$100 = (100 + 100)\,\frac{a_B}{9.81}$$

$$= 200 \frac{a_B}{9.81}$$

or

$$a_B - \frac{9.81}{2} = 4.905 \text{ m/s}^2$$

$\therefore$ Acc *n* of *A* $\quad a_A = \frac{a_B}{2} = 2.45 \text{ m/s}^2$

Acc of *B* $\quad a_B = a = 4.905 \text{ m/s}^2$ **Ans.**

7.9 : *Two blocks weighing 200 N and 140 N are attacked to the two ends of a rope of negligible weight, which is passing over a rough pulley free to rotate about its axis. (Fig. 7.25). The weight of the pulley is 60 N and it can be assumed as a disc of radius 0.5 m. Neglect friction of the bearings. Find the tension in the rope on both sides.*

Solution : Given data

$$W_1 = 200 \text{ N}$$

$$W_2 = 140 \text{ N}$$

$$W_3 = \text{weight of pulley}$$

$$= 60 \text{ N}$$

mass of pulley $\quad = m_3 = \frac{60}{9.81}$ kg

Radius of pulley $\quad R = 0.5$ m

Let the acceleration of weights be $= a$ m/s^2

$\because$ Angular acceleration of pulley

$$= \quad \alpha = \frac{a}{R}$$

$$= 2a \text{ rad/sec}^2$$

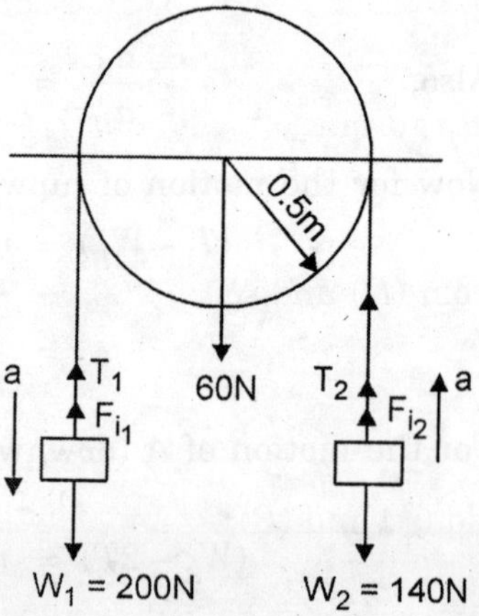

Fig. 7.25

1. Motion of Weight (W_1)

It is moving down Downward force

$$(W_1 - T_1) = \text{mass} \times \text{acc}$$

or $$(W_1 - T_1) = \frac{W_1}{8} \cdot a$$

or $$(200 - T_1) = \frac{200}{9.81} a \qquad ...(A)$$

For Motion of W_2

It is moving upward with acc = a

i.e., $$(T_2 - W_2) = \frac{W_2}{g} a$$

$$(T_2 - 140) = \frac{140}{9.81} \times a \qquad ...(B)$$

Rotary Motion of Pulley

$$I = m_3 \frac{R^2}{2} = \frac{60}{9.81} \times \frac{(.5)^2}{2} \qquad ...(C)$$

$\therefore$ Torque on pulley $= I\,\alpha$ [where angular acceleration $= \frac{a}{R}$]

$\therefore$ Torque on pulley $= \left[\frac{60}{9.81} \times \frac{(.5)^2}{2}\right] \cdot \frac{a}{R}$

$$= \frac{60}{9.81} \times \frac{(0.5)^2}{2} \times \frac{a}{(.5)}$$

$$T = \frac{15}{9.81} a \qquad ...(D)$$

Also torque on pulley $= (T_1 - T_2) R$

$$= (T_1 - T_2)\, 0.5 \qquad (E)$$

From (*D*) and (*E*), $$(T_1 - T_2)(0.5) = \frac{15}{9.81} \cdot a$$

$$T_1 - T_2 = \frac{30a}{9.81} \qquad ...(F)$$

Add (*A*) and (*B*),

$$(T_2 - T_1) + (200 - 140) = \frac{340}{9.81} \cdot a$$

$$(T_2 - T_1) + 60 = \frac{340}{9.81} \cdot a$$

EXAMPLE

E X A M P L E

For (*F*), $$-\frac{30a}{9.81}+60 = \frac{340}{9.81}\cdot a$$

$$\frac{370}{9.81}a = 60$$

$$a = \frac{60\times 9.81}{370}$$

$$= 1.5908 \text{ m/s}^2$$

∴ From equation (*A*), $$200 - T_1 = \frac{200}{9.81}\cdot a$$

or $$T_1 = 200-\frac{200}{9.81}\times 1.5908$$

$$= 200 - 32.43$$

$$= 1.67.57 \text{ N } \textbf{Ans.}$$

and from (*F*), $$T_2 = T_1-\frac{30a}{9.81}$$

$$= 167.57-\frac{30\times 1.59}{9.81}$$

$$= 162.71 \text{ N } \textbf{Ans.}$$

7.10 : *A cage weighing 2940 N is raised by a rope, one end of which is wrapped round a drum 1.2 m in diameter and weighing 735 N and having radius of gyration of 0.55 m. The drum is rotated by an electric motor, which exerts constant torque of 4000 N - m. If the rope is tight when the drum begins to rotate, determine,*

(i) The acceleration of cage

(ii) The tension in the rope

(iii) The time required to raise the cage 20 m from ground. Neglect loss of energy due to friction and weight of rope. Take g = 9.8 m / s².

(AMIE 1982)

Solution : Given data

Weight of cage = W = 2940 N

Diameter of drum = 1.2 m

∴ Radius of drum = $\frac{1.2}{2} = 0.6$ m

Weight of drum = 735 N

∴ Mass of drum = $\frac{735}{9.8}$ kg = 7.5 kg

Radius of gyration of drum = k = 0.55 m

$$\alpha = \text{Angular acceleration of drum}$$

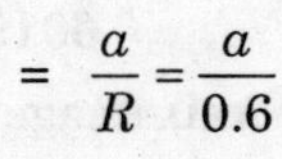

$$= \frac{a}{R} = \frac{a}{0.6}$$

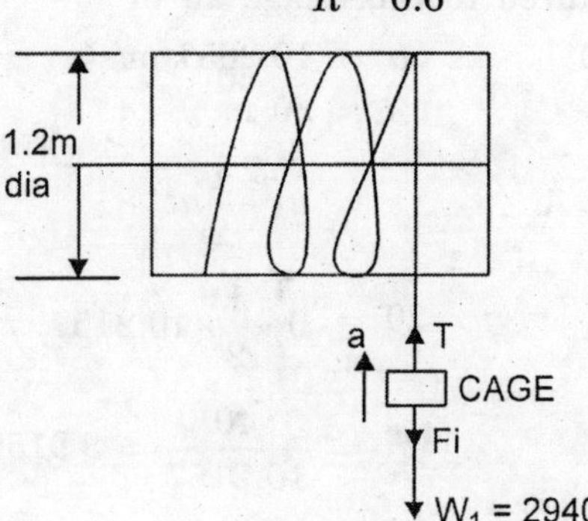

Fig. 7.26

Motion of Cage $T_1 - W_1 = \frac{W}{g} \cdot a$

$$(T_1 - 2940) = \frac{2940}{9.8} \times a = 300a$$

$$T_1 = 2940 + 300a \quad ...(A)$$

Moment of Inertia of drum $= MR^2$

$$= 75 \times (0.55)^2$$

$$= 22.6875 \text{ kg} - \text{m}^2$$

Rotational Motion of Drum

Torque applied = 4000 Nm

Resisting torque due to $T_1 = T_1 \times r$

$$= T_1 \times 0.6$$

$\therefore$ Net torque on drum

$$(4000 - T_1 \times .6) = I\,\alpha$$

where, I = moment of inertia of drum

an α = angular acceleration of drum

$$\therefore \quad (4000 - 0.06\, T_1) = 22.6875 \times \frac{a}{0.6}$$

or $$4000 - 0.6\, T_1 = 37.8125\, a \quad ...(B)$$

Substituting value of T_1 from equation (A) into (B),

$$4000 - 0.6\,[2940 + 300a] = 37.8125a$$

$$4000 - 1764 - 180a = 37.8125a$$

or $$2236 = 217.8125a$$

or $a = 10.2657 \text{ m/s}^2$ **Ans.**

Then from (A), $T_1 = 2940 + 300 \times 10.2657$

$$= 2940 + 3079.71$$

$$= \mathbf{6019.71\ Ans.}$$

(*c*) Let t = time required to raise cage 20 m

$$u = 0 \qquad a = 10.2657 \text{ m/s}^2$$

$$s = 20 \text{ m}$$

We have $$s = ut + \frac{1}{2}at^2$$

$$20 = 0 + \frac{1}{2} \times 10.2157 \times t^2$$

or $$t^2 = \frac{40}{10.2157} = 3.9155$$

$\therefore$ $$t = 1.978 \text{ sec}$$

EXAMPLES ON WORK AND ENERGY

7.11 : *A body weighing 2 kN is lying an a horizontal pavement. This body is to be shifted through 35 m on the road. Find the work done, if the resistance to motion is 15N per 1 kN weight of the body.*

Solution : The resisting force $= \frac{15 \text{ N}}{1 \text{ kN}} \times 2 \text{ kN} = 30\text{N}$

WD = against resistance $= 30 \times 35$

$= 1050$ N–m

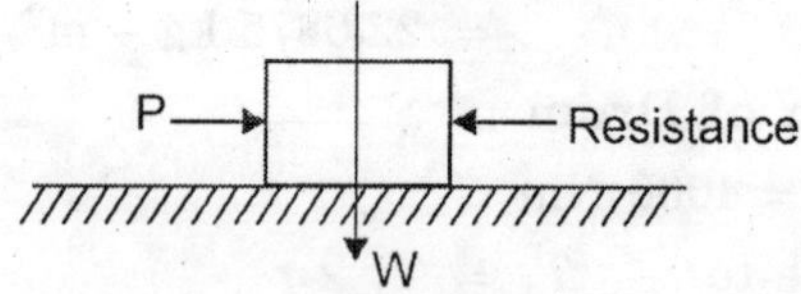

Fig. 7.27

7.12 : *A block weighing 1 kN is lying on a rough inclined plane. (30° to horizontal) A force P is applied along the plane on the body to move it up by 7 m. Find the work done. The co–efficient of friction between the body and plane is 0.3.*

Solution : Resolving forces along for plane against which P does the work.

$$P = \mu R + W \sin 30$$

Normal to Plane $$R = W \cos 30$$

$$= 1000 \times \frac{\sqrt{3}}{2}$$

$$= 500\sqrt{3}$$

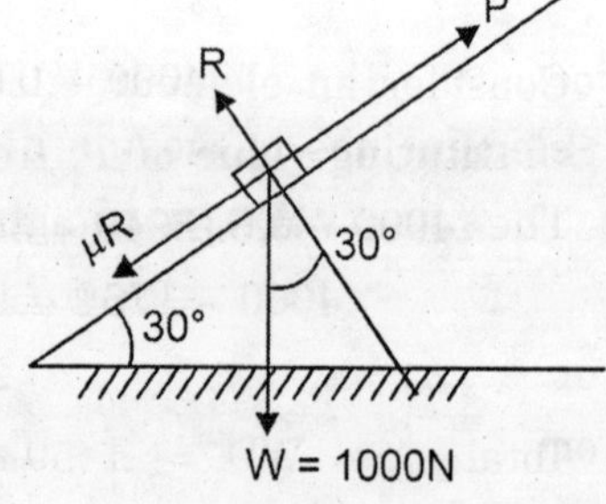

Fig. 7.28

$\therefore$

$$P = 0.3 \times 500\sqrt{3} + 1000 \times \frac{1}{2}$$

$$P = 259.8 + 500.0 = 759.8 \text{ N}$$

$\therefore$ W.D. by $\quad P = P \times s$

$= 759.8 \times 7$

$= 5318.6$ N–m **Ans.**

7.13 : *A train of weight 100 kN is pulled up, by a diesel locomotive an, inclined plane, slope 15° with horizontal. Its uniform speed is 36 km/hr. Determine the power to be consumed by the locomotive. Total resistance (air & friction) is 6N/kN weight of the train..*

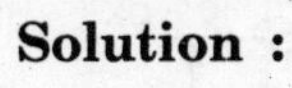

Solution : $W = 100$ kN

$v = 36$ km/hr

$$= \frac{36 \times 1000}{3600} = 10 \text{ km /s}$$

$W = 100000$ N

Resistance $= R_2 = 6 \times 100$

$= 600$ N

R1
P
R2
15°
15°
W = 10000N

Fig. 7.29

Since the velocity is uniform acceleration = 0

$\therefore$ Resolving forces along the plane,

$$P - R_2 - W \sin 15° = 0$$

$$P = R_2 + W \sin 15°$$

$$= 600 + 100000 \times 0.25889$$

$$= 600 + 258\text{i}0$$

$$= 26410 \text{ N}$$

$\therefore$ Power $= P \times v = 26410 \times 10$

$= 264100$ Nm/s

$= 264.1\ kW$

7.14 : *Determine the work done by electric motor in winding up a uniform cable which hangs from a hoisting drum, if its free length is 10 m and weighs 500 N. The drum is rotated by the motor.* (AMIE 1984)

Solution : Weight of cable per unit length

$$= \frac{500}{10} = 50 \text{ N/m}$$

Consider an elemental length dx at a height x from the end A. This element has to move a distance of $(10 - x)$ to be wound up.

The $\quad d(\text{W.D.}) = (\text{weight of CD})\ (10 - x)$

$= (50\ dx)\ (10 - x)$

Total $\quad \text{WD} = \int_0^{10} (50\,dx)\,(10 - x)$

or $$\text{WD} = \int_0^{10} (500 - 50x)\,dx$$

$$= \int_0^{10} 500\,dx - \int_0^{10} 50\,x\,dx$$

$$= [500x]_0^{10} - \left[50 \cdot \frac{x^2}{2}\right]_0^{10}$$

$$= 500(10-0) - \left[50\frac{x^2}{2}\right]_0^{10}$$

$$= 5000 - 25\ (100 - 0) = 2500 \text{ Nm}$$

Fig. 7.30

7.15 : *A train of weight 5000 kN is pulley by an electric engine on a level track at constant speed of 100 km/hr. The air resistance on train to be overcome by engine, is 12 N/kN of the train weight. Find the power of the engine.*

Solution : Velocity $= 100 \text{ k/m} = \dfrac{100 \times 1000}{3600}$ m/s

Resistance $= \dfrac{12}{1} \times 500 = 6$ kN

$= 6$ kN

Thus, if P is the tractive force developed by the engine

$$P = 60000 \text{ kN}$$

Power $= P \times$ velocity

$$= 60000 \text{ kN} \times \frac{10^5}{3600} \text{ kN m/s}$$

$$= \frac{10^7}{6} \text{ kN - m}$$

$$= 1.667 \times 10^9 \text{ N-m/s}$$

7.16 : *A train weighing 1000 kN is moving down an inclined plane (15° to horizontal) at 18 km/h when the locomotive develops a power of 15 kW. If the train is pulled up at the same speed, what power will be required to pull the train?*

Solution : Given data $v = \dfrac{18000}{3000} = 5$ m/s

Power $= 15$ kW

If P is the tractive force, then

$$Pv = 15 \text{ kW} = 15000 \text{ W}$$

$$\therefore \quad P = \frac{15000}{5} = 3000\text{N}$$

Let Fair be the resistance to train by air and other factors.

or $\quad$ Air resistance $= F_{air}$

Resolving forces along plane

$$\therefore \quad P + W \sin \alpha - F_{air} = 0$$

$$P - F_{air} + 1000 \times 10^3 \times \sin 15° = 0$$

$$P - F_{air} + 0.2588 \times 10^6 = 0$$

$$F_{air} = P + 0.2588 \times 10^6$$

$$= 3 \times 10^3 + 0.2588 + 10^6$$

$$= 261.588 \text{ kN}$$

$$\therefore \quad F_{air} = 261.58 \text{ kN}$$

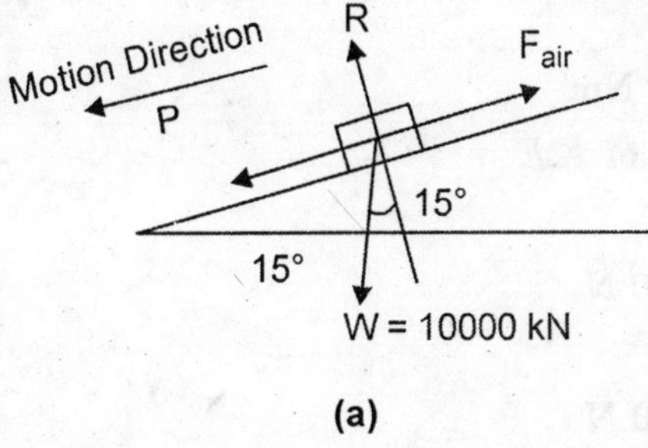

(a) (b)

Fig. 7.31

Which Moving up

Resolving forces along plane,

$$P_2 - F_{air} - W \sin 15° = 0$$

$$P_2 = F_{air} + W \sin 15°$$

$$= 261.588 + 0.2588 \times 10^6$$

$$= 261.588 + 258.8 \text{ kN}$$

$$= 261.588 \text{ kN}$$

$$v = 5 \text{ m/s}$$

$$\therefore \quad \text{Power} = 520.388 \times 5$$

$$= 2601.94 \text{ kW}$$

7.17 : *A bullet of mass 81 gm and moving with a velocity of 300 m/s is fired into block of wood and it penetrates to a depth of 10 cm. If the bullet moving with the same velocity were fired into a similar price of wood 5 cm then with what velocity would it emerge? Find also the force of resistance assuming it to be uniform.* (AMIE 1981)

E X A M P L E

Solution : Given data Mass of bullet

$= 0.081$ kg

Initial velocity of bullet $u = 300$ m/s

First Case :

Distance moved $s = 0.1$ m

The final velocity $v = 0$

Let F = Uniform resisting force in N

Therefore, $\text{WD} = F \times s = F \times 0.1$

Change of kinetic energy $= \frac{1}{2}mu^2 - \frac{1}{2}mv^2$

$= \frac{1}{2}m \times 300^2 - 0$

$= \frac{1}{2} \times 0.081 \times (300)^2$

$= 3645$ Nm

Since WD = Loss of $K.E$

$3645 = 0.1$ F

$\therefore$ $F = 36450$ N

Second Case:

Resisting force $= 36450$ N

$s = 5$ cm

$= 0.05$ m

$\therefore$ $\text{WD} = -36450 \times 0.05$

$= -1822.5$ Nm

(It is $-ve$ as it is lost by bullet that is in opposite direction)

Thus, if v_1 is velocity of bullet after emerging out.

Then $\frac{1}{2}mv_1^2 - \frac{1}{2}mu^2 = -1822.5$

$$\frac{1}{2} \times 0.081(v_1^2 - 300^2) = -1822.5$$

$$\frac{1}{2} \times 0.081 \times v_1^2 = \frac{1}{2} \times 0.081 \times 300^2 - 1822.5$$

or $v_1^2 = \frac{1792.5 \times 2}{0.081}$

or $v_1 = 210.38$ m/s **Ans.**

7.17 LAW OF CONSERVATION OF ENERGY

Law of conservation of energy states that energy can not be destroyed or created even through it can be transformed from one form to another form. Alternatively total energy possessed by a body remains constant provided no energy is added or taken out of it.

Consider a body of weight W at the top of a tower of height H above the ground level which is taken as the datum.

Fig. 7.32

Then Potential Energy of body

$$PE = WH$$

Since at the top the body is at rest its $K.E = 0$

$$\therefore \quad \text{Total energy} = WH$$

Let it fall down through a height h to reach the position B. Then as per equation of motion.

$$v_B^2 - v_A^2 = 2hg$$

or $$v_B = \sqrt{2hg} \quad (\text{as } v_A = 0)$$

Substitute v_B from (A) in equation for $K.E$,

$$\therefore \quad (K.E)_B = \frac{1}{2} m v_B^2$$

$$= \frac{1}{2}\left(\frac{W}{g}\right)(2hg)$$

$$= Wh$$

$$PE \text{ at } B, \text{ w.r.t datum} = W(H-h)$$

$$\therefore \text{ Total Energy at } B = Wh + W(H-h)$$

$$= WH$$

$$= \text{Same as at A}$$

EXAMPLE

7.18 : *A car, weighing 10 kN, is moving on a highway at a velocity of 60 km/hr. All of a sudden the driver applies brakes and it is brought to rest after moving 30 m. Determine the average resisting force acting on the car in this duration.*

Solution : Given data

$$\text{Weight} = 10 \text{ kN}$$

$$\therefore \quad \text{Mass} = \frac{10 \times 1000}{9.81} \text{ kg}$$

$$\text{Initial velocity} = 60 \text{ km/hr}$$

$$= \frac{60 \times 1000}{3600} = 16.667 \text{ m/s}$$

$$\text{Final velocity} = 0/\text{sec}$$

EXAMPLE

$\therefore$ Change in $KE = \frac{1}{2}mv^2 - \frac{1}{2}mu^2$

$$= 0 - \frac{1}{2} \times \frac{10^4}{9.81} \times (16.667)^2$$

$$\Delta KE = -14.158 \times 10^4 \text{ N–m}$$

(–*ve* sign indicates loses)

Thus, if average resistance is *F*,

Then $-Fs = -\Delta KE$

$$F \times 30 = -14.158 \times 10^4$$

$$F = \frac{14.158 \times 10^4}{30} = 0.472 \times 10^4$$

$$= 4720 \text{ N} \quad \textbf{Ans.}$$

7.19 : *A car weighing 6 kN accelerates from zero velocity to 50 km/hr in a distance of 50 m against an average resistance of 120 N. Determine the average driving force by the engine of the car, then find the greatest power developed by the engine.*

Solution : Given data :

$$v = 50 \text{ km/hr}$$

$$= \frac{50 \times 1000}{3600} = 13.89 \text{ m/s}$$

$$m = \frac{W}{g} = \frac{6 \times 1000}{9.81} = 611.62 \text{ kg}$$

$$s = 50 \text{ m}$$

If F = average resisting force = 120 N

$\therefore$ Change in $KE = \frac{1}{2}mv^2 - \frac{1}{2}mu^2$

$$= \frac{1}{2}mv^2 \qquad \text{(as } u = 0\text{)}$$

or $\Delta KE = \frac{1}{2} \times 611.62 \times (13.89)^2$

$$= 59000.5 \text{ Nm}$$

If P = Driving force of the engine

Then $(P - F)$ = Resultant driving force

$\therefore$ $(P - F)\,s = \Delta KE$

$$= 59000.5$$

$$P = F + \frac{59000.5}{s}$$

$$= 120 + \frac{59000.5}{50}$$

$$= 120 + 1180$$

$$= 1300 \text{ N} \quad \textbf{Ans.}$$

7.20 : *A flywheel weighing 9.810 N has radius of gyration 1.5 m is intially rotating at 440 rpm which reduces to 120 rpm in 1.5 minute. Calculate the retarding torque and the change in kinetic energy in the above duration.*

Take $\alpha = 9.81\ m/sec^2$

Solution : Given data

$$K = 1.5 \text{ m}$$

$$m = \frac{9810°}{9.81} = 10^3 \text{ kg}$$

$$\omega_1 = 440 \text{ rpm}$$

$$= \frac{440 \times 2\pi}{60} \quad 46.007 \text{ rad/sec}$$

$$\omega_2 = 120 \text{ rpm}$$

$$= \frac{120 \times 2\pi}{60}$$

$$= 12.56 \text{ rad/s}$$

$$\therefore \quad \alpha = \frac{\omega_1 - \omega_2}{t} = \frac{46.0 - 12.56}{1.5 \times 10}$$

$$= 6.3716 \text{ rad/s}^2$$

Weight of flywheel $= 9810$ N

Mass of $= \dfrac{9810}{9.81} = 1000$ kg

$$I = mk^2$$

$$= 1000 \times (1.5)^2$$

$$= 2250 \text{ kg m}^2$$

Then retardation torque $= I\,\alpha$

$$= 2250 \times 0.3716$$

$$= 836.1 \text{ N-m}$$

(*b*) Charge in kinetic energy $= \dfrac{1}{2} I w_1^2 - \dfrac{1}{2} I w_2^2$

$$= \frac{1}{2} \times 2250\,[46^2 - (12.56)^2]$$

$$= 2173.67 \text{ kJ}$$

7.21 : *A spherical steel ball of radius 0.1 m is welded to a steel rod AB, 1 m long and diameter 40 mm. The rod is hinged at point A and held in horizontal position as shown in Fig. 7.33. It is released from this position and rotates about A in the vertical plane about YY axis. Determine the angular velocity, when the rod passes through the vertical position. Also, find the angular acceleration at this position.)Take the density of steel as 7800 kg/m³.)*

EXAMPLE

Solution : Given: Diameter of rod = 40 mm = 0.04 m

length = 100 m

Vol. of Rod $AB = \dfrac{\pi(0.04)^2}{4} \times 1$

$= .0012566 \text{ m}^3$

Mass of rod $= M_1 = \text{vol} \times \text{density}$

$= 0.0012566 \times 7800$

$= 9.8017 \text{ kg}$

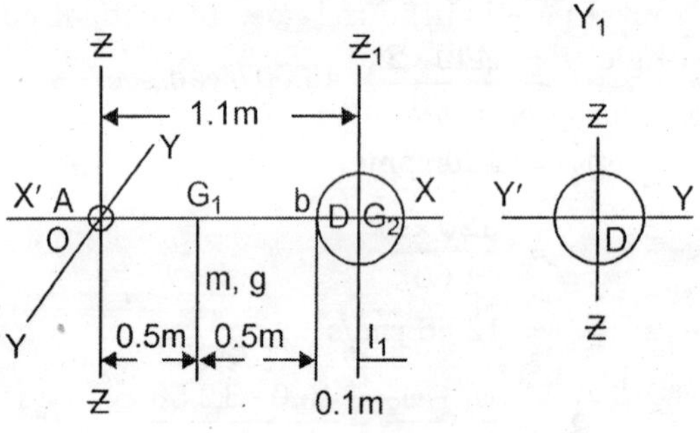

Fig. 7.33

Mass of steel ball = volume × density

$= \dfrac{4}{3}\pi(0.1)^3 \times 7800$

$M_2 = 32.6725 \text{ kg}$

Take A as origin and axes XOX and ZOZ' in the vertical plane as shown Moment of Inertia of rod

$$(I_{YY})_1 = \frac{ML^2}{3} \qquad ...(7.29)$$

$$= \frac{1}{3} \times 9.8017 \times 1^2$$

$$= 3.2672 \text{ kg}\theta\text{m}^2$$

Moment of Inertia of Ball about YY passing Theorem A (By theorem of parallel axes),

$$(I_{YY})_2 = \frac{2}{5} M_2 R^2 + M_2 (AD)^2$$

or $$(I_{YY})_2 = \frac{2}{5} \times 32.6725 \times (0.1)^2 + 32.6725 \times (1.1)^2$$

$$= 32.6725 \left[\frac{2}{5} \times (.1)^2 + (1.1)^2\right]$$

$$= 32.6725\ [0.004 + 1.21]$$

$$= 39.6644 \text{ kg-m}^2 \quad ...(A0$$

Now weight of rod which is concentrated at G_1

$$W_1 = M_1 g$$

$$= 9.8017 \times 9.81$$

$$= 96.0567 \text{ N}$$

Weight of ball $W_2 = M_2 g = W_2$

$$= 32.6725 \times 9.81$$

$$= 320.517 \text{ N}$$

∴ From the horizontal position the rod goes to vertical position, then it rotates through angle $\theta = \pi/2$.

∴ Applying work Energy equation,

$$WD = W_1\theta + W_2\theta$$

$$= 96.0567 \times \theta + 320.577 + \theta$$

$$= 416.5737\ \theta$$

Also, change in $\Delta K.E = \frac{1}{2}\frac{I_1}{1}(w^2 - 0^2) + \frac{1}{2}\frac{I_2}{2}(w^2 - 0^2)$

$$= \frac{1}{2}\omega^2[(I_{yy})_1 + (I_{yy})_2]$$

Where w is angular vol. at vertical position then,

∴ Applying work – Energy equation

$$416.57370\theta = \frac{1}{2}(I_{YY})_1 w^2 + \frac{1}{2}(I_{YY})_2 w^2$$

$$= \frac{1}{2}w^2[3.2672 + 39.6644]$$

Here, $= 21.4658\ w^2$

or $416.5737\ \theta = 21.4658 w^2 \quad ...(A)$

Here, $\theta = \pi/2$

∴ $w^2 = 416.5737 \times \frac{\pi}{2} \times \frac{1}{21.4658}$

$$= 30.4835$$

or $w = 5.521$ rad/sec

Differentiating, quation (A) w.r.t time

$$416.5737\frac{d\theta}{dt} = 21.4658 \times 2\frac{dw}{dt}w$$

Since $w = \frac{d\theta}{dt}$

$$\therefore \qquad 416.5737 = 21.4651 \times (2\alpha)$$

or
$$\alpha = \frac{416.5737}{214658 \times 2}$$

$$= 9.703 \text{ rad/s}^2$$

7.22 : *The pendulum of an Izod impact testing machine is made by welding two 70 mm thick and 250 mm diameter discs to a metre long bar of 40 mm dia. The discs are separated by welding a square rod of 40 mm side 100 mm long piece of steel as shown in Fig. (7.34). The rod AB is hinged at A and is released from rest in the horizontal position to strike a notched cantilever specimen diameter discs specimen placed vertically below A at a distance of 1.12 m. Find the velocity of the hammer with which it will strike the specimen. Also calculate the change in KE of the hammer and its energy contant at this time of striking. Take density of steel 7800 kg/m³*

Solution : Mass of rod $M_1 = \frac{\pi}{4}(.04)^2 \times 1 \times 7800 = 9.8017$ kg

Weight $W_1 = 9.8017 \times 9.81$

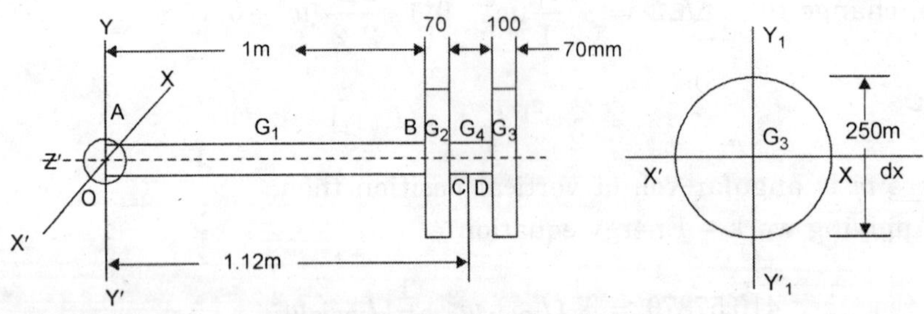

Fig. 7.34

Take origin at A moment of inertia of rod about X o X' from equation (7.29).

$$= \frac{m_1 L^2}{3} = \frac{9.8017 \times 1^2}{3} = 3.2672 \text{ kg} - \text{m}^2$$

Mass of each disc – vol × density,

$$m_2 = m_3 = \left[\frac{\pi}{4}(0.25)^2 \times .07\right] \times 7800$$

Weight of each disc $= 26.8$ kg

$$W_2 = W_3 = 26.8 \times 9.81 = 262.91 \text{ N}$$

Moment of inertia of each disc about an axis parallel to XOX and passing though it own CG *i.e.*, G_1 or G_2 as the case may be.

$$(I_{G_2})_{XX} = (I_{G_3})_{XX}$$

$$= \frac{m_2 r_2^2}{4} \text{ (equation 7.25)}$$

Now mass of each disc $= m_2 = m_3$

$= \text{vol} \times \text{density}$

$$= \left[\frac{\pi}{4}(0.250)^2 \times .07\right] \times 7800$$

$= 26.8$ kg

Weight of each disc $= W_2 = W_3$

$= 26.8 \times 9.81$

$= 262.91$ N

$$(I_{XX})_{G_2} = (I_{YY})_{G_2} = (I_{XX})_{G_3}$$

$$= \frac{m_2 r_2^2}{4}$$

$$= \frac{26.8 \times (0.125)^2}{4}$$

$= 0.105$ kg-m^2

Mass of central price of square section 40 mm side & length 100 mm

$$m_4 = \frac{(40 \times 40)}{10^6} \frac{100}{1000} \times 7800$$

$= 12.48$ kg

Weight of central piece $= 12.48 \times 9.81$

$= 122.43$ N

Moment of within of central piece about centroidal axis parallel XX

$$(I_{G_4})_{YY} = (I_{G_4})_{XX} = \frac{m_4 a^2}{12} \quad \text{(equation 7.30 } a\text{)}$$

$$= \frac{12.48 \times (0.04)^2}{12}$$

$= 0.001664$ kg-m^2

W.D. when the pendulum shifts from horizontal position through angle θ.

$$\text{W.D.} = M\theta + M_2\theta + M_3\theta + M_n\theta - \theta[M_1 - M_2 + M_3 + M_4]$$

When M_1, M_2, M_3 and M_4 are moments of respective weights about A.

Here, $M_1 = W_1 \times OG_1 = 96.0567 \times 0.5 = 48.028$ N–m

$M_2 = W_2 \times OG_2 = 262.91 \times 1.035 = 272.112$ N–m

$M_3 = W_3 \times OG_3 = 262.91 \times 1.205 = 316.807$ N–m

$M_4 = W_4 \times OG_4 = 122.43 \times 1.120 = 137.12$ N–m

For rotation through θ radians,

$$\therefore \quad \text{WD} = \theta[48.28 + 272.172 + 316.807 + 137.12] = 774.127\,\theta \quad ...(B)$$

EXAMPLE

E X A M P L E

Now moment of Inertia of each component about XoX passing through O, can be obtained by applying theorem of parallel axes.

$$(I_1)_{XoX} = 39.6644 \text{ kg m}^2 \quad \text{(from } A\text{)}$$

Disc 1 :

$$(I_2)_{XoX} = (I_{G_2})_{XX} + m_2(OG_2)^2$$

$$= 0.105 + 26.8\ (1.035)^2$$

$$= 0.105 \quad 28.709$$

$$= 28.813 \quad \text{kg m}^2$$

$$(I_3)_{XoX} = (I_{G_3})_{XX} + m_3(OG_3)^2$$

$$= 0.105 + 26.8 \times (1.205)^2$$

$$= 0.105 + 38.914$$

$$= 39.019 \text{ kg m}^2$$

$$(I_4)_{XoX} = (I_{G_4})_{XX} + m_4(OG_4)^2$$

$$= 0.001664 + 1248\ (1.12)^2$$

$$= 0.001664 + 15.655$$

$$= 15.6565 \text{ kg-m}^2$$

Total KE = Σ KE of each added up

$$= \frac{1}{2}I_1(w^2 = 0) + \frac{1}{2}I_2(w^2 - 0) + \frac{1}{2}I_3(w^2 - 0) + \frac{1}{2}I_4(w^2 - 0)$$

$$= \frac{1}{2}w^2[I_1 + I_2 + I_3 + I_4]_{XoX}$$

whereas in the angular speed at the time of striking

$$= \frac{1}{2}w^2[39.6644 + 28.313 + 39.019 + 15.65]$$

$$= \frac{1}{2}w^2[122.6529] = 61.326\omega^2$$

Then from (B) 61326 ω^2 = 774.127 θ

Here, it is found that $\theta = \pi/2$ ...(B)

$\therefore$ Using work energy equation we get

$$774.127 \times \frac{\pi}{2} = 61.326\ w^2 \quad ...(C)$$

$$\therefore \quad w^2 = \frac{774.127}{61.326} \times \frac{\pi}{2} = 19.828$$

or $\quad w = 4.453$ rad/s

Differentiating (B) w.r.t time

$$61.236 \times 2\omega \frac{dw}{dt} = 774.127 \frac{d\theta}{dt}$$

$$= 774.127\, w$$

or $$\alpha = \frac{d\omega}{dt} = \frac{774.127}{61.236 \times 2}$$

$$= 6.3208 \text{ rod/s}^2$$

Velocity of striking the test price

= Peripheral velocity of the hammer point

$= w(OG_4)$

$= 2.835 \times 1.120$

= 3.175 m/s **Ans.**

From (*C*) total energy content

$$= 774.127 \times \frac{\pi}{2}$$

= 1216 *J* **Ans.**

7.23 : *A body A weighing 5 kN is lying on a rough inclined plane inclined at 45° to the horizontal. The body A is connected to another body B weighing 1.4 kN with a nor extensible string passing over a frictionless pulley of diameter 130 mm and weight 500 N as shown in the figure (7.35). If the co–efficient of friction between the plane and body A is 0.3, determine the distance named by of block A, when it achieves the velocity of 2 m/s. Also, find the acceleration of body. Use work energy method.*

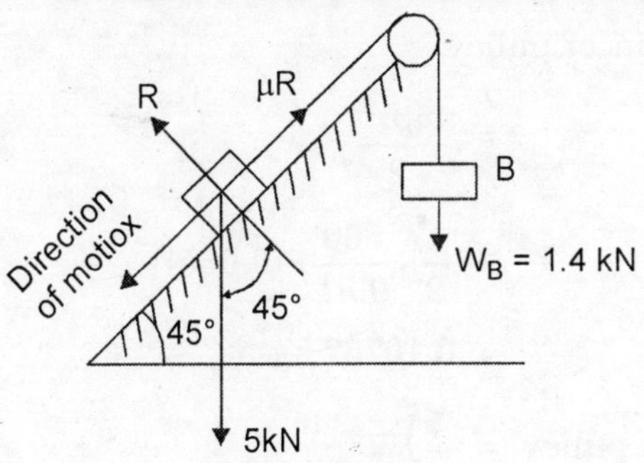

Fig. 7.35

Solution : Since the movement of body *A* is downward, the frictional force with the in the upward direction.

Component of W_A down the plane

$= W_A \sin 50°$

$$= \frac{5000}{\sqrt{2}} \text{ N} = 3535.53 \text{ N}$$

Resolving normal the plane we get,

E X A M P L E

$$R = W_A \sin 50°$$

$$= \frac{5000}{\sqrt{2}} = 3535.\sqrt{3} \text{ N}$$

$\therefore$ Frinctional force $\mu R = \frac{5000}{\sqrt{2}} \times 0.3 = 1060.66$ N

$\therefore$ Now downward force along the plane

$$= W_A \sin 45° - \mu R$$

$$= 3535.53 - 1060.66$$

$$= 2474.87 \text{ N}$$

$\therefore$ W.D. by W_A = time $\times$ s

When s is the distance moved, then velocity becomes 2m/s

$$v_A = 2\text{m/s}$$

$$= v_B = wr$$

where w is the rpm of pulley of radius r

Here, $r = \frac{130}{2} = 65 \text{ mm} = 0.065 \text{ m}$

w_p = 500, mm of pulley

$$= m_p = \frac{500}{9.81}$$

Net workdone by moving weights

$(W_A - F)_s - W_B s$ = kE or pulley + kE of A + kE of B

Moment of inertia of pulley

$$= \frac{mpr^2}{2} \quad ...(A)$$

$$= \frac{1}{2} \times \frac{500}{9.81} \times (0.065)^2$$

$$= 0.10767 \text{ kg-m}^2$$

$\therefore$ kE of pulley $= \frac{1}{2} IW^2$

Now $w = \frac{v}{r} = \frac{v}{(0.065)}$

$$kE\ P\ eg \cdot \frac{I}{2} w^2 = \frac{0.10767}{2} \times \left(\frac{v}{.065}\right)^2$$

$$= 12.742\ v^2$$

$$kE \text{ of } A = \frac{1}{2} mAv^2$$

$$= \frac{1}{2}\left(\frac{5000}{9.81}\right)v^2$$

$$= 254.84\ v^2$$

$$kE \text{ of } B = \frac{1}{2} m_B v^2$$

$$= \frac{1}{2} \times \frac{1400}{9.81} v^2$$

$$= 71.35\ v^2$$

∴ Substituting values in equation (A),

$$2474.87s - 1400s = 12.742v^2 + 2541.84v^2 + 71.35v^2$$

$$1074.87s = 2625.9v^2$$

But $s = 2$ $\qquad = 2625.9 \times (2)^2$ ...(B)

∴ $\qquad s = 9.77$ m **Ans.**

Differentiating equation (B),

$$1074.89\frac{ds}{dt} = 2625.9 \times 2v\frac{dv}{dt}$$

or
$$1074.89 = 2625.9 \times 2\frac{dv}{dt}$$

or
$$\frac{dv}{dt} = \frac{1074.89}{2625.9 \times 2}$$

$$= 0.2046 \text{ m/s}^2$$

E X A M P L E

7.19 IMPULSE AND MOMENTUM

So far we have learnt that D'Alembert's principle is applied to the problems involving force and acceleration. Similarly work-energy equation is used to deal with the problems where forces velocity and displacement are involved. However, in situations where one has to deal with forces time and velocity the concept of impulse and momentum is brought in. We now proceed to define these terms.

7.19.1 Lirear Impulse and Momentum

If R is the resultant force acting on a body of mass m having acceleration a, then we have from Newton's law,

$$R = m \cdot a$$

$$= m\frac{dv}{dt}$$

or
$$R\ dt = m\ dv \qquad ...(7.64)$$

Suppose at time = 0, the initial velocity is u and after time t, it is v then integrating equation (7.64) we get,

$$\int_0^t R\,dt = m\int_u^v dv = m(v-u) \qquad ...(7.65)$$

The term $\int_0^t R\,dt$ is known as IMPULSE. Thus, if R is in Newton and time in sec, the units of Impulse are $N - s$.

In general R may be function of t for variable force system. However, if in the interval zero to t, sec, R is constant, then,

$$\text{Impulse} = Rt$$

The right hand side of equation 7.65 in the difference of the terms mv and mu.

The product of mass and velocity is called the momentum.

If mass is in kg, and velocity in m/s then

$$\text{Units of momentum} = \frac{\text{weight}}{g} \times velocity$$

$$= \frac{N}{\text{m/s}^2} \cdot \text{m/s}$$

$= N - s$ same as those of impouse

The equation (7.65) we see that

$$\text{Impulse} = \text{Final momentum} - \text{Initial momentum}$$

Equation (7.66) is known as impulse momentum equation.

Note : We know that R, the resultant force is a vector quantity and so is the velocity. Therefore, we conclude that Impulse – momentum equation is valid only, when R, u and v are in the same direction.

Thus, the statement of the impulse – momentum equation is "The component of the resultant linear impulse m any direction is equal the charge in the component of momentum in that direction."

EXAMPLE 7.24 : *A rigid body weighing 2 kN is lying on a rough horizontal plane. If a force P of 500 N is applied along the plane, after how much time its velocity will be 5 m/s? Take the co–efficient of friction between the body and plane surface as 0.2.*

Solution : Let P be the force applied.

Frictional force $= \mu R = 0.2\ R$

But $R = W = 2000\ N$

$\therefore \quad F = 0.2 \times 2000$

$= 400$ N

Therefore, resultant driving force,

$$= P - F$$

Then from Impulse – momentum equation

$$(P - F)\,t = m\,(v - u)$$

Here, $u = 0,\quad v = 5$ m/s

$$(500 - 400)\,t = \frac{W}{g}(5-0)$$

$$100\,t = \frac{2000}{9.81}\times 5$$

$$t = \frac{2000\times 5}{9.81\times 100}$$

$$= \textbf{10.18 sec Ans.}$$

Fig. 7.36 (*a*)

7.25 : *In example 7.24 after the block attains the velocity of 5 m/s, if the force P is removed, how long the body will continue moving before consing to rest?*

Solution : In this case the motion will be against the frictional force *F*. Applying the impulse momentum equals.

$$-Ft = m(0 - 5)$$

$$-\,0.2 \times 2000 = t - \frac{2000}{9.81}\times 5$$

$$t = \frac{5}{9.81\times .2}$$

$$= \textbf{2.548 sec}$$

Fig. 7.36 (*b*)

7.26 : *A body weighing 20 N has its velocity increased from 15 m/s to 25 m/s in the same direction in 60 secs. Find the change in momentum and the force that caused the change.*

Solution : Given data $u = 15$ m/s

$$v = 25 \text{ m/s}$$

$$t = 60 \text{ sec}$$

$$W = 20 \text{ N}$$

$$m = \frac{20}{9.81}$$

Therefore, using the Impulse – momentum equation we have,

$$F \times t = \frac{W}{g}(v-u)$$

$$F \times 60 = \frac{20}{9.81}(25-15)$$

$$\therefore \quad F = \frac{20\times 10}{60\times 9.81} = 0.3398 \text{ N}$$

$$\text{Change in momentum} = mv - mu$$

$$= \frac{m}{9.81}(v-u)$$

$$= \frac{20}{9.81}(25-15) = 20.39 \quad \textbf{Ans.}$$

7.27 : *A car weighing 2kN is moving at 72 km/hr when due to some emergency the driver had to apply brakes and the car came to rest after 10 secs. Determine the resisting force offered by the brakes.*

Solution : Given data weight of car

$$W = 2 \text{ kN}$$

$$= 2000 \text{ N}$$

$$u = 72 \text{ km/hr}$$

$$= \frac{72 \times 1000}{3600}$$

$$= 20 \text{ m/s}$$

Using impulse, momentum equation we have,

$$-F \times t = m(v - u)$$

or

$$-F \times 10 = \frac{2000}{9.81}(0-20)$$

or

$$F = \frac{2000 \times 20}{9.81 \times 10}$$

$$= \textbf{407.74 N Ans.}$$

7.28 : *A car weighing 2kN is moving down a hilly road whose inclination is 20° with the horizontal. Its initial velocity down the incline is 3 m/s, what will be its velocity after 10 sec? The resistance to the motion of car is 0.1 N per 100 N reaction of the road on the car.*

Solution : Given data

$$\theta = 20°$$

$$W = 2000 \text{ N}$$

$$R = W \cos 20°$$

$$= 2000 \times 0.951°$$

$$= 1902.1 \text{ N}$$

$$\therefore \text{ Resisting force} \quad F = 0.1 \times R$$

$$= 190.21 \text{ N}$$

$$\text{Now initial vel} \quad u = 3 \text{ m/s}$$

$$t = 10 \text{ sec}$$

$$v = ?$$

Fig. 7.37

Therefore, using impulse momentum equation,

$$F \times t = \frac{W}{g}(v-u)$$

$$190.21 \times 10 = \frac{2000}{9.81}(v-3)$$

or $$(v-3) = \frac{190.21 \times 10 \times 9.81}{2000} = 9.33$$

$\therefore$ $$v = \mathbf{3 \times 9.33 = 12.33\ m/s}$$

7.29 : *Two blocks A an B weighing 2kN and 1.8 kN are connected by an unstretchable cord which passes over a smooth pulley as shown in Fig. 7.38. Block B is hanging vertically while A is lying on a rough plane inclined at 30° with the horizontal. If the co–efficient of friction between A and plane is 0.2, find the time that will lapse when the blocks will attain a velocity of 10 m/sec. Also, find the tension in the cord.*

Solution : Given data

$$W_A = 2000 \text{ N}$$
$$W_B = 1800 \text{ N}$$
$$\mu = 0.2$$
$$\theta = 30°$$

Since the block A moves up, the frictional force will be acting downward.

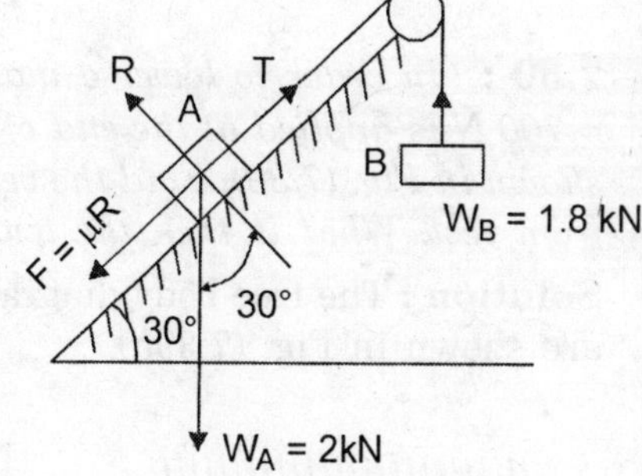

Fig. 7.38

Thus, Net Tractive force

$$= T - F - W_A \sin 30$$

$$= T - \mu R - \frac{W_A}{2}$$

But $$R = W \cos 30$$

$$= 2000 \times \frac{\sqrt{3}}{2} = 1000\sqrt{3}$$

Applying Impulse – momentum equation to Block A,

$$(T - \mu R - W_A \sin 30)\, t = \frac{W_A}{g}(10-0)$$

or $$\left(T - 0.2 \times 1000\sqrt{3} - 2000 \times \frac{1}{2}\right) t = \frac{2000}{9.81} \times 10$$

or $$(T - 364.41 - 1000)\, t = 20.38.73$$

or $$(T - 1364.41)\, t = 2038.73 \quad ...(A)$$

Again applying Impulse – momentum equation to Block B,

$$(W_B - T)\, t = \frac{W_B}{g}(10-0)$$

E X A M P L E

$$(1800 - T)\,t = \frac{1800}{9.81}\times 10 \qquad \ldots(B)$$

Dividing (A) by (B),

$$\frac{(T-1364.1)}{(1800-T)} = \frac{2000}{1800} = \frac{10}{9}$$

or $$9T - 9 \times 1364.1 = 18000 - 10T$$

or $$19T = 18000 + 12276.9$$

$$= 30277$$

$\therefore$ From equation (B),

or $$T = \frac{30277}{19}$$

$$= 1593 \text{ N } \textbf{Ans.}$$

$$(1800 - 1593.0)\,t = \frac{1800\times 10}{9.81} = 1834.86$$

$$t = \frac{1834.86}{207}$$

$$= 8.86 \text{ sec } \textbf{Ans.}$$

7.30 : *In order to lower a machine of weight W = 2100 N an effort P = 700 N is applied at the end of the rope in pulley block arrangement as shown in Fig. (7.39). Find the velocity of the machine after 4 secs starting from rest. What is then the tension in the rope?*

Solution : The free body diagram of machine as well as the pulley block are shown in Fig. (7.39*b*).

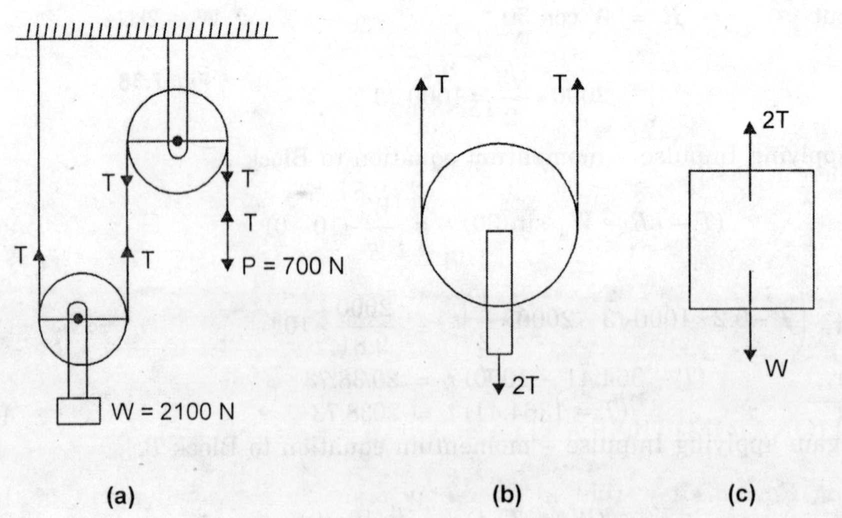

Fig. 7.39

Given data $u = 0$, $v = ?$, $t = 4$ sec

Applying the impulse – momentum equation to the machine

$$(2100 - 2T)\,t = \frac{2100}{9.81}(v - u)$$

$$(2100 - 2T)\,4 = \frac{2100}{9.81}v$$

or $$(2100 - 2T) = \frac{2100}{4 \times 9.81}v \qquad ...(A)$$

When machine moves a distance s, P moves $2s$ thus vel. of machine = $\frac{1}{2}v$. Again applying the impulse momentum equation to effort. Here, initial vol of P is given as zero and after 4 secs 2 v.

$$(T - 270)\,t = \frac{2700}{9.81}(2v - 0)$$

$$T - 270 = \frac{270}{4 \times 9.81}(2v) \qquad ...(B)$$

or $$2T - 540 = \frac{270 \times 2}{4 \times 9.81}(2v)$$

or $$2T - 540 = \frac{270.0v}{9.81} \qquad ...(B)$$

Radius (A) and (B), $$2100 - 540 = \frac{2100v}{4 \times 9.81} + \frac{540}{9.81}v$$

$$1560 = \frac{v}{9.81}(525 + 540)$$

$$= \frac{965v}{9.81}$$

$\therefore$ $$v = \frac{9.81 \times 1560}{965}$$

$$= \mathbf{15.85\ m/s\ Ans.}$$

Then from (B), $$T = 270 + \frac{270 \times v}{2 \times 9.81}$$

$$= 270\left(1 \times \frac{15.85}{2 \times 9.81}\right)$$

$$= 476.4 \text{ N } \mathbf{Ans.}$$

EXAMPLE

7.20 ROTATIONAL IMPULSE AND ANGULAR MOMENTUM

We have from Newtons law that a Torque T acting an body will result in the angular acceleration α given by the relation.

$$T = I\alpha \qquad \text{...(7.67)}$$

Where, I = Mass moment of inertia about the axis of rotation or that of T

Equation (7.66) can be written in the form

$$T = I\frac{dw}{dt}$$

or

$$T \text{ at} = I\, d\omega \qquad \text{...(7.68)}$$

Integrating (7.68),

$$\int_0^t T\,dt = \int_{W_0}^{W} I\,dv = I(\omega - \omega_0)$$

The term $\int_0^t T\,dt$ is called the angular impulse. Also, ω_0 = is the initial velocity at time 0 and ω after t sec.

E X A M P L E

7.31 : *The moment of Inertia of a rotor of a motor is 200 kg m². If the torque generalised by the electric current is 50 N – m, find the speed in rpm of the rotor after 10 sec starting from rest.*

Solution : Given data $\omega_0 = 0$

$$\omega = ?$$

$$T = 50 \text{ N} - \text{m}$$

$$I = 200 \text{ Kg} - \text{m}^2$$

We have from impulse and angular momentum equation,

$$Tt = I(\omega - \omega_0)$$

$$(50 \text{ Nm})\, t \text{ (sec)} = 200 \text{ (Kg} - \text{m}^2)\omega \text{ (rad/sec)}$$

$$\left(50 \times \frac{1}{9.81} \text{Kgm}\right) t\text{(sec)} = w\, 200\left(\text{Kg}\, m^{2w} \times \frac{\text{rad}}{\text{sec}}\right)$$

$$\left(\frac{50 \times 1}{9.81}\right) 10 = 200 \times \omega$$

$$\therefore \quad \omega = \frac{500}{9.81 \times 200}$$

$$= \mathbf{0.2548 \text{ rad/sec Ans.}}$$

7.21 CONSERVATION OF MOMENTUM

From equation (7.66), we know that,

$$\int_0^t R\,dt = m(v - u) \qquad \text{...(7.66)}$$

In case the resultant force R on a body is zero, then from equation (7.66) we have

$$m(v - u) = 0$$

or

$$mv = mu \qquad \text{...(7.67)}$$

That is Initial momentum is equal to the final momentum.

Important Note : It is re-iterated that the resultant force on the body or system must be zero and equation (7.69) is appliable only when the entire body is considered. The body cannot be subdivided into smaller elements and after drawing the free body diagram of each element, the equation (7.67) can not be applied to individual element. For example, consider a case where two persons are sitting on bicycle one turning the pedal and other sitting on the carrier. This velocity is u and all of a sudden the person sitting on the carrier jumps out, then the initial momentum will be,

$$= \left(\frac{W_1 + W_2}{g}\right)u$$

and final momentum will be $\frac{W_1}{g} v_1 + \frac{W_2}{g} v_2$

where v_1 is velocity with which the person sitting on carrer has jumped on v_2 is the velocity of the person still sitting on the sad be,

is $$\left(\frac{W_1 + W_2}{g}\right)u = \frac{W_1}{g} v_1 + \frac{W_2}{g} v_2 \qquad ...(7.68)$$

EXAMPLE

7.32 : *A gun shell weighing 240 N is fired from a cannon weighing 8500 N in the horizontal direction. The velocity of shell is 450 m/s. Find the initial recoil velocity of the cannon. If the recoil is resisted by a constant force, Find its value if the recoil distance is 700 mm.*

Solution : Let v to the recoil velocity of the gun.

Initial momentum = 0 since all the items is gun, shall an at rest.

Total momentum after the shell left the cannon

$$= \frac{8500}{9.81} \times v + \frac{240}{9.81} \times 450 = 0$$

$$\therefore \quad 8500\, v = -\, 240 \times 450$$

$$\therefore \quad v = \frac{-240 \times 400}{8500}$$

$$= -12.7 \text{ m/s}$$

Negative sign indicates that the connon moves in the opposite to that of shell.

The connon comes to rest after moving 700 mm.

Therefore, using work-energy equation we get,

$$-\frac{700}{1000} \times F = \left[0 - \frac{1}{2}\left(\frac{8500}{9.81}\right)(12.7)^2\right]$$

$$F = \frac{1}{2} \times \frac{8500}{9.81} \times \frac{(12.7)^2}{0.7}$$

$$= 99822.7 \text{ N}$$

$$= 99.822 \text{ kN } \textbf{Ans.}$$

E X A M P L E

7.33 : *A car A weighing 80 N is moving along OX direction at a velocity of 45 km/m and collides with another car B weighing 98/N moving in the reaction CA as shown in Fig. (7.40a) with a velocity 60 km/hr. Find the velocity of two cars after the collision.*

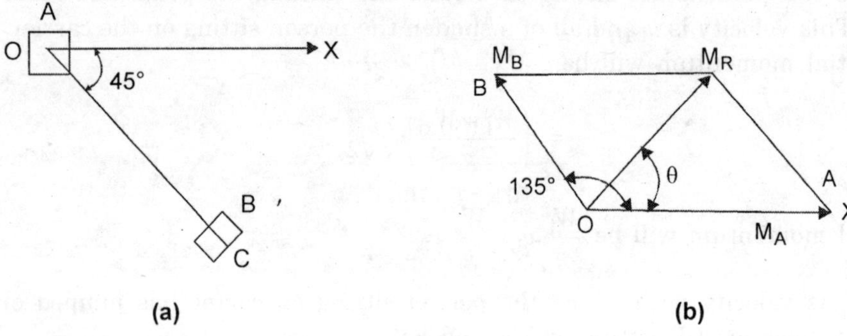

Fig. 7.40

Solution : Initial momentum of

$$A = \frac{80}{9.81} \times \frac{4500}{3600} \text{ along } OX$$

$$= 101.94 \text{ kg} - \text{m/sec}$$

$$\text{Initial momentum of } B = \frac{981}{9.81} \times \frac{6000}{3600} \text{ along } CO$$

$$= 166.67 \text{ kg} - \text{m/sec}$$

From the vector diagram Fig. (7.38*b*),

Final or Resultant movement

$$= \overline{OR}$$

$$= \sqrt{OA^2 + OB^2 + 2(OA)(OB)\cos 135°}$$

$$= \left[(101.94)^2 + (166.67)^2 - 2(101.94)(166.67)\frac{1}{\sqrt{2}}\right]$$

$$= [10391.76 + 27778.9 - 24028]^{\frac{1}{2}}$$

$$= [35762.66]^{\frac{1}{2}}$$

$$= 189.1 = M_e \text{ (say)}$$

Then from the conservation of momentum equation,

$$M_R = v \times \left(\frac{80 + 981}{9.81}\right) \text{ where } v \text{ is the velocity of two cars}$$

or $\quad 189.1 = 108.15\, v = 189.1$

$$\therefore \quad v = \frac{189.1}{108.15}$$

$$= 1.748 \text{ m/s}$$

$$= \frac{1.748 \times 3600}{1000} = 6.29 \text{ Km/h}$$

The direction of movement after collision is found by the relation.

$$\tan \theta = \frac{M_B \sin 135}{M_A + M_B \cos 135}$$

$$= \frac{166.67 \times \frac{1}{\sqrt{2}}}{101.94 - 166.67 \times \frac{1}{\sqrt{2}}}$$

$$= \frac{117.85}{101.94 - 117.85}$$

$$= -\frac{117.85}{15.91} = -61.70$$

$$\theta = 180 - 98.97°$$

$$= 81.03°$$

7.34 : *A body A of mass 9 Kg, moving at a velocity of 4 m/s collides with a body B of mass 7 kg moving at a velocity of 2 m/s (a) in the same direction (b) in the opposite direction. After impact both the bodies A and B merge into one another and move as one body C. Find the velocity of the body C. Also, find if there is any loss of kinetic energy.*

Solution : (*a*) Let the velocity of the body (*C*) after *A* has merged into *B* be = *v*.

Then for the conservation of momentum equation we have,

$$9 \times 4 + 7 \times 2 = (9 + 7)\, v$$

or

$$36 + 14 = 16\, v$$

$$v = \frac{50}{16} = 3.125 \text{ m/s}$$

(*b*)

$$\Delta KE = \left(\frac{1}{2}m_1v_1^2 + \frac{1}{2}m_2v_2^2\right) - \frac{1}{2}(m_1 + m_2)v^2$$

$$= \frac{1}{2} \times 9 \times (4)^2 + \frac{1}{2} \times 7 \times (2)^2 - \frac{1}{2}(9+7)(3.125)^2$$

$$= 72 + 14 - 8 \times (3.125)^2$$

$$= 86 - 78.125$$

$$= \mathbf{7.8751 \text{ N - m Ans.}}$$

(*c*) When *A* and *B* are moving in opposite directions,

$$m_1v_1 - m_2v_2 = (9 + 7)\, v$$

$$9 \times 4 - 7 \times 2 = 14\, v$$

$$36 - 14 = 16v$$

E X A M P L E

or $$v = \frac{32}{16} = 1.375\,\text{m/s}$$

$$\Delta KE = \left[\frac{1}{2}m_1v_1^2 + \frac{1}{2}m_2(-v_2)^2\right] - \frac{1}{2}(m_1 + m_2)\,v^2$$

$$= \frac{1}{2}\times 9\times(4^2) + \frac{1}{2}\times 7\times(-2)^2 - \frac{1}{2}(9+7)1.325^2$$

$$= 72 + 14 - 8\,(1.375)^2$$

$$= 86 - 15.125$$

$$= 70.875\ \text{N} - \text{m}$$

7.35 : *A body of mass 18 Kg is moving with a velocity of 15 m/s. While moving at due to internal explosion, it splits into two parts of masses 10 kg and 8 kg and start moving in the same direction. If due to splitting this is increase in kinetic energy by 100 Nm, find the value of the velocities of the the portions.*

Solution : Let the masses of two portions be m_1 and m_2 and their respective velocities be v_1 and v_2.

Give data $\rightarrow m_1 = 10$ Kg,

$m_2 = 8$ Kg

Original mass of body $= 10 + 8$

$= 18$ Kg

Original vel of body 15 m/s

Increase in $KE = 100$ N – m

From conservation of momentum equation we have,

$$m_1v_1 + m_2v_2 = (m_1 + m_2)\,v$$

$$10v_1 + 8v_2 = 18 \times 15$$

$$= 270$$

or $$v_1 + 0.8\,v_2 = 27 \qquad \text{...}(A)$$

Also, increase in KE = 100 Nm

$$\therefore \quad \frac{1}{2}m_1v_1^2 + \frac{1}{2}m_2v_2^2 = \frac{1}{2}(m_1 + m_2)v^2 + 100$$

$$\frac{10}{2}v_1^2 + \frac{8}{2}v_2^2 = \frac{1}{2}\times 18\times 15^2 + 100$$

or $$5v_1^2 + 4v_2^2 = 2025 + 100$$

$$= 2125 \qquad \text{...}(B)$$

Substituting value of v_1 from (A) into equation (B),

$$5(27 - 0.8v_2)^2 + 4v_2^2 = 2125$$

or $$\left(729 + 0.64v_2^2 - 43.2v_2\right) + 0.8v_v^2 = 425$$

EXAMPLE

or $1.44 v_2^2 - 43.2\, v_2 + 304 = 0$

$$\therefore \quad v_2 = \frac{43.2 \pm \sqrt{(43.2)^2 - 4 \times 1.44 \times 304}}{2 \times 1.44}$$

$$= \frac{1}{2.88}\left[43.2 \pm \sqrt{1866.24 - 1751}\right]$$

$$= \frac{1}{2.88}\left[43.2 \pm \sqrt{115.24}\right]$$

$$= \frac{1}{2.88}[43.2 \pm 10.73]$$

$$= \frac{1}{2.88}[53.93] \text{ or } \frac{1}{2.88}[32.47]$$

$$= 18.65 \quad \text{or} \quad 11.27$$

Then corresponding value of v_1 is obtained from equation A,

$$v_1 = 27 - 0.8\, v_2$$
$$= 27 - .8 + 18.65$$
$$= 27 - 14.92$$
$$= 12.08 \text{ m/s corresponding}$$
$$v_2 = 18.65$$

For $v_2 = 11.27$

$$v_1 = 27 - .8 \times 11.27$$
$$= 27 - 9.02$$
$$= 17.08 \text{ m/s} \qquad \text{corresponding to } v_2 = 11.27$$

Since $v_1 - v_2 > 0$

The valid answer is $\left.\begin{matrix} v_1 = 17.08 \text{ m/s} \\ v_2 = 11.27 \text{ m/s} \end{matrix}\right]$ **Ans.**

7.22 VIRTUAL WORK

In article 7.11 the concept of work done by a force has been discussed, the equation (11.53) has been derived. For the W.D. by force P, when it moves through a distance ds in a direction to which P is inclined at θ. It can be re-written as below,

$$\text{W.D.} = (P \cos \theta)\, S \qquad ...(7.53)$$

where, the component of P moves a distance S.

Before the definition of virtual work is taken up, we proceed to define the virtual displacement in the following paragraph.

Virtual Displacement may be defined as infinitely small imaginary or (hypothetical or virtual) given to the body or to a system of bodies in equilibrium consistant with the end constraints. It may be either rectilinear or angular. Similarly virtual work is the product of force P and the virtual displacement ds in the direction of the force. Thus, if δu is the virtual work.

i.e.,
$$\delta U = P \cos \theta \, ds \qquad ...(7.71)$$

Let there be a body A lying at 0, and be acted upon by a large number of forces P_1, P_2, P_3 ... P_n, which are inclined at angles δ_1, $\delta_2 = \delta_n$ respectively with the axis OX shown in Fig. 7.41.

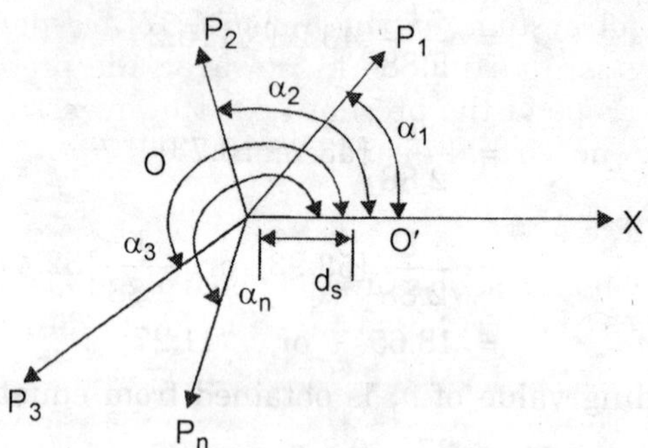

Fig. 7.41

Let δs be the virtual displacement given to the body along X-axis so that the distance moved by the body at 0 is equal is,

$$00' = ds$$

The initial work done by each of the forces P_1, P_2 ... P_n will be the sun of individual vertical done by each

i.e.,
$$\delta U = P_1 \cos \delta_1 \, \delta s + P_2 \cos d_2 \, \delta B + P_3 \cos \delta_3 \, \delta s + ... + P_n \cos \delta n \, ds$$
$$=(P_1 \cos \delta_1 + P_2 \cos \delta_2 + ... + P_n \cos \delta_3) \, ds \qquad ...(7.72)$$

But we know that,

$$P_1 \cos \delta_1 + P_2 \cos \delta_2 + P_n \cos \delta_3 = R_X$$

When R_X is the improvement of the resultant of the force P_1, $P_2 = P_n$ along the axis OX'_1 X.

$$R_X = (P_1 \cos \delta_1 + P_2 \cos \delta_2 + ... + P_n \cos \delta_n) \qquad ...(7.73)$$

Where, R_X = Algetive sum of the component of forces along XOX' axis

Further we know that if the body is m equilibrium be,

$$R_X = 0 \qquad ...(7.74)$$

From equation (7.73) and (7.74) we conclude that,

$$\delta U = 0 \qquad ...(7.75)$$

Thus, it is concluded that in the case of a rigid body, the principle of virtual work states that if a rigid body is in equilibrium the total virtual work of the external forces acting on the body is zero for any virtual displacement of the body.

Note: However, in the case of system of connected bodies during the virtual displacement, only the virtual work of forces external to the system are to be considered because the total work of the internal forces of various connection is zero.

7.22.1 Active Forces

The external forces which are acting on the body are known as *Active Force*. While reactions on the supports and all external forces at fixed reactions an called *Reactive Force*. It may be noted that the reactive forces do not do any work during any virtual displacement which are consistant with the constaints of these reaction.

It is further pointed out that the number of virtual displacement is equal to the number of degrees of freedom which is defined as the number of independent – co-ordinates which would have to be specified in order to fix completely the location of each and every part of the system. "For example for system, having *n*-degrees of freedom, it is essential to solve *n*-equations (each expressing zero virtual work. However, the problems involving n = degrees of freedom an beyond the scope of the present text. However, such problem will require are equation no (7.74) for the one virtual displacement.

7.22.2 Ideal Systems

Ideal systems are those in which two or more rigid bodies are fastened together by mechanical connections, where there is no friction so that the joints do not consume any energy.

Under the above mentioned conditions, we need not draw the free body diagrams of individual members. The relation between the active forces is determined directly without considering the reactive forces.

EXAMPLE

7.36 : *A machine of weight 3000 N being lifted by a system of pulleys as shown in Fig. 7.42 using the method of virtual work, determine the value of effort P which will hold the machine in equilibrium.*

Solution : Here, W = 3000 N, P = ?

Let the effort P move downward by a small distance δy. Then the upward movement of machine will be $\frac{\delta y}{2}$.

$$\therefore \quad \delta U = P \cdot \delta y - W\frac{\delta y}{2} = 0$$

Therefore,

$$P = \frac{W}{2} = \frac{3000}{2} = 1500 \text{ N}$$

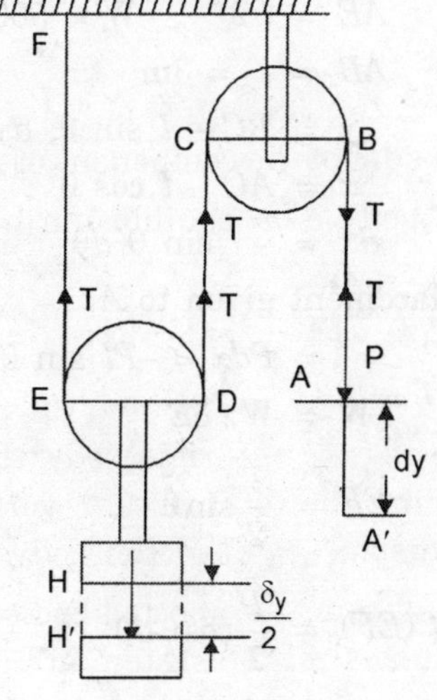

Fig. 7.42

EXAMPLE

7.37 : *A ladder (weightless) rests against a smooth wall and on smooth floor. (Fig. 7.43). It is held in equilibrium with the help of the forces F_1 and F_2. If the inclination of ladder with horizontal is θ, find the ratio F_1/F_2.*

Solution : Let the length of ladder be = l

$$AC = x = l \cos\theta \quad ...(A)$$

$$BC = y = l \sin\theta \quad ...(B)$$

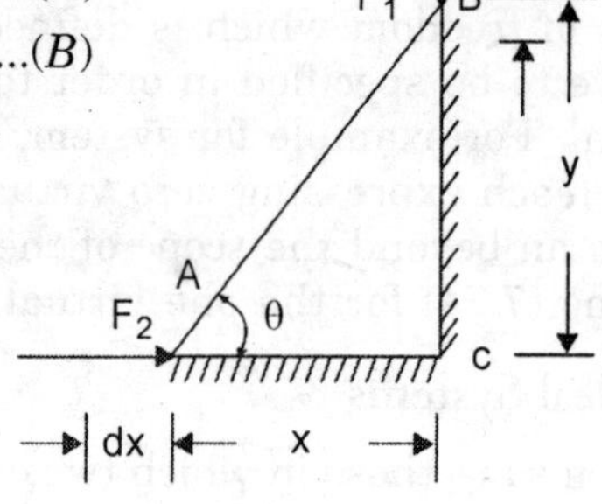

Fig. 7.43

Let a small displacement δx be given to A.

Differentiaing equation (A) and (B),

$$dx = -l \sin\theta\, d\theta$$

and $$dy = l \cos\theta\, d\theta$$

Vertical work done by,

$$F_1 = F_1\, dx$$

$$= -F_1\, l \sin\theta\, d\theta$$

Vertical work done by,

$$F_2 = F_2\, dy$$

$$= F_2 l \cos\theta\, d\theta$$

From equation (7.74),

$$\delta U = -F_1\, \theta\, \delta_n \theta\, d\theta + F_2\, l \cos\theta\, d\theta = 0$$

or $$\frac{F_1}{F_2} = \frac{\cos\theta}{\sin\theta}$$

$$= \cot\theta \textbf{ Ans.}$$

7.38 : *A ladder weight w N and length l m is resting against a wall and standing at floor making an angle θ. It is kept in equilibrium in this position by applying force P as shown in Fig. (7.74). If W = 500 N and l = 3 m, find the value of P. Take $\theta = 60°$.*

Soution : $AE = l/2 \qquad W = 500$

$$AB = l = 3\text{m}$$

$$y = BC - l \sin\theta,\ dy = 1 \cos\theta$$

$$x = AC - l \cos\theta$$

$$dx = -\, l \sin\theta\, d\theta$$

Let δx be a small displacement given to A.

Virtual work done by P $= Pdx = -Pl \sin\theta\, d\theta$

Initial work done by $W = W \cdot \delta E$

and $$EF = \frac{l}{2} \sin\theta$$

$$d\,(EF) = \frac{l}{2} \cos\theta\, d\theta$$

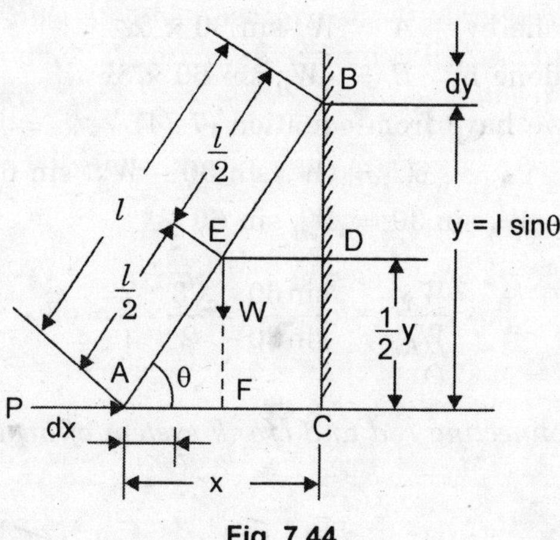

Fig. 7.44

$\therefore$ Vertical work by $\quad W = W(d\ EF) = W\dfrac{l}{2}\cos\theta\ d\theta$

$\therefore$ The equation (7.74),

$$-Pl\sin\theta\ d\theta + W\frac{l}{2}\cos\theta\ d\theta = 0$$

i.e.,

$$P\sin\theta = \frac{W\cos\theta}{2}$$

$$\frac{W}{P} = 2\tan\theta$$

$$P = \frac{W}{2}\cot\theta \equiv \frac{500}{2}\times\frac{1}{\tan 60}$$

$$= \frac{250}{1.376} = 181.6\text{ N}$$

7.39 : *Two blocks A and B weighing W_A and W_B respectively rest on frictionless planes as shown in Fig. 7.45. They are connected by an inextensible string passing over a frictionless pulley. Find by using principle of virtual work, the ratio of W_A/W_B for equilibrium of the system.*

Solution : Since the string is inextensible, if the displacement δx in B will be equal to the displacement δx in A.

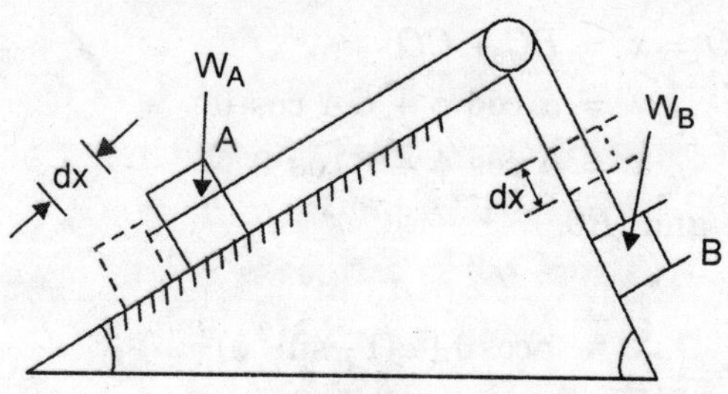

Fig. 7.45

$\therefore$ Vertical workdone by $A = W_A \sin 30 \times \delta x$

Vertical workdone by $B = -W_B \sin 60 \times \delta x$

For equilibrium we have from equation (7.74),

$$\delta U = (W_A \sin 30 - W_B \sin 60)\, \delta x = 0$$

i.e.,
$$W_A \sin 30 = W_B \sin 60$$

or
$$\frac{W_A}{W_B} = \frac{\sin 60}{\sin 30} = \frac{\sqrt{3}}{2} \times \frac{2}{1} = \sqrt{3}$$

7.40 : *Piston, connecting rod and crank system of an Engine is shown in Fig. 7.46.*

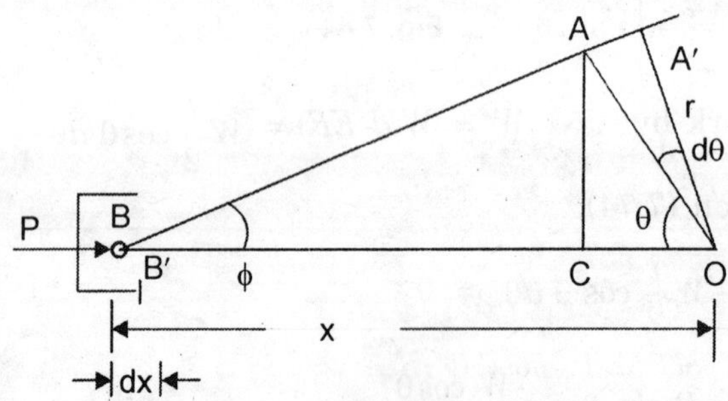

Fig. 7.46

During explosion stroke the gases exert force P on the piston at any instant and the torque on the crank shaft is Qr, when crank has turned θ from T.D.C, *where r is the crank radius. If the length of connecting rod is 'a', then using the method of virtual work find the ratio of Q/P at this instant.*

Solution : Let the crank be given a shall angular displacement $d\theta$ so that A goes to A' and B goes & B'.

Let
$$BB' = dx,$$
$$AA' = rd\theta$$

From triangles ABC and CAO,

$$CA = a \sin \phi = r \sin \theta$$

$$\therefore \quad \sin \phi = \frac{r}{a} \sin\theta \qquad ...(A)$$

Also, let
$$BO = x = BC + CO$$
$$= a \text{ cod } \phi + OA \cos \theta$$
$$x = a \cos \phi + r \cos \theta \qquad ...(B)$$

From equation (A) and $(B0,$

$$x = r\cos\theta + a(1-\sin^2\phi)^{\frac{1}{2}}$$

$$= r\cos\theta + a\left[1 - \frac{r^2}{a^2}\sin^2\theta\right]^{\frac{1}{2}}$$

or $$x = r\cos\theta + [a^2 \cdot r^2\sin^2\theta]^{\frac{1}{2}} \quad ...(C)$$

Differentiate (C) w.r.t θ

$$\therefore \quad dv = -r\sin\theta\, d\theta + \frac{1}{2}\left[a^2 - r^2\sin^2\theta\right]^{-\frac{1}{2}} \times \left[-r^2 2\sin\theta\cos\theta\right] d\theta$$

$$= -\left[r\sin\theta + \frac{r^2\sin\theta\cos\theta}{(a^2 - r^2\sin^2\theta)^{\frac{1}{2}}}\right] d\theta \quad ...(D)$$

$\therefore$ Virtual work done by

$$P = P\, dv$$

And vertical work done by

$$Q = Q\,(rd\,\theta)$$

$$\therefore \quad \delta U = P\,dx + Q(rd\theta)$$

$$= -P\left[r\sin\theta + \frac{r^2\sin\theta\cos\theta}{(a^2 - r^2\sin^2\theta)^{\frac{1}{2}}}\right] d\theta + Q\,rd\theta \quad ...(E)$$

Therefore, using equation (7.47),

$$\delta U = 0 \quad ...(F)$$

Therefore, from (E) and (F),

$$P\left[r\sin\theta + \frac{r^2\sin\theta\cos\theta}{(a^2 - r^2\sin^2\theta)^{\frac{1}{2}}}\right] d\theta = Q\,rd\,\theta$$

$$\therefore \quad \frac{Q}{P} = \sin\theta + \frac{r\sin\theta\cos\theta}{\sqrt{a^3 - r^2\sin^2\theta}}$$

$$= \sin\theta\left(1 + \frac{r\cos\theta}{\sqrt{a^2 - r^2\sin^2\theta}}\right)$$

$$\frac{Q}{P} = \sin\theta\left[1 + \frac{\cos\theta}{\sqrt{\frac{a^2}{r^2} - \sin^2\theta}}\right]$$

EXAMPLE

PROBLEM FOR PRACTICE

Velocity and Acceleration

1. A man has to catch a lens swtich leaves with an acceleration 0.3 m/s^2, when the man was still 15 m from the bus. Seeing that the law has started, he starts runing at 3m/s. Will he be able to catch the bus, if yes, after how many seconds of running? (**Ans.** Yes, After 10 sec)

2. A stone is projected vertically upwards from the point *A*, at the floot of tower. It passes point *B* in 3 sees and moves up and returns point *A* after 4 more secs. Determine the height of *B* from the foot of the tower and the velocity of stone, when it passes the point M_C which is midway between *A* and *B* while going up. (**Ans.** 58.8 m, 24.6 m/s)

D'Alembert's Principle

3. Two blocks *A* and *B* weighing 2000 N and 1200 N respectively are held with a system of frictionless pulleys as shown in Fig. 7.47. Neglecting the weight of pulleys, find the value of tension in the string and acceleration of each weight.

(**Ans.** 1058.8 N, a_A = 1.1154 m/s, a_B = 0.577 m/s^2)

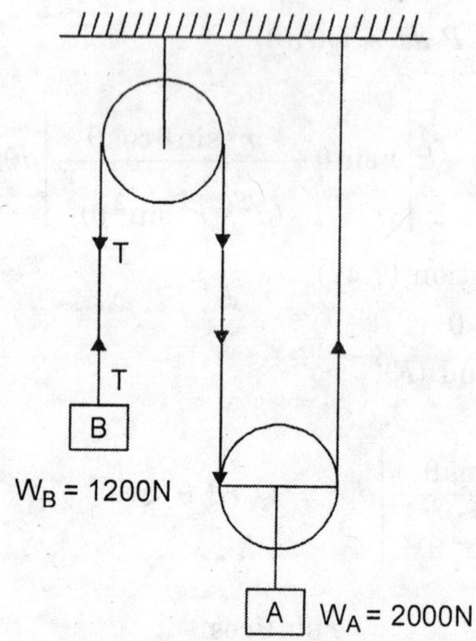

Fig. 7.47

4. A block weighing 150 N in lying on a horizontal rough plane. the co–efficient of friction between the block and the plane is 0.15. A push of 70 N acts on the block which is inclined with the horizontal at 30°, find the acceleration of the block. (**Ans.** 1.898 m/s^2)

5. A block *A* weighing 1000 N is resting on an inclined smooth plane as shown in Fig. 7.48. The block *A* is attached to another block *B*, weighing *W* with a inextensible string passing over a frictionless pulley. The block *A* starts moving down the plane with a velocity of 4 m/s. Find the value of *W* so as to bring the block *A* rest in 2 m. (**Ans.** 1532.6 *N*)

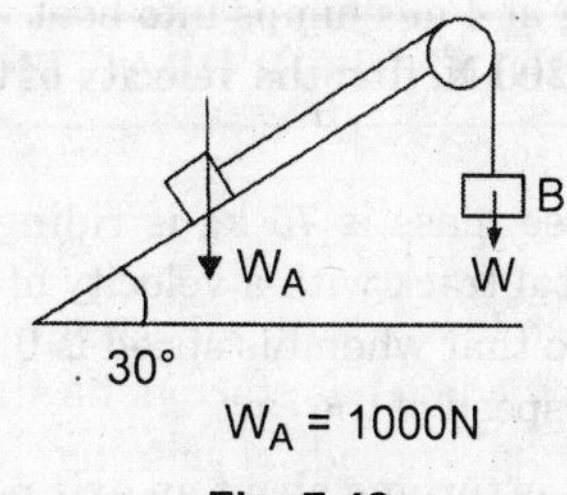

Fig. 7.48

6. A man weighing 0.5 kN is standing in an elevator. Find the reaction on the floor of the elevator in the following cases:

(*a*) Elevator is moving down with an acceleration of 2 m/s^2.

(*b*) Elevator is moving down with constant velocity.

(*c*) Elevator is moving down with retardation of 2 m/s^2. (**Ans.** 398.1 *N*, 500 *N*, 611.9 *N*)

Work and Energy

7. A loaded trolly is being pulled up an inclined rough plane as shown in Fig. 7.49 with the help of weight W_2, 30 kN. Total weight of the trolly is $W_1{}^2$ 50000 N and the frictional resistance offered by it is 5N/kN. The pulleys are frictionless. If the initial velocity of the trolly is 12 m/s, find the distance travelled by the trolly before coming to rest.

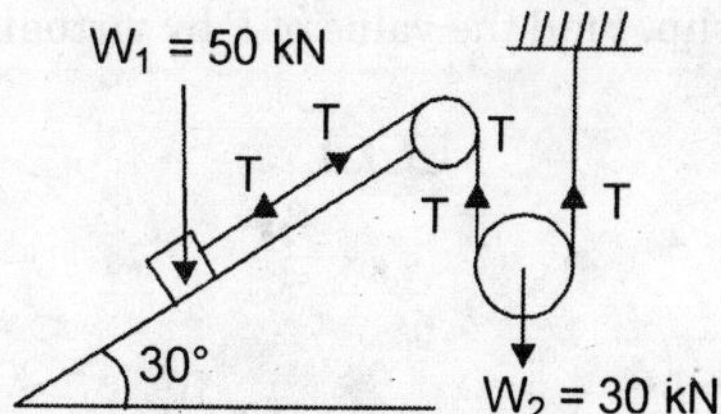

Fig. 7.49

8. (*a*) An electric train is moving up an inclined plane with inclination 1 m 120 at a velocity of 56 Km per m. The frictional resistance is 5N/kN. The weight of the train is 420 kN. Find the power of the locomotive.

(*b*) When the train is moving at 56 km/hr velocity, the electric supply to the locomative is cut off. Find how far it will move before coming to rest. (**Ans.** 87.11 km, 9.25 km)

9. A car weighing 20 kN is running at a speed of 70 km/hr when the brakes are suddenly applied. If the co–efficient of function of tyres with road is 0.75, find the time taken when the car with come to rest. (**Ans.** 2.64 s)

10. A system of two pulleys is shown in Fig. 7.50. Find the velocity of each weight when W_1 has moved a distance of 5 m from rest. Take W_1 = 1 kN and W_2 = 500 kN. Also, find the tension in the rope. (**Ans.** 8.1 m/s, 4.05 m/s, 1668.8 N)

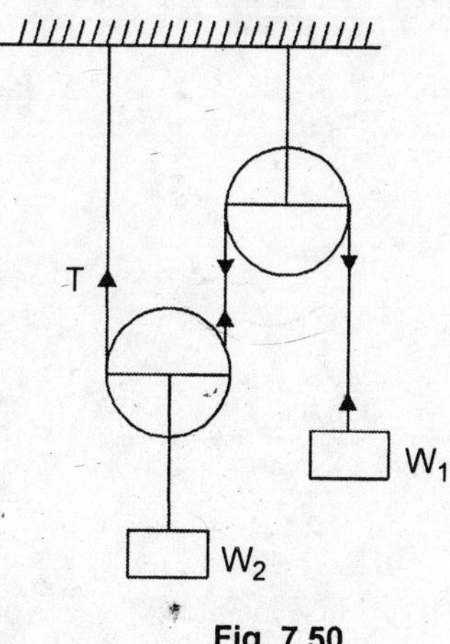

Fig. 7.50

Impulse and Momentum

11. A mass of 2 kg initially at rest explodes into three parts of masses in the ratio 1:1:3. Two pieces of equal mass move after explosion at right angles to each other with velocity of each equal to 25 m/s. Find the velocity of the third piece. (**Ans.** 11.785 m/s)

12. A man weighing 500 N running at 4 m/s jumps into boat which is coming towards him with vel., 5 m/s. If the weight of boat is 1200 N, find the velocity of the boat after the man jumps into it. **(Ans.** 2.235 m/s)

13. An Inspector of Rail road whose mass is 75 kg is riding a rail cast of mass 150 kg. which is moving to the right on horizontal track with a velocity of 1.5 m/s. The Inspector starts running to the left, in still on the cart, so that when his speed is 0.5 m/s relative to the cart, he jumps off the cart. What will then be the speed of the cart. **(Ans.** 1.33 m/s)

14. A round turn table 2.5 m radius is turning about an axis normal to the table and passing through is centre with an angular speed of 30 rpm. There are two blocks, each of mass 5 kg placed at the ends of a diameter on the table near the periphery of the table. The moment of Inertia of the table alone is 20 kg m^2. If the two blocks are suddenly made to fall off, what will then be the speed of the table? **(Ans.** 123.75 rpm)

Virtual Work

15. In the system of pulleys shown in Fig. (7.49) weight W_2 is to be lefted with the help of W_1 find the ratio of W_2/W_1 by method of virtual work. **(Ans.** $W_2/w_1 = 2$)

16. A ladder *AB* of length 3.9 m has to do at a distance of 1.3 m from *A*. The lower end of the ladder is on the ground and upper end *B* rests against the wall making angle of 60° with the ground. The weight of the ladder is 400 N. A horizontal force *P* is applied at the end *A* directed towards the wall so that it does not slip. Find the value of *P* by virtual work method. **(Ans.** 96.87 N)

CHAPTER 8

SIMPLE STRESS AND STRAIN

8.1 INTRODUCTION

In the present day world, the production and consumption of steel is considered to be index of development of some country. Not only the items that are made of steel and its alloys but also for the non-metal items that we see around indirectly lead us to the fact that for their production, steel has extensively been used in the form of machinery which is made of steel. Be it paper, a book, plastic toys, household goods, furniture as a matter of fact any item we see around, tools and machinery made of steel have been employed for their manufacture.

Steel which is extracted from iron ore is subjected to various manufacturing processes to ultimately acquire the shape of some component or the other so as to form part of an assembly which is called a machine. Although as part of a particular machine, these components are subjected to various types of loads but they do not change their configuration and size so that the machine continues to perform its assigned duty for which it has been designed. This is possible only if the behaviour of the engineering materials used, is thoroughly investigated and understood by the designer. This objective is fulfilled by the subject called "Mechanics of Solids "for which more popular prevalent name is "Strength of Materials"

Before topics related to 'Mechanics of solid' is taken up, it is considered essential to review the basic knowledge already gained by the reader during his pre-engineering course particularly about, free body digarms, equations of static equilibrium and other associated terms in the following paragraphs:

8.2 FREE BODY DIAGRAM

Free body diagram of a member of a structure, or of an element of a member in static equilibrium, is the diagram of only that member or the element, as if made free from the rest of the structure, with all the external forces acting on it duly marked. These external forces will include those due to the presence of adjacent members in contact with the one in question, the gravitational effect and the inertia force, if the member is in motion. Fig. 8.1(*b*) shows the free body diagram of a weight hanging from the end of a crane rope, the gravitational force is balanced by the tension T in the rope. The Fig. 8.1(*c*) shows the free body diagram, of the two elements of a tension test specimen, held in tensile testing mechine, Fig. 8.1(*a*), under the action of two equal and opposite forces P, obtained by taking a section CC. The force, F_1 on the element AC_1 is due to the cohesion of the plane C_2, while F_2 on the element C_2B is due to the plane C_1. Since the test specimen AB is in equilibrium, any portion of that will also be in equilibrium. For the equilibrium of individual elements AC_1 and C_2B, F_1 and F_2 should both be equal to P.

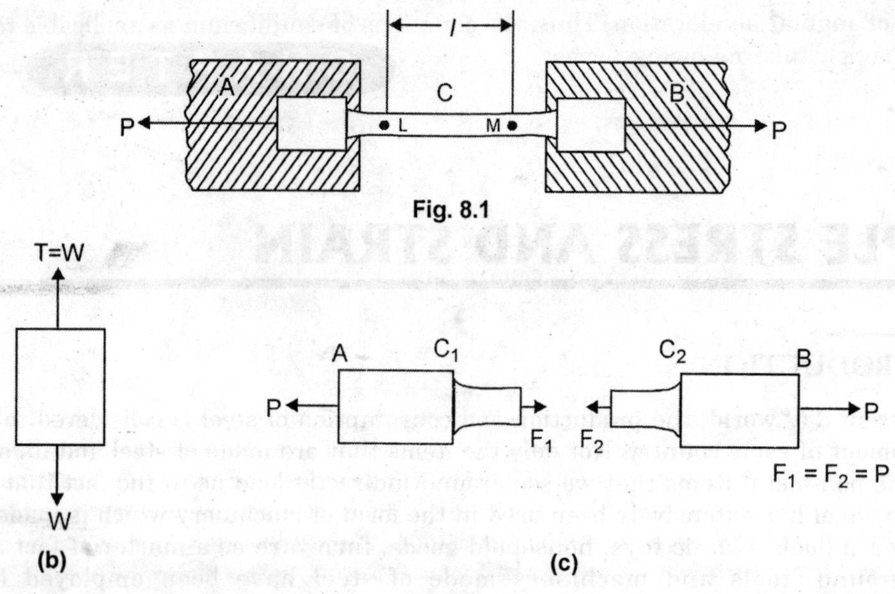

Fig. 8.1

Fig. 8.1

8.2.1 Equations of Static Equilibrium

Despite the fact that the reader would have been familiar with the equations of static equilibrium, still it is felt necessary to re-write these equations here so that the reader is not at loss to understand the basic concepts involved in this subject properly. Whenever a solid body under the action of various forces is in static equilibrium or in the state of rest then the algebraic sum of the components of these forces along any chosen three perpendicular axes of reference are separately zero. In addition, the algebraic sum of the moments of these forces about any point is also zero, if it is not rotating. Mathematically, these results are expressed in the form of following four equations:

$$\Sigma X = 0 \quad ...(A)$$

$$\Sigma Y = 0 \quad ...(B)$$

$$\Sigma Z = 0 \quad ...(C)$$

$$\Sigma M = 0 \quad ...(D)$$

where ΣX, ΣY and ΣZ represent the algebraic sum of the components of forces along X, Y and Z direction respectively and ΣM the algebraic sum of moments of these individual forces about any point.

8.2.2 Equations of Static Equilibrium and Body in Motion

In article 8.2.1, the equations of equilibrium have been introduced and it was pointed out that these equations are derived for the state of rest of the solid body. However, D' Alembert pointed out that these equations can be applied to a solid body in motion (linear or rotory) as well, provided the inertia force in the case of linear motion or inertia torque in the case of rotary motion is superimposed on the body in addition to the given forces. The inertia force is equal to the product of mass and acceleration and is directed opposite to the direction of acceleration. Similarly in the case of rotory motion the inertia torque is the product of mass moment of inertia of body and the angular acceleration and is directed opposite to the direction of angular acceleration. Thus, the equations of equilibrium as applicable to the solid body in motion take following forms:

$$\Sigma X + (F_i)_x = 0 \quad \text{...(E)}$$

$$\Sigma Y + (F_i)_y = 0 \quad \text{...(F)}$$

$$\Sigma Z + (F_i)_z = 0 \quad \text{...(G)}$$

$$\Sigma M + T_i = 0 \quad \text{...(H)}$$

where,

F_i = Inertia force = $m \times a$

$(Fi)_x$ = Component of F_i along X axis

$(Fi)_y$ = component of F_i along Y-axis

$(Fi)_z$ = component of F_i along Z-axis

T_i = Inertia torque = $I_m \alpha$

M = Mass of the solid body

a = Linear acceteration of body

I_m = Mass moment of Inertia

α = Angular acceleration of body

These equations are known as *D'* Alemberts equations of equilibrium for solid bodies in motion and this is known as *D'* Alembert's Principle.

As a special case, if the body is having uniform linear motion or uniform angular motion then both linear acceleration '*a*' and angular acceleration 'α' will be zero and the equations (*E*), (*F*), (*G*) and (*H*) reduce to (*A*), (*B*), (*C*) and (*D*) respectively. Thus, we conclude that when a solid body is at rest or in the state of uniform motion the equations (*A*) through (*H*) apply. However, when it is accelerating or deaccelerating then equation (*E*) through (*H*) will apply.

The utility of the *D'* Alemberts principle as applicable to bodies in motion can better be understood with the help of following example:

Let a passanger train consisting of 16 bogies be pulled by an electric locomotive. If it is assumed that the mass of each bogie be *M* and the air resistance experienced by each bogie be F_a, then it is desired to calculate the stress in each coupler which connects two adjacent bogies. The area of cross-section of each coupler is 'A'. For evaluating the stress in the last coupler connecting 16th bogie with 15th bogie, we take a section any where along the length of the coupler and draw the free body diagram of the last bogie. Two situations may arise, firstly, the train be moving with uniform speed and secondly, when the train is accelerating.

In first case, there will be no inertia force as the acceleration is zero. Thus, the free body diagrams of the last two bogies will be as given in figure 8.2. Then according to *D'* Alemberts' principle considering the equalibrium of bogie number 16, we have for the horizontal forces.

$$F_a = P_{16} \quad \text{...(F)}$$

Similarly, for the equalibrium of 15th bogie we will get,

$$P_{16} + F_a = P_{15} \quad \text{or} \quad P_{15} = 2F_a$$

and so on

$$P_{15} + F_a = F_{14} \quad \text{or} \quad P_{14} = 3F_a$$

$$\vdots \qquad\qquad P_2 = 15F_a$$

$$P_2 + F_a = P_1 \quad \text{or} \quad P_1 = 16F_a$$

Thus, the first coupler which connects the first bogie with the locomotive will be stressed maximum and it will be equal to $\frac{16F_a}{A}$. While the last coupler will be stressed least $\left(= \frac{F_a}{A}\right)$.

The maximum stress is $\frac{16F_a}{A}$.

In the second case, when the train is accelerating with an acceleration equal to 'a', then according to *D'* Alembert's principle, each bogie will be accelerating and on each the inertia force which is equal to $(M \times a)$ will be acting and it will be in the opposite direction to that of the acceleration. The free body diagrams of a few bogies is given in Fig (8.3)

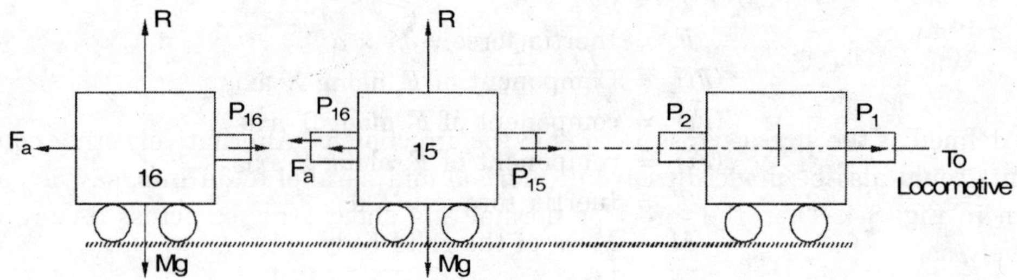

Fig. 8.2

Then for the equilibrium of bogies number 16,

$$F_a + F_i = P_{16}$$

or
$$P_{16} = F_a + M.a$$

Similarly, for bogie no. 15,

$$M.a + F_a + P_{16} = P_{15}$$

or
$$(M.a + F_a) + (F_a + M.a) = P_{15}$$

or
$$2M.a + 2F_a = P_{15}$$

and so on
$$F_i + F_a + P_2 = P_1$$

or
$$16F_a + 16M.a = P_1$$

The stress in each coupler can then be evaluated by dividing the force transmitted by each coupler with area of cross–section A.

Thus, the max stress will occur in the first coupler which will be equal to $(16F_a + 16M.a)/A$ and likewise the least stressed coupler will be the 16th coupler connecting the 15th and 16th bogies and will be equal to $(F_a + M.a)/A$.

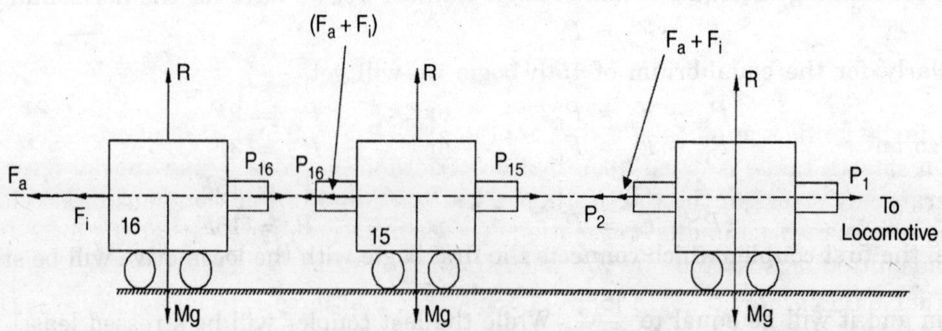

Fig. 8.3

8.3 DIRECT STRESS

Let O be a point in a solid body which is in equilibrium under the action of a number of forces. Let δA be a small plane area with ON normal to it. Across area δA a force δF, say, is being transmitted or sustained by the body to give a quantity,

$$\sigma = \underset{\delta A \to 0}{Lt}\left(\frac{\delta F}{\delta A}\right)$$

which is defined as the stress across δA at O in the direction δF. Alternatively other equivalent statements could also be made. Resolve dF normal and parallel to δA into, say δF_n and δFs as shown in Fig. 8.4. Then the direct or normal and shear stresses across dA are defined respectively as,

$$\sigma_n = \underset{\delta A \to 0}{Lt}\left(\frac{\delta F_n}{\delta A}\right)$$

$$\tau = \underset{\delta A \to 0}{Lt}\left(\frac{\delta F_s}{\delta A}\right)$$

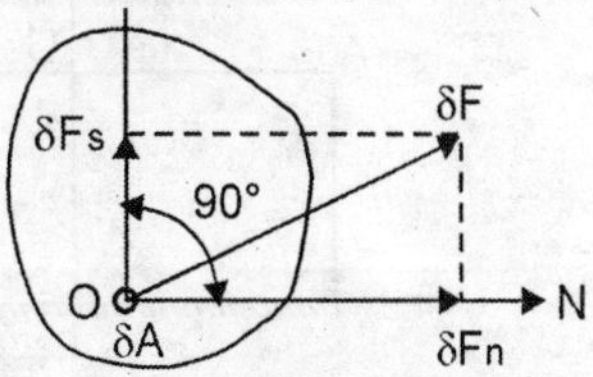

Fig. 8.4

In case δF is acting along ON, that is normal to δA, the stress σ itself will be the normal or direct stress and the shear stress τ will be zero. Thus, the direct or normal stress at a point of a plane section is defined as the force sustained per unit area of cross-section normal to it.

It is once again emphasized that the solid body must be in equilibrium and it should sustain or transmit force. Evidently to generate stress in a solid body, there must be two equal and opposite forces acting on the body and these forces must have their respective points of application separated by some distance. The intervening material of the solid body will then be stressed since that portion of the body will be sustaining the force. As an example in the test specimen of Fig. 8.1(*b*), the direct stress over the section CC will be F_1/A_0 (= P/A_0), where A_0 is the area of cross-section of the specimen since A_0 is the area which is sustaining the normal force P. The stress is usually measured in Newton/Metre Square (N/m^2) or Pascal (Pa). Since Pascal is a very small unit it is not uncommon to use Mega Pascal (MPa) or Giga Pascal (GPa) as units of stress, where,

$$1 \text{ MPa} = 10^6 \text{ Pa}$$

and $$1 \text{ GPa} = 10^9 \text{ Pa}$$

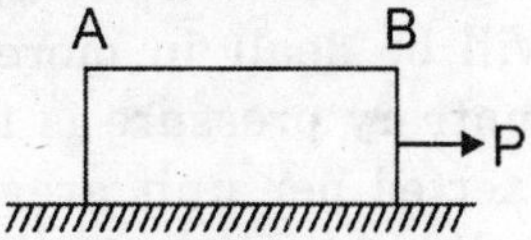

Fig. 8.5

As a further illustration take the case of a rod AB Fig. 8.5 of area of cross-section, A_0, lying on a smooth table, is acted upon by a force P along its axis. This external force P will not generate any stress in the rod. Rather it will set the rod in motion causing acceleration according to Newton's second law of motion. The point to note here is that the force P is not being transmitted or sustained by the rod AB.

In order to generate stress in it, would require equal force P to be applied at section A in the opposite direction. In that event the rod would have not been set into motion but would have been at rest or in state of equilibrium. It may further he noted that a solid body need not necessarily be in static equilibrium for it to be in the state of stress. It could be stressed while in motion as well. For example, if the rod AB, Fig. 8.5, is acted upon by the force $F(<P)$ on the section A towards left, then force F will be sustained by the rod AB to generate normal stress F/A_0 in the rod. However, the force $(P - F)$ will set the rod into motion according to Newtons second law of motion.

In order to elaborate this point, further let the rod AB be acted upon by an other force P acting on the section B but in leftward direction as shown in fig. 8.7 and we intend to find the stress at any section CC lying in between A and B. Then taking a section CC at the point, the free body diagrams of the two elements AC_1 and C_2B are drawn as shown in Fig. 8.5. Evidently if the body AB is in static equilibrium, so will be the elements AC_1 and C_2B. If we consider the static equilibrium of individual elements AC_1 and C_2B and resolve the forces, along the direction of P we have,

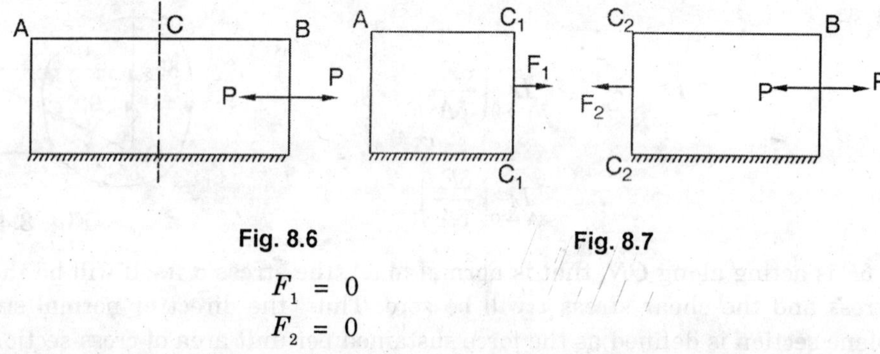

Fig. 8.6

Fig. 8.7

$$F_1 = 0$$

and
$$F_2 = 0$$

where F_1 is the reaction of C_2 on C_1 and F_2 the reaction of C_1 on C_2 and $F_1 = F_2 = 0$

Thus, the force sustained by any cross-section lying between A and B will be zero. In other words, the rod material will not be stressed. This proves the statement made above in the previous paragraph *i.e.,* to generate stress, two equal and opposite forces must be applied at points at finite distance apart.

A beginner in the subject is sometimes at loss to distinguish between stress and pressure as both have the same units. It is to be noted that stress is encountered in solids and is associated with the direction of the sustained force with respect to the plane passing through the point under consideration. The magnitude of stresses will be different on different planes passing through the point under consideration. This aspect will be dealt in more detail later, when complex stresses will be considered. On the contrary pressure is the term associated with fluids (liquids and gases) and the force exerted per unit area due to the impact of the moving molecules of the fluid on the walls of the container or on the body immersed in the fluid and is same at a point in the fluid, in all directions (Pascal's Law).

8.3.1 Direct Strain

Once again referring to the Fig 8.1(*b*), let there be two points L and M on the specimen which are at a distance l_0 called the gauge length, before the application of the force P. After the force is applied, the specimen will stretch and the distance between L and M will increase to say l. The change in length per unit gauge length $(l - l_0) / l_0$, is called the axial or direct strain.

8.4 THE TENSION TEST

Consider a tension test specimen of round section as shown in Fig. 8.1(*b*) being stretched in a testing machine. The load recorded by the universal testing machine is the reaction of the test specimen. At room temperature, the load extension diagram in general, at strain rate less than 2×10^{-3}/sec, will tend to follow the curve of Fig. 8.8(*a*), which is typical for an annealed metal like aluminium or copper.

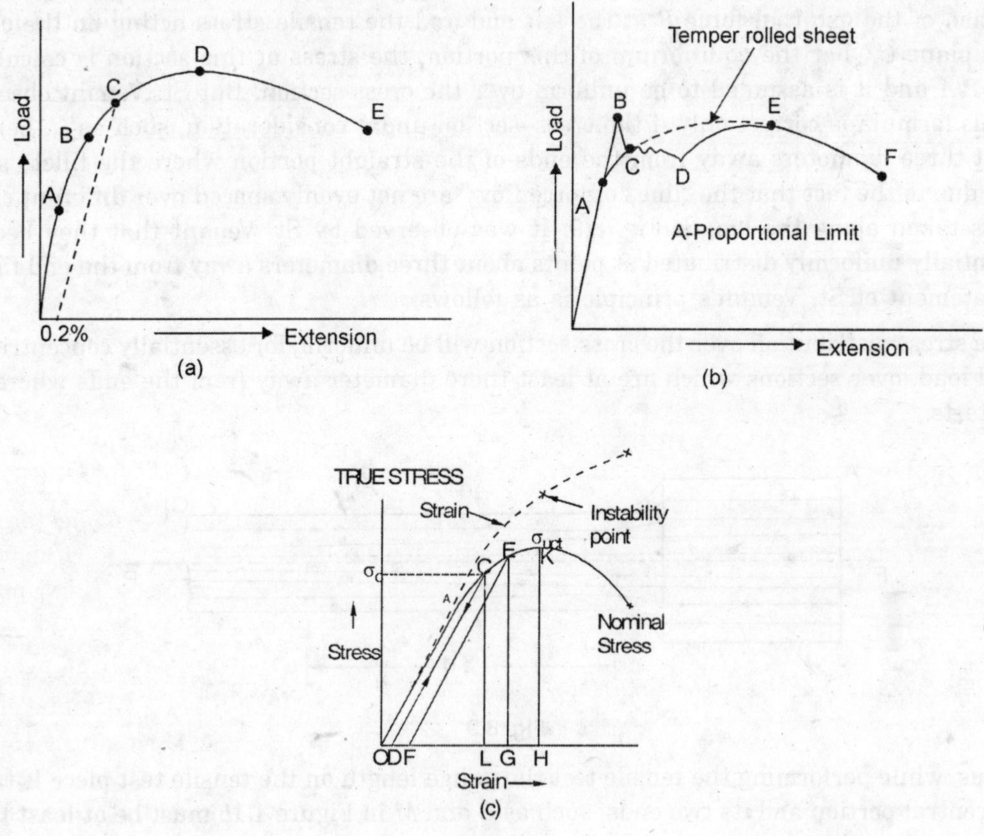

Fig. 8.8

Initially, the relation between the load and extension is essentially linear—that is, portion *OA* of the curve, where *A* defines the limit of proportionality. On further straining, the relation between the load and extension no longer remains linear but the material is still elastic. That is, if the load is released, the specimen will revert back to its original length. The maximum load which can be applied without causing permanent deformation is known as the 'elastic limit' represented by point *B* in Fig. 8.8(*a*). Usually the points *A* and *B* are very near to each other and the real difference can be achieved only by using highly sensitive measuring devices. At the point *B*, ends the elastic straining and the 'inelastic' or plastic deformation ensues. Beyond the point B, there is a general permanent extension of the test piece until the load attains a maximum value, point *D*, corresponding to which the stress is called the 'ultimate tensile stress' σ_{ult} $(= P_{max} / A_0)$. At this load, the specimen necks down locally rendering the deformation non-uniform and the load falls off with increasing extension, thereby causing the fracture at a load, point E, which is less than that corresponding to the ultimate tensile stress point. In addition another term, called the *'proof stress'* is usually quoted in structural design. This refers to the stress, σ_c (= P_c/A_0) required to produce a small strain of 0.2%. This is represented by point C in the Fig. 8.8(*b*).

8.4.1 St. Venant's Principle of Uniform Stress

In figure 8.1 is illustrated the equilibrium of the left portion AC_1 of the test specimen under the action of the external force P at the left end and the tensile stress acting on the cross-section plane C_1. For the equilibrium of this portion, the stress at this section is calculated as $\sigma = P/A$ and it is assumed to be uniform over the cross-section. But St. Venant observed that this formula is correct only if the cross–section under consideration, such as C_1 here, is at least three diameters away from the ends of the straight portion where the fillets start. This is due to the fact that the "lines of force flow" are not evenly spaced over different cross-sections taken along the length Fig. 8.9. It was observed by St. Venant that they become substantially uniformly distributed at points about three diameters away from the end fillets. The statement of St. Venant's principle is as follows:

The stress distribution over the cross-section will be uniform, for essentially concentrically applied load, over sections which are at least there diameter away from the ends where the fillet starts.

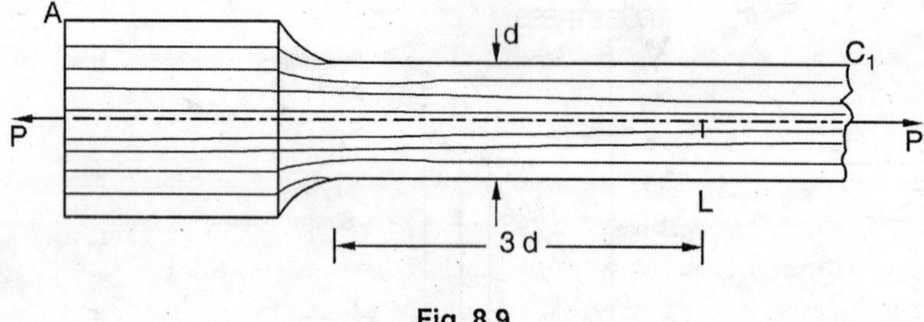

Fig. 8.9

Thus, while performing the tensile test the gauge length on the tensile test piece is taken in the central portion and its two ends, such as L and M in Figure 1.1*b* must be at least three diameters from the respective ends, where the fillet starts and on both sides the machine grips should be at equal distance from the ends. Then and then only the stress on any cross-section taken any where along the length LM will be uniform.

8.4.2 The Yield Point in Low Carbon Steel

Typical load extension curve after Murray (1) for annealed low carbon steel (earlier called mild steel) is shown in Fig. 8.8 (*b*). The marked difference between this and two diagrams of the materials of the type discussed above is the exhibition of pronounced yield point. Upto the point B, the load rises steadily and the material remains elastic. The extension per unit guage length at this point is of the order 10^{-3}. At low carbon or nitrogen levels in mild steels, at the point B the load suddenly drops and then fluctuates about some constant value of load. The specimen extends considerably at this almost constant load. This sudden extension in mild steels is localized to certain zones in the metal and surface markings, known as "Luder bands" or "Strecteher strains" are produced. Under this situation, we have some material which has yielded and some which has not. As the stretching continues upto the point *D* almost the entire specimen gets uniformly deformed. The extension *CD* is of the order of 10^{-2} per unit gunge length. Beyond the ponit *D,* the load rises again as the specimen work hardens.

The stress at the point *B* at which the load suddenly drops is called the 'upper yield point' and the average constant load in the range CD to which the load has suddenly dropped is called the 'Lower yield point'. This gives rise to a peculiar feature that for low carbon steel (mild steels) that the stress required to maintain plastic flow immediately after yielding has started, is lower than that required to start it. Cottrell (1) (2) has put forth his theory of "Formation of cottrell's Atmosphere" around the dislocation to explain the occurrence of this phenomenon in low carbon steels. According to this theory, carbon or nitrogen atoms, in *BCC* ion *i.e.,* in Ferrite phase, diffuse the positions of low energy. The dislocations are pinned or locked with these interstitial solutes (carbon or nitrogen) and additional stress, over that normally required for the movement of dislocations, is needed in order to tear some of the dislocations away from the restraining impurity atoms. This results in increase in stress. When the dislocation line is pulled free from the influence of the solute atoms, it can slip to a lower stress called lower yield point stress. The significant point is that a very small concentration of solute is required to lock or pin along the whole length of dislocation. This phenomenon gives rise to rough surface during drawing or stamping processes.

In the field of metal working, such as in deep drawing of steel blanks, for making steel utensils etc, the presence of Luder Bands is not desirable. The modern method to suppress this phenomenon is to "temper roll" it. The soft annealed sheet of mild steel is subject to light cold rolling (½ to 2 %) reduction in thickness) not more than 24 hours before deep drawing because the yield point suppression is not permanent. Lightly cold rolled sheet, deep drawn after standing for sufficient length of time, does show Luder Bands. Apart from this, slow hardening occurs and results in loss of much desired ductility. This process of age-hardening is usually checked by adding aluminium into steel and producing the so called killed ingot.

For temper rolled sheet the stress-strain curve assumes the shape of the one shown in broken lines in Fig. 8.8 *b*.

Modern practice is to add a small percentage of metal like Titanium or Molybdenum into mild steel, which forms carbide with the carbon and the phenomena of lower yield point is eliminated completely and no Luder Bonds are produced in the finished product.

Due to wide use of mild steel in industry, where permanent change in shape in brought about, the complete stress-strain diagram for the metal is invariably considered. On the other hand, its use in structures and machine elements is limited only upto the limit of proportionality point, where the emphasis is only on small strains.

---

1. Murry 6, "The yield point in mild steel, Engineering, Vol. 178, PP. 366, 1954
2. Cotrell; Physical Metallurgy, *Mc Graw* hill.
3. Vijendra Singh, Physical Metallurgy, Standard Publishers and Distributers, New Delhi.

8.4.3 True Stress Strain Curve

If we now divide the load by the original area of cross-section A of the test specimen and the extension by the original length to (which is usually called the gauge length), then the curves shown in Fig. (8.8) will be called the nominal or Engg. stress-strain curve. These are called nominal or Engg. stress and strain because both the parameters *i.e.* stress and strain are obtained by dividing with constant quantities respectively. Actually as the load is slowly or gradually increased, the area of cross-section will continuously go on reducing (poissons ratio effect, see chapter 2). Similarly at any instant we take the extension with respect to the original length ℓ_o. But actually the stretch at any particular instant occurs over the current length. Thus, if we consider the current area of cross-sections and divide the current load with it, we will get the "TRUE STRESS". Likewise the change in length over the current length should be taken into account for calculating the TRUE STRAIN.

In the light of above, if P is the current load and 'd' is the current diameter of the specimens then,

$$\text{True stress } \sigma = \frac{4P}{\pi d^2}$$

Similarly, if dl is the elemental elongation in current length l, at which point the current load is P, then the incremental strain $d\varepsilon$ is defined as,

$$d\varepsilon = \frac{dl}{l}$$

Integrating this equation we get the total current strain ε,

$$\int_0^{\varepsilon} d\varepsilon = \int_{l_0}^{l} \frac{dl}{l}$$

i.e.,

$$\varepsilon = \ln\left(\frac{l}{l_0}\right)$$

Thus, at any current load P the true strain will be,

$$\varepsilon = \ln\left(\frac{l}{l_0}\right)$$

In view of the logarithm value of (l/l_0) true strain is also called logarithmic strain or natural strain and associated with this are the names True stress or natural stress of the value $\frac{4P}{\pi d^2}$

The true stress-strain curve will be a monotonically rising curve and it will be higher than the nominal or Engg. Stress strain curve as shown in Fig. (8.8C)

Upto the elastic limit point, these curves are very near to each other. However, after elastic limit point, the true stress-strain curve is much higher than nominal stress-strain curve.

8.4.4 Yield Envelope and Bauschinger Effect

In Fig. (8.8*c*), portion *AK* of the stress-strain curve is known as the plastic portion of the curve. If the stress on specimen corresponding at any point *C* in this portion of curve σ_C is reduced gradually to zero value, this reverse stress-strain curve will be a straight line *CD* parallel to the original elastic line *OA*. This phenomenon was observed by Bauschinger and is known as *Bauschinger effect*. At the point *D* the stress will be zero but residue strain *OD* will be left in the specimen. This strain *OD* is also known as the 'plastic strain' component, ε_p, of the total strain *EC* at *C* which is equal to *OL* the portion *DL* is called the elastic component, ε_e, of total strain.

Thus,
$$\varepsilon_C = OL = OD + DL$$
$$= \varepsilon_p + \varepsilon_e$$

Once again if we re–start increasing the stress on the specimen, it will follow the path ***DC*** and then along the curve *EK*. In other words *C* becomes the new yield point and the corresponding stress, σ_c, the new yield stress. Similarly, if this process is repeated when the stress level is σ_E then *E* will yet become another new yield point stress, This will continue upto the point *K*, which is the ultimate stress point.

Therefore, the portion *AK* is also known as "yield envelope". Another name assigned to the curve *AK* is strain hardening portion of the stress-strain curve.

8.4.5 Ductility and Malleability

Ductility hitherto, had been defined as the property of the material due to which it could be drawn into wire. It was very well known to the people that gold is more ductile than copper, Copper is more ductile than aluminium and so on. That is, this term was known qualitatively and had not been difined quantitatively. It was in late fifties of the last century that during an international meet, the scientists put their stamp on the proper definition of this term. According to this, the strain to which a uniaxially stressed specimen can be subjected prior to instability point or ultimate stress point is called the ductility of the material. That is *OH* (Fig. 8.8 *c*) is the measure of ductility.

Malleability

Similarly, the malleability which is the property due to which the material can be deformed into plate in quantitatively defined as the strain at instability point for that particular material.

8.5 UNIFORM STRESS

To generate a uniform stress over a section, the force must be applied at its centroid. To prove this statement, assume a section shown in Fig.. 8.10, with the centroid at C, under a uniform stress σ. Let the total force acting on the section be P.

If *dA* be an elemental area at a distance *r* from *C*.

Then, we have $\int \sigma \, dA = P$ ...(8.1)

Taking moments of elemental forces about the centroid, we have, if *P* is at a distance *a* from *C*

$$P.a = \int r\sigma \, dA = \sigma \int r \, dA \quad \text{(as } \sigma \text{ is constant)}$$

$= \sigma \times$ First moment of area about centroid.

$= 0.$

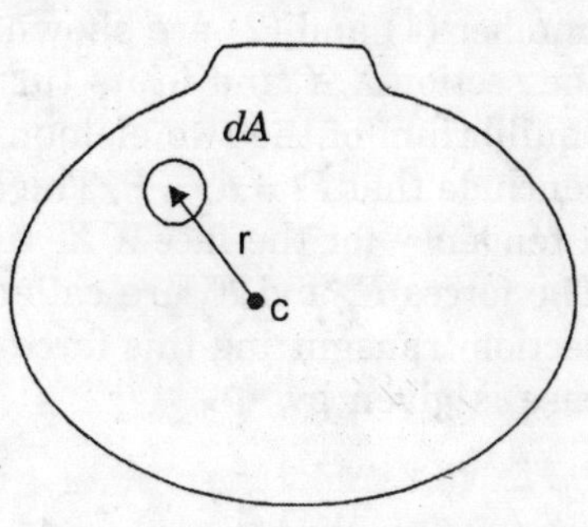

Fig. 8.10

Now since P is not zero, therefore,

$$a = 0.$$

Thus, the line of action of P must pass through the centroid of the section.

8.6 SHEAR STRESS AND STRAIN

Consider a rectangular block $ABCD$ (Fig. 8.11) of dimensions $a \times b$ and thickness t. Let it be subjected to two equal and opposite forces P on the faces AB and CD as shown. Taking a section XX at a height 'y' from the face DC the free body diagrams of the two elements

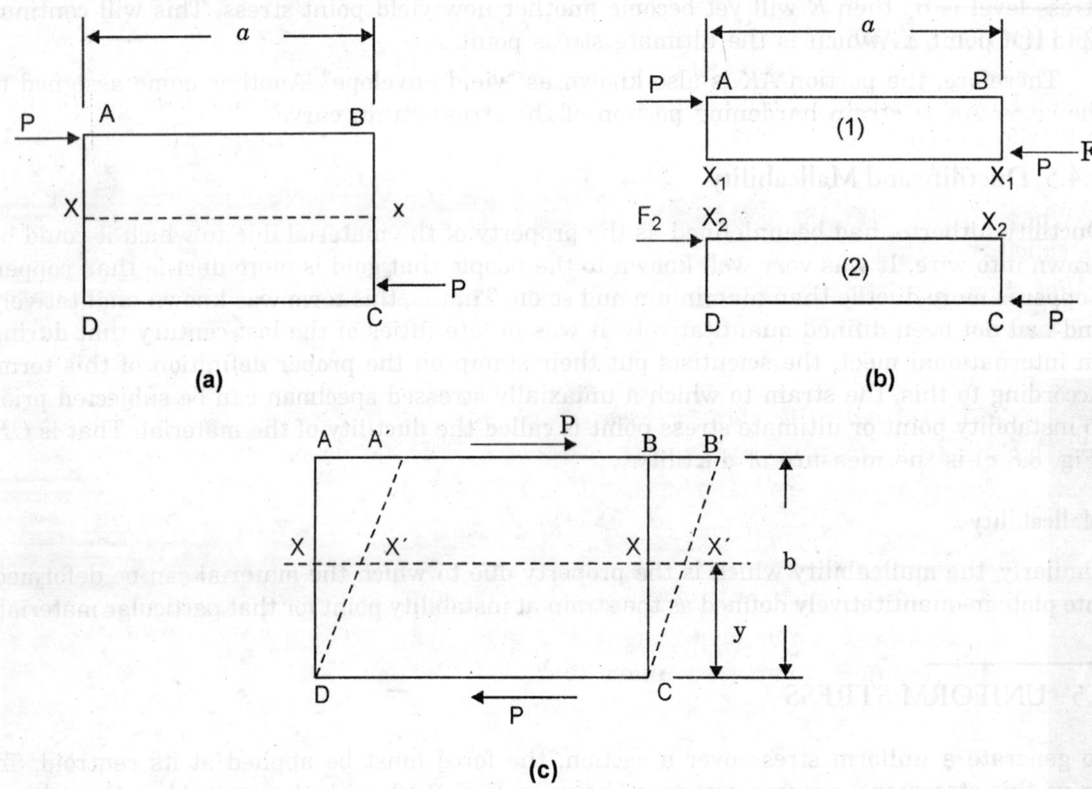

Fig. 8.11

number (1) and (2) are shown in Fig. 8.11(b). The face X_1X_1 offers a cohesive force F_2 on to the section X_2X_2 and in its turn the face X_2X_2 gives a reaction F_1 onto the face X_1X_1. For the equilibrium of the two elements separately which follows from that of the original block, we conclude that $F_1 = F_2 = P$. The effect of the force F_1 and F_2 on the respective faces is to induce a tendency for the face X_1X_1 to slide over the face X_2X_2 in the direction of the arrowhead F_2. The forces F_1 and F_2 are called the shearing forces and the force per unit area of the cross-section transmitting this force is known as the shear stress. It is represented by τ and in this case is given by,

$$\tau = F_1 / \text{at} \ (= P/at) \qquad ...(8.2)$$

It is evident that the shear stress, in this case, is independent of the distance of the section XX from CD. That is shear stress on any horizontal section will be constant and that section will have a tendency to slide to the right with respect to its adjacent plane immediately below it. The cumulative effect will be to displace the face *AB* to the position *A'B'* prime with respect to *DC*. Likewise then, section *XX* will occupy the position *X'X'* such that at any height *y* from *DC* the displacement at right angles to *y* is *XX'*. This displacement *XX'* over unit height is called the shear strain and is represented by symbol ϕ Fig. 8.11(*c*) such that,

$$\phi = \frac{XX'}{Y} = \tan ADA' \simeq \text{angle } ADA' \qquad ...(8.3)$$

That is, the shear strain is an angular displacement measured in radian and is constant for any value of *y* between zero and *b*.

8.7 HOOKE'S LAW

Hooke's Law states that within elastic limit stress is proportional to strain. Mathematically, for tensile or compressive stress,

$$\sigma/\varepsilon = E \qquad ...(8.4)$$

Where *E* is constant for material and is known as the Young's modulus of elasticity and ε is called the strain.

For shear stress this ratio is denoted by *G* or *C* and is called the modulus of rigidity.

Thus,

$$G = \tau/\phi \qquad ...(8.5)$$

EXAMPLE

8.1 : *A bar of uniform cross-sectional area 100 mm² carries forces in Newtons as shown in Fig. 8.12. Calculate the relative movement of end A with respect to D. E = 200 × 10³ MN/m².*

Solution : Taking a section anywhere in between *A* and *B*, it is found by drawing the free body diagram and considering the equilibrium of their element *AX*, or X'_1BCD and taking the algebraic sum of the forces along *X*-direction and equating it to zero, we get (Fig. 8.12*b*) where *F*, is then assumed compressive force transmitted by the section *X, X,*

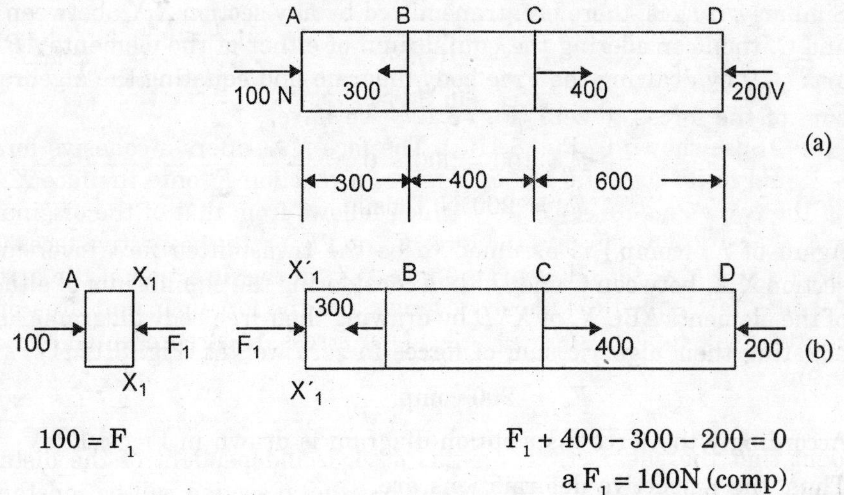

$100 = F_1$ $\qquad$ $F_1 + 400 - 300 - 200 = 0$

a F_1 = 100N (comp)

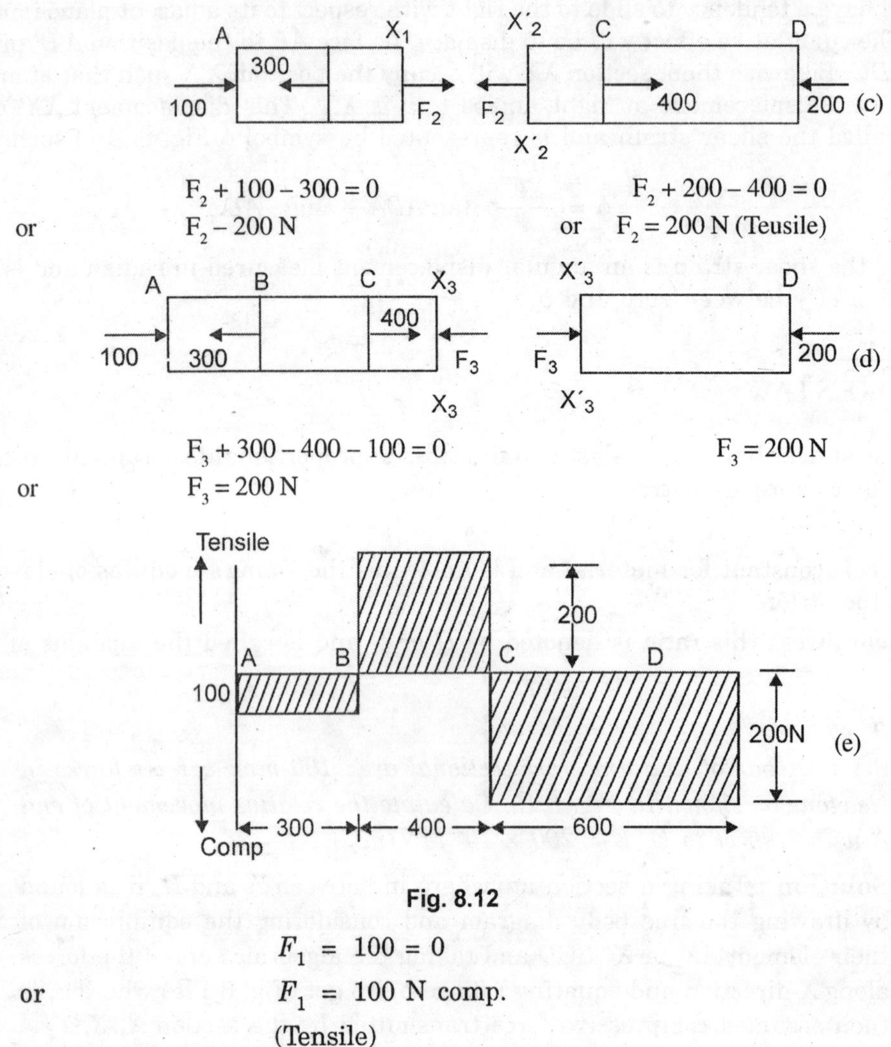

Fig. 8.12

$$F_1 = 100 = 0$$

or $$F_1 = 100 \text{ N comp.}$$

(Tensile)

Similarly, if F_2 is, therefore, transmitted by any section X_2X_2, between B and C, then considering the equilibrium of either of the elements ABX_2 or X'_2CD by drawing the Free body diagram and equating the algebraic sum of the forces to zero (Fig. 8.12*c*) we have,

$$F_2 + 100 - 300 = 0$$

$$F_2 = 200 \text{ N Tensile}$$

Again of F_3 (comp.) is assumed to be the transmitted force over any section X_3X_3 between C and D, then considering the equilibrium of either of the elements ABC X_3 or X'_3D by drawing their free body diagrams and equating their algebric sum of forces to zero we get (Fig. 8.12*d*),

$$F_3 = 200 \text{ comp.}$$

Accordingly the force distribution diagram is drawn in Fig. 8.12(*e*).

Thus, the respective deformations are,

$$\delta l_{\text{AB}} = \frac{1}{E}\sigma_{AB}.l_{AB} = \frac{1}{200\times10^3}\times\frac{100}{100}\times300$$

$$= 0.0015 \text{ mm decrease}$$

$$\delta l_{\text{AB}} = \frac{1}{E}\sigma_{BC}.l_{BC} = \frac{1}{200\times10^3}\times\frac{200}{100}\times400$$

$$= 0.004 \text{ mm increase}$$

$$\delta l_{\text{AB}} = \frac{1}{E}\sigma_{CD}.l_{CD} = \frac{1}{200\times10^3}\times\frac{200}{100}\times600$$

$$= 0.006 \text{ mm decrease}$$

The movement of 'A' w.r.t. 'D'

$$= .004 - (.0015 + .006) = -0.0035 \text{ mm}$$

That is, the points A and D will come closer to each other by 0.0035 mm.

8.2 : *A turn buckle is used to connect two steel bars each effectively 3 m long. One of the bars is 50 mm dia and has* $1\frac{4}{5}$ *threads per cm over its end, while the other is 75 mm dia with* $1\frac{2}{5}$ *threads per cm on its end. If there is a small tension initially in the rods, find the increase in tension when the turn buckle is tightened through half a turn.*

Take $E = 200\ GPa = 200 \times 10^9\ N/m^2$.

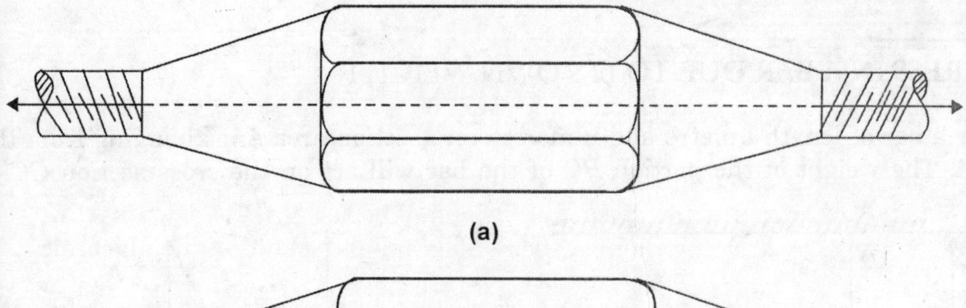

(a)

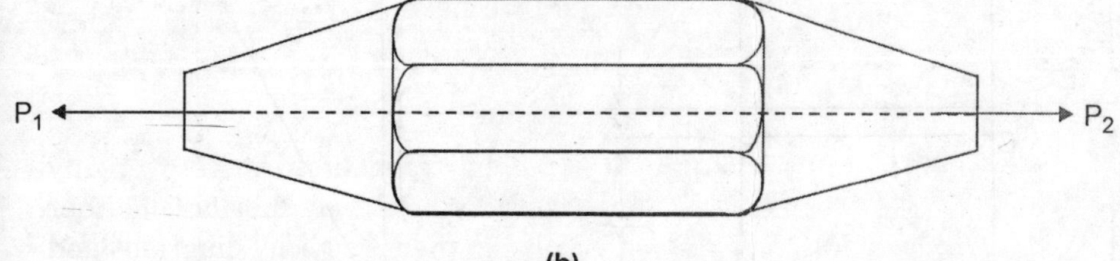

(b)

Fig. 8.13

Solution : Fig. 8.13(b) shows the free body diagram of the turn buckle with P_1 and P_2 respectively as the reactions of the two bars. Evidently for the equilibrium of the turn buckle,

$$P_1 = P_2 = P(\text{say}) \quad ...(8.6)$$

Where P is the increase in tension required to be evaluated.

Total stretch in the bars = sum of the movements of the ends due to rotation of turn buckle

$$= \left(\frac{5\times 10}{9}\times\frac{1}{2}\times\frac{5\times 10}{7}\times\frac{1}{2}\right) \text{ mm}$$

$$= \frac{25\times 16}{63} \text{ mm} \quad ..(8.7)$$

If ε_1 and ε_2 are the respective strains in the two bars, the total stretch is also given by $(\varepsilon_1 l_1 + \varepsilon_2 l_2)$, where l_1 and l_2 are the respective lengths of the bars. Thus, with the help of equations (1.6) and (1.7), we have

$$\frac{25\times 16}{63} = \varepsilon_1 l_1 + \varepsilon_2 l_2$$

$$= \frac{P_1}{A_1 E} l_1 + \frac{P_2}{A_2 E} l_2$$

$$= \frac{P}{E} l_1 \left(\frac{1}{A_1} + \frac{1}{A_2}\right) \quad \text{as } l_1 = l_2$$

$$= \frac{P\times 3000}{200\times 10^3}\left[\frac{4}{\pi(50)^2} + \frac{4}{\pi(75)^2}\right]$$

Thus, $P = 0.575$ MN.

EXAMPLE

8.8 STRESS IN A BAR DUE TO ITS OWN WEIGHT

Consider a bar of length l metre and uniform cross-section area A m² hanging from the top, Fig. 8.14. The weight of the portion BC of the bar will act on the cross section CC.

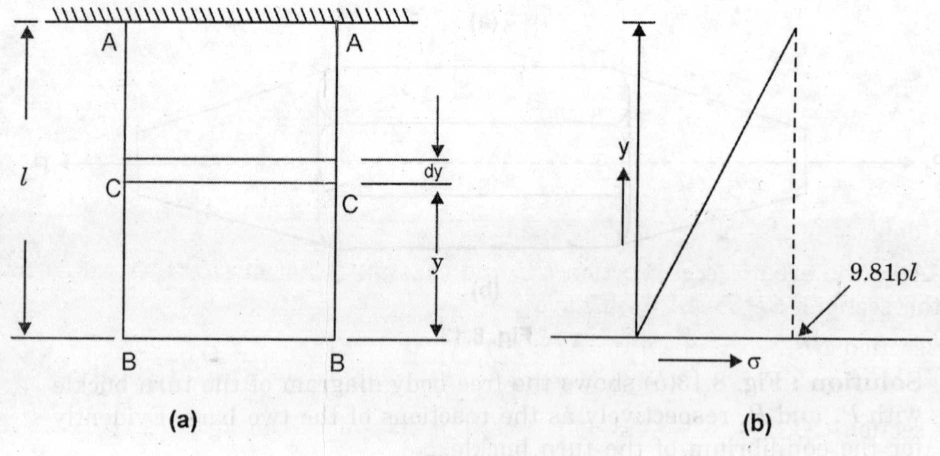

Fig. 8.14

If ρ kg/m³ is the density of the material, the stress at a height y metre from B is given by,

$$\sigma = \frac{9.81 \times Ay\rho}{A}$$

$$= 9.81\ \sigma y\ \text{N/m}^2 \quad \text{...(8.8)}$$

Evidently the stress rises linearly with y and

$$\sigma_{m\ ax} = \sigma_A = 9.81\ \sigma l\ \text{N/m}^2 \text{ or Pa} \quad \text{...(8.9)}$$

The stretch dl in the length dy will be given by,

$$dl = \frac{\sigma}{E} dy \quad \text{...(8.10)}$$

Where E is the Young's modulus of elasticity.

Total elongation in the bar

$$= \int_0^l (\sigma / E)\, dy = \int_0^l \frac{9.81}{E} \rho y dy$$

$$= \frac{9.81}{2E} \rho l^2 \quad \text{...(8.11)}$$

8.8.1 Tie Bar of Uniform Strength

A tie bar of uniform strength carrying a load of 'P' Newton is shown hanging from the top in the Fig. 1.8. Let the stress everywhere by σ N/m².

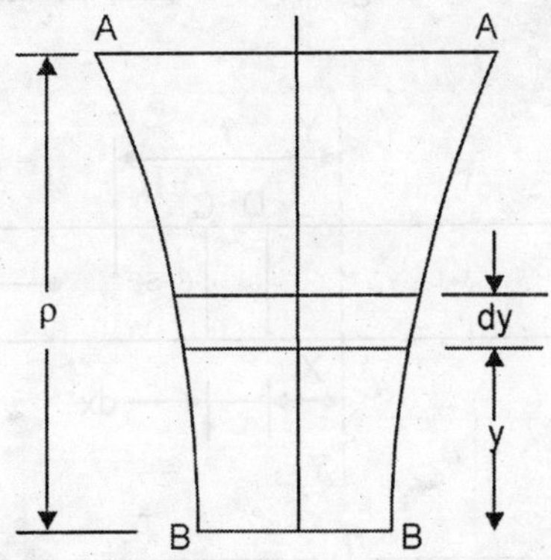

Fig. 8.15

The density of the material is ρ kg/m³.

Let A be the area of cross-section CC and $(A + dA)$ that of DD. If A varies from A_1 to A_2 for the sections BB to AA ; we have

For section BB, $\quad A_1\sigma = P$

or $\quad \sigma = P/A_1 \quad$ ...(8.12)

For section CC, $\quad A\sigma = P + 9.81 \times$ mass of BC in kg $\quad$ ...(8.13)

For section DD,

$$(A + dA)\,\sigma = P + 9.81 \text{ mass of portion } BC \text{ in kg} + 9.81(\text{mass of portion } CD \text{ in kg})$$

$$= A\sigma + 9.81\ (\text{mass of portion } CD \text{ in kg})$$

...from equation (8.13)

That is $$A\sigma + dA\sigma = A\sigma + 9.81 \times (Ady)\rho$$

or $$\frac{dA}{A} = \frac{9.81\,\rho}{\sigma}dy \qquad ...(8.14)$$

Integrating $$[\ln A]_{A_1}^{A} = \left[\left(\frac{9.81\,\rho}{\sigma}\right)y\right]_0^y$$

or $$\ln(A/A_1) = 9.81\ \rho y/\sigma$$

or $$A = A_1\,e^{9.81\ \rho y/\sigma}$$

At $$y = l\ ,\ A = A_2$$

Therefore, $$A_2 = A_1 e^{\left[\frac{9.81\,\rho \ell}{\sigma}\right]} \qquad ...(8.15)$$

8.9 STRESS IN A BAR DUE TO ROTATION

Fig. 8.16 shows a bar of length l revolving about Y-axis at an angular speed of omega rad/sec.

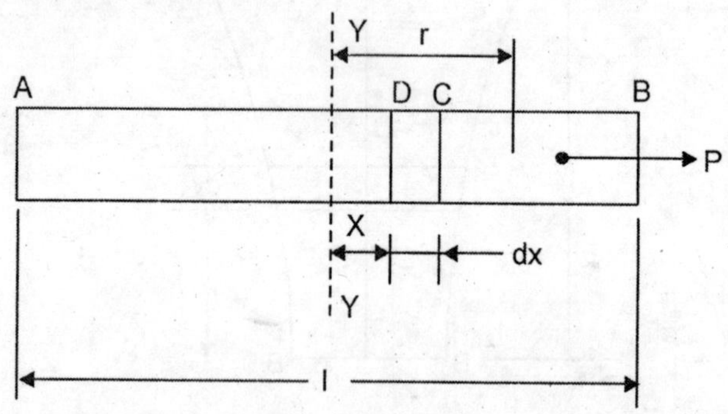

Fig. 8.16

Tensile force on the element DC = Centrifugal force P on the part CB = $\omega^2\,rM$, where M is the mass of part CB.

Thus, if A is the area of cross-section (in m^2) of the bar and ρ the density of the material (in kg/m^3) then

$$P = \rho A\left(\frac{\ell}{2} - x\right)\omega^2\left[x + \frac{1}{2}\left(\frac{l}{2} - x\right)\right]$$

or $$P = \frac{1}{2}\rho A\omega^2\,(l^2/4 - x^2)\text{ Newton}$$

i.e. $$\sigma_c = \frac{1}{2}\rho\omega^2\,(l^2/4 - x^2)\text{ N/m}^2 \qquad ...(8.16)$$

Maximum stress will occur at

$$x = 0$$

i.e. $$\sigma_{max} = \frac{1}{8}\rho\omega^2 l^2 \qquad ...(8.17)$$

Extension of the element $DC = \frac{\sigma_c}{E}dx$ ...(8.18)

Thus, substituting the value of σ_c from equation (1.16) and integrating, the total extension in the bar is given by,

$$dl = 2\int_0^{\frac{\ell}{2}} \frac{1}{2E}\rho\omega^2\,(l^2/4 - x^2)\,dx$$

$$= \frac{\rho\omega^2}{E}\left[\frac{l^2}{4}x - \frac{x^3}{3}\right]_0^{\ell/2}$$

$$= \rho\omega^2\, l^3/(12\,E)\text{ metre.}$$

8.10 COMPOSITE BARS

A load carrying member which is made up of two different materials is called a composite or compound bar.

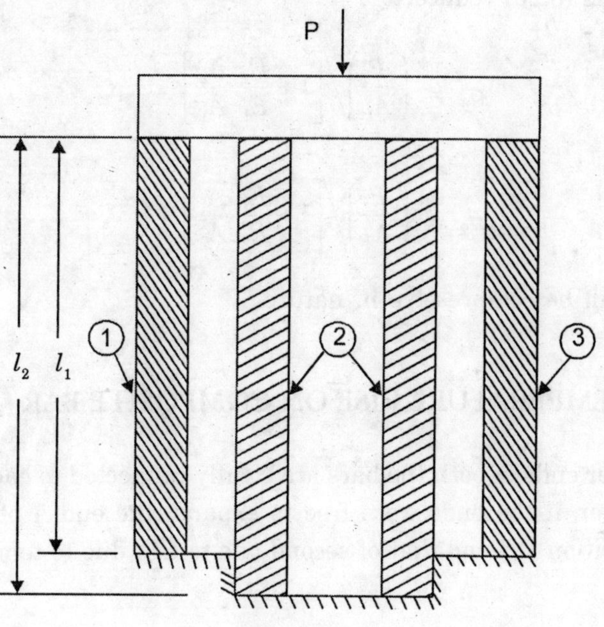

Fig. 8.17

In Fig. 8.17 the two materials having the cross-sectional areas A_1 and A_2, the corresponding moduli of elasticity as E_1 and E_2 constitute a composite bar.

8.10.1 Composite Bar under the Effect of External Load

If the upper ends are rigidly connected to each other, the compression in each will be same,

i.e. $$\delta l_1 = \delta l_2$$

or $$\frac{\sigma_1}{E_1} l_1 = \frac{\sigma_2}{E_2} l_2 \qquad \text{...(8.19)}$$

Also, the applied load will be equal to the sum of loads sustained by each part

i.e. $$P_1 + P_2 = P$$

or $$\sigma_1 A_1 + \sigma_2 A_2 = P \qquad \text{...(8.19)}$$

Then σ_1 and σ_2 can be evaluated by solving the simultaneous equations (8.18) and (8.19) to give,

$$\sigma_1 = \left(\frac{P}{A_1}\right) \Big/ \left[1 + \frac{E_2 l_1}{E_1 l_2} \frac{A_2}{A_1}\right] \qquad \text{...(8.20}a\text{)}$$

$$\sigma_2 = \left[\frac{P}{A_2}\right] \Big/ \left[1 + \frac{E_1 l_2}{E_2 l_1} \cdot \frac{A_1}{A_2}\right] \qquad \text{...(8.20}b\text{)}$$

If $l_1 = l_2$, equations (8.20) reduce to

$$\sigma_1 = \left(\frac{P}{A_1}\right) \Big/ \left[1 + \frac{E_2}{E_1} \frac{A_2}{A_1}\right] \qquad \text{...(8.21}a\text{)}$$

and $$\sigma_2 = \left(\frac{P}{A_2}\right) \Big/ \left[1 + \frac{E_1}{E_2} \frac{A_1}{A_2}\right] \qquad \text{...(8.21}b\text{)}$$

Both σ_1 and σ_2 will be compressive in nature.

8.11 EFFECT OF TEMPERATURE RISE ON COMPOSITE BAR

To start with, the upper ends of both the bars are rigidly connected to each other so that they are at level *A*. However, if the ends were free to expand, the end *A* of the first bar would have moved to the position *BB* and that of second one to *CC*, due to temperature rise of *t*°C, (Fig. 8.18) such that,

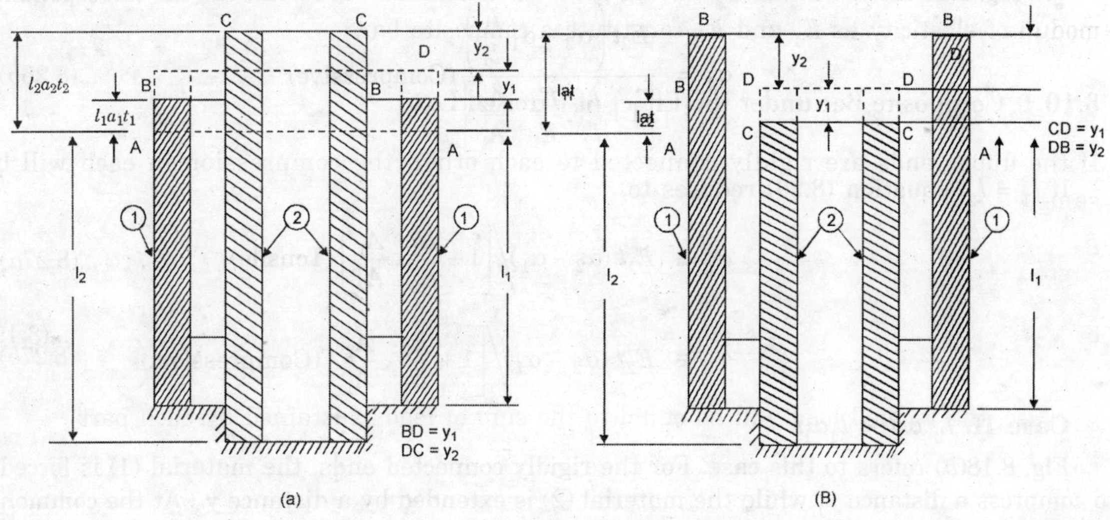

Fig. 8.18

$$AB = l_1 \alpha_1 t \qquad \text{...(8.22)}$$

$$AC = l_2 \alpha_2 t \qquad \text{...(8.23)}$$

where α_1 and α_2 are the respective coefficients of linear thermal expansion.

Case I. $l_2 \alpha_2 t > l_1 \alpha_1 t$

Since the two ends are rigidly connected to each other material (1) is forced to extend through a distance y_1 to the level D while the material (2) is forced to compress through a distance y_2 to the level D Fig. 8.11(a)

Geometrically, from the Fig. (8.11a),

$$y_1 + y_2 = (l_2\alpha_2 - l_1\alpha_1)t$$

or

$$\frac{\sigma_1}{E_2} l_1 + \frac{\sigma_2}{E_2} l_2 = (l_2\alpha_2 - l_1\alpha_1) \qquad \text{...(8.24)}$$

Where α_1 and α_2 are respectively the stresses in the two materials.

Also, as no external force is applied at the level D when equilibrium is attained :

Tensile force in material (1) = Compressive force in material (2)

i.e.

$$P_1 = P_2$$

$$\sigma_1 A_1 = \sigma_2 A_2 \qquad \text{...(8.25)}$$

Solving the simultaneous equations (8.24) and (8.25), we get,

$$\sigma_1 = \frac{E_1 t(\alpha_2 - \alpha_1 l_1 / l_2)}{\left[\frac{l_1}{l_2} + \frac{E_1}{E_2} \cdot \frac{A_1}{A_2}\right]} \text{(Tensile)} \qquad \text{...(8.26}a\text{)}$$

$$\sigma_2 = \frac{E_2 t\left(\alpha_2 - \alpha_1 \frac{l_1}{l_2}\right)}{\left[1 + \frac{l_1}{l_2} \cdot \frac{E_2}{E_1} \cdot \frac{A_2}{A_1}\right]} \text{(Compressive)} \qquad \text{...(8.26}b\text{)}$$

If $l_1 = l_2$, equation (8.26) reduces to,

$$\sigma_1 = E_1 t\,(\alpha_2 - \alpha_1) \Big/ \left[1 + \frac{E_1}{E_2} \cdot \frac{A_1}{A_2}\right] \text{(Tensile)} \qquad \text{...(8.27}a\text{)}$$

$$\sigma_1 = E_2 t\,(\alpha_2 - \alpha_1) \Big/ \left[1 + \frac{E_2}{E_1} \cdot \frac{A_2}{A_1}\right] \text{(Compressive.)} \qquad \text{(8.27}b\text{)}$$

Case II. $l_2\ \alpha_2 t < l_1 \alpha_1 t$

Fig. 8.18(*b*) refers to this case. For the rigidly connected ends, the material (1) is forced to compress a distance y_1 while the material (2) is extended by a distance y_2. At the common level *DD* the tensile force experienced by material (2) will be equal to the magnitude of the compressive force in material (1).

Thus, $$y_1 + y_2 = t\,(l_1\ \sigma_1 - l_2\ \alpha_2\ )$$

i.e. $$\frac{\sigma_1 l_1}{E_1} + \frac{\sigma_2 l_2}{E_2} = t\,(l_1\alpha_1 - l_2 a_2) \qquad \text{...(8.28)}$$

Equation (1.25) holds good here also. Solving the equations (1.25) and (1.28) for σ_1 and σ_2 we get,

$$\sigma_1 = \frac{E_1 t\left(\alpha_1 - \frac{l_2}{l_1}\alpha_2\right)}{\left[1 + \frac{E_1 l_2 A_1}{E_2 l_1 A_2}\right]} \text{(Compressive)} \qquad \text{...(8.29}a\text{)}$$

$$\sigma_1 = \frac{E_2 t\left[\alpha_1 - \left(\frac{l_2}{l_1}\alpha_2\right)\right]}{\left[1 + \frac{E_1 A_2}{E_2 A_1}\right]} \text{(Tensile)} \qquad \text{..(8.30}a\text{)}$$

If $l_1 = l_2$ these reduce to

$$\sigma_1 = \frac{E_1 t(\alpha_1 - \alpha_2)}{\left[1 + \frac{E_1}{E_2} \cdot \frac{A_1}{A_2}\right]} \text{(Compressive)} \qquad \text{...(8.29}b\text{)}$$

$$\sigma_1 = \frac{E_2 t(\alpha_1 - \alpha_2)}{\left[1 + \frac{E_1 A_2}{E_2 A_1}\right]} \text{(Tensile)} \qquad \text{...(8.30}b\text{)}$$

8.12 EFFECT OF BOTH THE TEMPERATURE RISE AND EXTERNAL LOAD

The stresses due to individual effects are evaluated as per the method explained above. The resultant stresses are then calculated by adding the stresses algebraically in each material.

EXAMPLE 8.3 : *A steel bar of uniform diameter of 40 mm is heated to 80°C and then clamped at the ends with the help of two fixtures 4 m apart, and left to cool down to room temperature of 20° C at which temperature the distance between the fixtures was found to be 1.0 mm shorter than that at 80° C. Determine the stress in the bar, when it has cooled down to room temperature and the reaction at the fixtures.*

Take $E = 200\ GN/m^2$ and

$\alpha = 1.1 \times 10^{-5}$/deg.C.

Solution : $l = 4000$mm

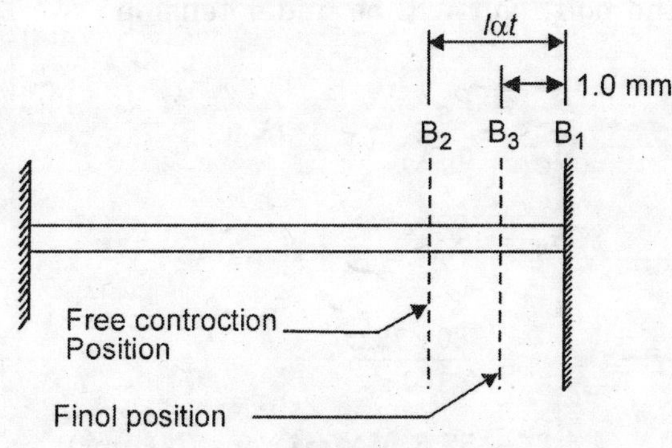

Fig. 8.19

Temperature drop $t = 80 - 20 = 60°C$

B_1B_2 = Free contraction $= l\ \alpha\ t$

$= 4000 \times 1.1 \times 10^{-5} \times 60 = 2.64$ mm

Therefore, B_2B_3 = Stretch in the bar

$= B_1B_2 = - B_1B_3$

$= 2.64 - 1.0 = 1.64$ mm

Thus, the stress in the bar is given by,

$$\sigma = \frac{E}{l} \times 1.64$$

$$= \frac{200 \times 10^3}{4000} \times 1.64$$

$= 82\ MN/m^2$ or N/mm^2 (Tensile)

Reaction of Supports $= \sigma \times A = 82 \times \frac{\pi}{4}(40)^2$

$= 103 \times 10^3$ N

$= 103$ kN.

8.4 : *A copper sleeve, 21 mm internal and, 27 mm external diameter, surrounds a 20 mm steel bolt, one end of the sleeve being in contact with the shoulder of the bolt. After putting a rigid washer on the other end of the tube, a nut is screwed on the bolt until the compressive stress in the sleeve is 80 MN/m². Find the range of the external axial load that can be applied to this assembly, if the stress in the bolt is never to be compressive and in the sleeve never to be zero. Take* $E_S = 200\ GN/m^2$ *and* $E_C = 90\ GN/m^2$.

Solution :

$$A_S = \frac{\pi}{4} \times 20^2$$

$$= 10\pi \text{ mm}^2;$$

$$A_C = \frac{\pi}{4}(27^2 - 21^2)$$

$$= 72\ \pi\ \mu\mu^2$$

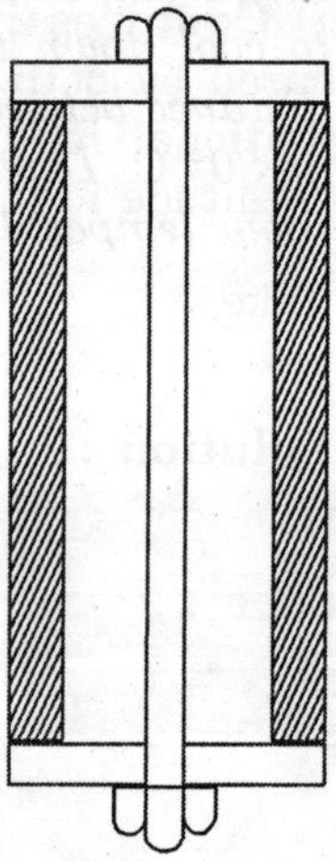

When the tube is in compression due to the tightening of the bolt, bolt will be under tension end

$$P_C = P_S$$

or

$$\sigma_C A_C = \sigma_S A_S$$

or

$$\sigma_S = \frac{\sigma_C A_C}{A_S}$$

$$= \frac{80 \times 72\pi}{100\ \pi}$$

$$= 57.6 \text{ MN/m}^2 \text{ or } (\text{N/mm}^2) \qquad ...(8.31)$$

$$P_S = P_C = 57.6 \times 100 \times \pi$$

$$= 18.1\text{kN} \qquad ...(8.32)$$

$$\varepsilon_S = \frac{\sigma_S}{E_S}$$

$$= \frac{57.6}{200 \times 10^3}$$

$$= 2.88 \times 10^{-4} \text{ (Tensile)} \qquad ...(8.33)$$

$$\varepsilon_C = \frac{\sigma_C}{E_C} = \frac{80}{90 \times 10^3}$$

$$= 8.89 \times 10^{-4} \text{ (Compressive)} \qquad ...(8.34)$$

Fig. 8.20

(*b*) Two cases can arise depending on whether the load, compressive or tensile, is applied on the bolt or on the tube.

(*i*) **Tensile load on the bolt :** In this case the limiting condition will be when the compressive stress already existing in the tube is just brought down to zero. This will happen when the bolt is stretched further to the extent that the additional stretch or strain induced in it is equal to the

E X A M P L E

compression or compressive strain in tube as calculated in equation (8.34) (as these lengths are equal)

$$\varepsilon_{S_2} = \varepsilon_C = 8.89 \times 10^{-4} \quad ...(8.35)$$

$$P_2 = P_{S_2} = A_S E_S \quad ...(8.36)$$

$$= 100\,\pi \times (200 \times 10^{-4})\,(8.89 \times 10^{-4})\text{N}$$

$$= 5.9 \text{ N (Tensile)} \quad ...(8.37)$$

(*ii*) **Compressive load on the bolt :** Here the applied load will be shared by both bolt and tube and the limiting case will be when the additional induced compressive stress in the bolt becomes equal in magnitude to the existing tensile stress in the bolt as given by equation (8.31).

i.e.
$$\sigma_{S_3} = \sigma_S \quad ...(8.38)$$

$$P_3 = P_{S_3} + P_{C_3} \quad ...(8.39)$$

and
$$\varepsilon_{S_3} = \varepsilon_{C_3}$$

or
$$\frac{\sigma_S}{E_S} = \frac{p_{C_3}}{A_C E_C}$$

i.e.
$$P_{C_3} = \sigma_S A_C E_C / E_S \quad ...(8.40)$$

Thus, substituting values in equation (8.39),

$$P_3 = \sigma_S A_S + \sigma_S A_C / E_S$$

$$= \sigma_S [A_S + A_C E_C / E_S]$$

$$= 57.6 \left[100\pi + 72\pi \frac{90}{200}\right]$$

i.e
$$P_3 = 24 \text{ kN (Compressive)} \quad ...(8.41)$$

(*iii*) **Tensile load on tube :** Here again the load will be shared by both the bolt and the tube and the stress in the tube will be the limiting case and the additional tensile stress induced in the tube will become equal to 80 N/mm^2 or 80 MN/m^2.

i.e.
$$\sigma_{C_4} = 80 \text{ N/mm}^2 \quad ...(8.42)$$

$$P_4 = P_{S_4} + P_{C_4} \quad ...(8.43)$$

$$\varepsilon_{54} = \varepsilon_{C_4}$$

or
$$\sigma_{S_4} / E_S = \sigma_{C_4} / E_C$$

Thus,
$$P_4 = \sigma_{S_4} A_S + \sigma_{C_4} A_C$$

$$= \sigma_{C_4}\left[A_C + \frac{E_S}{E_C}A_S\right]$$

$$= 80\left[72\pi + \frac{200}{90}\times 100\pi\right] \text{N}$$

i.e. $P_4 = 73.9$ kN (Tensile) ...(8.44)

(*iv*) **Compressive load applied on tube :** In this case the load with be born only by tube and the limiting condition will be when the stress in bolt reduces to zero. This will happen when the compressive strain induced in tube due to external load is just equal in magnitude to the strain ε_S in bolt as given in equation (8.33).

Thus, $\varepsilon_{C_5} = \varepsilon_S$ ...(8.45)

$$P_5 = A_{C_5} A_C$$

$$= 72 \times (2.88 \times 10^{-4}) \times (90 \times 10^3) \text{ N}$$

i.e. $P_5 = 5.86$ kN (Compressive) ...(8.46)

From equations (8.37), (8.41), (8.44) and (8.46), it is evident that the given conditions to be fulfilled the allowable range of the external load is from 5.86 kN (Compressive) to 55.9 kN (tensile).

8.5 : *If instead of applying the external load to the bolt and tube assembly of Example 8.4, the temperature is raised through 50°C, find the resultant stresses in the bolt and the tube.*

Take $\alpha_{steel} = 11 \times 10^{-6}/°C$

$\alpha_{copper} = 18 \times 10^{-6}/°C$.

Solution : In this case since the lengths are equal, we have from equation (8.24),

$$\frac{\sigma_S}{E_S} = \frac{\sigma_S}{E_S} + \frac{\sigma_C}{E_C} = t\,(\alpha_C - \alpha_S) \quad ...(8.47)$$

Where t is the temperature rise.

Also, in equilibrium state the compressive load in copper will be equal to tensile load in bolt, due to the temperature rise effect.

i.e $\sigma_S A_S = \sigma_c A_c$

or $\sigma_S = (72/100)\sigma_c$ ...(8.48)

Substituting equation (8.48) in equation (8.47),

$$\sigma_c\left[\frac{0.72}{E_S} + \frac{1}{E_C}\right] = t(\alpha_c - \alpha_S)$$

EXAMPLE

E X A M P L E

or $\sigma_c\left(\dfrac{0.72}{200\times10^3}+\dfrac{1}{90\times10^3}\right) = 50\ (7 \times 10^{-6})$

$= 23.8\ \text{N/mm}^2$ or 23.8 MPa (Comp.) ...(8.49*a*)

and from eqn. (1.48),

$$\sigma_S = .72 \times 23.8$$

$= 17.2\ \text{MN/m}^2$ or 17.2 MPa (Tensile) ...(8.49*b*)

Thus, from equations (8.31) and (8.49)

Resultant stress in bolt

$$= 57.6 + 17.2 = 74.8\ \text{MN/m}^2 \text{ (Tensile)}$$

Resultant stress in tube

$$= 80.0 + 23.8 = 103.8\text{MPa (Comp.)}$$

8.6 : *Two bars one of aluminium and the other of steel, joined end to end, are placed lengthwise between two rigid supports 600 mm apart. Their axes are horizontal and are along the same straight line. Aluminium bar is 200 mm long and steel bar is 400 mm long. Their cross-sectional areas are 1000 mm² and 200 mm² while Young's modulii are 0.6 × 10⁶ kg/cm² (588.6 MN/m²) and 2 × 10⁶ kg/cm² (1962 MN/m²) respectively. The whole system is in un-stressed condition at 10°C. The system is then heated to a temperature of 60°C. Find the stresses developed in the two materials if,*

$$a_{\text{steel}} = 12 \times 10^6/°C$$

and $\alpha_{al} = 24 \times 10^{-6}/°C$ (Panjab University)

Solution : If the composite bar were free to expand, the corresponding increase in length of the two portions will be added up to give the total increase in length δl as

$$\delta l = \delta l_s + \delta l_a$$
$$= l_s\, \alpha_s\, (T_2 - T_1) + l_a\, \alpha_a\, (T_2 - T_1)$$
$$= (T_2 - T_1)\,[\alpha_s l_s + \alpha_a l_a]$$
$$= (60 - 10)\,(12 \times 10^{-6} \times 400 + 24 \times 10^{-6} \times 200)$$
$$= 50 \times 9600 \times 10^{-6}$$
$$= 0.48 \text{ mm.}$$

Since both the metals have the same axis, they will carry the same compressive axial load, P, if the free expansion is checked by rigid supports. Thus, the compression in both the metals due to the load P will add upto *dl*

i.e., $$\frac{P}{A_s S_s} l_s + \frac{P}{A_a E_a} l_a = \delta l$$

or $$P\left[\frac{400\times10^2}{(200)\,(2\times10^6)}+\frac{200\times10^2}{(1000)\,(0.6\times10^6)}\right] = 0.48$$

or $$P = 11076.9 \text{ kg}$$

Therefore, $$\sigma_s = \frac{P}{A_S} = \frac{110076.9}{200} \times 10^2$$

$$= 553.85 \text{ kg/ cm}^2 \text{ (comp.)}$$

and $$\sigma_a = \frac{P}{A_a} = \frac{11076.9}{1000} \times 10^2$$

$$= 1107.70 \text{ kg/cm}^2 \text{ (comp.)}$$

8.7 : *A brass rod of 6 mm diameter and 1 m long is joined at one end to a rod of steel 6 mm diameter and 1.3 m long. The compound rod is placed in a vertical position with the steel rod at the top and connected top and bottom to rigid fixings in such a way that it is carrying a tensile load of 3.5 kN.*

An attachment is fixed at the junction of the two rods and to this a vertical axial load of 1.3 kN is applied downwards. Calculate the stresses in the steel and brass. The temperature is then raised through 30°C. What are the final stresses in steel and brass? Take,

$$E_s = 200 \text{ GN/m}^2$$
$$E_b = 85 \text{ GN/m}^2$$
$$\alpha_s = 12 \times 10^{-6}/°C,$$
$$\alpha_b = 19 \times 10^{-6}/°C.$$

(Pnnjab University)

Solution : (*a*) In the first instance, when the composite bar *ABC* is held in fixtures, both steel and brass portions of the rod carry the tensile load of 3500 *N*. Thus the initial tensile stresses in each of the metals will be equal, as their areas of cross sections are equal.

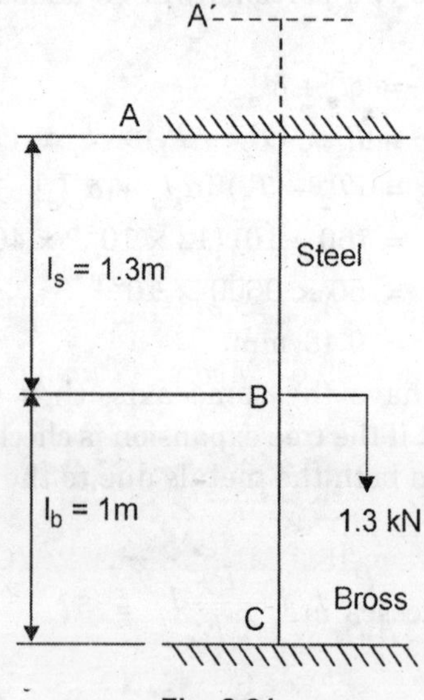

Fig. 8.21

E X A M P L E

$$A_s = A_b = \frac{\pi}{4}(6)^2$$

$$= 28.274\text{mm}^2$$

$$\sigma_{s_1} = \sigma_{b_1} = \frac{3500}{28.274}$$

$$= 123.79\text{N/mm}^2$$

or MN/m² (Tensile)

(*b*) The load of 1300 *N* applied at the junction point *B* (Fig. 8.14), will induce compressive stress in brass and tensile stress in steel. The magnitudes of the loads P_s and P_b due to the action of 1300 *N* load at *B*, will be such that the stretch in steel will be equal in magnitude to the compression in brass, because the total length *AC* remains constant.

That is $$P_s + P_b = 1300$$

and $$\delta l_s = \delta l_b$$

or $$\frac{P_s l_s}{A_s E_s} = \frac{P_b l_b}{A_b E_b}$$

or $$P_s = \frac{E_s l_b}{E_b l_s} P_b \qquad (\because A_s = A_b)$$

$$= \frac{200}{85} \times \frac{1}{1.3} P_b = 1.81\, P_b$$

Therefore, $$1.81\, P_b + P_b = 1300$$

or $$P_b = 461 \text{ N}$$

Then $$P_s = 1300 - 461$$

$$= 839 \text{ N}$$

Thus, the corresponding additional stresses in steel and brass, due to the application of 1300 N load are,

$$\sigma_{s2} = \frac{P_s}{A_s} = \frac{839}{28.274}$$

$$= 29.67 \text{ MN/m}^2 \text{ (Tensile)}$$

$$\sigma_{b2} = \frac{P_b}{A_b} = \frac{461}{28.274}$$

$$= 18.3 \text{ MN/m}^2 \text{ or MPa (Comp.)}$$

Then the resultant stresses are,

$$\sigma_b = \sigma_{b1} - \sigma_{b2}$$

$$= 123.79 - 18.3$$

$$= 107.49 \text{ MN/m}^2 \text{ or MPa (Tensile)}$$

$$\sigma_s = \sigma_{s1} + \sigma_{s2}$$

$$= 123.79 + 29.67$$

$$= 153.46 \text{ MN/m}^2 \text{ or MPa (Tensile)}$$

(c) In the absence of fixtures, the rod AC would have expanded freely due to temperature rise to CA'. But the presence of fixtures effectively compresses the rod AC by an amount AA' which is equal to ($l_s \alpha_s t + l_b \alpha_b t$). Thus, if σ N/m² is the compressive stress in steel and brass (it will be same as the cross section areas are same) due to temperature rise, then we have,

$$\frac{\sigma}{E_s} l_s + \frac{\sigma}{E_b} l_b = l_s \alpha_s t + l_b \alpha_b t$$

or
$$\left[\frac{1.3}{200 \times 10^9} + \frac{1}{85 \times 10^9}\right] = 30 \, (1.3 \times 12 \times 10^{-6} + 1 \times 9 \times 10^{-6})$$

or
$$\sigma = 56.9 \times 10^6 \text{ N/mm}^2$$

$$= 56.9 \text{MN/m}^2 \text{ or MPa}$$

Therefore, the final stresses in steel and brass are,

$$\sigma_{s3} = \sigma_s - \sigma$$

$$= 143.46 - 56.9$$

$$= 98.56 \text{ MN/m}^2 \text{ or MPa (Tensile)}$$

$$\sigma_{b3} = \sigma_b - \sigma$$

$$= 107.49 - 56.9$$

$$= 50.59 \text{MN/m}^2 \text{ or MPa (Tensile)}$$

8.8 : *A bar AD (Fig. 8.22) is pinned at A, and supported by a copper rod 40 mm² in area and 1.5 m long at B, also supported at C by a steel rod of 30 mm² in area and 1 m length. The bar AD is rigid and horizontal before applying a load of 1000 kg (9810 N) at the end D. Determine the stresses and elongations produced in copper and steel rods after the application of above load. Take*

$$E_c = 10 \times 10^3 \text{ kg/mm}^2 \text{ (} 98.1 \text{GN/m}^2\text{)}$$

$$E_s = 20 \times 10^3 \text{ kg/mm}^2 \text{ (198.2 GN/m}^2\text{)}.$$

(Punjab University)

Solution : Let the tensile loads sustained by copper and steel rods be P_c and P_s respectively. Then for the equilibrium of the rod **AD**, taking moments of P, P_c and P_s about A we have,

$$P_c \times 1 + P_s \times 2 = P \times 2.5$$

or
$$P_c + 2P_s = 9810 \times 2.5 = 24525 \quad \text{...(A)}$$

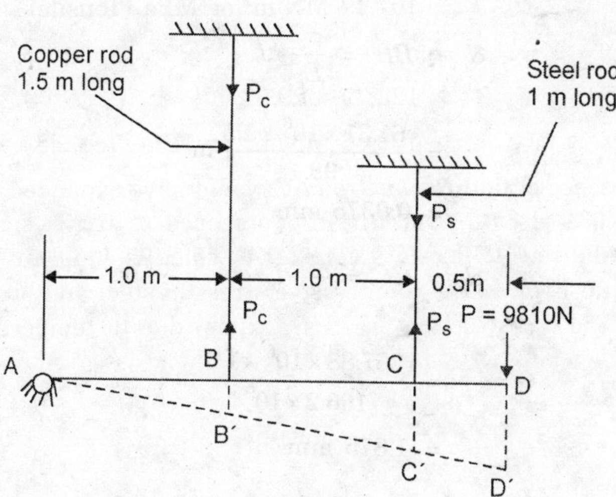

Fig. 8.22

Further due to the loads P_c and P_s the copper and steel rods will get stretched by BB' and CC' respectively. However, the extensions will be such that A, B', C' and D' will lie in one straight line, as the rod AD is rigid. Therefore, from similar triangles ABB' and ACC', we have,

$$\frac{CC'}{BB'} = \frac{AC}{AB} = \frac{2}{1}$$

or $$CC' = 2BB'$$

or $$\frac{P_S}{A_S E_S} l_s = 2 \frac{P_C}{A_C E_C} l_c$$

or $$P_s = 2\left(\frac{A_s}{A_C}\right)\left(\frac{E_s}{E_C}\right)\left(\frac{l_c}{l_s}\right) P_c$$

$$= 2\left(\frac{30}{40}\right)\left(\frac{196.2}{98.1}\right)\left(\frac{1.5}{1.0}\right) P_c$$

or $$P_s = 4.5\, P_c \qquad ...(B)$$

Substituting (B) into (A), we get,

$$P_c = 2452.5 \text{ N}$$

and $$P_s = 11038.25 \text{ N}$$

Therefore, $$\sigma_c = \frac{2452.5}{40}$$

$$= 61.31 \text{ N/mm}^2 \text{ or MPa (Tensile)}$$

and $$\sigma_s = \frac{11036.25}{30}$$

$$= 367.88 \text{N/mm}^2 \text{ or MPa (Tensile)}$$

Also, $\delta l_c = BB' = \frac{\sigma_c}{E_c} \times l_c$

$$= \frac{61.31 \times 10^6 \times 1.5}{98.1} \text{ m}$$

$$= 0.9375 \text{ mm}$$

Similarly, $\delta l_s = CC' = \frac{\sigma_s}{E_s} l_s$

$$= \frac{367.88 \times 10^6 \times 1.0}{196.2 \times 10^9} \text{ m}$$

$$= 1.875 \text{ mm.}$$

EXAMPLE

8.9 : *Fig. 8.24 shows a round steel rod supported in a recess and surrounded by a coaxial brass tube. The upper end of the rod is 0.1 mm below that of the tube and an axial load is applied to a rigid plate resting on the top of the tube.*

(a) Determine the magnitude of the maximum permissible load, if the compressive stress in the rod is not to exceed 110 MN/m² and that in the tube is not to exceed 80 MN/m².

(b) Find the amount by which the tube will be shortened by the load, if the compressive stress in the tube is the same as that in the rod,

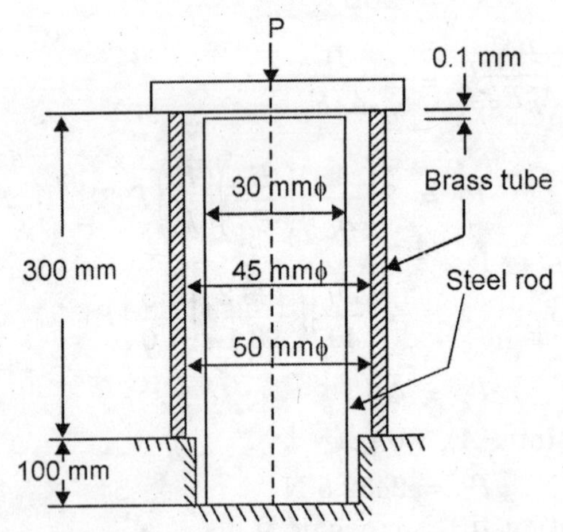

Fig. 8.23

Take $E_s = 200 \text{ GN/m}^2 = 200 \times 10^3 \text{ MPa}$

$E_b = 100 \text{ GN/m}^2 = 100 \times 10^3 \text{ MPa}$ *(Punjab University)*

Solution. (*a*) $A_s = \frac{\pi}{4} \times 706.86 \text{ mm}^2$

$$A_b = \frac{\pi}{4}(50^2 - 45)^2 = 373.07 \text{ mm}^2$$

E X A M P L E

Let σ_s and σ_b (MN/m^2 or N/mm^2) be the stresses induced in the steel rod and brass tube respectively. Since the sum of the loads carried by the steel and brass is equal to the total load

$$P = \sigma_s A_s + \sigma_b A_b$$

$$= 708.86\ \sigma_s + 373.07\sigma_b \text{ N} \qquad ...(A)$$

When the top plate makes contact with steel and compresses it, the compression of the brass will exceed that of steel by 0.1 mm.

That is $\delta l_b = \delta l_s + 0.1$

or $$\frac{\sigma_b}{E_b} l_b = \frac{\sigma_s}{E_s} l_s + 0.1$$

or $$\frac{\sigma_b}{100 \times 10^3} \times 300 = \frac{\sigma_s}{200 \times 10^3} \times 400 + 0.1$$

or $$3\sigma_b = 3\sigma_s + 100 \qquad ...(B)$$

The maximum compressive stress of 110 MN/m^2 in rod and 80 MN/m^2 in tube will not occur simultaneously, rather the magnitudes of induced stresses in the two materials will be related to each other by the equation (*B*). Thus, if it is assumed that the stress in steel is the limiting case, then from equation (*B*), for $\sigma_s = 110$ MN/m^2,

$$\sigma_b = \frac{2 \times 110 + 100}{3}$$

$$= 108.7 \text{ MN/m}^2$$

which is more than the allowable stress in brass leading to the conclusion that the assumption is wrong. Then again assuming the stress in brass to be the limiting case, then from (*B*) for $\sigma_b = 800$ MN/m^2,

$$\sigma_s = \frac{3 \times 80 - 100}{2} = 70 \text{ MN/m}^2$$

which is safe being less than the maximum allowable stress in steel. Thus, substituting these values of σ_b and σ_g in equation (*A*), we get the maximum permissible load *P* as,

$$P = (706.86 \times 70 + 373.07 \times 80) \text{ N}$$

$$= 79.325 \text{ kN}$$

(*b*) In this case $\sigma_s = \sigma_b$

Then from equation (*B*),

$$\sigma_b = \sigma_s = 100 \text{ MN/m}^2$$

Therefore, $$\delta l_b = \frac{\sigma_b}{E_b} \times l_b$$

$$= \frac{100}{100 \times 10^3} \times 300 \text{ mm}$$

$$= 0.3 \text{ mm.}$$

8.10 : *A reinforced concrete column is 300 mm in diameter and has 4 steel bars of 12 mm diameter embedded in it. If the allowable stresses in steel and concrete are 65.0 MN/m² and 4 MN/m² respectively, calculate the safe axial load which the column can carry. Take* $E_s/E_c = 15$.

Solution : Total area of column $= \frac{\pi}{4}(300)^2 = 70686 \text{ mm}^2$

$$A_s = \text{Area of steel} = 4 \times \frac{\pi}{4}(12)^2 = 452.4 \text{ mm}^2$$

Therefore, area of concrete portion

$$= A_c = 70686 - 452.4$$

$$= 70233.6 \text{ mm}^2$$

Since both steel and concrete portions will be compressed by the same amount and also their original lengths are same, the compressive strains produced by the corresponding loads sustained by each will be same.

i.e., $$\varepsilon_c = \varepsilon_s$$

or $$\frac{P_c}{A_c E_c} = \frac{P_s}{A_s E_s} \quad \text{...}(A)$$

Also, from (*A*), $$\frac{\sigma_c}{E_c} = \frac{\sigma_s}{E_s}$$

or $$\frac{\sigma_s}{\sigma_c} = \frac{E_s}{E_c} = 15 \quad \text{...}(B)$$

Thus, if it is assumed that the stress in concrete is the limiting case, then for 4 MN/m² stress to be induced in concrete the stress induced in steel will be from (*B*),

$$\sigma_s = 15 \times \sigma_c$$

$$= 15 \times 4 = 60 \text{ MN/m}^2$$

(which is less than the allowable stress in steel)

Thus, our assumption of limiting stress in concrete is valid. Corresponding loads sustained by concrete and steel are,

$$P_c = \sigma_c A_c = 4 \times 70233.6$$

$$= 280934 \text{ N}$$

$$= 280.934 \text{ kN}$$

$$P_s = \sigma_s \times A_s = 60 \times 452.4 = 27144 \text{ N}$$

$$= 27.144 \text{ kN}$$

Thus, total load which the column can carry

$$= P_c + P_s$$

$$= 280.934 + 27.144$$

$$= 308.078 \text{ kN}.$$

EXAMPLE

E X A M P L E

8.11 : *Two steel plates, each 25 mm thick, are held together with the help of 20 mm diameter steel bolts. If the pitch of the threads is 2.5 mm, find the increase in stress in the shank of the bolt when the nut is turned through $\frac{1}{200}$ of a turn with respect to the head of the bolt, assuming that the plates do not deform. Take E_s = 205 GN/m².*

Solution : Length of the bolt shank = 25 + 25 = 50 mm.

Axial movement of the nut corresponding to rotation of $\frac{1}{200}$ of a turn

$$= \frac{2.5}{200} = 0.0125 \text{ mm.}$$

Thus, that rotation of the nut will cause 0.0125 mm stretch in 50 mm length of the shank giving rise to strain is given by,

$$\varepsilon = \frac{0.0125}{50}$$

Therefore $$\sigma = \varepsilon E = \frac{0.0125}{50} \times (205 \times 10^3)$$

$$= 51.25 \text{ MN/m}^2.$$

8.12 : *A copper sleeve, 21 mm internal and 27 mm external diameter, surrounds a 20 mm steel bolt, one end of the sleeve being in contact with the shoulder of the bolt. The sleeve is 60 mm long. After putting a rigid washer on the other end of the sleeve, a nut is screwed on the bolt through 10 degrees. If the pitch of the threads is 2.5 mm, find the stresses induced in the copper sleeve and steel bolt. Take E_s = 200 GN/m² and E_c = 90 GN/m². (Panjab University)*

Solution : Due to 10° rotation of the nut, it will move along the axis of the bolt through a distance δ (say), such that,

$$\delta = \frac{10}{360} \times \text{pitch}$$

$$= \frac{10}{360} \times 2.5 = 0.0694 \text{ mm}$$

Further the axial movement of the nut, δ, will be equal to the compression in the copper sleeve plus stretch in the steel bolt.

i.e. $$\delta = \varepsilon_c l_c + \varepsilon_s l_s$$

or $$0.0694 = 60(\varepsilon_c + \varepsilon_s) \qquad [\because l_c = l_s = 60 \text{ mm}]$$

or $$\frac{0.0694}{60} = \frac{\sigma_c}{E_c} + \frac{\sigma_s}{E_s} \qquad ...(A)$$

Also, the tube and the bolt, in equilibrium, will carry the same load.

i.e., $P_c = P_s$

$$\sigma_c A_c = \sigma_s A_s .$$

or $$\sigma_c = \frac{\sigma_s A_s}{A_c}$$

$$= \sigma_s \frac{\frac{\pi}{4}(20)^2}{\frac{\pi}{4}(27^2 - 21^2)} = 1.39\sigma_s \quad ...(B)$$

From (*A*) and (*B*), we get,

$$\frac{0.0694}{60} = \left(\frac{1.39}{90 \times 10^3} + \frac{1}{200 \times 10^3}\right)\sigma_s$$

$$= 0.2044 \times 10^{-4}\ \sigma_s$$

or $$\sigma_s = \frac{0.0694}{60} \times \frac{1}{0.2044 \times 10^{-4}}$$

$$= 58.59 \text{ (Tensile)}$$

From (*B*), $$\sigma_c = 1.39 \times 58.59$$

$$= 78.66 \text{ MN/m}^2 \text{ (compressive)}$$

EXAMPLE

8.13 MECHANICAL PROPERTIES OF MATERIALS AND THEIR CLASSIFICATION

8.13.1 Homogeneous and Isotropic Material

A homogeneous material is one in which the physical properties like the density, yield stress, modulus of elasticity. Poisons ratio, modulus of rigidity etc. are same at each and every point in the body. Isotropic means that these poperties do not change with the change in direction *e.g.* the yield point will have the same value at a point even if it is measured in any direction. In the case of rolled metallic sheet, the value of youngs modulus will be more, if the test piece is cut out of sheet along the diection of rolling as compared to the test piece cut along the direction perpendicular to its direction of rolling which will mean that the rolled sheet is not isotropic.

8.13.2 Elastic Plastic Materials

A material whose stress strain curve is linear upto elastic limit point and plastic beyond elastic limit point is called the elastic plastic material, steel falls in this category and its stress-strain curve is shown in Fig. 8.8. These are also called a elastic-strain hardening materials.

8.13.3 RIGID PERFECTLY PLASTIC MATERIAL

Rigid material will have stress-strain curve on shown in Fig. 8.24 (*a*) such material will not be strained upto yield point. It will be a vertical line along y-axis and then horizontal at σ_y level which will correspond to perfectly plastic region where there will be constant stress level but strain will increase.

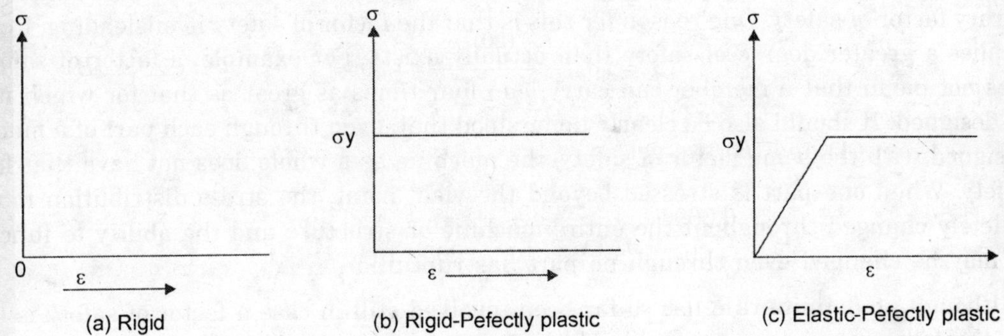

(a) Rigid (b) Rigid-Pefectly plastic (c) Elastic-Pefectly plastic

Fig. 8.24

8.13.4 Ductile and Brittle Materials

Ductile materials under tensile load show large permanent strain upto the instability point. Thus, these materials, can be drawn into thin wives. There are two thoughts to quantify ductility. According to one, it is defined as the % elongation of the specimen after it breaks. The length l of the specimen after the specimen has fractured is measured between the two points, which had been used to originally measure l_0 before yielding, by placing the two fractured pieces together.

Then percentage elongation,

$$= \frac{l - l_0}{l_0} \times 100$$

However, the latest thinking which (has now come up) is to stretch the specimen upto instability point, when the neck formation has just started, the length l is measured. Then the total stress upto instability point which is defined below, gives the measure of ductibility,

$$(\varepsilon_t)_{wet} = \frac{l - l_0}{l_0}$$

8.13.5 Working Stress or Allowable Stress

Working stress is the maximum stress to which, it is considered safe to subject a member in service. The term "allowable stress" is preferable to "working stress", since the latter often is used to indicate the actual stress in a material, when in service. Allowable stresses for different materials for different conditions of service are specified by different authorities on the basis of tests and on experience. In general, for ductile materials, allowable stress is considerably less than elastic limit.

8.13.6 Factor of Safety

Factor of safety is ratio of ultimate strength of the material to allowable stress. The terms was originated for determining the allowable stress. The ultimate strength of a given material is divided by an arbitrary factor, called the safety factor, dependent on material and the use to which it has to be put, gives the allowable stress. In present day design practice, it is customary to use allowable stress as specified by recognised authorities, or codes rather than arbitrary factor of safety. One reason for this is that the factor of safety is misleading, in that, it implies a greater degree of safety than actually exists. For example, a factor of safety of 4 does not mean that a member can carry load four times as great as that for which it has been designed. It should also be clearly understood that, even through each part of a machine is designed with the same factor of safety, the machine as a whole does not have that factor of safety. When one part is stressed beyond the yield point, the stress distribution may be completely changed throughout the entire machine or structure and the ability to function, then may be changed even through no part has ruptured.

Although no definite rule has so far been evolved still in case a factor of safety is to be used, the following circumstances should be taken into account in its selection:

(*a*) When the ultimate strength of the material is known within narrow limits, as for structural steel for which tests on samples have been made, when the load is entirely a steady one of known value, and there is no reason to fear the deterioration of metal by corrosion, the lowest factor of safety should be taken as 3.

(*b*) When the circumstances of (a) are modified by a portion of the load being variable, as in floors of warehouses, the factor of safety should not be less than 4.

(*c*) When the whole or nearly whole load is likely to be put on and taken off, as the suspension rods of floors of bridges, the factor of safety should be 5 or 6.

(*d*) When the stresses are reversed in direction from tension to compression as in bridges and parts of machines, the factor of safety should not be less than 6.

(e) When the part or component is subjected to repeated shocks, the factor of safety should not be less than 10.

(f) When there is probability of property loss caused by failure of the part, the factor of safety should be very large even as high as 50.

In mathematical terms,

$$\text{Safety factor} = \frac{\text{Ultimate stress}}{\text{Working stress}}$$

8.13.7 Hardness

It is the resistance to indentation or to seratehing, abrasion or cutting. Indentation hardness is usually measued by indentation with hardened steel balls, Diamond cone or Diamond pyramid. Most popular tests are Brinell hardness test or Rockwell hardness test.

8.13.8 Fatigue

Fatigue of metals is a phenomenon leading to the fracture under the repeated cyclic or fluctuating stresses which are much below the elastic limit of the material. It is a progressive phenomenon and initiates as a minute crack developing under the fluctuating stresses.

(*a*) **Endurance limit:** It is the maximum stress level below which the material will not fracture or fail even infinite number of cycles the stress may be repeated. Fig. 8.26 shows the plot on log-log graph paper of σ v_s N, are number of cycles to cause failure.

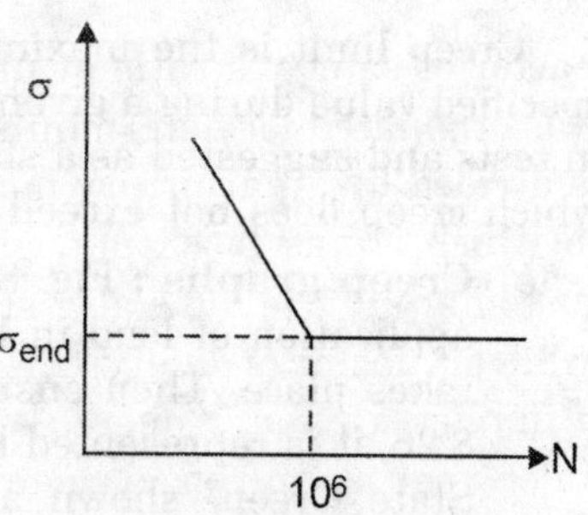

Fig. 8.25. No. of cycles to cause failure

It is observed that as the stress level decreases the number of cycles requied to cause failure increased. However, when the stress level is σ_e it is noticed that the curve becomes horizontal. The knee of the curve corresponds to 10^6 cycles and the stress level is σ_e, called Endurance limit. Evidently at σ_e stress, the number of cycles may even be very large still the failure does not occur. That is why, the limit if 10^6 cycles is in a way pertains to infinite life and σ_e the endurance limit for the particular type of loading. (That is when the stress is due to bending, direct stress, repeated or cycle etc.)

8.13.9 Fatigue life

At any particular stress level which is less than elastic limit, the number of cycles of application of stress required to cause failure is called the fatigue life at that particular stress level. If the stress level is equal to the endurance limit, then the life is called the infinite life.

8.13.10 Fatigue or Endurance Ratio

The ratio of endurance limit to the tensile strength or ultimate stress is called the Endurance or Fatigue ratio.

8.13.11 Fatigue Strength

The limiting stress below which a material will withstand a specified number of cycles of stress without fracture, is called the fatigue strength for the given life in number of cycles.

8.13.12 Creep

Creep or flow of metals in a phase of plastic or inelastic action. Some solids, such as asphalt or paraffin, flow specially at room temperature index extremely small stress, zinc, lead and tin show signs of creep at room temperature under extremely small stresses; zinc. lead and tin show signs of creep at room temperature under moderate stresses. At sufficiently high temperatures, practically all metals creep under stresses which vary with temperature, the higher the temperatures lower being the stress at which creep takes place. The deformation due to creep continues to increase indefinitely and becomes the extreme importance in members subjected to high temperatures, as parts in turbines, boilers, super heaters etc.

Creep limit is the maximum unit stress under which unit distortion will not exceed a specified value during a given period of time at a specified temperature. A value is much used in tests and suggested as a standard for comparing materials, is the maximum unit stress at which creep does not exceed 1% in 100,000 hours.

(*b*) **Creep graphs :** Fig. 8.26 shows the type of the creep strain vs time curve. After the application of load in laboratory test, the instantaneous elastic deformation $OA = E_i$ takes place. Then ensures the creep. At first it is "primary transient creep". In Fig. 8.26, it is represented by the portion *AB* of the curve which is followed by the "Steady State Creep" shown as *BC* and thirdly by what is called the "Tertiary Creep" or "Accelerated Creep" which is shown as CD in the Fig. 8.26.

The minimum creep rate $\frac{\Delta\varepsilon}{\Delta t}$ is determined by the slope of the secondary creep portion of the curve is of portion *BC*

i.e. $$\frac{\Delta\varepsilon}{\Delta t} = \tan\phi = \phi \qquad ...(8.560)$$

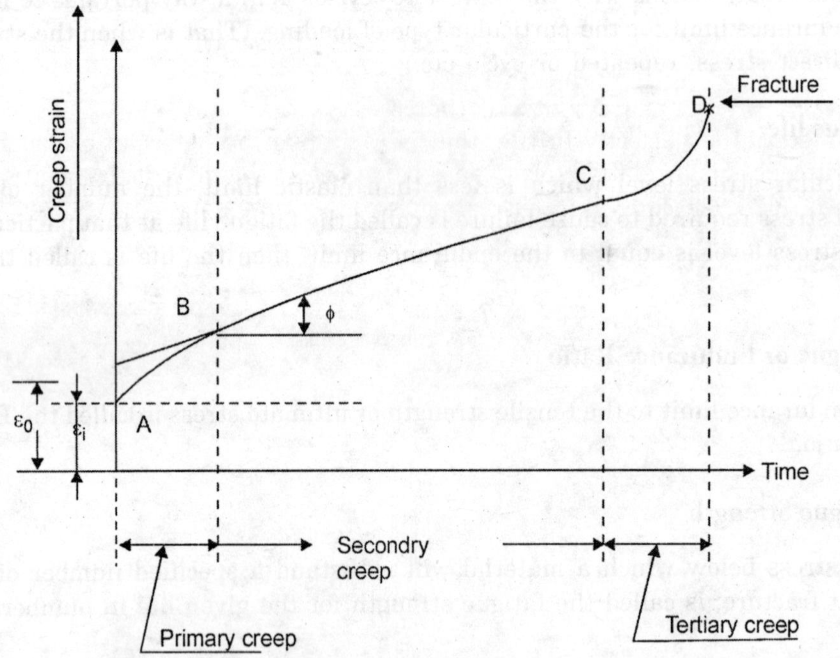

Fig. 8.26

Then at any time *t*, with in the secondary creep range the total creep strain ε_t.

$$\varepsilon_t = \varepsilon_0 + t\tan\phi$$

$$\simeq \varepsilon_0 + t\,\phi \qquad ...(8.51)$$

where, $\varepsilon_0 = OE$

= intercept on *y*-axis of the line *CB* produced

(*c*) **Transient Creep:** Again primary creep or cold creep is a transient phenomenon as the work hardening of the material due to deformation, starts side by side and therefore, the primary creep has a decreasing rate *i.e.* the slope of the curve *AB* goes on decreasing.

Because of its nature, it is also called the transient creep.

(*d*) **Secondary Creep:** Secondary creep curve BC is a straight line and has steady rate or constant slope. Therefore, it is also called steady state or hot or viscous creep. The slope remains constant due to the equilibrium attained between the work hardening processes and annealing effect due to high temperature.

(*e*) **Tertiary Creep** or **Accelerated creep:** This is the final stage where the creep is at an accelerating rate leading to fracture which represents the process of progressive damage which ends up into a fracture.

(*f*) **Creep Ductility :** Elongation at rupture, of the creep specimen is known as creep ductility.

(*g*) **Creep Strength :** It is defned as the highest stress which the material can withstand for a specified length of time without excessive strain.

(*h*) **Creep Rupture Strength :** It is defined as the highest stress which a material can withstand for a specified length of time without rupture.

(*i*) **Creep Limit :** It is the maximum stress which a metal can sustain indefinitely without deforming which is generally taken as 0.001% strain after between the 25th and 35th hours and 0.2% strains after 45 hours.

8.13.13 Brittleness

Brittleness is the property of a material that permits it to be slightly deformed without rupture. No material is perfectly brittle, that is capable of no deformation before rupture. Therefore, brittceness is a relative term, glass being one of the most brittle material. Brittle materials have relatively short stress-strain curves the common structural materials, cast iron , brick and stone are brittle in comparison with steel.

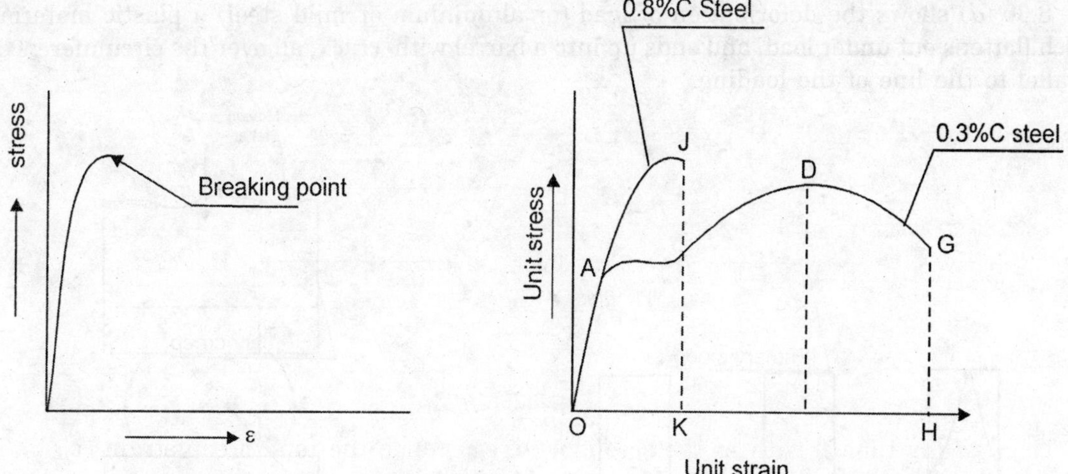

Fig. 8.27. *Stress-strain curve of brittle material* **Fig. 8.28.** *Thoughness comparison.*

8.13.14 Toughness

Toughness is ability of the material to withstand high unit stress together with great unit strain without complete fracture. The area *OAGH* or *OJK*, under the curve of stress-strain

(Fig. 8.28) is the measure of the toughness of the material.

8.13.15 Difference between Toughness and Ductility

The distinction between toughness and ductility is that ductility deals only with the ability to deform and is represented by the strain at instability point, whereas toughness considers both the ability to deform and the stress developed during deformation that is the energy content.

8.14 TYPES OF FRACTURE

8.14.1 Under Tension

A rod of brittle material, as of cast iron will rupture in a tension test in a clean sharp fracture with very little reduction of cross-sectional area and very little elongation. The plane of rupture is normal to the axis of the load (Fig. 8.30 a).

In a ductile material, as in low carbon steels, the reduction of area and elongation are greater. There is a neck formation at the ultimate stress point, at which three dimensional stresses start acting and the fracture is in the form of cup and cone.

(a) (b)

Fig. 8.29

8.14.2 Under Compression

A prism of brittle material, in which compression will break by shearing along oblique planes. Greater the brittleness of the material, the more nearly will these planes be parallel to the direction of applied force Fig. 8.30 (*a*) and 8.30 (*a*) and 8.30 (*c*) for brick, concrete and timber respectively in descending order of brittleness Fig. 8.30 (*d*) shows the deformation of lead (or aluminium or mild steel) a plastic material which flattens out under load, and ends up into a barrel with crack, all over the circumference, parallel to the line of the loading.

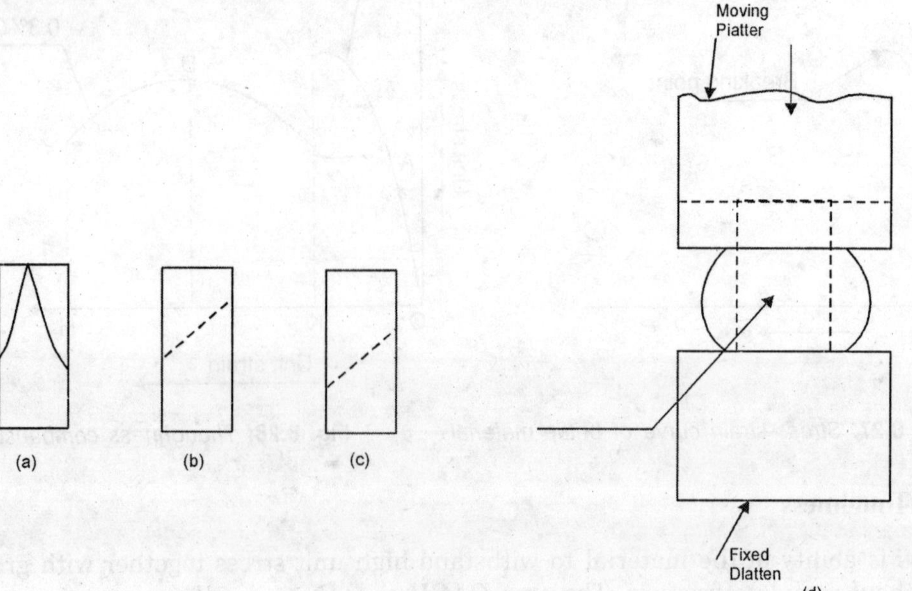

Fig. 8.30

8.14.3 Under Twisting Moment

(a) When a circular shaft of ductile material like low carbon steel is subjected to twisting moment, the shearing along the cross-sectional plane, normal to the axis of the twisting moment, will take place Fig. 8.31 (*a*).

(*b*) In the case of brittle material like *C.I.* put to Twisting Moment, the fracture takes place along the helix inclined to the cross-sectional plane at 45°. The failure is actually due to tensile stress acting normal to plane which is inclined to cross-sectional plane at 45°. Fig. 8.31 (*b*).

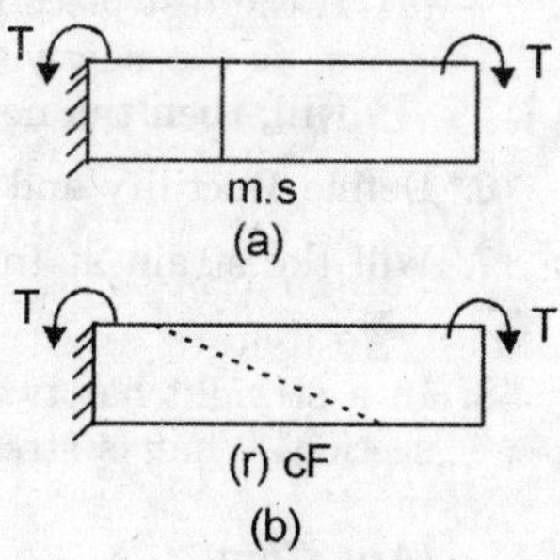

Fig. 8.31

REVIEW QUESTIONS

1. Define the terms "stress", "pressure".
2. Explain the difference between Nominal Strain and True Strain in a Simple Tension Test.
3. Explain the difference between Nominal stress and True Stress in a uniaxial stress, test price.
4. State whether it is true or false : "If the tensile test piece while in the plastic range of the curve, is unloaded, it follows a straight path parallel to the elastic line. (**Ans.** True)
5. (*a*) State whether true or false :

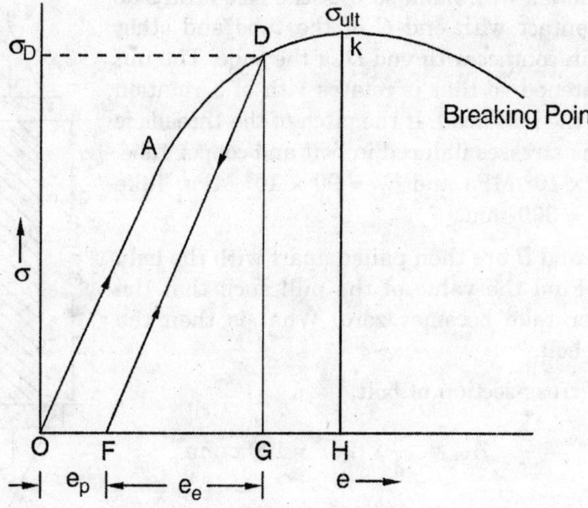

Fig. 8.32

When the stress in the simple tension test is at level D *i.e.*, σ_D (Fig. 8.32) and the specimen is unloaded to no stress state, the left over strain OF is equal to the plastic strain component of the total strain of D and FG is the elastic strain component or elastic recovery.

(**Ans.** True)

(b) If the test piece is re–loaded or re–stressed from the point F, it follows the path FD. Then σ_D is the new yield stress. Similarly, if this process is repeated with reference to point L will, then the new yield point corrspond to σ_L? (Yes)

6. Define Ductility and Malleability.

7. Will the strain at Instability point in the case of Aluminium be more than that for steel? (Yes)

8. In a straight bar two equal and opposite forces of 100 kN each are acting at the mid length section. What is stress in the cross section at ¾th length? Take the area of x-section as 0.01 m^2.
(**Ans.** Zero)

9. (a) State D' Alembert's Principle.

(b) Can a body in linear motion be in a state of stress ? State Yes or No.

If yes, give example. (Yes)

10. Can the equilibrium equations for a body in static equilibrium be applied to the body in motion? If so, explain the phenomenon. (Yes)

11. State Hooke's law.

12. Define Proof stress.

13. Define quantitatively the terms ductility and malleability.

14. What is meant by "Homogeneous materials" and "Isotropic materials"?

15. (a) A copper tube 25 mm internal diameter and 30 mm external diameter surrounds a 20 mm steel bolt Fig. (8.33). After putting rigid washers A and B on two sides of the tube, the nut is just tightened with hand so that the bolt head side washer is in contact with end C of the tube and other washer is just in contact with end D of the tube. The nut is further tightened so that it rotates ¼th of a rotation with respect to the bolt head. If the pitch of the threads is 1.5 mm, find the stresses induced in bolt and copper tube. Take E_S = 200 × 10^3 MPa and E_C = 90 × 10^3 MPa. Take length of tube = 300 mm.

(b) The washers A and B are then pulled apart with the help of wire ropes. Find the value of the pull such that the stress in copper take becomes zero. What is then the stress in steel bolt ?

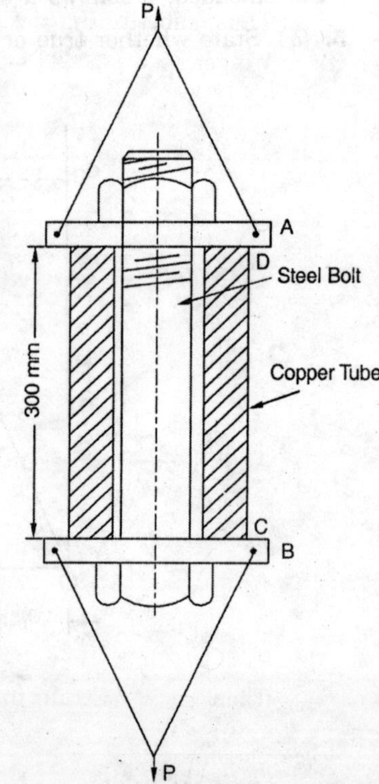

Fig. 8.33

Solution : Area of cross-section of bolt,

$$A_S = \frac{\pi}{4} \times (20)^2 = 100\pi \text{ mm}^2$$

$$= 314.159 \text{ mm}^2$$

Area of cross-section of the tube,

$$A_C = \frac{\pi}{4}(30^2 - 25^2)$$

$$= \frac{\pi}{4} \times 275 = 215.98 \text{ mm}^2$$

Free movement of nut with respect of head

= Pitch × no of terms

$$= 1.5 \times \frac{1}{4} = 0.375 \text{ mm}$$

$$= \text{compression in copper + stretch in steel} \quad ...(A)$$

If W is the axial compressive load in copper,

Then compression in Cu = $\varepsilon_C l_C$

$$= \left(\frac{W}{A_C}\right)\frac{l_C}{E_C}$$

$$= \frac{W}{215.98} \times \frac{300}{90 \times 10^3} \text{ mm}$$

$$= 0.0154 \times 10^{-3} \text{ W mm}$$

At equilibrium tensile load in steel = compressive load in Cu = W

Therefore, Stretch in steel $= \frac{W}{A_S}\frac{1}{E_S} l_s$

$$= \frac{W}{314.119} \times \frac{1}{200 \times 10^3} \times 300 \text{ mm}$$

$$= 0.00477 \times 10^{-3} \text{ W mm}$$

Thus, substituting the values in equation (*A*),

We get, $0.375 = 0.154 \times 10^{-3} \text{ W} + .00477 \times 10^{-3} \text{ W}$

or $W = \frac{0.375 \times 10^3}{0.02017}$

$$= 18.59 \times 10^3 \text{N} = 18.59 \text{ kN}$$

Therefore, σ_s **stress in steel** $= \frac{W}{A_S} = \frac{18.59 \times 10^3}{314.159}$

$$= 0.0592 \times 10^3 \text{ N/mm}^2$$

$$= 59.2 \text{ MPa (Tensile)}$$

And **stress in copper** $= \sigma_c = \frac{W}{A_c}$

$$= \frac{18.59 \times 10^3}{215.98} = 0.08614 \times 10^3 \text{ N/mm}^2$$

$$= 88.14 \text{ MPa (comp.)}$$

Then **Strain in copper** $= \frac{\sigma_C}{E_C} = \frac{86.14 \text{ MPa}}{90 \times 10^3 \text{ MPa}}$

i.e., $\varepsilon_c = \frac{86.14}{90 \times 10^3} = 0.957 \times 10^{-3}$

$$\delta l_c = 300 \times 0.957 \times 10^{-3}$$

$$= 0.287 \text{ mm}$$

Solution : (*b*) In the part (*b*) the load *P* will act as pull only on the bolt. Its value is to be found such that the resultant stress in copper is zero. In other words, if the bolt stretches further by an amount 0.287 mm then resultant compression stress in copper will become zero because when additional stretch in steel equals 0.287 mm at that time the washers will just leave contact with tribe.

Thus, additional stress in steel will be due stretch of 0.287 mm

$$\delta'_S = \varepsilon'_S\, l_s$$

i.e.,
$$\varepsilon'_S = \frac{\delta'_S}{l_s} = \frac{0.287}{300}$$

Therefore,
$$\sigma'_S = \varepsilon'_S E_S$$

$$= \frac{0.287}{300} \times 200 \times 10^3 \text{ MPa}$$

$$= 191.33 \text{ MPa (tensile)}$$

That is additional load
$$P'_S = \sigma'_S A_s = \left(\frac{0.287}{300} \times 200 \times 10^3\right) 314.159$$

$$= 60 \times 10^3 \text{ N} = 60 \text{ kN}$$

Thus, we conclude that 60 kN pull applied in the bold will bring the stress in copper to zero.

Then the resultant stress in steel bolt

$$= 59.2 + 191.33$$

$$= 250.53 \text{ MPa (Tensile)}$$

16. A brass tube 20 mm internal dia and 20 mm external diameter surrounds a 18 mm shank of a steel bolt. After putting rigid washers on the two sides of the tube the nut is rotated such that it is just tight and no load is exerted on the tube by the washers. The length of the tube is 200 mm. The temperature is then raised to 270°C. If the room temperature is 20°C, find the increase in the stresses in brass and steel.

Take
$$\alpha_b = 16.7 \times 10^{-6}/°\text{C}$$

$$\alpha_s = 6.6 \times 10^{-6}/°\text{C}$$

$$E_b = 100 \text{ GPa}$$

$$E_S = 200 \text{ GPa}$$

Solution : Rise in temperature $= t$

$$= 220 - 20 = 200°\text{C}$$

Free expansion of brass $= l_b \alpha_b t = \delta_b$

or
$$\delta_b = 200 \times 16.7 \times 10^{-6} \times 200$$

$$= 0.748 \text{ mm}$$

Free expansion of steel $= l_S \alpha_S t = \delta_S$

or
$$\delta_S = 200 \times 6.6 \times 10^{-6} \times 200$$

$$= 0.464 \text{ mm}$$

Because of this difference in the free expressions, steel will get pulled up while brass will get compressed due to temperature rise.

Thus, if σ_S and σ_b are the stresses induced in two materials then

$$(\delta_b - \delta_s) = \frac{\sigma_b}{E_b} l_b + \frac{\sigma_s}{E_S} l_s$$

But $l_b = l_s$

$$(0.748 - 0.464) = \left(\frac{\sigma_b}{E_b} + \frac{\sigma_S}{E_S}\right) l_S$$

or

$$0.284 = \left[\frac{\sigma_b}{E_b} + \frac{\sigma_S}{E_S}\right] 200$$

or

$$\frac{\sigma_b}{E_b} + \frac{\sigma_S}{E_S} = \frac{.254}{200} = 0.00142 \quad ...(A)$$

Also, at equilibrium,

$$P_S = P_b$$

$$\sigma_S A_S = \sigma_b A_b$$

or

$$\sigma_S = \left(\frac{A_b}{A_S}\right) \sigma_b \quad ...(B)$$

But

$$A_b = \frac{\pi}{4}(25^2 - 20^2)$$

$$= \frac{\pi}{4} \times 227 = 178.71 \text{ mm}^2$$

$$A_S = \frac{\pi}{4} \times 18^2 = 254.469 \text{ mm}^2$$

$\therefore$

$$\sigma_S = \frac{176.71}{254.469} \times \sigma_b = 0.6938\ \sigma_b$$

Thus, from equations (*A*) and (*B*) we get,

$$\frac{\sigma_b}{(100 \times 10^3)} + \frac{0.6938\ \sigma_b}{(200 \times 10^3)} = 0.00142$$

or

$$\sigma_b\ [2 + 0.6938] = 0.00142 \times 200 \times 10^3$$

$$\sigma_b = \frac{0.00142 \times 200 \times 10^3}{2.6938}$$

$$= 0.528 \text{ MPa}$$

Then

$$\sigma_S = 0.6938\ \sigma_b$$

$$= 0.6938 \times 0.528$$

$$= 0.3663 \text{ MPa}$$

PROBLEM FOR PRACTICE

1. A rigid cross bar is supported horizontally by two vertical bars, A and B of equal length and hanging from their tops. The bars A and B are 0.6 m apart. The cross bar stays horizontal even after a vertical force of 6000 N is applied to it at a point 0.4 m from B. If the stress in A is 200 MN/m², find the stress in B and the areas of cross–section of the two rods.

Take $E_A = 200$ GN/m²

and $E_B = 130$ GN/m²

(**Ans.** : 130 MN/m², 20.0 mm² ; 15.385 mm²)

2. A rigid bar AD, 1 m long, is hinged at A and is held horizontal by supporting it at B and C as shown in the Fig. 8.34 The area of cross section of the steel rod BE is 16 mm² and that of copper rod CF is 12 mm². The bar AD is acted upon by a vertical force of 500 N at D. Find the stresses induced in the rods BE and CF. Take

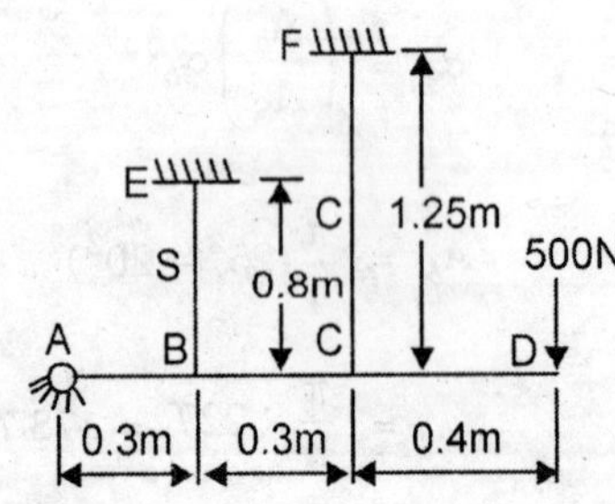

Fig. 8.34

$E_S = 200$ GN/m²

and $E_C = 90$ GN/m². (**Ans.** 56.0 MN/m², 32.2 MN/m²)

3. A steel rod, 20 mm diameter, passes centrally through a steel tube 25 mm internal diameter and 40 mm external diameter. The tube, 750 mm long, is closed by two rigid washers of negligible thickness which are fixed onto the tube by tightening the nuts on the threaded rod. If the pitch of the threads is 2.5 mm, find the increase in the stresses in the tube and the rod when one nut is turned through one quarter of a turn with respect to the nut on the other end.

Take $E_{steel} = 205$ GN/m². (**Ans.** 49.8 MN/m² ; 121.2 MN/m²)

4. A cast iron bracket as shown in Fig. 8.35 is held fast to the foundation with the help of 4 steel bolts. Find the increase in tensile stresses of the four bolts due to the load P of magnitude 1000 N. The bolts are all 12 mm in diameter.

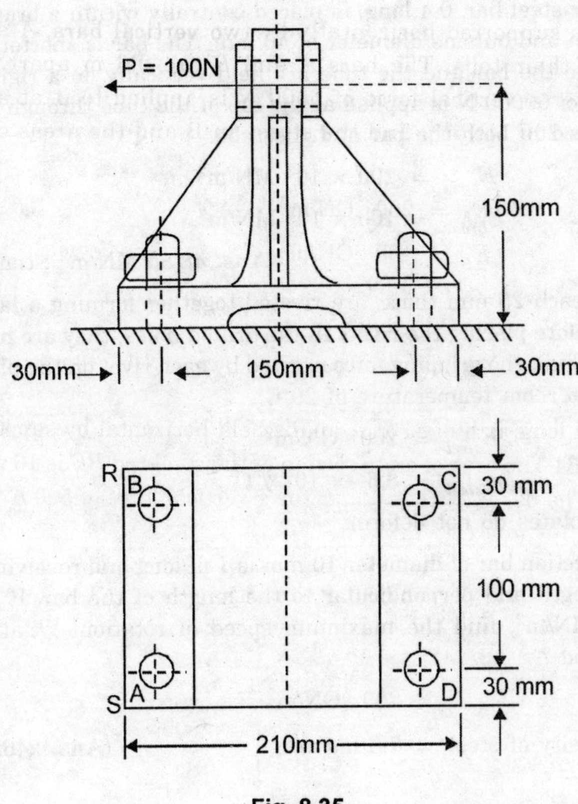

Fig. 8.35

(**Ans.** For A and B 0.596 MN/m^2 ;
For C and D 3.576 MN/m^2)

[**Hint** : The bracket will have a tendency of turning about the line RS and the bolts will have a tendency to stretch proportional to their distances from RS].

5. A 25 mm diameter brass bar is enclosed in a steel tube of 25 mm internal diameter and 50 mm external diameter. Both of them are 1m long at room temperature and fastened rigidly to each other at the ends.

If the room temperature is 20°C, find to what temperature the assembly should be heated so as to generate a compressive stress of 48.7 MN/m^2 in brass. What is then the stress in steel?

If the composite bar is then pulled with a force of 50000 N, find the resulting stresses in both the metals.

Take

$$E_{steel} = 200 \text{ GN / m}^2$$

$$E_{brass} = 100 \text{ GN / m}^2$$

$$\alpha_{steel} = 8.6 \times 10^{-6} \text{ /°C ;}$$

$$\alpha_{brass} = 18.7 \times 10^{-6}\text{/°C .}$$

(**Ans.** 80°C, 18.23 MN/m^2 tensile ;
34.17 MN/m^2 compressive ;
45.3 MN/m^2 tensile)

6. A 28 mm diameter steel bar, 0.4 long, is placed centrally within a brass tube having an inside diameter of 30 mm and outside diameter of 40 mm. The bar is shorter in length than the tube by 0.12 mm. While the bar and the tube are held vertically on a rigid horizontal platform, a compressive force of 60000 N is applied at the top of the tube through a rigid plate. Determine the stresses induced in both the bar and the tube.

Take $E_{steel} = 200 \times 10^3$ MN/m^2

$E_{brass} = 100 \times 10^3$ MN/m^2.

(**Ans.** 48.85 MN/m^2 ; comp. 54.42 MN/m^2 comp.)

7. Two steel plates, each 25 mm thick, are riveted together forming a lap joint by using 25 mm dia steel rivets. Before placing the rivets in the drilled holes, they are heated to 600°C and then the heads formed. Find the gripping force exerted by each rivet on the plates, when the assembly has cooled down to room temperature of 20°C.

Take $E_{steel} = 200$ GN/m^2

$\alpha_{steel} = 8.6 \times 10^{-6}$/°C .

Assume that the plates do not deform. (**Ans.** 660.52 *kN*)

8. A uniform cross–section bar of diameter 10 mm is 1 m long and revolving at a high speed about an axis at mid-length and perpendicular to the length of the bar. If the maximum stress is restricted to 65 MN/m^2, find the maximum speed of rotation. What will then be the total extension of the rod ?

Take $E = 200$ GN/m^2

Density of steel = 7.8 mg/m^3. (**Ans.** 2465.7 r.p.m. ; 0.2167 mm)

CHAPTER 9

ELASTIC CONSTANTS, COMPLEX STRESSES AND STRAIN ENERGY

9.1 INTRODUCTION

By elastic constants of a material we mean four constants of the material namely, Youngs Modulus or Modulus of elasticity E, Modulus of rigidity or Shear Modulus G, Bulk Modulus K and Poisson ratio ν.

While discussing the Hooke's law, the Youngs Modulus E has been introduced in Chapter 8. It deals with the direct stress or normal stress and the corresponding direct strain. The stress and the corresponding strain may be tensile or compressive. In case it is tensile stress, it will give rise to positive strain or alternatively the dimension will increase, (+ sign) . Similarly in the case of compressive stress the dimension along the direction of the stress will decrease (– ve sign). Accordingly, for tensile and compressive direct stresses, the Young Modulus of elasticity, E, had been introduced as per Hooke's Law. Similarly for the shear stresses, the Modulus of rigidity or Shear Modulus G, had been defined and introduced. In the following paragraphs other two constants namely the Poisson's ratio and Bulk modulus also one introduced and the relationship between these constants has been derived.

9.2 POISSON'S RATIO

When a body is acted upon by a direct stress, tensile or compressive, the dimensions will increase or decrease along the line of action of the stress applied. The change in length per unit length along this direction is known as longitudinal or primary strain. Accompanied with the strain in the longitudinal direction, there are strains in the lateral directions (*i.e.* along both y and z directions if the primary strain is along x-axis) known as lateral or secondary strains but these strains are of the opposite sign. That is if their is tensile stress σ_x acting the x-axis, then there will be positive strain, in the primary direction and negative strains in the lateral directions and vice versa. For a material stressed within elastic limit, the ratio of the magnitudes of lateral strains to primary strain is constant. This ratio is known as Poisson's ratio which is represented by ν.

Thus, if ε_x is the strain along the direction of the stress σ_x (Fig. 9.1), which is along x-axis or the longitudinal axis of the circular bar AB, then longitudinal strain ε_x, along x-axis is given by Hooke's law.

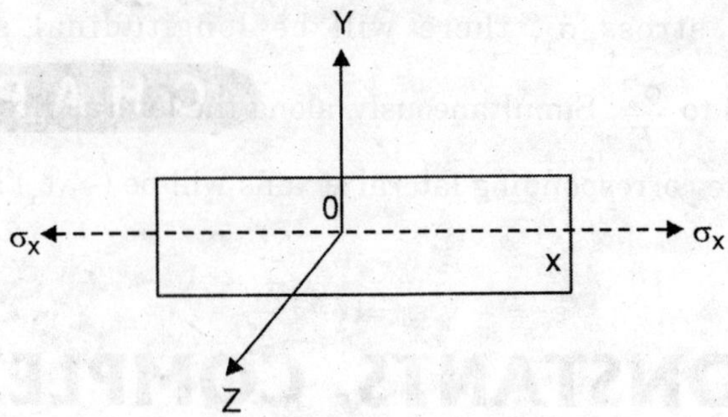

Fig. 9.1

$$\varepsilon_x = \frac{\sigma_x}{E} \qquad ...(9.1)$$

when E = Young's Modulus of the material and if along the lateral directions YY and ZZ' the lateral strains are ε_y and ε_z. These lateral strains, will be negative if ε_x is positive.

The Poisson's ratio ν is given by,

$$\nu = \frac{\text{(Lateral strain)}}{\text{(Longitudinal strain)}}$$

$$\nu = \left(\frac{-\varepsilon_y}{\varepsilon_x}\right) = -\frac{\varepsilon_y}{\varepsilon_x} \qquad ...(9.2a)$$

Also,

$$\nu = \left(\frac{-\varepsilon_z}{\varepsilon_x}\right) \qquad ...(9.2b)$$

or

$$\varepsilon_y = -\nu\,\varepsilon_x \qquad ...(9.3a)$$

and

$$\varepsilon_z = -\nu\,\varepsilon_n \qquad ...(9.3b)$$

From equation (9.2), we observe that ν is – ve but this negative sign is not reported in the data books, where only the numerical value is recorded for different materials. But certainly while calculating the lateral strains the negative sign must be taken into account.

9.2.1 Volumetric Strain

Consider a rectangular parallelepiped *ABCDEFGH*, (Fig. 9.2) under the action of three direct stresses, σ_x, σ_y and σ_z acting along the three axes, *xx*, *yy* and *zz* normal to the plane, *ABCD*, *ABGF* and *ADEF* respectively.

On these planes there is no shear stress acting that is these planes have only normal or direct stresses acting on them. Those planes on which there is no shear stress are called principal planes. The corresponding direct stresses on these planes are called the principal stresses. Let the Young's Modulus or Modulus of elasticity and Poisson's ratio of the material are E and ν respectively.

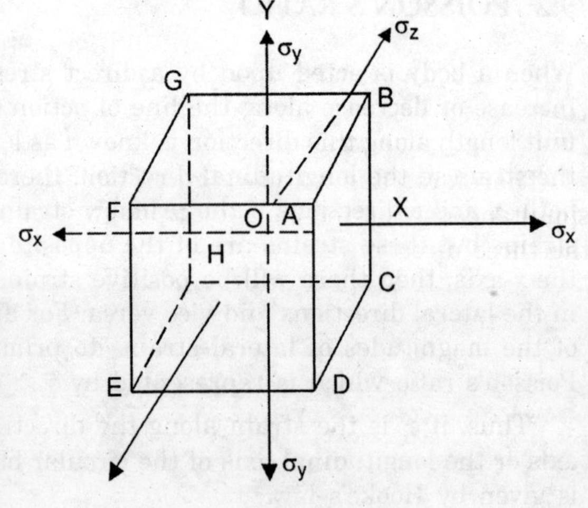

Fig. 9.2

Due to the tensile stress σ_x, there will be longitudinal strain ε_n along the x-axis which will be equal to $\frac{\sigma_x}{E}$. Simultaneously, along the lateral directions yy and zz there will be contraction and the corresponding lateral strains will be $(-\nu\varepsilon_x)$ along both YY and ZZ axis.

i.e.

$$\varepsilon_y = -\nu\,\varepsilon_x = -\nu\frac{\sigma_x}{E}$$

and

$$\varepsilon_z = -\nu\,\varepsilon_x = -\upsilon\frac{\sigma_x}{E}$$

Thus, due to σ_x, σ_y and σ_z the strains along the three directions are as under:

Due to σ_x

$$\varepsilon_x = \frac{\sigma_x}{E}$$

$$\varepsilon_y = -\nu\frac{\sigma_x}{E}$$

$$\varepsilon_z = -\nu\frac{\sigma_x}{E} \quad ...(9.4a)$$

Due to σ_y

$$\varepsilon_x = -\nu\frac{\sigma_y}{E}$$

$$\varepsilon_y = \frac{\sigma_y}{E}$$

$$\varepsilon_z = -\nu\frac{\sigma_y}{E} \quad ...(9.4b)$$

Due to σ_z

$$\varepsilon_x = -\nu\frac{\sigma_z}{E}$$

$$\varepsilon_y = -\nu\frac{\sigma_z}{E}$$

$$\varepsilon_z = \frac{\sigma_z}{E} \quad ...(9.4c)$$

Therefore, due to the combined effect of all the three principal stresses acting simultaneously on the parallelopiped, we get by the principle of superposition, the strains in the three direction as,

$$\varepsilon_x = \frac{\sigma_x}{E} - \frac{\nu}{E}(\sigma_y + \sigma_z) \quad ...(9.5a)$$

$$\varepsilon_y = \frac{\sigma_y}{E} - \frac{\nu}{E}(\sigma_z + \sigma_x) \quad ...(9.5b)$$

$$\varepsilon_z = \frac{\sigma_z}{E} - \frac{\nu}{E}(\sigma_x + \sigma_y) \quad ...(9.5c)$$

Let x, y and z be the original lengths of the three sides of the parallelopiped along the xx, yy and zz directions respectively

Then volume V is given by,

$$V = xyz \qquad ...(9.6)$$

Again due to the three stresses σ_x, σ_y and σ_z let three sides increase respectively by dx, dy and dz due to which the corresponding increase in volume be dV. Then differentiating equation (9.6) w.r.t. V we get,

$$1 = xy\frac{dz}{dV} + yz\frac{dx}{dV} + zx\frac{dy}{dV}$$

or
$$dV = xy\,dz + yz\,dx + zx\,dy \qquad ...(9.7a)$$

Dividing equation (9.7) with equation (9.6) we get,

$$\frac{dV}{\nu} = \frac{dx}{x} + \frac{dy}{y} + \frac{dz}{z} \qquad ...(9.7b)$$

But
$$\frac{dV}{V} = \text{Volumetric strain} = \varepsilon_V$$

$$\frac{dx}{x} = \text{Strain along x direction} = \varepsilon_x$$

Similarly
$$\frac{dy}{y} = \varepsilon_y$$

or
$$\frac{dz}{z} = \varepsilon_z$$

Therefore, equation (9.7) reduces to

$$\varepsilon_V = \varepsilon_x + \varepsilon_y + \varepsilon_z \qquad ...(9.8)$$

Therefore, substituting equation (9.5) into equation (9.8) we get,

$$\varepsilon_V = \left[\frac{\sigma_x}{E} - \frac{\nu}{E}(\sigma_y + \sigma_z)\right] + \left[\frac{\sigma_y}{E} - \frac{\nu}{E}(\sigma_z + \sigma_x)\right] + \left[\frac{\sigma_z}{E} - \frac{\nu}{E}(\sigma_x + \sigma_y)\right]$$

$$= \frac{1}{E}(\sigma_x + \sigma_y + \sigma_z) - \frac{2\nu}{E}(\sigma_x + \sigma_y + \sigma_z)$$

$$\varepsilon_V = \frac{dV}{V} = \frac{(1 - 2\nu)}{E}(\sigma_x + \sigma_y + \sigma_z) \qquad ...(9.9)$$

9.2.2 HYDROSTATIC STRESS AND BULK MODULUS AND RELATION BETWEEN E, K AND ν

As special case when the parallelopiped in Fig. (9.2) is acted upon in all the directions by the same stress σ

i.e. when $\sigma_x = \sigma_y = \sigma_z = \sigma$

then σ is known as hydrostatic stress as it is synonymous with hydrostatic pressure p acting on it when the body is immensed deep in water.

Then in that case

$$\frac{dV}{V} = \varepsilon_V = \frac{(1-2\nu)}{E}(3\sigma) \qquad \text{...(9.10)}$$

The ratio $\dfrac{\sigma}{\left(\dfrac{dV}{V}\right)}$ *i.e.* hydrostatic stress divided by volumetric strain is called the Bulk Modulus of the material represented by K and is constant for any material

i.e.

$$K = \frac{\sigma}{\left(\dfrac{dV}{V}\right)} \qquad \text{...(9.11)}$$

Therefore, from (9.10) and (9.11) we get,

$$E = 3\,(1-2\nu)\,\frac{\sigma}{\left(\dfrac{dV}{V}\right)}$$

or

$$E = 3\,(1-2\nu)\,K \qquad \text{...(9.12)}$$

Equation (9.12) gives the relationship between the Young's Modulus, Poisson's ratio and Bulk Modulus.

9.3 COMPLEMENTARY SHEAR STRESSES

Consider a rectangular block of length x, height y and thickness z under the action of shear stresses τ and τ_1 as shown in Fig. 9.3 *i.e* on the plane $BCHG$ normal to x axis the shear stress is τ and that on the plane normal to y axis ($ABGF$) is τ_1. For the static equilibrium of the block and equating,

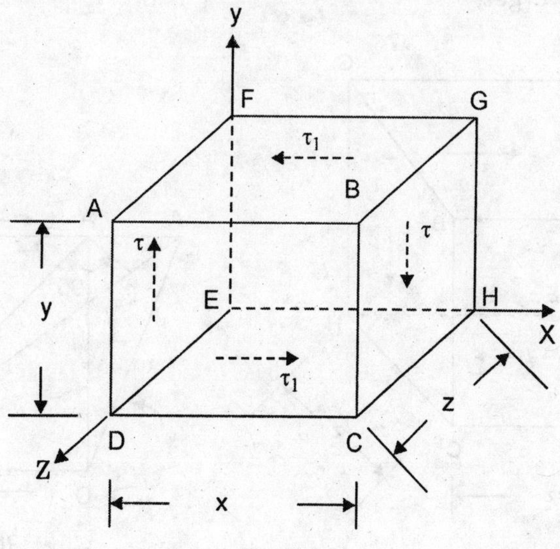

Fig. 9.3

$$\Sigma M = 0$$

i.e. Algebraic sum of the movements due to the forces must be zero.

Let the corresponding forces due to the shear stress τ and τ_1 be P and P_1

Then $$P = \tau \text{ (area } BCHG) = \tau \times (yz) \qquad ...(9.13)$$

and $$P_1 = \tau_1 \text{ (area } DEHC) = \tau_1 \text{ } (xz) \qquad ...(9.14)$$

Taking moments due to forces P acting on faces $BCHG$ and $DEFA$ at a distance x which are equal and opposite

$$= P \times x \text{ (clockwise)}.$$

Similarly, the moment of face P_1 equal and opposite acting on faces $ABGF$ and $DCHE$.

$$= P_1 \, y \text{ (anticlockwise)}$$

$\therefore$ For $$\Sigma M = 0$$

$$P_x - P_1 y = 0$$

or $$(\tau \, yz) \, x = (\tau_1 \, xz) \, y$$

or $$\tau = \tau_1 \qquad ...(9.15)$$

Equation ((9.15) shows that a shear stress τ must be accompanied by another shear stress τ_1 of equal magnitude but of opposite sense, acting on planes at right angles to the planes of original shear stress τ. Then shear stress τ_1 is called the complementary shear stress. Thus system is known as system under pure shear stress.

9.3.1 Cube Under Pure Shear

Consider a parallelopiped $ABCOEPG$ of square section under the action of complementary shear stress as shown in the Fig. 9.4 and it is in static equilibrium. If the entire block is in static equilibrium, then a part of it will also be in equilibrium. Therefore, taking a section along the diagonal plane $ACHF$ and assuming the direct and shear stress in the plane $ACHF$ of which the side view is shown in Fig. (9.4 *b*), are σ_{n_1} and τ_1 resolving the faces along the normal to the face AC we get,

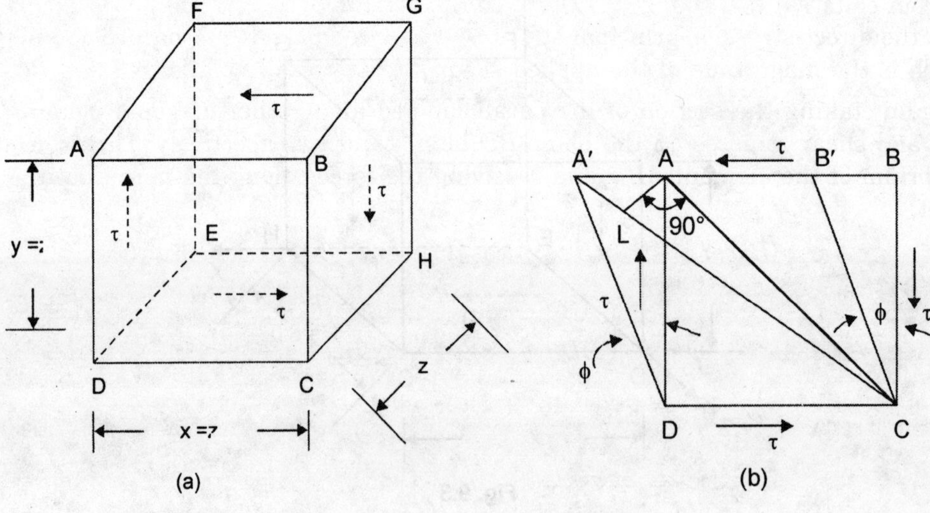

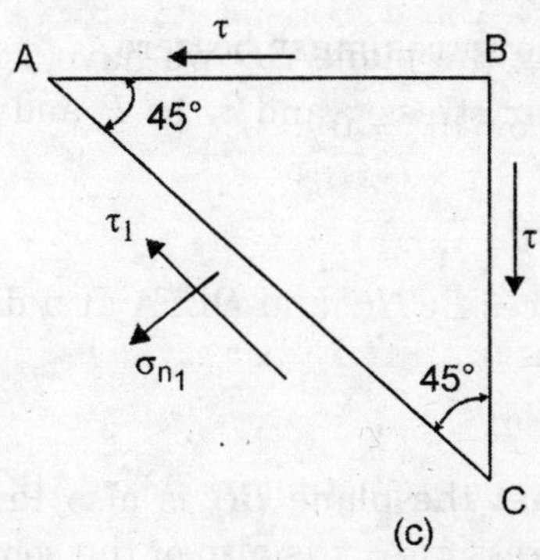

(c)

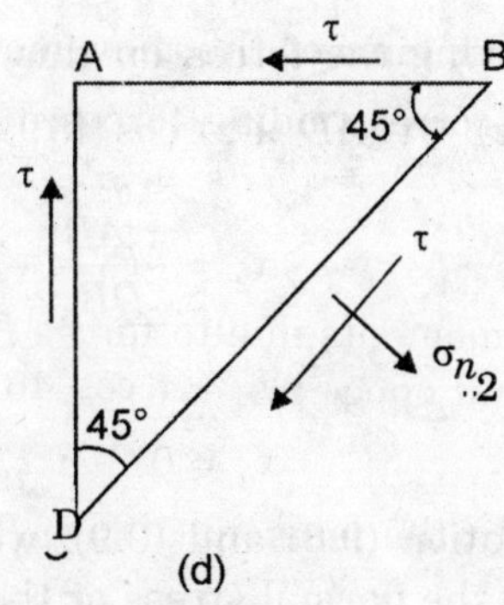

(d)

Fig. 9.4

$$\sigma_{n_1} (AC \times z) + \tau (BC \times z) \cos 45° + \tau (AB \times z) \cos 45° = 0$$

or
$$\sigma_{n_1} = -\tau \frac{BC}{AC} \cos 45° - \tau \frac{BC}{AC} \cos 45°$$

$$= -\tau (\cos 45)^2 - \tau (\cos 45)^2$$

$$= -2\tau \left(\frac{1}{2}\right) = -\tau$$

That is $\sigma_{n_1} = -\tau$ ...(9.16)

Similarly, resolving the forces along the plane CA

$$\tau_1 (CA \times z) + \tau (AB \times z) \cos 45 - \tau (BC \times z) \cos 45° = 0$$

Since $AB = BC$ being side of square it reduces to

$$\tau_1 (CA \times z) = 0$$

or $\tau_1 = 0$ ...(9.17)

From equation (9.16) and (9.17), we conclude that the plane AC is a principal plane on which the direct stress or principal stress is $-\tau$ *i.e.* compressive in nature and of magnitude τ which is the magnitude of the applied shear stresses on the planes AB and BC as shown.

Again taking the section of the parallolopiped along other diagonal plane BD. Let the direct and shear stresses on the plane BD be σ_{n_2} and τ_2 respectively. Then considering the equilibrium of the element ABD and resolving the forces along the normal to the plane BD we get,

$$\sigma_{n2} (BD \times z) - \tau (AB \times z) \cos 45° - \tau (AD \times z) \cos 45 = 0$$

or
$$\sigma_{n2} = \tau \left(\frac{AB}{BD}\right) \cos 45° + \tau \left(\frac{AD}{BD}\right) \cos 45°$$

$$= \tau \cos^2 45° + \tau \cos^2 45°$$

$$= 2\tau \times \frac{1}{2} = \tau$$

or $$\sigma_{n_2} = \tau \qquad ...(9.18)$$

Now resolving the forces on this element along the plane BD we have,

$$\tau_2\,(BD \times z) - \tau\,(AD \times z)\cos 45° + \tau\,(AB \times z)\cos 45° = 0$$

or $$\tau_2 = \frac{\tau\,AB}{DB} - \frac{\tau\,AB}{DB}$$

$$= \tau \cos 45 - \tau \cos 45° = 0$$

i.e. $$\tau_2 = 0 \qquad ...(9.19)$$

From equation (9.8) and (9.9), we conclude that the plane BC is also Principal plane having $+\tau_x$ as the normal stress or the Principal stress $\sigma_{n2} = \tau$ is also of the some magnitude as $\sigma_{n1} = \tau$ but it is tensile as against σ_{n1} which is compressive. The directions of σ_{n1} and σ_{n2} are along DB and AC respectively.

$$\therefore \quad \text{Strain along } CA = \frac{\sigma_{n_2}}{E} - \nu\,\frac{\sigma_{n_1}}{E}$$

or $$\varepsilon_{CA} = \frac{\tau}{E} - \frac{\nu\,(-\tau)}{E} = \frac{\tau}{E} + \frac{\nu\,\tau}{E}$$

or $$\varepsilon_{CA} = \frac{\tau}{E}\,(1 + \nu) \qquad ...(9.20)$$

And $$\text{Strain along } BD = \frac{\sigma_{n_1}}{E} - \nu\,\frac{\sigma_{n_2}}{E}$$

$$= \frac{(-\tau)}{E} - \frac{\nu}{E}\,\tau$$

$$= -\frac{\tau}{E}\,(1 + \nu) \qquad ...(9.11)$$

Again considering the effect of shearing stresses the parallelopiped. The shear stress will distort the square section to shape $A'B'CD$ (if we assume the force AB moves relative to CD) Then angle A DA' will be the shear strain angle ϕ as per its definition,

$$\tan\phi = \frac{AA'}{AD} \simeq \phi \text{ (as } \phi \text{ is small)} \qquad ...(9.21)$$

where G is shear modulus or modulus of rigidity

Also, if AC is perpendicular on CA'

Shear $$AA' = \frac{AL}{\cos(AA'L)}$$

Now $$\angle AA'L \simeq 45°$$

Also, $$\angle ACD = 45°$$

$\therefore$ $$LA' = AA' \cos 45 = \frac{AA'}{\sqrt{2}}$$

Now $$\varepsilon_{CA} = \frac{AL - AC}{AC} = \frac{A'L}{DA\sqrt{2}}$$

$$= \frac{AA'}{\sqrt{2}} \times \frac{1}{DA\sqrt{2}}$$

$$= \frac{AA'}{2\,DA}$$

$$= \frac{\phi}{2} \quad \text{...(9.21)}$$

$$\varepsilon_{CA} = \frac{1}{2}\left(\frac{\tau}{G}\right) \quad \text{...(9.22)}$$

Therefore, from equations (9.10) and (9.13),

We get, $$\varepsilon_{CA} = \frac{\tau}{E}(1 + \nu) = \frac{\tau}{2G}$$

or $$E = 2(1 + \nu)\,G \quad \text{...(9.23)}$$

Thus, from equation (9.21) and (9.23), we have the final relationship between the four static constants E, K, G and ν as,

$$E = 3(1 - 2\nu)\,K = 2(1 + \nu)\,G \quad \text{...(9.24)}$$

Again from (9.24), $$\frac{E}{3K} = 1 - 2\nu$$

or $$\nu = \frac{1}{2}\left[1 - \frac{E}{3K}\right]$$

and $$\frac{E}{2G} = 1 + \nu$$

or $$\nu = \frac{E}{2G} - 1$$

$\therefore$ $$\nu = \frac{1}{2}\left[\frac{3K - E}{3K}\right]$$

$$= \left(\frac{E - 2G}{2G}\right)$$

By cross-multiplication we have,

or $$3\,GK - GE = 3\,KE - 6\,KG$$

$$3\,GK + 6\,KG = E\,(3K + G)$$

$$E = \frac{9\,GK}{3K + G} \qquad \text{...(9.25a)}$$

9.4 LIMITING VALUES OF POISSON'S RATIO

For an isotropic (a material which does not have a preferred orientation) perfectly plastic material the stress-strain curve will be horizontal. Thus, the value of the Young's Modules, E, which is equal to the slope of the stress-strain curve will be zero.

Therefore, from equation 9.4, we get from the first part of this equation,

$$\nu = -1$$

From the second part of this equation,

$$\nu = 0.5$$

Thus, the limiting values of n are – 1 to 0.5. In other words, 0.5 is the maximum value of the Poisson's ratio.

Since ordinary materials contract laterally when stretched and expand laterally when compressed axially, Poisson's ration for such materials is positive. It is 0.45 for lead, 0.33 for aluminium and 0.27 to 0.33 for common steels. For cellular solids, such as typical foams its value lies within the range 0.1 to 0.4.

For cork the value of Poisson's ratio is nearly zero. It is due to this property that cork stopper is used while bottling champagne. The cork must be easily inserted and removed, yet it also must withstand the pressure from within the bottle. Rubber, with a Poisson's ratio of nearly 0.5 could not be used for this purpose because it would expand when compressed in to the neck of the bottle and would jam. Cork, by contrast with Poisson's ratio of nearly zero, is ideal in this application.

Biological tissues and various rubbery materials have values of Poisson's ratio approaching 0.5. Therefore, from equation (9.24), it is seen that,

$$G = \frac{3K\,(1 - 2v)}{[2(1 + v)]} \qquad \text{...(9.25b)}$$

and when $\nu \rightarrow 0.5$ the value of G approaches zero.

Thus, for rubbery materials the shear modulus G, is much less than the bulk modulus, K. In physical sense, these materials readily undergo shear deformation but resist volumetric (or bulk) deformation.

On the other extreme, a solid with Poisson's ratio close to – 1 would have the value of K from equation (2.71) approaching to zero because,

$$K = \frac{2G\,(1 + v)}{[3(1 - 2v)]} \qquad \text{...(9.26)}$$

When $\nu \to -1, K \to 0$

In physical sense their behaviour will be opposite of rubber. That is they are very difficult to shear but easy to deform volumetrically. Mathematically for such solids the value of shear modulus, G will be much greater than the bulk modulus K. Therefore, these materials may be called ANTI-RUBBER materials.

9.4.1 Negative Poisson's Ratio Materials

Till 1968 it was believed by many that materials with negative values of Poisson's ratio were unknown (Young 1968). (Love 1944) had reported a single example of "Single Crystal" pyrite in having a Poisson's ratio of –0.14 in his book published in 1944. It was further suggested that this effect might have been result from a twinned crystal. Analysis of the tensorial elastic constants (R. Hill) of an isotropic single crystal cadmium suggests Poisson's ratio may attain negative values in some directions (Li 1976). An isotropic macroscopic, two-dimensional flexible models of certain honeycomb structures (not materials) may be used in such applications as sponges, shock absorbing materials, air filters, fasteners, car and air craft seats. For having additional information the reader is advised to refer to the article entitled "Advances in negative Poisson's ratio materials" published in "Advanced Materials, 5, pp 229–296 (1993).

9.5 SHEARING FORCE AND BENDING MOMENT IN BEAMS

9.5.1 Beams and their Classification

Any member of a machine or structure under the action of lateral forces in the axial plane is called a beam. The effect of these forces will be to cause bending of the member. Depending upon the end conditions there are different types of beams. A beam with both the ends rigidly fixed is known as a fixed beam. If one of the ends is rigidly fixed and the other is absolutely free, it is known as a cantilever beam. If both the supports at the ends are such that the beam is capable of taking any angular position at the ends and not the lateral movement, the beam is called to be a freely or simply supported beam ; in actual practice this requirement is met by providing a well lubricated hinge at one end and a support on rollers at the other.

If such a beam has ends extending beyond the supports, it is called an overhanging beam. A beam which is supported at more than two supports is called a continuous beam.

As beams are invariably horizontal, and the external forces are weights, the beams will be taken horizontal and the external forces as vertical, although the same conclusions will hold good in other cases. Members of structures and machines transmit axial loads in addition to lateral forces acting upon them. However, in the present chapter only transverse forces will be considered.

A concentrated or point load is one which is applied to the beam through a knife edge. A uniformly distributed load is that which is spread evenly over a certain length of the beam so that the load carried by a portion of the beam is proportional to the length of that portion. A distributed load will be called a varying load, when the rate of loading varies from one point to another.

9.6 MOMENT OF A FORCE, PLANE OF MOMENT AND AXIS OF MOMENT

Moment of a force about a point is the product of the force and the perpendicular distance of the point from the force vector. Plane containing the force vector and the point about which moment taken is called in PLANE OF MOMENT. And the axis passing through the point and normal the plane of moment is called the axis of the moment. Since force is a vector quantity, the moment will also be a vector quantity as the distance is a scalar quantity. The force vector is represented by an arrowhead the length of which to some chosen scale represents the magnitude of the force and the arrowhead indicates the direction of the force. Similarly to represent a moment vector with an arrowhead right handed screw rule is followed. The length of the arrowhead will represent the magnitude of the moment to some scale. However, before we describe the method to decide the placement of the arrow and the direction to which it is to point we have to see whether the moment about the given point is clockwise moment or a anticlockwise moment depending upon the location of the point with respect to the direction of the arrowhead representing the force vector. If the point lies on the left side of force vector arrowhead as seen along the arrowhead, then the moment will be reckoned as anticlockwise. Similarly if the point in question lies on the right hand side of the force vector, then it will be called clockwise moment because this moment will tend to rotate the minutes arm of the clock, placed at the point where the moment is taken, in the clockwise direction or vice versa Fig. (9.5). After having decided whether the moment is clockwise or anticlockwise, we proceed to apply the right hand screw rule. The right handed screw rule states that with respect to a fixed nut placed in the plane of the moment, the direction of axial movement of the right handed screw when acted upon by the given moment, will be the direction of the arrowhead representing the moment vector. Evidently the axial movement of the screw with respect to fixed nut which is placed in the plane of the moment, will be normal to the plane of the moment. Thus, if the plane of the moment is vertical and is visible in elevation, the normal will be visible in the plane view and will be normal to plane of the moment. It may be noted that the arrowhead will be normal to the plane of the moment and it will be parallel to the axis of the moment.

(a) Clockwise moment (b) Anticlockwise moment

Fig. 9.5

9.7 CLASSIFICATION OF MOMENT : BENDING MOMENT AND TWISTING MOMENT

The next question arises as to whether the moment acting on any section of a beam or shaft is bending moment or twisting moment. When the plane of the moment is normal to the cross–sectional plane of the beam or shaft on which the moment is acting, the moment is called Bending moment, as it will tend to bend the beam or shaft. Alternately moment will be a Bending Moment, if the axis of the moment is parallel to or lies in the cross–sectional plane of the beam.

Similarly if the plane of the moment is parallel to or contains the cross–sectional plane of the beam or shaft, it will be called twisting moment as it tends to twist the shaft. Alternately the moment will be a twisting moment, if the axis of the moment is normal to the cross-sectional plane of the beam or shaft or if the axis of the moment coincides with the longitudinal axis of the shaft.

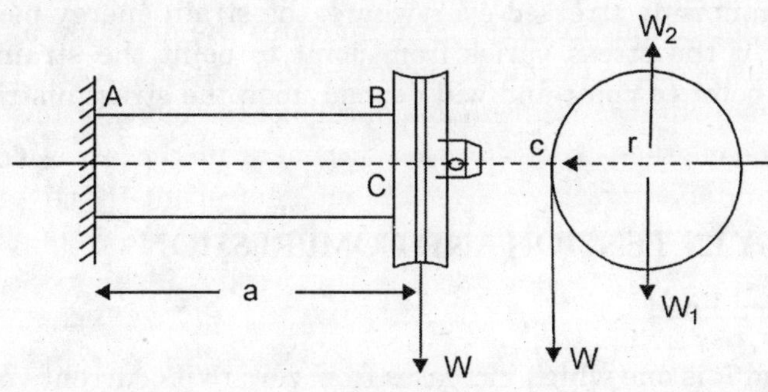

Fig. 9.6

As an illustration, Fig. 9.6 shows a circular bar *AB* with end *A* fixed in wall and at the end *B* is attached a pulley over which a rope passes around its circumference. Let a load *W* be hanging from the B end of the rope.

That is load *W* is applied vertically along the tangent to the circumference of the pulley at *C*. Let the radius of the pulley be *r*.

From our own side we apply equal and opposite vertical loads W_1 and W_2 at the point *B*. Also, both W_1 and W_2 are equal to *W*.

That is $W_1 = W_2 = W$, then they will not alter the given load condition.

(*a*) **Effect of W and W_2.** These loads constitute a momemt of magnitude $(W \times r) = T$ (say). The plane of the moment *T* is the plane of the pulley and is visible in the side view. Evidently the axis of the moment, which is normal to the plane of the movement will coincide with the axis of the rod. Thus, moment *T* is the Twisting Moment" as per the definition. It will try to twist the axis *B* of the rod with respect to the end *A*.

(*b*) Effect W_1 will give rise to the moment '$W_1 \times a = W_1$ a' about the end *A* of this rod which is a cantilever beam. The plane of the moment is the vertical plane passing through the points *A* and *B*. Evidently the axis of the rod *AB* lies in the plane of the moment itself or the axis of the moment which is normal to this plane of moment will be normal to axis of rod. Thus, as per definition this moment $W_1 a = M$ (say) will be the Bending moment and it will tend to bend the rod.

9.8 STRAIN ENERGY AND RESILIENCE

When a bar is acted upon by a tensile or compressive load, there is a change in the length and the point of application of the load moves against the resistance offered by the material which gets strained under the action of the load. In so doing the work is done on the bar. If the bar had been loaded within the elastic limit of the material, the work done on the bar will be recovered when the load is removed. In other words, the work done on the bar, while loading, is stored in the bar as strain energy which can be recovered wholly or partially depending upon whether the bar was loaded upto or beyond its elastic limit. The strain energy stored per unit volume is called the resilience of the body. Further the '*Proof Resilience*' is the strain energy per unit volume stored in the body at its elastic limit point. Strain energy is always a positive quantity and, being work units, will be expressed in Nm (or Joules).

If the material is uniformly stressed everywhere, the strain energy per unit volume will be constant. However, if the stress varies from point to point the strain energy per unit volume will vary from point to point and will depend upon the stress distribution and shape of the body.

9.9 STRAIN ENERGY IN TENSION AND COMPRESSION

9.9.1 Gradually Applied Load

A gradually applied load P is one which increases from zero to its current value P in infinitely large number of very small steps. In actual practice, if the strain rate is less than 2×10^{-3}/sec, it is called a gradually applied or static load.

For a gradually applied tensile load the load-extension (or load-contraction for compressive load) curve for a bar of uniform section will be a straight line upto the limit of proportionality point (Fig. 9.7). Corresponding to the load P the stretch will be x and the work done will be represented by the shaded area. Thus, the strain energy U is given by,

$$U = \frac{1}{2} Px \qquad \text{...(9.27)}$$

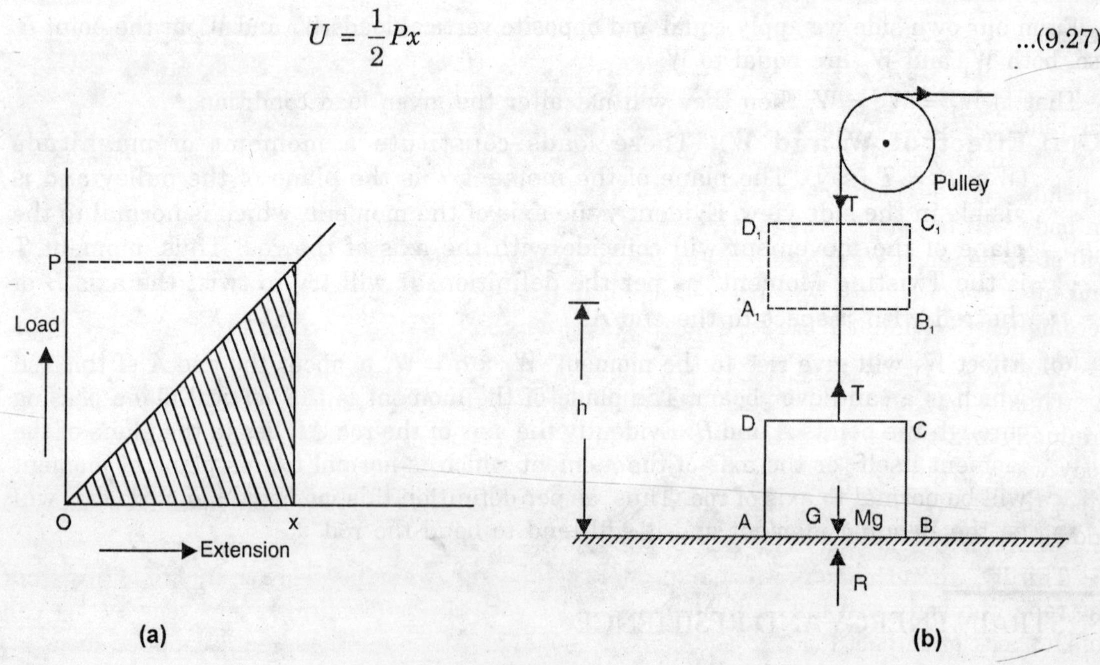

Fig. 9.7

(*i*) Bottom of body AB, just touching table and

$$R = 0$$

$$T = \text{Mg}$$

Rope snapped, then sudden load on table

(*ii*) Rope tight but carries tension T, (< Mg)

$$R + T = \text{Mg}$$

(*iii*) Rope just loose so that R = Mg and Tension = 0 gradually applied load.

(*iv*) *Bottom of body, AB at height h above table rope snapped → shock on impact load.*

If the length of the bar is l and its area of cross-section A, then the strain energy can be expressed in terms of stress σ and Young's modulus E of the material,

$$U = (A\sigma)\frac{1}{2}\left(\frac{\sigma}{E}l\right) = \frac{\sigma^2}{2E}(Al)$$

or $$U = \frac{\sigma^2}{2E}(\text{Volume}) \qquad ...(9.28)$$

For equation (9.28), the resilience or the strain energy per unit volume in simple tension or compression is $\sigma^2/2E$. Further if σ_e is the elastic limit of the material, the proof resilience will be $\sigma_e^2/(2E)$ which is constant for a material.

9.9.2 Suddenly Applied Load

A suddenly applied load P_1 is one which is applied in one step of value P_1, that is, it remains constant during the period the stretch or contraction attains the final value. Consider a body $ABCD$ of mass M hanging from the end of a rope as shown in Fig. 9.7*b* and is being lowered slowly by releasing the rope with the help of a worm and worm wheel arrangement attached to the other end of the rope. When the body is completely in air and not touching the table, the tension T in the rope will be equal to Mg. When the body is lowered slowly its lower face AB will come close to the top face of the table and it will be just above the table surface at a height zero but not touching the table then again $T = Mg$ and the reaction, R of the table on body will be euqal to zero. But when the rope is further released slowly, the reaction R will start increasing and the tension T will simultaneously start decreasing at the same rate and after some time R will become equal Mg and T will reduce to zero and the rope will become loose. Evidently the load on the table will be equal to R which in turn will be equal to Mg.

Thus, a load of 'Mg' has been applied which has increased from zero to current value Mg gradually and hence, it will be called gradually applied load. Secondly, when the bottom of the body was just above the table at zero height but $T = Mg$ and $R = 0$ at that time we cut the rope, the entire load Mg would have come on table in one go. It would then be called a suddenly applied load.

Thirdly, when the body was hanging in the air and at height '*h*' above the table, if we cut the rope, the body will fall and come in contact with the table with certain velocity. We call this a shock or impact load. In the light of above, in the case of a suddenly applied load, P_1, the strain energy is given by,

$$U = P_1 x \qquad ...(9.29)$$

If the deformation x is same as in the gradually-applied-load case the bar will experience the same strain and will, therefore, be under the action of same stress which in turn will result in the same strain energies in the two cases. Then comparing equations (9.28) and (9.29),

$$\frac{1}{2}Px = P_1x$$

or $$P = 2P_1 \qquad ...(9.30)$$

From equation (9.30), it is observed that in order to produce the same strain and same stress a gradually applied load should be twice the value of the suddenly applied load. Alternatively a suddenly applied load will induce stress in the bar which is twice the one induced by same load on the bar but applied gradually. It may be noted that in the case of the suddenly applied load, the strain and stress considered are instantaneous maximum ones obtained immediately after the application of the sudden load. As a matter of fact in the case of suddenly applied load, the system is set into free damped vibrations and the stress in the bar changes in magnitude and nature from tension to compression to tension and so on. The maximum value of the stress on each side progressively goes on diminishing till it ultimately settles down to a value equal to that due to gradually applied load* when the vibrations die out.

9.9.3 Impact Load

An impact load is one which strikes the body with certain velocity, as in the case of a falling mass M through height h (Fig. 9.8) above the collar attached to the lower end of the rod AB of length l.

It is assumed that the mass M falls freely without any loss of energy in impact and produces an instantaneous extension x and due to it instantaneous stress s is induced in this rod. Then the loss of potential energy of the mass will be stored in the form of strain energy in the rod.

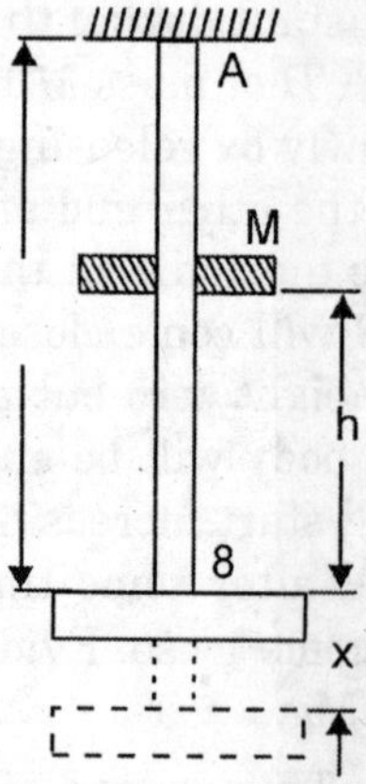

$$Mg\ (h + x) = \frac{\sigma^2}{2E}(Al) \qquad ...(9.31)$$

or $$Mg\left(h + \frac{\sigma l}{E}\right) = \frac{\sigma^2}{2E}(Al)$$

or $$\sigma^2 - \sigma\left(\frac{2Mg}{A}\right) - \frac{2\ MghE}{Al} = 0 \qquad ...(9.32)$$

A

M

h

8

x

Fig. 9.8

Taking only positive root of equation (9.32),

$$\sigma = \frac{Mg}{A}\left[1 + \sqrt{1 + \frac{2AEh}{Mgl}}\right] \qquad ...(9.33)$$

Case I.

When $h = 0$, the impact velocity will be zero and the load will be a suddenly applied load.

Then the equation (9.33) gives $\sigma = \frac{Mg}{A} = 2\frac{P}{A}$ which is same as obtained in article 9.9.2.

☛ The reader is urged to refer to any good book on vibrations.

Case II.

When x is very small as compared to h, then from equation (9.31),

$$Mgh = \frac{\sigma^2}{2E}(Al) \qquad ...(9.31)$$

or

$$\sigma = \sqrt{\frac{2\,Mgh\,E}{Al}}$$

EXAMPLE

9.1 : *A round bar of steel 1 m long is 20 mm diameter for half its length and 25 mm diameter over the rest. A gradually applied axial load of 1 kN acts on it. Find the total strain energy and strain energy per unit volume. E = 200 GN/m².*

Solution :

$$A_1 = \frac{\pi}{4}(0.02)^2 = \pi \times 10^{-4}\ \text{m}^2$$

$$\sigma_1 = \frac{1000}{\pi \times 10^{-4}} = 3.18 \times 10^6\ \text{N/m}^2$$

$$A_2 = \frac{\pi}{4}(.025)^2 = 4.91 \times 10^{-4}\ \text{m}^2$$

$$\sigma_2 = \frac{1000}{4.91 \times 10^{-4}} = 2.035 \times 10^6\ \text{N/m}^2$$

$$\text{Total strain energy} = \frac{\sigma_1^2}{2E}(A_1 l_1) + \frac{\sigma_2^2}{2E}(A_2 l_2)$$

$$= \frac{l_1}{2E}[\sigma_1(\sigma_1 A_1) + \sigma_2(\sigma_2 A_2)]$$

$$= \frac{0.5}{2E}(P\sigma_1 + P\sigma_2)$$

[Because $P = \sigma_1 A_1 = \sigma_2 A_2 = 1000$ N and $l_1 + l_2 = .5$]

$$= \frac{100 \times 0.5}{2 \times 200 \times 10^9}(3.18 \times 10^6 + 2.035 \times 10^6)$$

$$= 0.006515\ J$$

$$\text{Total volume} = A_1 l_1 + A_2 l_2 = 0.5\,(p \times 10^{-4} + 4.91 \times 10^{-4})$$

$$= 4.025 \times 10^{-4}\ \text{m}^3$$

Therefore, average strain energy per unit volume,

$$U/V = \frac{0.006515}{4.025 \times 10^{-4}} = 16.2\ \text{J/m}^3.$$

9.2 : *A wagon, of mass 5 t, is attached to a steel wire rope, the other end of which is wound round a brake drum. The wagon is descending a slope at a uniform speed of 5 km/h and the rope is taut. By the action of the brake the drum is suddenly stopped and the wagon is brought to rest. If the length of the rope between the wagon and the drum at that instant was 250 m and its cross-sectional area is 300 mm², find the maximum stress in the rope. E = 200 GN/m².*

Solution : Strain energy of rope = Kinetic energy of wagon

$$\frac{\sigma^2}{2E} Al = \frac{1}{2} Mv^2$$

or $$\sigma^2 = 5000 \times \left(\frac{5000}{3600}\right)^2 \times \frac{200 \times 10^9}{(300 \times 10^{-6}) \times 250}$$

or $$\sigma^2 = \frac{(10)^{10}}{36\sqrt{3}}$$

i.e., $$\sigma = 160 \times 10^6 \text{ N/m}^2 = 160 \text{ MN/m}^2$$

9.3 : *A bar of certain material, 40 mm diameter and 1.2 m long, has a collar securely fitted to one end. It is suspended vertically with the collar at the lower end and a mass of 2000 kg is gradually lowered on to the collar, producing an extension in the bar of 0.25 mm. Find the height from which this load could be dropped onto the collar, if the maximum tensile stress in the bar is to be 100 MN/m².*

Solution. In the first case when the load is applied gradually,

$$\sigma_1 = \frac{(2000 \times 9.81) \times 4}{\pi \times (0.040)^2} = 15.6 \times 10^6 \text{ N/m}^2$$

and $$\varepsilon = \frac{0.25 \times 10^{-3}}{1.2}$$

Therefore, $$E = \frac{\sigma_1}{\varepsilon} = \frac{15.6 \times 10^6 \times 1.2}{(0.25 \times 10^{-3})} = 75.0 \times 10^9 \text{ N/m}^2$$

For the second case,

$$\sigma_2 = 100 \times 10^6 \text{ N/m}^2$$

Corresponding extension,

$$x = \frac{(100 \times 10^6)}{75 \times 10^9} \times 1.2 = 1.6 \times 10^{-3} \text{ m}$$

If h is the height through which the mass has to be dropped, then from equation (9.31), we have,

$$(2000 \times 9.81)(h + 1.6 \times 10^{-3}) = \frac{(100 \times 10^6)}{75 \times 10^{-3}} \times 1.2$$

from which $$h = 3.52 \text{ mm.}$$

EXAMPLE

9.10 STRAIN ENERGY DUE TO SIMPLE SHEAR

Referring to Fig. 9.9, if a, b and t are the length, breadth and thickness of the rectangular block under the action of gradually applied shearing force P due to which the movement of the face AB is AA'. Then the work done, which is stored in the block as shear strain energy is given by the expression given below, where ϕ is the shear strain angle.

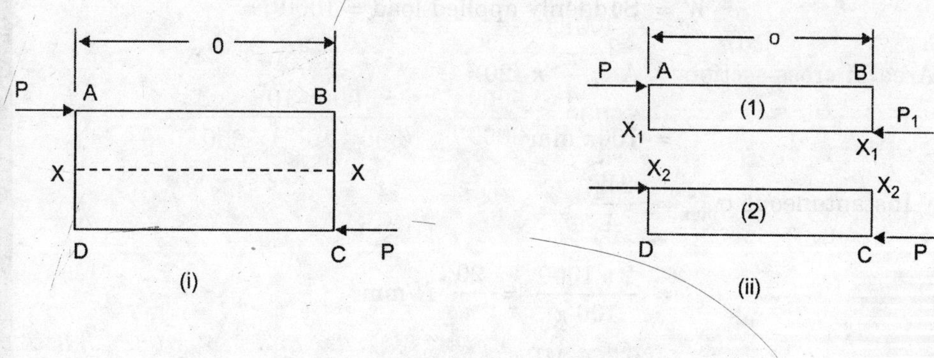

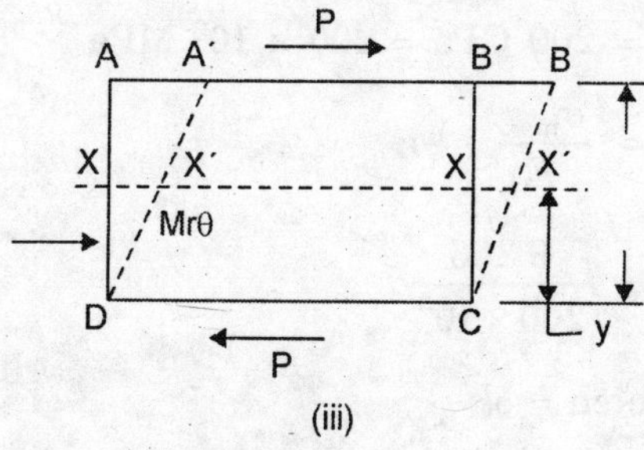

Fig. 9.9

$$U = \frac{1}{2} P(AA')$$

$$= \frac{1}{2} (\tau \,.\, a \,.\, t)(b\,\phi)$$

$$= \frac{1}{2} \tau\phi \text{ (volume)}$$

$$= \frac{1}{2} \tau \left(\frac{\tau}{G}\right) \text{volume}$$

i.e.

$$U = \frac{\tau^2}{2G} \text{ (volume)} \qquad \ldots(9.34)$$

9.4 : *On A bar of steel 1m long 20 mm diameter, a load of 1 kN is applied suddenly. Calculate the instantaneous maximum stress and the elongation. What will be stress and elongation after some time when the initial vibrations have died out? Take E = 200 GPa.*

Solution :

$$d = 20 \text{ mm}$$
$$l = 1 \text{ m}$$
$$E = 200 \text{ GPa}$$
$$W = \text{Suddenly applied load} = 1000 \text{ N}$$

$$\text{Area of cross-section} = A = \frac{\pi}{4} \times (20)^2$$
$$= 100\pi \text{ mm}^2$$

$$\text{Instantaneous } \sigma_{max} = \frac{2W}{A}$$
$$= \frac{2 \times 1000}{100\,\pi} = \frac{20}{\pi} \; N/\text{mm}^2$$
$$= 6.366 \text{ MPa}$$

Given $E = 200 \text{ GPa} = 200 \times 10^3 \text{ MPa}$

$$\text{Corresponding } \varepsilon_{max} = \frac{\sigma_{max}}{E}$$
$$= \frac{6.366}{200 \times 10^3}$$

$\therefore$ Instantaneous stretch $= \delta l$

$$= l\,(\varepsilon_{max})$$
$$= (1000) \times \frac{6.366}{200 \times 10^3} = 0.0318 \text{ mm}$$

(*b*) When the sudden load is applied the system is set into vibration mode. These vibrations die out after some time and the system comes to static state.

$$\text{Then the stress } \sigma = \frac{W}{A}$$
$$= \frac{1000}{100\,\pi} = \frac{10}{\pi} \text{ N/mm}^2 = 3.183 \text{ MPa}$$

$$\text{Corresponding stretch} = \left(\frac{\sigma}{E} \times l\right)$$
$$= \frac{3.183}{200 \times 10^3} \times 1000 \text{ mm} = 0.0159 \text{ mm}$$

EXAMPLE

E X A M P L E

9.5 : *A circular bar 50 mm diameter, 0.5 m long is subjected to a gradually applied load of 60 kN. Find the strain energy and the stain energy per unit volume.*

Take $E = 200\ GPa$

Solution :

$$l = 500 \text{ mm}$$

$$d = 50 \text{ mm}$$

$$W = 60 \times 1000\ N$$

$$A = \frac{\pi}{4} \times (50)^2 \text{ mm}^2$$

$$\sigma = \frac{60 \times 1000 \times 4}{\pi \times (50)^2} \text{ N/mm}^2$$

$$= 30.5577 \text{ MPa}$$

$$U = \text{Strain energy} = \frac{\sigma^2}{2E} \times \text{Vol}$$

$$= \frac{(30.5577)^2}{2E} \times \left(\frac{\pi}{4}(50)^2 \times 500\right)$$

$$= \frac{(30.5577)^2 \times \pi \times 2500 \times 500}{8 \times 200 \times 10^3}$$

$$= 2291.8 \text{ N-mm}$$

$$= 22918 \text{ N-m or Joules}$$

$$\text{Vol.} = \frac{\pi}{4}(50)^2 \times 500 = 981.74 \text{ mm}^3$$

$$\text{U/unit vol.} = \left[\frac{2.2918}{(981.74 \times 10^{-9})}\right] \text{J/m}^3$$

$$= 2.3344 \times 106 \text{ J/m}^3$$

$$= 2334.4 \text{ k/m}^3 \text{ or } 2.3344 \text{ kJ/m}^3.$$

9.6 : *A mild steel test piece was tested in tension and the following readings were obtained :*

Diameter of specimen = *20 mm*

Length of specimen = *20 cm*

Extension under 30 kN load = *0.08 mm*

Load at yield point = *150 kN*

Maximum load = *225 kN*

Length of specimen after fracture = *25 cm*

Calculate the values of the following :

(a) Young's Modulus

(b) Yield point stress

(c) Ultimate strengths

(d) Percentage elongation *[P.T.U. 2001]*

Solution : Diameter of specimen

$$d = 20 \text{ mm}$$

$$l = 200 \text{ mm}$$

$$dl = \text{extension } 0.08 \text{ mm at load}$$

$$P = 30 \text{ kN}$$

$$\therefore \quad \varepsilon = \frac{\delta l}{l} = \frac{0.08}{200}$$

and
$$\sigma = \frac{30 \times 1000 \times 4}{\pi \times (20)^2}$$

$$= \frac{300}{\pi} \text{ N/mm}^2$$

$$\therefore \quad E = \frac{\sigma}{\varepsilon} \times \frac{300}{\pi} \times \frac{200}{0.08}$$

$$= 238732.4 \text{ N/mm}^2 \text{ or MPa}$$

$$= 238.73 \text{ MPa or}$$

(*b*) Yield stress $= \sigma_y = \dfrac{\text{Load at yield}}{\text{Area}}$

$$\sigma_y = \frac{150 \times 10^3 \times 4}{\pi \times (20)^2}$$

$$= 477.46 \text{ N/mm}^2$$

$$= 477.46 \text{ MPa}$$

(*c*) Ultimate strength $= \dfrac{\text{Maximum load}}{\text{Area}}$

$$= \frac{225 \times 10^3 \times 4}{\pi \times (20)^2} \text{ N/mm}^2$$

$$= 716.2 \text{ N/mm}^2 \text{ or MPa}$$

(d) % elongation $= \dfrac{25 - 20}{20} \times 100$

$$= \frac{5}{20} \times 100 = 25\%$$

EXAMPLE

EXAMPLE

9.7 : *A cylindrical test piece of diameter 30 mm was put to tension. When the load was 60 kN the measured extension was 0.08 mm over the gauge length of 200 mm and the diameter reduced by 0.0045 mm. Calculate the Young's Modulus and Poisson's ratio of the material. Also, calculate the value of the Bulk Modulus and Modulus of rigidity.*

Solution : $d = 30$ mm

$$A = \frac{\pi}{4} \times (30)^2 = 225\ \pi\ \text{mm}^2$$

$$P = 60\ \text{kN}$$

$$\therefore \quad \sigma = \frac{P}{A} = \frac{60 \times 1000}{225\ \pi}\ \text{N/mm}^2$$

$$\varepsilon = \frac{dl}{l} = \frac{0.08}{200} = 0.0004$$

E = Young's Modulus

$$= \frac{\sigma}{\varepsilon} = \frac{60 \times 1000}{225\ \pi} \times \frac{1}{0.0004}$$

$$= 0.212 \times 106\ \text{N/mm}^2 \text{ or MPa}$$

$$= 212\ \text{GPa}$$

$$\text{Lateral stress} = \frac{\delta d}{d} = \frac{0.0045}{30}$$

$$\therefore \quad \text{Poisson's ratio} = \nu = \left(\frac{\frac{\delta d}{d}}{\varepsilon}\right)$$

$$= \frac{0.0045}{30} \times \frac{1}{0.0004}$$

$$= 0.375$$

(b) From equation (9.24) we have,

$$E = 3k(1 - 2\nu)$$

or
$$k = \frac{E}{3(1 - 2\nu)}$$

$$= \frac{212 \times 10^9}{3(1 - 2 \times 0.375)}\ N/\text{mm}^2$$

$$= 282.6 \times 10^9\ N/\text{mm}^2$$

$$= 282.6\ \text{GPa}$$

(c) Again from equation (9.24) we have,

$$E = 2G(1 + \nu)$$

or
$$G = \frac{E}{2(1+\nu)}$$

$$= \frac{212\times10^{9}}{2(1+0.375)} = 77.09 \times 10^{9}\ N/mm^{2}$$

$$= 77.09\ GPa$$

9.8 : *A bar 4 m long and 1200 mm² area of cross-section is subjected to a sudden load which produces an instantaneous stretch of 1 mm. Find the magnitude of the sudden load and the maximum stress induced in the bar. Take E = 200 GPa*

Solution : Given data

$$\delta l = 1.0\ mm$$
$$l = 4000\ mm$$

Then
$$\varepsilon_{max} = \frac{1.0}{4000} = 0.25\times10^{-3}$$

Therefore,
$$\sigma_{max} = E\,\varepsilon_{max}$$
$$= (200 \times 10^{3})\,(0.25 \times 10^{-3})$$
$$= 50\ MPa$$

(b) Let the suddenly applied load = W N

Then the max stress $= \sigma_{max} = \frac{2W}{A}\ N/mm^{2}$

or
$$W = \frac{A}{2}\,\sigma_{max}$$
$$= \frac{1200}{2}\times50$$
$$= 30000\ N$$
$$= 30\ kN$$

9.9 : A steel rod 4m long and 1000 mm² area of cross-section is hanging from the top and has a rigid collar at the lower end. A 10kg moving mass falls on the collar at the lower end from a height of 1cm. Calculate the instantaneous maximum stress in the rod. Take E = 200 GPa.

Solution : Given data

$$M = 10\ kg$$
$$h = 10\ mm$$
$$E = 200 \times 10^{3}\ MPa$$
$$l = 400\ mm$$

EXAMPLE

$$A = 1000 \text{ mm}^2$$

$$g = 9.827 \text{ m/sec}^2$$

$$= 9827 \text{ mm/sec}^2$$

We have from equation (13.32),

$$\sigma = \frac{Mg}{A}\left(1+\sqrt{1+\frac{2AEh}{Mgl}}\right)$$

$$= \frac{0 \times 98.27}{1000}\left[1+\sqrt{1+\frac{2\times 1000 \times (200\times 10^3)\times 10}{10\times 9827\times 400}}\right]$$

$$= 98.27\left[1+\sqrt{102.76}\right]$$

$$= 1094.4 \text{ N/mm}^2 \text{ or MPa}$$

9.10 : *A steel bar of square section 10 × 10 mm, 1 m long, when subjected to a gradually applied load of 200 N stretches 0.01 mm and the sides of the square section reduce by 2.7 × 10^{-5} mm each. Determine the Poisson's ratio, Young's Modulus, Bulk Modulus and Shear Modulus of the material.*

Solution : $A = 10 \times 10 = 100 \text{ mm}^2$

$$\varepsilon = \text{Longitudinal strain} = \frac{\delta l}{l} = \frac{0.01}{1000} = 1 \times 10^{-5}$$

$$\text{Applied stress } \sigma = \frac{W}{A} = \frac{200}{100}$$

$$= 2 \text{ N/mm}^2 \text{ or MPa}$$

$$\therefore \quad E = \frac{\sigma}{\varepsilon} = \frac{2}{1\times 10^{-5}}$$

$$= 2 \times 10^5$$

$$= 200 \times 10^3 \text{ MPa}$$

$$= 200 \text{ GPa}$$

$$\text{Lateral strain} = \frac{\delta a}{a}$$

where a = side of the square section

$$= 10 \text{ mm}$$

and $\delta a = 2.7 \times 10^{-5}$

$$\therefore \quad \text{Lateral strain} = \frac{\delta a}{a} = \frac{2.7\times 10^{-5}}{10}$$

$$= 2.7 \times 10^{-6}$$

$$\therefore \qquad \text{Poisson's ratio } \nu = \frac{\left(\dfrac{\delta a}{a}\right)}{\left(\dfrac{\delta l}{l}\right)}$$

$$= \frac{2.7 \times 10^{-6}}{10 \times 10^{-5}}$$

$$= 0.27$$

From equation (9.24), $E = 3K(1 - 2\nu) = 2G(1 + \nu)$

$$K = \frac{E}{3(1 - 2\nu)}$$

$$= \frac{200 \times 10^3}{3(1 - 2 \times .27)} = \frac{200 \times 10^3}{3 \times 0.46}$$

$$= 144.9 \times 10^3 \text{ N/mm}^2$$

$$= 144.9 \times 10^3 \text{ MPa}$$

$$= 144.9 \text{ GPa}$$

Again from equation (9.24),

$$E = 2G(1 + \nu)$$

or $$G = \frac{E}{2(1 + 0.27)} = \frac{200 \times 10^3}{2 \times 1.27} \text{ N/mm}^2$$

$$= 0.78 \times 10^5$$

$$= 78 \times 10^3 \text{ MPa}$$

$$= 78 \text{ GPa} \quad \textbf{Ans.}$$

9.11 : *A gradually applied tensile load of 1500 N acts on the specimen of shape and give shown in Fig. 9.10.*

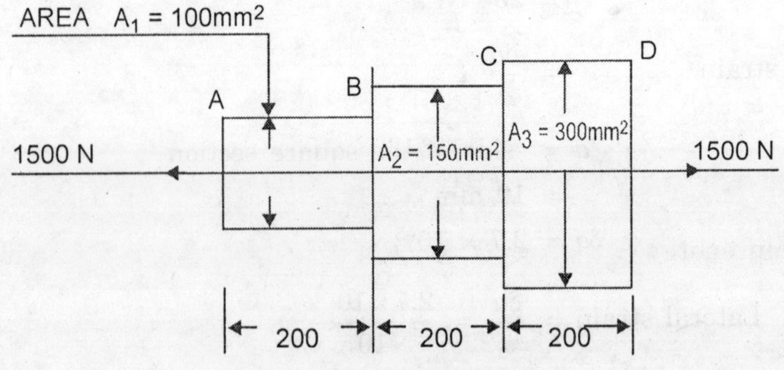

Fig. 9.10

EXAMPLE

E X A M P L E

Calculate the strain energy stored in the bar AD

Take $E = 200\ GPa$

Solution : Stress in $AB = \sigma_1 = \dfrac{1500}{100} = 15\ \text{N/mm}^2$

Strain in $BC = \sigma_2 = \dfrac{1500}{150} = 10\ \text{N/mm}^2$

Stress in $CD = \sigma_3 = \dfrac{1500}{300} = 5\ \text{N/mm}^2$

U_1 = Strain Energy in $AB = \dfrac{1}{2}\dfrac{\sigma_1^2 V_1}{E}$

where V_1 = volume of portion AB

$\therefore \quad U_1 = \dfrac{1}{2E}\sigma_1^2 A_1 l_1$

$$= \frac{1}{2E}(15)^2 \times 100 \times 200$$

$$= \frac{225}{E} \times 10^4 \text{ N-mm}$$

$$= \frac{225 \times 10^4}{200 \times 10^3} = 11.25 \text{ N-mm}$$

Similarly $U_2 = \dfrac{1}{2E}\sigma_2^2 A_2 l_2$

$$U_2 = \frac{(10)^2 \times 150 \times 200}{2 \times 200 \times 10^3}$$

$$= 7.5 \text{ N-mm}$$

and $U_3 = \dfrac{1}{2E}\sigma_3^2 A_3 l_3$

$$= \frac{1 \times (5)^2 \times 300\ (200)}{2 \times 200 \times 10^3}$$

$$= 3.75 \text{ N-mm}$$

Total strain energy $= U_1 + U_2 + U_3$

$= 11.25 + 7.50 + 3.75$

$= 22.50$ N-mm

$= 0.025$ N-mm or J

9.12 : *A low carbon steel test piece 'A' 1 m long is 20 mm diameter over 600 mm length and 15 mm diameter over 400 mm length. Another test piece of the same material and of same length is 20 mm diameter over 400 mm length and 15 mm diameter over 600 mm length. Compare the energies stored in A and B, if the same gradually applied tensile load is applied on them.*

Solution. For A,
$$l_1 = 600 \text{ mm}$$
$$d_1 = 20 \text{ mm}$$
$$l_2 = 400 \text{ mm}$$
$$d_2 = 15 \text{ mm}$$

For B,
$$L_1 = 400 \text{ mm}$$
$$D_1 = 20 \text{ mm}$$
$$L_2 = 600 \text{ mm}$$
$$D_2 = 15 \text{ mm}$$

Let the tensile gradually applied tensile load be P

For test piece *A*

$$\sigma_1 = \frac{4P}{\pi d_1^2} \; ; \; l_1 = 600 \; d_1 = 20$$

$$V_1 = \frac{\pi}{4} d_1^2 l_1$$

$$U_1 = \frac{\sigma_1^2}{2E} \times V_1$$

$$= \frac{1}{2E}\left[\frac{4P}{\pi (d_1)^2}\right]^2 \left(l_1 \frac{\pi}{4} d_1^2\right)$$

$$= \frac{1}{2E} \frac{16P^2}{(\pi d_1^2)^2} \times l_1 \times \frac{\pi d_1^2}{4}$$

$$= \frac{1}{2E} \frac{4P^2}{\pi d_1^2} l_1$$

$$= \frac{2P^2 l_1}{\pi d_1^2 E} = \frac{2P^2 (600)}{\pi (20)^2 E}$$

Similarly
$$U_2 = \frac{2P^2 l_2}{\pi d_2^2 E} = \frac{2P^2 (400)}{\pi (15)^2 E}$$

$$\therefore \quad U_A = \frac{2P^2}{\pi E}\left[\frac{l_1}{d_1^2} + \frac{l_2}{d_2^2}\right]$$

E X A M P L E

$$= \frac{2P^2}{\pi E}\left[\frac{600}{20^2} + \frac{400}{15^2}\right]$$

$$= \frac{2P^2}{\pi E}[1.5 + 1.7777]$$

$$= \frac{2P^2}{\pi E}(3.2777)$$

For the test piece B

$$D_1 = 20 \text{ mm}$$

$$D_2 = 15 \text{ mm}$$

$$U_B = \frac{2P^2}{\pi E}\left[\frac{L_1}{D_1^2} + \frac{L_2}{D_2^2}\right]$$

$$= \frac{2P^2}{\pi E}\left[\frac{400}{(20)^2} + \frac{600}{(15)^2}\right]$$

$$= \frac{2P^2}{\pi E}[1 + 2.6666]$$

$$= \frac{2P^2}{\pi E}[3.6666]$$

$$\therefore \quad \frac{U_A}{U_B} = \frac{\frac{2P^2}{\pi E}(3.2777)}{\frac{2P^2}{\pi E}(3.6666)}$$

$$= 0.8939$$

9.13 : *Two bars A and B each 0.6 m long are subjected to the same axial blow. 'A' is 40 mm over 0.4 m length and 20 mm over 0.2 m length and B is 40 mm diameter over 0.2 m length and 20 mm diameter over 0.4 m length. If the material of the both the rods is same, find the ratio of the maximum stress in A to maximum stress in B.*

Solution. Since the axial flow is same as both the rods. The energy emitted by the blow for both the rods will be same. Thus, the energy contents of the rods after they receive the same blow will be same.

i.e. $U_A = U_B$

Let σ_a be the maximum stress induced in rod A and σ_b that in the rod B. Since the maximum stress induced in any bar will occur at that portion at which the area, of cross-section is least.

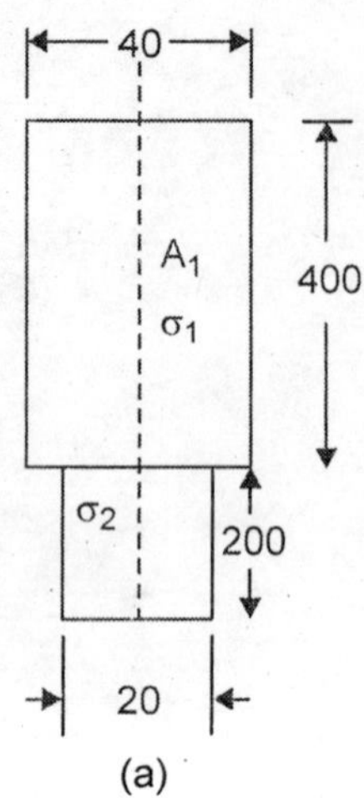

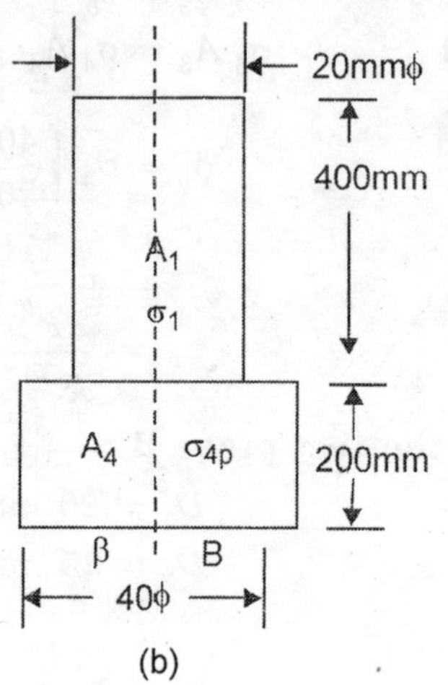

Fig. 9.11

For bar A $\quad U_A = U_1 + U_2$

$$= \frac{\sigma_1^{\ 2}}{2E} A_1 l_1 + \frac{\sigma_2^{\ 2}}{2E} A_2 l_2 \qquad \text{(since } s_a = s_1\text{)}$$

Now $\quad \sigma_1 A_1 = \sigma_2 A_2$

$$\therefore \qquad \sigma_2 = \frac{\sigma_1 A_1}{A_2} = \frac{\sigma_a \frac{\pi}{4} d_1^{\ 2}}{\frac{\pi}{4}(d_2)^2} = \frac{\sigma_a \cdot (20)^2}{40^2}$$

$$= \frac{\sigma_a}{4} = \frac{\sigma_1}{4}$$

R\ $\quad U_A = \frac{\sigma_1^{\ 2}}{2E} A_1 l_1 + \frac{\sigma_2^{\ 2}}{2E} A_2 l_2$

$$= \frac{1}{2E}\left[\sigma_a^{\ 2} \frac{\pi}{4}(20)^2 \, 200 + \left(\frac{\sigma_a}{4}\right)^2 \left(\frac{\pi}{4}\right) \times (40)^2 \times 400\right]$$

$$= \frac{\sigma_a^{\ 2}}{2E}\left(\frac{\pi}{4}\right)(80000 + 40000)$$

$$= \frac{\sigma_a^{\ 2}}{2E}\left(\frac{\pi}{4}\right)(120000)$$

For bar B $\quad U_B = \frac{\sigma_3^{\ 2}}{2E} A_3 l_3 + \frac{\sigma_4^{\ 2}}{2E} A_4 l_4$

E X A M P L E

E X A M P L E

Since $\sigma_3 = \sigma_b$

and $\sigma_3 A_3 = \sigma_4 A_4$

$$\sigma_b = \sigma_4 \left(\frac{40}{20}\right)^2$$

or $\sigma_4 = \frac{\pi}{4}$

$$U_B = \frac{1}{2E}\left[\sigma_b^2 \frac{\pi}{4}(20)^2 \times 400 + \left(\frac{\sigma_b}{4}\right)^2\left(\frac{\pi}{4}\right)(400)^2\,(200)\right]$$

$$= \left(\frac{\sigma_b^2}{2E}\right)\left(\frac{\pi}{4}\right) \times (20 \times 400)\left[20 \times 10 + \left(\frac{1}{4}\right)^2 \times 8000 \times 5\right]$$

$$= \frac{\sigma_b^2}{2E}\left(\frac{\pi}{4}\right)(8000)\left[20 + \frac{40}{16}\right]$$

$$U_B = \frac{\sigma_b^2}{2E}\left(\frac{\pi}{4}\right)(8000)\,(22.5)$$

Since $U_A = U_B$

Therefore, substituting the values of U_A and U_B,

$$\frac{\sigma_b^2}{2E}\left(\frac{\pi}{4}\right)(120000) = \frac{\sigma_b^2}{2E}\left(\frac{\pi}{4}\right)(8000)\,(22.5)$$

or
$$\left(\frac{\sigma_a}{\sigma_b}\right)^2 = \frac{8000 \times 22.5}{120000} = \frac{3}{2}$$

$$\frac{\sigma_a}{\sigma_b} = \sqrt{\frac{3}{2}}$$

or $\sigma_a : \sigma_b = \sqrt{3} : \sqrt{2}$

9.14 : *A steel bar of area of cross-section 1200 mm² and length 1.0 m is subjected to a gradually applied of load 120 kN. Find (a) the strain energy stored in the bar (b) If the elastic unit of steel is 24 MPa. Find the proof resilience.*

Take $E = 200\ GPa$

Solution : $A = 1200\ \text{mm}^2$

$l = 1000\ \text{mm}$

$W = 120000\ \text{N}$

$$\sigma = \frac{120000}{1200} = 100 \text{ N/mm}^2$$

$$\therefore \quad U = \frac{\sigma^2}{2E} \times (A)(l) = \frac{(100)^2 \times 1200 \times 1000}{2 \times (200 \times 10^3)}$$

$$= 3 \times 10^5 \text{ N-mm}$$

$$= 300 \text{ Nm (J)}$$

$$\sigma_e = 240 \text{ MPa or N/mm}^2$$

$$E = 200 \times 10^3 \text{ N/mm}$$

$$\text{Proof resilience} = \frac{\sigma_e^{\ 2}}{2E} \times (A)(l)$$

$$= \frac{(240)^2 \times 1200 \times 1000}{2 \times (200 \times 10^3)}$$

$$= (240)^2 \times 3$$

$$= 172800 \text{ N-mm}$$

$$= 172.8 \text{ N-m or J}$$

$$\text{Modulus of resilience} = \frac{\sigma_e^{\ 2}}{2E} = \text{Proof resilience per unit vol.}$$

$$= \frac{(240)^2}{2 \times 200 \times 10^3}$$

$$= \frac{240 \times 240}{400 \times 10^3}$$

$$= 144 \times 10^{-3} \text{ N/mm}^2$$

$$= 0.144 \text{ N/mm}^2$$

9.15 : *Three bars A, B and C of the same material and of same length have their areas of cross sections in the ratio 1 : 2 : 3. Find the ratio of the strain energies of A, B and C when:*

(a) the three bars are subjected to same gradually applied load

(b) when the maximum stress induced in all the three rods is same.

(c) If these bars receive the same below, find the ratio of the maximum stresses induced in the three bars.

Solution. (*a*) Gradually applied load is same let the areas of cross-sections are A_a, A_b are A_c

Then $\quad A_a : A_b : A_c :: 1 : 2 : 3$

i.e. $\quad A_b = 2\,A_a$

$\quad A_e = 3\,A_a$

E X A M P L E

and $$l_a = l_b = l_c = l \text{ (say)}$$

$$\sigma_a = \frac{W}{A_a}$$

$\therefore$ $$U_a = \frac{1}{2}\frac{\sigma_a^{\,2}}{E} \times A_a\, l_a = \frac{1}{2}\frac{W^2}{EA_a} \,.\, l$$

$$U_b = \frac{1}{2}\frac{\sigma_b^{\,2}}{E} A_b\, l_b$$

$$= \frac{1}{2E}\frac{W^2}{A_b}\, l = \frac{lW^2}{2E} \cdot \frac{1}{A_b}$$

$$= \frac{lW^2}{2E\,(2A_a)}$$

Similarly, $$U_c = \frac{1}{2}\frac{W^2}{A_c}\, l = \frac{lW^2}{2E}\frac{1}{A_c}$$

$$= \frac{lW^2}{2E(3Aa)}$$

$\therefore$ $$\frac{U_b}{U_a} = \frac{\dfrac{lW^2}{2E}\left(\dfrac{1}{2A_a}\right)}{\dfrac{lW^2}{2E}\dfrac{1}{A_a}} = \frac{1}{2}$$

and $$\frac{U_c}{U_a} = \frac{\dfrac{lW^2}{2E}\left(\dfrac{1}{3A_a}\right)}{\dfrac{lW^2}{2E}\left(\dfrac{1}{A_a}\right)} = \frac{1}{3}$$

$\therefore$ $$U_a : U_b : U_c = 1 : \frac{1}{2} : \frac{1}{3}$$

(*b*) Let the loads be such that,

$$\sigma_a = \sigma_b = \sigma_c = \sigma$$

But $$A_a : A_b : A_c : : 1 : 2 : 3$$

Then $$U_a = \frac{\sigma_a^{\,2}}{2E} A_a\, l$$

$$U_b = \frac{\sigma_b^{\,2}}{2E} A_b\, l$$

$$U_c = \frac{\sigma_c^{\ 2}}{2E} A_c\, l$$

$$\frac{U_b}{U_a} = \frac{A_b}{A_a} = 2 \qquad \text{(given } A_b = 2\, A_a\text{)}$$

and $$\frac{U_c}{U_a} = \frac{A_c}{A_a} = 3 \qquad \text{(given } A_c = 3\, A_a\text{)}$$

$\therefore$ $$U_a : U_b : U_c : : 1 : 2 : 3$$

(*c*) When the three bars receive the same below

$$U_a = U_b = U_c$$

Let the three corresponding maximum stresses be σ_a, σ_b, σ_c

Then $$U_a = \frac{\sigma_a^{\ 2}}{2E} A_a\, l$$

$$U_b = \frac{\sigma_b^{\ 2}}{2E} A_b\, l = \frac{\sigma_a^{\ 2}}{2E} A_a\, l$$

Since $$U_b = U_a$$

$\therefore$ $$\frac{U_b}{U_a} = \frac{2\sigma_b^{\ 2}}{2E} A_a\, l \,.\, \frac{2E}{(\sigma_a^2 A_a l)}$$

$$= 2\,\frac{\sigma_b^{\ 2}}{\sigma_a^{\ 2}} \quad [\text{Because } \mathrm{A_a : A_b : A_c : : 1 : 2 : 3}]$$

or $$\sigma^2 = 2\, s_1^{\ 2}$$

or $$\sigma_a = \sqrt{2}\ s_1$$

and $$U_c = \frac{\sigma_c^{\ 2}}{2E} A_c\, l$$

$$= \frac{\sigma_c^{\ 2}}{2E}\, 3\, A_a\, l$$

$\therefore$ $$\frac{\dfrac{\sigma_c^{\ 2}}{2E}\, 3\, A_a l}{\dfrac{\sigma_2^{\ 2}}{2E}\, A_a\, l} = 3\,\frac{\sigma_c^{\ 2}}{\sigma_a^{\ 2}}$$

or $$\sigma_a^{\ 2} = 3\,\sigma_c^{\ 2}$$

or $$U_c = U_c$$

or $$\sigma_a = \sqrt{3}\ \sigma_c$$

$\therefore$ $$\sigma_a : \sigma_b : \sigma_c : : 1 : \sqrt{2} : \sqrt{3}$$

9.16 : *Three test pieces of the same material and of same length have their dimensions as shown in Fig. 9.12. They are subjected to the same gradually applied load, find the ratio of the strain energies stress in them.*

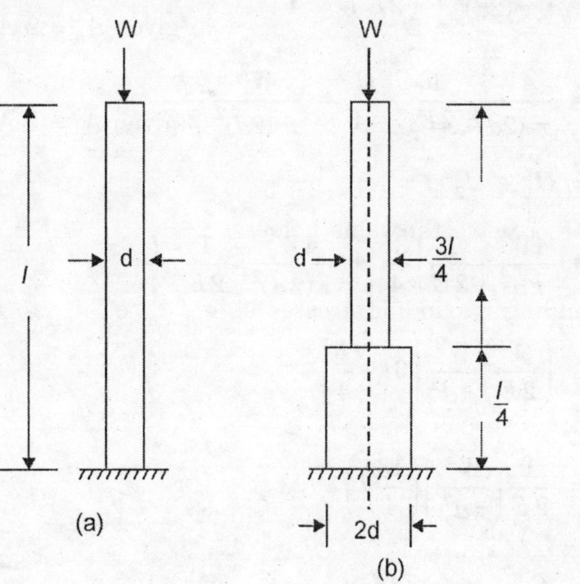

Fig. 9.12

Solution. $\sigma_a = \frac{W}{A}$

For test piece *A*

$$U_a = \frac{\sigma^2}{2E} \times \left(\frac{\pi}{4} d^2 l\right)$$

$$= \left(\frac{4W}{\pi d^2}\right)^2 \frac{1}{2E}\left(\frac{\pi}{4} d^2 l\right)$$

$$= \frac{1}{2E}\left[\frac{W^2 \times 4}{\pi d^2} l\right] = \frac{W^2}{2E(\pi d^2)}(4l)$$

For test piece *B*

$$U_b = U_1 + U_2$$

For the portion of dia 'd'

$$U_1 = \left(\frac{4W}{\pi d^2}\right)^2 \frac{1}{2E} \times \frac{\pi}{4} d^2 \left(\frac{3l}{4}\right)$$

$$= \frac{4W}{\pi d^2} \frac{1}{2E} \frac{3l}{4}$$

For portion of diameter = $2d$

$$U_2 = \left[\frac{4W}{\pi\,(2d)^2}\right]^2 \left(\frac{1}{2E}\right) \times \frac{\pi}{4}\,(2\,d)^2\,\frac{l}{4}$$

$$= \frac{4W^2}{\pi\,(2d)^2}\,\frac{1}{2E} \times \frac{l}{4} = \frac{W^2}{\pi\,(2d)^2}\,\frac{l}{2E}$$

$$\therefore \quad U_b = U_1 + U_2$$

$$= \frac{4W^2}{\pi d^2}\left(\frac{1}{2E}\right)\frac{3l}{4} + \frac{4W^2}{\pi(2d)^2}\cdot\frac{1}{2E}\cdot\frac{l}{4}$$

$$= \left(\frac{1}{2E}\right)\frac{W^2}{\pi d^2}\left[3l + \frac{l}{4}\right]$$

$$= \frac{1}{2E}\left(\frac{W^2}{\pi d^2}\right)\left(\frac{13}{4}\,l\right)$$

$$\therefore \quad \frac{U_B}{U_A} = \frac{\left(\frac{1}{2E}\,\frac{W^2}{\pi d^2}\right)\left(\frac{13l}{4}\right)}{\left(\frac{1}{2E}\,\frac{W^2}{\pi d^2}\right)(4l)}$$

$$= \frac{13}{4 \times 4} = \frac{13}{16}$$

For test piece *C*

For portion of diameter = d

$$U_1 = \frac{\sigma^2}{2E} \times \frac{\pi}{4}\,d^2\left(\frac{l}{3}\right)$$

$$= \frac{1}{2E}\left(\frac{4W}{\pi d^2}\right)^2 \times \left(\frac{\pi}{4}\,d^2\right)\frac{2l}{3}$$

$$= \frac{1}{2E}\left(\frac{W^2}{\pi d^2}\right)\left(\frac{8l}{3}\right)$$

For portion of diameter = $3d$

$$U_2 = \frac{1}{2E}\,\sigma^2 \times \frac{\pi}{4}\,(3d)^2\left(\frac{l}{3}\right)$$

EXAMPLE

E X A M P L E

$$= \frac{1}{2E}\left[\frac{4W}{\pi(3d)^2}\right]^2\left[\frac{\pi}{4}(3d)^2\right]\frac{l}{3}$$

$$= \frac{1}{2E}\frac{4W}{\pi(3d)^2}\frac{l}{3}$$

$$= \frac{1}{2E}\left(\frac{W^2}{\pi d^2}\right)\frac{4}{9}\cdot\frac{l}{3}$$

$$= \frac{W^2}{2E(\pi d^2)}\cdot\frac{4l}{27}$$

Now, $$U_c = U_1 + U_2 = \frac{1}{2E}\left(\frac{W^2}{\pi d^2}\right)\left(\frac{8l}{3}\right) + \frac{1}{2E}\left(\frac{W^2}{\pi d^2}\right)\left(\frac{4l}{27}\right)$$

$$U_c = \frac{1}{2E}\left(\frac{W^2}{\pi d^2}\right)\left[\frac{2l}{3} + \frac{4l}{27}\right]$$

$$= \frac{1}{2E}\left(\frac{W^2}{\pi d^2}\right)\left(\frac{22l}{27}\right)$$

$$\therefore \quad \frac{U_C}{U_A} = \frac{\frac{1}{2E}\left(\frac{W^2}{\pi d^2}\right)\frac{22}{27}l}{\frac{1}{2E}\left(\frac{W^2}{\pi d^2}\right)(4l)}$$

$$= \frac{22}{27 \times 4} = \frac{11}{54}$$

or $$U_A : U_B : U_C :: 1 : \frac{13}{16} : \frac{11}{54}$$

9.17 : *A round bar 2 m long and 25 mm diameter hangs vertically from one end. At the lower end a rigid collar is attached. A mass of 2000 kg falls from a height of 100 mm on the rigid collar. Find (a) the instantaneous maximum stress induced in the bar. (b) If the stretch in the bar is neglected, then what will be its value?*

Take $E = 200$ *GPa*

Solution : $g = 9.827$ m/sec^2

$= 9827$ mm/sec^2

$M = 2000$ kg

$$h = 100 \text{ mm}$$

$$d = 25 \text{ mm}^2$$

$$\therefore \quad A = \frac{\pi}{4} \times (25)^2 = 490.874 \text{ mm}^2$$

$$l = 2000 \text{ mm}$$

From equation (9.32),

$$\sigma = \frac{Mg}{A}\left[1 + \sqrt{\frac{1 + 2\,A\,E\,h}{Mgl}}\right]$$

$$= \frac{2000 \times 98.27}{490.874}\left[1 + \sqrt{\frac{1 + 2 \times 490.874 \times 200 \times 10^3 \times h}{2000 \times 9827 \times 2000}}\right]$$

$$= 40.038 \times 10^3 \left[1 + \sqrt{1.07067}\right]$$

$$= 40.038 \times 10^3 \, (2.034)$$

$$= 81.557 \times 10^3 \text{ N/mm}^2$$

(*b*) When the sketch in the bar is neglected, then we have from equation (9.33),

$$\sigma = \left[\frac{2\,Mg\,h\,E}{Al}\right]^{\frac{1}{2}}$$

$$= \left[\frac{2 \times 2000 \times 9827 \times 100 \times 200 \times 10^3}{490.874 \times 2000}\right]^{\frac{1}{2}}$$

$$= 28.297 \text{ N/mm}^2 \text{ or MPa}$$

9.18 : *A load W acts at end of a cantilever. AB of length L (Fig. 9.13) State whether the moment of this load W at the section A is bending movement or twisting moment. Explain why ?*

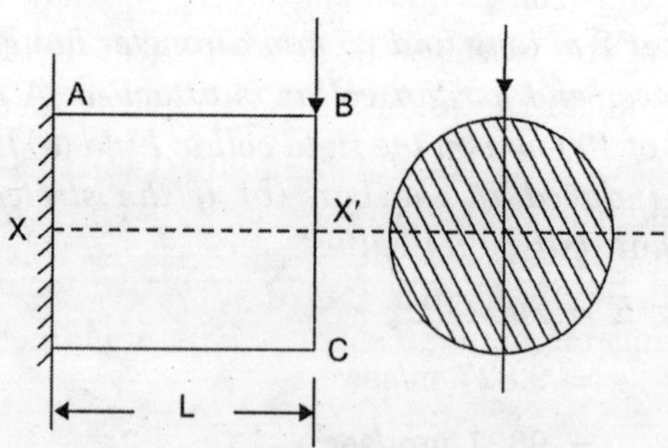

Fig. 9.13

EXAMPLE

Solution : $W \times L = M$ is the bending movement because the plane of the moment is the plane of the paper. Then the axis will be normal to the plane. Since longitudinal axis of the leam is in the plane of the paper, the axis of the moment is the axis of the beam, therefore, it is a bending moment.

9.19 : *In Figure 9.14 is shown we circular section bar AB with its end A fixed in the wall so that AB acts like a cantilever. At the end B is fixed a wheel of radius 'r' and at the periphery of the wheel is fixed a rope from which a load W is hanging. State whether at the section A of the cantilever, there is a BM or twisting moment and what are their values?*

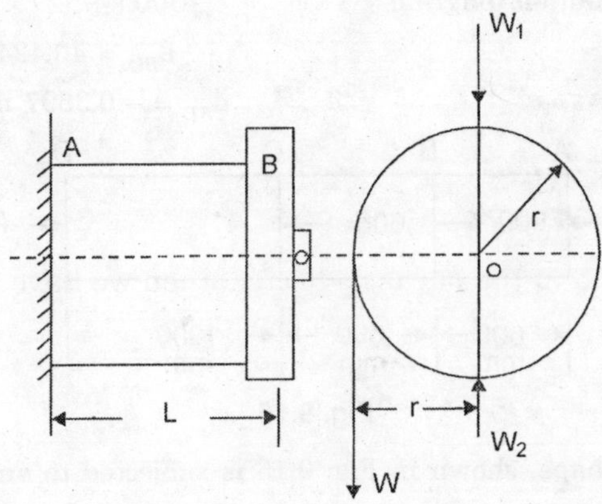

Fig. 9.14

Solution. The system is in static equilibrium condition. At the point O, centre of the wheel from our own side, we apply two equal and opposite loads. W_1 and W_2

Such that $W_1 = W_2 = W$

Now W and W_2 being equal and opposite in direction and are at a distance equal to 'r' they constitute a couple the moment of which is ($W \times r$) with the plane of the wheel as its plane. Naturally the axis of this moment will be normal to the plane of the wheel as seen in side view. Thus, it axis will coincide with axis of the bar AB. Thus, (Wr) moment is the twisting moment.

In addition the load W, at the end B will give rise to moment $W \times L$ which will be bending moment at section A as explained in. Example 9.18.

PROBLEM FOR PRACTICE

1. Define and explain direct stress, shear stress and hydrostatic stress.
2. Define and explain the terms (*a*) Longitudinal strain (*b*) Lateral strain (*c*) Poisson's ratio.
3. What are the limiting values of Poisson's ratio?
4. What type of material exhibit negative Poisson's ratio ? Where do such materials find their utility ?
5. Explain the 'Rubbery Material' and "Anti Rubber "Materials". Where are these used ?
6. A low carbon steel bar of diameter 60 mm is subjected to axial forces, as shown in Fig. 9.15. Find the change in length of each segment of the bar and then calculate the relative movement of the end *D* with respect to end A of the bar.

   Take $E = 200$ GPa

   Draw the force distribution diagram. (**Ans.** $d_{AB} = 0.53.5$ mm, $d_{BC} = 40.4244$ mm, $d_{CD} = -0.3536$ mm, $d_{AD} = -0.3697$ mm, *i.e.* *A* and *D* conclusion)

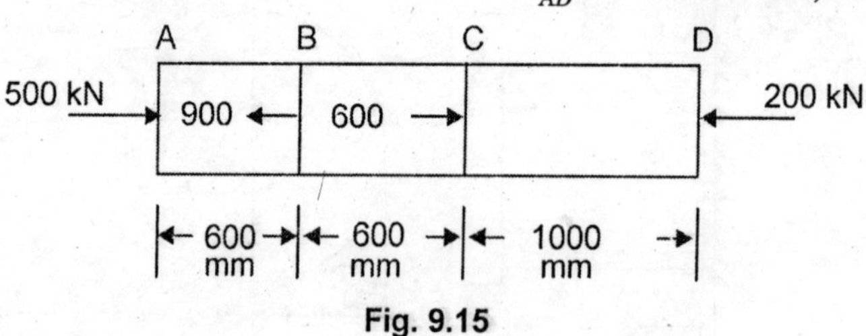

Fig. 9.15

7. A steel piece of the shape, shown in Fig. 9.16 is subjected to an axial pull of 10 kN.

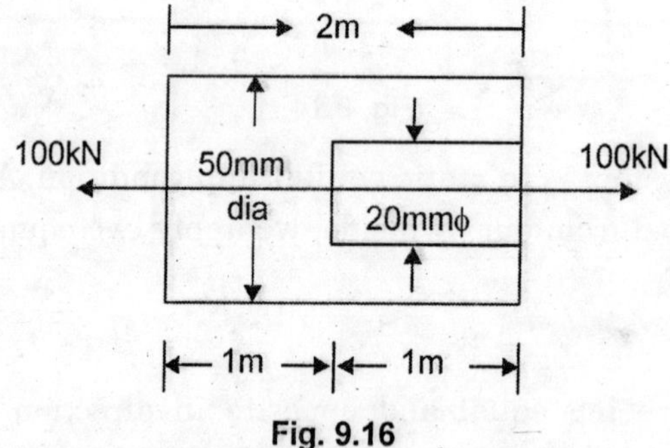

Fig. 9.16

   Find the extension of the piece *AC*, Take E = 200 GPa

8. A hole of 20 mm diameter is to be punched in a plate of 25 mm thickness (*i*) If the ultimate shear stress of the plate material is 360 MPa find the force required to punch the hole. (**Ans.** 565.5 kN)
9. A steel bar 1 m long and 20 mm diameter is subjected to an axial pull of 25 kN. Find the change is length, change in diameter and volumetric strain.

   Take $E = 200$ GPa

   and Poisson ratio = 0.25 (**Ans.** $dl = 3.979 \times 10^{-4}$ mm, $d = 2.4864 \times 10^{-6}$ mm, $\frac{dV}{V} = 5.9655 \times 10^{-7}$)

10. A cylindrical test piece of diameter 25 mm was subjected to a tensile load of 40 kN. The measured extension was 0.08 mm over a gauge length of 200 mm and the reduction in diameter was 0.0025 mm. Find Young's Modulus, Poisson's ratio, Bulk Modulus and Shear Modulus.

(**Ans.** E = 203.72 × 10^3 MPa n = 0.25,
K = 135.8 × 10^3 MPa G = 81.49 × 10^3 MPa)

11. A steel ball of 100 mm diameter is immersed in sea water to a depth where the hydrostalic pressure is 100 MPa. If the value of E = 200 GPa and n = 0.27 for steel, calculate the change in volume (decrease) of steel ball. (**Ans.** 1.021 mm^3)

12. A steel bar AB, 2m long is fixed at the end A and at end B, a rigid circular disc. of diameter 400 mm is keyed. At the periphery of the disc a force of 10 kN is applied in the vertically downward direction as shown in the figure (9.17). Find the values of bending moment and twisting moment at the fixed end A of the bar. (**Ans.** BM = 20 kN-m, T = 2 kN-m)

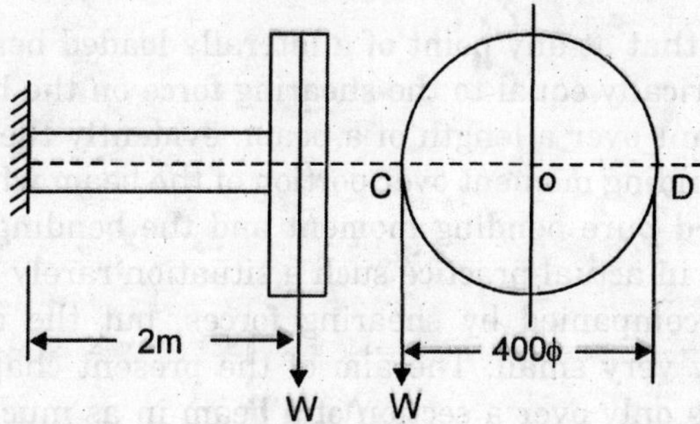

Fig. 9.17

13. Two bars A and B, each 0.5 m long are subjected to same axial blow. A is 50 mm diameter over 0.3 m length and 25 mm diameter over 0.2 m length and 25 mm diameter over 0.2 m length and B is 50 mm diameter over 0.2 m length and 25 mm diameter over 0.3 m length. If the material of both the bars is same, find the ratio of maximum induced in A to that induced in the bar B.

$\left(\textbf{Ans.}\ \sqrt{\frac{14}{11}}\right)$

13. A bar 10 mm diameter and 3 m long is fixed the top and hangs vertically. At the lower and is fixed is rigid disc and on this disc a mass of 1 kg is allowed to be fall freely through a distance of 50 mm. Calculate the maximum stress induced in the rod.

Take E = 200 GPa. (**Ans.** 92.5 MPa)

❑❑❑

CHAPTER 10

PURE BENDING OF BEAMS

10.1 SIMPLE OR PURE BENDING

It was seen in article 3.4 that at any point of a laterally loaded beam, the rate of change of bending moment is numerically equal to the shearing force on the beam at that point. If the bending moment is constant over a length of a beam, evidently the shearing force over that length will be zero. The bending moment over portion of the beam where it is not accompanied by shearing force, is called pure bending moment and the bending of the beam as pure or simple bending. However, in actual practice such a situation rarely arises and invariably the bending moments are accompanied by shearing forces, but the distortion caused by the shearing action is usually very small. The aim of the present chapter will be to study the effect of bending moments only over a section of a beam in as much as to how the material is stressed at various points of a section of the beam.

10.2 THEORY OF SIMPLE BENDING

If an initially straight beam is acted upon by a pure bending moment 'M', the beam will get bend as shown in Fig. 10.1. Evidently the fibres in the upper portion will be extended and those in the lower part will get compressed. Alternatively the fibres which are nearer to the centre of the curvature of the bent beam will be under compression and those away in tension. In between there will be some surface NS the fibres along which will neither extend nor shorten. This surface is known as the neutral surface and the intersection of this surface with any cross–sectional plane is called the *neutral axis* of the section. It is represented by NA in Fig. 10.1. In evaluating the tensile or compressive stresses and the curvature of the beam following assumptions will be made :

1. The beam is initially straight and the radius of curvature of the bent beam is very large as compared to the dimensions of the cross–section.
2. The material of the beam is homogeneous and isotropic, that is, its properties are alike in all directions and at all points.
3. The material is not stressed beyond the elastic limit and obeys Hooke's law.
4. The values of a Young's modulii of elasticity in tension and compression are equal.
5. Transverse plane sections remain plane even after bending.
6. Each layer is free to expand or contract independently without being constrained by the adjoining fibres.

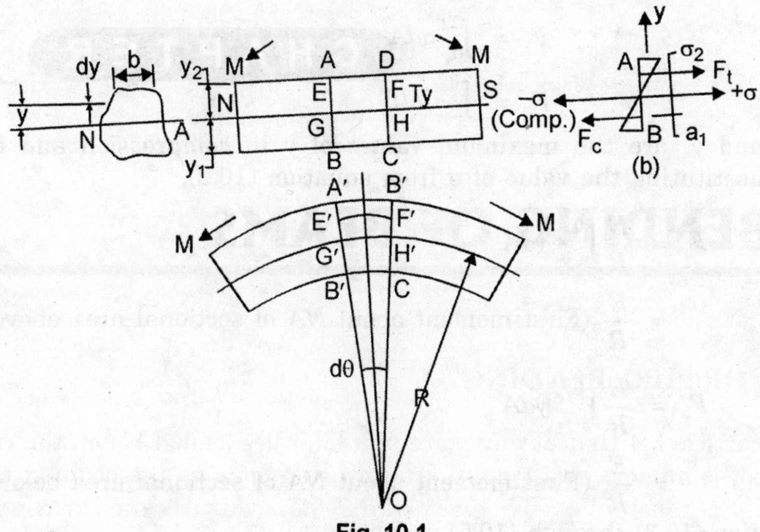

Fig. 10.1

Let the radius of curvature of the neutral surface of the beam at the section AB, be R such that $EF = GH = G'\ H' = R\ (d\ \theta)$. Considering the change in length of the fibre EF at a distance 'y' from the Neutral Surface the tensile strain along EF is,

$$\varepsilon = \frac{E'F' - EF}{EF} = \frac{[(R+y)\,d\theta - Rd\theta)]}{Rd\theta}$$

$$= \frac{y}{R} \qquad ...(10.1)$$

If σ is the direct stress along the fibre EF and E the Young's modulus of the Material, then equation (10.1) gives,

$$\frac{\sigma}{E} = \frac{y}{R}$$

or

$$\frac{\sigma}{y} = \frac{E}{R} \qquad ...(10.2)$$

From equation (10.2), it is evident that the direct stress at a point in the cross–section increases with y linearly. Its value will be zero at any point lying on the neutral axis. Similarly if y is measured from the neutral axis to any point in the region nearer to the centre of curvature the direct stress, there will be compressive and increase linearly with y. Since the value of the Young's Modulus E has been assumed to be same for tension and compression, the stress distribution curve in the two regions will have the same slope, rather extension of the same line in the two regions Fig. 10.1(b).

Position of the Neutral Axis

If there is no externally applied axial force, for the equilibrium of the portion of the beam lying on the left of the section AB the algebraic sum of the forces in the axial direction will be zero. That is, if the total force in the tension region is F_t and that in the compression region F_c then,

$$F_t + F_c = 0 \qquad ...(10.3)$$

Now if an elemental area dA is taken at a distance y from the neutral axis

$$F_t = \int_0^{y_2} \sigma dA$$

and
$$F_c = \int_{-y_1}^{0} a dA$$

Where y_1 and y_2 are the maximum values of y in compression and tension regions respectively. Substituting the value of σ from equation (10.2),

$$F_t = \frac{E}{R}\int_{-y_1}^{0} ydA$$

$$= \frac{E}{R} \text{ (First moment about } NA \text{ of sectional area above } NA) \quad ...(10.4)$$

and
$$F_c = \frac{E}{R}\int_{-y_1}^{0} ydA$$

$$= \frac{E}{R} \text{ (First moment about } NA \text{ of sectional area below } NA) \quad ...(10.5)$$

From equation (10.3) through (10.5) we obtain,

$$\text{F}_t + \text{F}_c = \frac{E}{R} \text{ (First moment about } NA \text{ of the cross–section)} = 0 \quad ...(10.6)$$

But E/R cannot be zero, hence, from equation (10.6), the first moment about NA of the cross–section must be zero. This will only be true if the neutral axis NA passes through the centroid of the section.

Evaluation of Radius of Curvature

From equation (10.3), it is evident that the two forces F_t and F_c will be equal and opposite. The couple produced by these forces acting at a distance $(a_1 + a_2)$ apart, Fig. 10.1(*b*), will balance the moment M for the equilibrium of the portion of the beam to the left of the section AB. The moment $F_t\ (a_1 + a_2)$ is called the resisting moment of the section.

Thus,
$$M = F_t\ (a_2 + a_1)$$

$$= F_t a_2 + F_c a_1 \quad \text{(as } F_t = F_c)$$

$$= \text{Moment of the tensile forces about NA + Moment of compressive force about } NA$$

$$= \int_0^{y_2} (\sigma dA)\, y + \int_{-y_1}^{0} (\sigma dA)\, y$$

$$= \int_0^{y_2} \left(\frac{yE}{R} dA\right) y + \int_{y_1}^{0} \left(\frac{E}{R} ydA\right) y \quad \text{(From Eq. 10.2)}$$

$$= \frac{E}{R}\left[\int_0^{y_2} (dAy)^2 + \int_{-y_1}^{0} (dAy^2)\right]$$

$$= \frac{E}{R}\int_{-y_1}^{y_2} (y^2 dA)$$

$$= \frac{E}{R} \times \text{(Second moment of the area about NA which passes through the centroid)}$$

$$= \frac{E}{R}.I$$

or
$$M/I = E/R \quad ...(10.7)$$

Where I is the second moment of the area of cross–section about a centroidal axis NA, which is also parallel to the axis of the moment M.

The equations (10.2) and (10.7) can also be written in the usual form as,

$$\frac{\sigma}{y} = \frac{M}{I} = \frac{E}{R} \quad ...(10.8)$$

Knowing the values of M and I at any section of a beam the radius of curvature R of the neutral surface at that section can be found with the help of equation (10.8).

Again from equation (10.8),

$$\sigma_{max} = \frac{M}{I / y_{max}} = \frac{M}{Z}$$

Where $Z = I/y_{max}$. is known as section modulus.

10.2.1 Second Moment of Area, Theorem of Perpendicular Axes

It is defined as the summation of the product of elemental areas and the square of their respective distances from a given axis, for the whole area.

Let 'O' be an elemental area at a distance y from x-axis. The integral of $\oint y^2\, dA$ for the entire area is called the second moment of area and represented by I_{xx}.

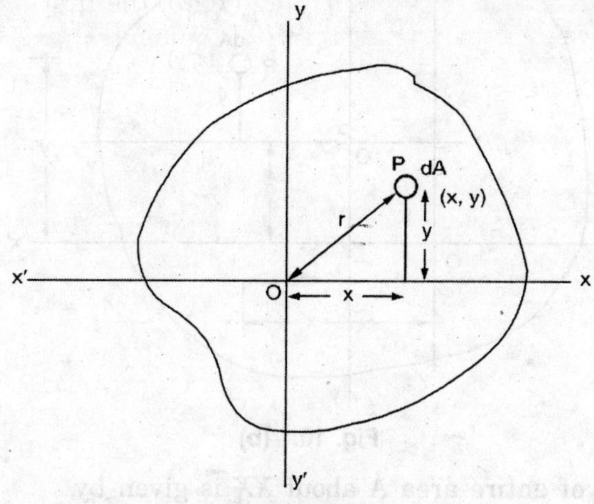

Fig. 10.1 (a)

Thus, $$I_{xx} = \oint y^2 dA \quad ...(A)$$

Similarly if the distance of dA is x from y-axis, then second moment of area about y-axis will be,

$$I_{yy} = \oint x^2 dA \quad ...(B)$$

However, if we take Z–axis which is normal to the plane $X - Y$ and passing through origin then I_{zz}, which is called the polar second moment of area is defined as,

$$I_{zz} = \oint r^2 dA \quad ..(C)$$

When r is the distance of dA from zz axis

Expression (C) can also be written as,

$$I_{zz} = \oint (x^2 + y^2)\, dA$$

$$= \oint y^2 dA + \oint x^2 dA$$

$$= I_{xx} + I_{yy} \qquad ...(D)$$

Equation (*D*) is known as theorem of perpendicular axis. That is the polar second moment of area of any section about an axis *OZ*, which is normal to the a in equal to the sum of second moment of areas about two perpendicular axis *XX* and *YY* lying in the cross-sectional plane.

Till the middle of twentieth century the term I had been called as moment of Inertia. The scientists from all over the world collectively decided to call this as second moment of area and the Mass Moment of Inertia for the second moment of the elemental mass. But still some of the authors continue with and misnomer the call it moment of inertia rather than second moment of area.

10.2.2 Theorem of Parallel Axis

Let 0 be the centroid of certain area and *OX* and *OY*, two axis of reference figure [10.1 (*b*)]. The elemental area of *dA* around *P* has co-ordinates (*x*, *y*).

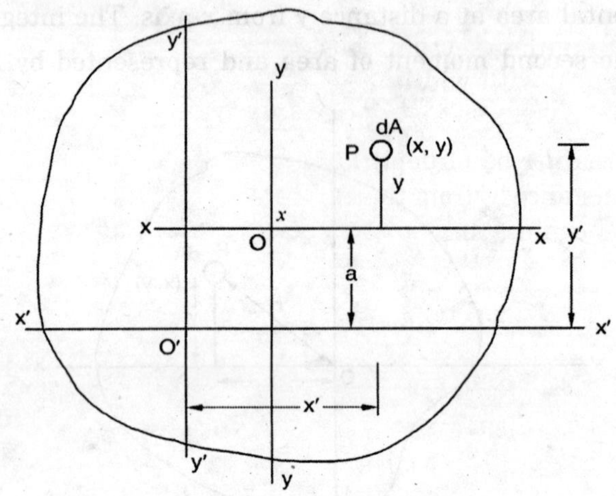

Fig. 10.1 (b)

The second moment of entire area Å about *XX* is given by,

$$I_{XX} = \oint y^2 dA \qquad ...(D)$$

Let $X'O'Y'$ be another set of axes such that the co-ordinates of point *P* with respect to $X'O'Y'$ axes are (x', y')

Then $$I_{x'x'} = \oint (y')^2 dA \qquad ...(E)$$

$$= \oint [y + (y' - y)]^2 dA$$

Let $$y' - y = a$$

Then $$I_{x'x'} = \oint (y + a)^2 dA$$

$$= \oint (y^2 + a^2 + 2ay)\, dA$$

$$= \oint y^2 dA + a^2 A + 2a \oint y dA$$

But $\oint ydA$ is first moment of the area about centroid 'O' of the section. It must be zero as X-axis passes through centroid of the section

Thus, $$I_{X'X'} = I_{XX} + Aa^2 \qquad \text{...(F)}$$

Equation (F) is known as theorem of parallel axes. It states that second moment of area of any section about X'X' which is parallel to XX passing through centroid, is equal to I_{XX} about the centroidal axis plus area times square of the distance between these axes.

10.2.3 Second Moment of Areas of Standard Sections

In the following paragraphs, expressions for the second moment of area of various commonly used sections, have been derived.

10.2.4 Rectangular Section

Let *ABCD* be the rectangular section with its centroid at 'O'. [Fig. 10.1(*c*)] with width as 'b' and depth 'd'.

Let the elemental area *dA* be of depth *dy* and width 'b' and taken at a distance 'y' from the axis *XX* and I_{XX} is intended to be evaluated.

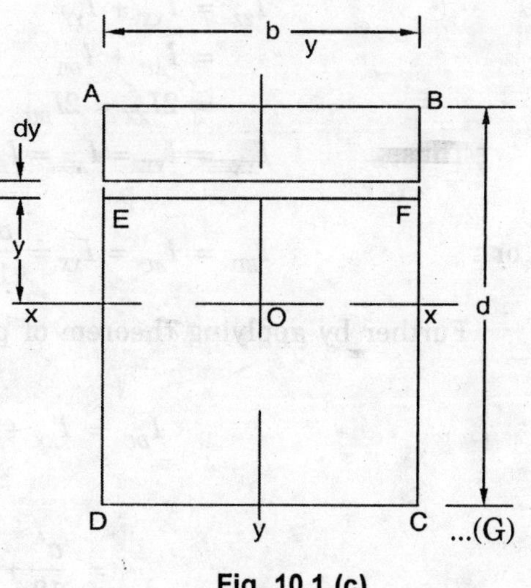

Fig. 10.1 (c)

Then $$I_{XX} = \int_{-d/2}^{d/2} y^2 dA = \int_{-d/2}^{d/2} y^2 b dy$$

$$= \left[\frac{y^3}{3}\right]_{-d/2}^{d/2} = \frac{bd^3}{12} \qquad \text{...(G)}$$

Similarly, by taking the elemental area strip parallel to *y* axes at a distance *x*, Fig. (10.1*d*) we have,

$$I_{YY} = \int_{-b/2}^{b/2} x^2 dA$$

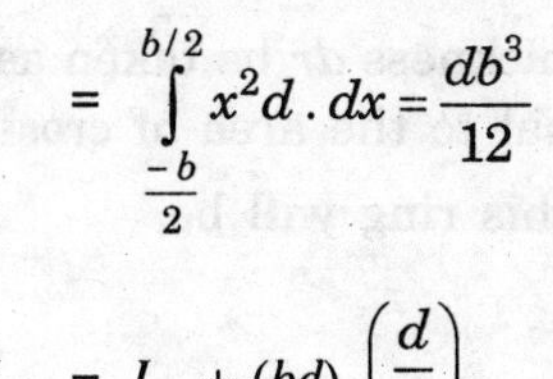

$$= \int_{-b/2}^{b/2} x^2 d \,.\, dx = \frac{db^3}{12}$$

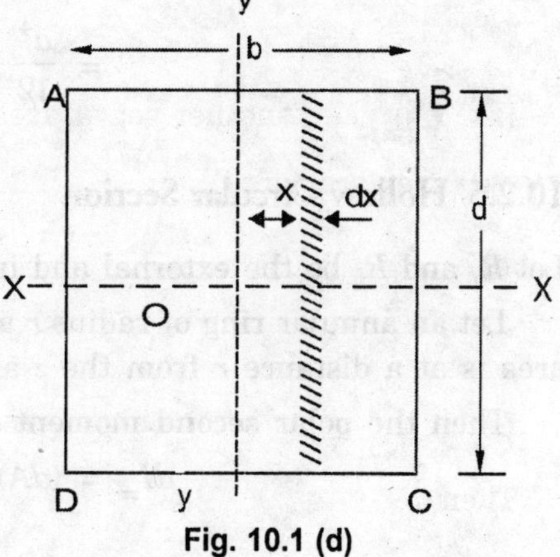

Fig. 10.1 (d)

Also, $$I_{DC} = I_{XX} + (bd)\left(\frac{d}{2}\right)$$

$$= \frac{bd^3}{12} + \frac{bd^3}{4} = \frac{bd^3}{3} \quad \text{...(H)}$$

It is important to note that is the expressions (*g*) and (*h*) the side of the rectangle which is parallel to the axis under consideration appears with exponent as unity and the perpendicular side with exponent 3.

10.2.4.1 Second Moment of Area of Square Section

If the two sides of the rectangular section are both equal to '*a*', '*say*' Fig (10.*1e*)

Then $$I_{XX} = I_{YY} = \frac{a \times a^3}{12} = \frac{a^4}{12}$$

Further since the two diagonals of square section are at right angles to each other, applying. Theorem of perpendicular axes, we get,

$$I_{ZZ} = I_{XX} + I_{YY}$$
$$= I_{AC} + I_{DB}$$
$$= 2I_{XX} = 2I_{DC}$$

Thus, $$I_{XX} = I_{YY} = I_{AC} = I_{BD}$$

or $$I_{BD} = I_{AC} = I_{XX} = \frac{a^4}{12} \quad \text{...(I)}$$

Fig. 10.1 (e)

Further by applying theorem of parallel axes we get,

$$I_{DC} = I_{XX} + (\text{Area } ABCD) \cdot \left(\frac{1}{2}BC\right)^2$$

$$= \frac{a^4}{12} + a^2\left(\frac{a}{2}\right)^2$$

$$= \frac{a^4}{12} + \frac{a^4}{4} = a^4\left(\frac{1}{12} + \frac{1}{4}\right) = \frac{a^4}{3} \quad \text{...(J)}$$

10.2.5 Hollow Circular Section

Let R_1 and R_2 be the external and inner radii of a hollow circular section [Fig (10.1 *f*)]

Let an annular ring of radius *r* and radial thickness *dr* be taken as elemental area. This area is at a distance *r* from the *z*-axis O_z normal to the area of cross-section.

Then the polar second moment of area of this ring will be,

$$dI_{zz} = (dA)z^2$$

or $$I_{zz} = \int_{R_2}^{R_1} r^2 dA$$

But $$dA = 2\pi r\, dr$$

$\therefore$ $$I_{zz} = \int_{R_2}^{R_1} r^2 (r\pi r dr)$$

$$= 2\pi \int_{R_2}^{R_1} r^3 dr$$

$$= 2\pi \left[\frac{r^4}{4}\right]_{R_2}^{R_1}$$

$$= 2\pi \left(\frac{R_1^4}{4} - \frac{R_2^2}{4}\right)$$

Fig. 10.1 (f)

or $$I_{zz} = \frac{\pi}{2}\left(R_1^4 - R_2^2\right) \qquad \text{...}(K-1)$$

$$= \frac{\pi}{32}\left(D_1^4 - D_2^4\right) \qquad \text{...}(K-2)$$

For the solid shaft the diameter D the value of I_{zz} is obtained by putting $D_2 = 0$ and replacing D_1 with D,

$$I_{zz} = \frac{\pi}{32} D^4 \qquad \text{...}(L)$$

But we know, with the help of theorem of perpendicular axes that,

$$I_{ZZ} = I_{XX} + I_{YY}$$

Since the section is symmetrical about XX and YY axes,

$$I_{XX} = I_{YY}$$

Therefore, we get, $$I_{ZZ} = 2I_{XX}$$

or $$I_{XX} = \frac{1}{2} I_{ZZ} = \frac{1}{2}\left(\frac{\pi D^4}{32}\right)$$

i.e., $$I_{XX} = I_{YY} = \frac{\pi D^4}{64} \qquad \text{...}(M)$$

10.2.6 Triangular Section

For a triangular section ABC Fig. (10.*1g*) with base $BC = b$ and height $AD = h$, the centroid G is at a height $\frac{h}{3}$ from the base.

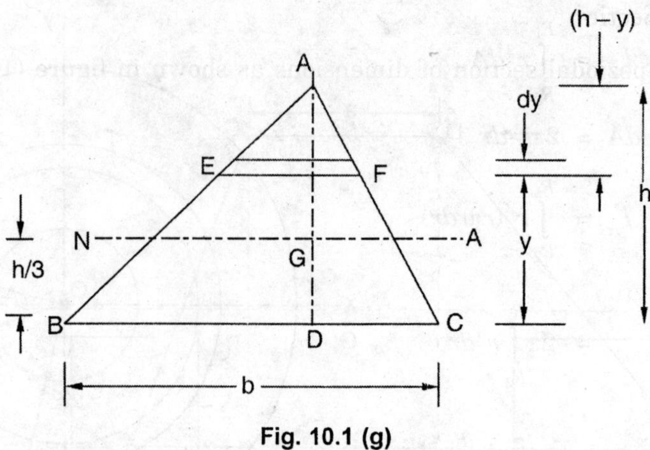

Fig. 10.1 (g)

Let an elemental area dA be at a height 'y' from the base and it thickness be dy

Then $$I_{BC} = \int_O^h (dAy^2) = \int_O^h y^2 (EF)\, dy$$

But $$EF = \frac{BC}{h} . (h-y) = \frac{b}{h}(h-y)$$

Therefore, $$I_{BC} = \int_0^h y^2 \frac{b}{h}(h-y)\, dy$$

or $$I_{BC} = b\int_0^h \left(y^2 - \frac{y^3}{h} \right) dy$$

$$= b\left[\frac{y^3}{3} - \frac{y^4}{4h} \right]_0^h$$

$$= b\left[\frac{h^3}{3} - \frac{h^3}{4} \right] = \frac{bh^2}{12} \quad ...(N)$$

Neutral axis passes through G, the centroid of the section Invoking theorem of parallel axes we have,

$$I_{BC} = I_{NA} + (\text{Area } ABC) \left(\frac{h}{3} \right)^2$$

$$\therefore \quad I_{NA} = I_{BC} - \left(\frac{1}{2} bh \right) \left(\frac{h}{3} \right)^2$$

$$= \frac{bh^3}{12} - \frac{bh^3}{18} = \frac{bh^3}{36} \quad ...(O)$$

10.2.7 Trapeziodal Section

Let *ABCD* be the trapeziodal section of dimensions as shown in figure (10.1*h*)

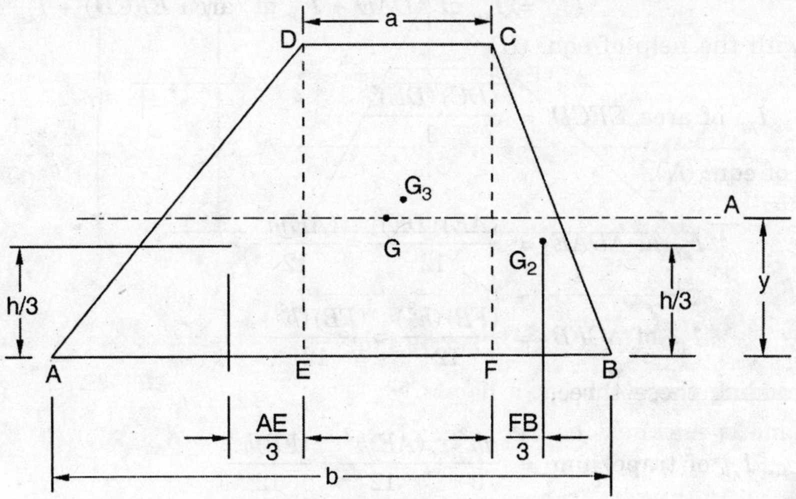

Fig. 10.1 (h)

i.e., $AB = b$

$CD = a$

and height $= DE = h$

For finding the location or specifically the height y of centroid of the trapezium from the base *AB*, we take the moments of areas *ADE, DEFC* and *CFB* about the base *AB* (since the height of G_1 and G_2, the centroids of Δs *DEA* and *BFC* respectively, from base *AB* is $h/3$ and centroid G_3 of rectangle *DCFE*, is at height $h/2$ from *AB*)

Then
$$(\text{Area } ABCD)\,\bar{y} = \text{Area }(AED)\frac{h}{3} + (\text{Area } DEFC)\frac{h}{2} + (\text{area } CFB)\frac{h}{3}$$

or
$$\left[\frac{1}{2}(a+b)h\right]\bar{y} = \left(\frac{1}{2}hAE\right)\frac{h}{3} + (ah)\frac{h}{2} + \frac{1}{2}h\,(FB)\frac{h}{3}$$

$$= \frac{h^2}{6}(AE + FB) + \frac{ah^2}{2}$$

$$= \frac{h^2}{6}(b - a) + \frac{ah^2}{2}$$

$$= \frac{h^2}{6}(b - a + 3a) = \frac{h^2}{6}(b + 2a)$$

Therefore,
$$\bar{y} = \frac{h^2}{6}(b + 2a)\frac{2}{(a+b)\,h}$$

or
$$\bar{y} = \frac{h}{3}\left(\frac{b + 2a}{b + a}\right) \quad \text{...(P)}$$

For finding the value of I_{NA} which passes through the centroid G, we first find the value of the second moment of area about the base BC and then transfer to centroidal axis NA

$$I_{AB} = I_{AB} \text{ of } \Delta DAE + I_{AB} \text{ of (area } EFCD) + I_{AB} \text{ of } \Delta FBE$$

We have with the help of eqn. (H),

$$I_{AB} \text{ of area } EFCD = \frac{(DC)(DE)^3}{3}$$

with the help of eqn. (N),

$$I_{AB} \text{ of } \Delta DAE = \frac{(AE)(DE)^3}{12} = \frac{(AE)h^3}{12}$$

and $$I_{AB} \text{ of } \Delta CFB = \frac{(FB)(h^3)}{12} = \frac{(FB)(h^3)}{12}$$

Therefore adding these three,

$$I_{AB} \text{ of trapezium} = \frac{ah^3}{3} + \frac{(AE)h^3}{12} + \frac{(FB)h^3}{12}$$

$$= \frac{ah^3}{3} + \frac{h^3}{12}(AE + FB)$$

$$= \frac{ah^3}{3} + \frac{h^3}{12}(b - a) \qquad (\because b = AE + FB + EF)$$

or $$I_{AB} = \frac{h^3}{12}(4a + b - a)$$

or $$I_{AB} = \frac{h^3}{12}(3a + b) \qquad ...(Q)$$

Now from theorem of parallel axes,

$$I_{AB} = I_{NA} + \text{Area } ABCD \times (\bar{y})^2$$

where $\bar{y}$ distance of centroid G from AB,

or $$I_{NA} = I_{AB} - \text{Area } ABCD \times (\bar{y})^2$$

$$= \frac{h^3}{12}(3a + b) - \frac{1}{2}(a + b)\,h\left[\frac{h}{3}\frac{(b + 2a)}{(b + a)}\right]^2$$

$$= \frac{h^3}{12}(3a + b) - \frac{(a + b)\,h^3}{18}\frac{(b + 2a)^2}{(b + a)^2}$$

$$= \frac{h^3}{12}(3a + b) - \frac{h^3}{18}\frac{(b + 2a)^2}{(b + a)}$$

$$= \frac{h^3}{36}\left[3(3a + b) - \frac{2(b + 2a)^2}{(b + a)}\right]$$

$$= \frac{h^3}{36(b+a)} [3(3a + b)(b + a) - 2(b + 2a)^2]$$

$$= \frac{h^3}{36\,(b+a)} [3(3ab + 3a^2 + b^2 + ab) - 2(b^2 + 4a^2 + 4ab)]$$

$$= \frac{h^3}{36\,(b+h)} [9ab + 9a^2 + 3b^2 + 3ab - 2b^2 - 8a^2 - 8ab]$$

$$= \frac{h^2}{36\,(b+a)} [4ab + b^2 + a^2]$$

$$= \frac{h^3}{36\,(b+a)} [(b+a)^2 + 2ab]$$

or
$$I_{NA} = \frac{h^3}{36}\left[(a+b) + \frac{2ab}{(a+b)}\right] \quad ...(R)$$

Equation (*R*) finds its wide application while determining the stresses in crane Hooks of Trapezoidal section.

EXAMPLE **10.1 :** *A beam of I–section shown in Fig. 10.2 is freely supported at its ends. Over what span can a uniform load of 5000 N/m run be carried, if the maximum stress is limited to 50 MN/m²?*

Solution :
$$I_{NA} = \frac{300 \times 760^3}{12} - \frac{280 \times 700^3}{12}$$

$$= 0.297 \times (10)^{10}\,(\text{mm})^4$$

Now
$$M_{max} = \frac{\sigma_{max}}{y_{max}}.$$

$$I_{NA} = \frac{50 \times 2}{760} \times 0.297 \times 10^{10}$$

$$= 391 \times 10^8 \text{ N-mm}$$

$$= 3.91 \times 10^5 \text{ N-m}$$

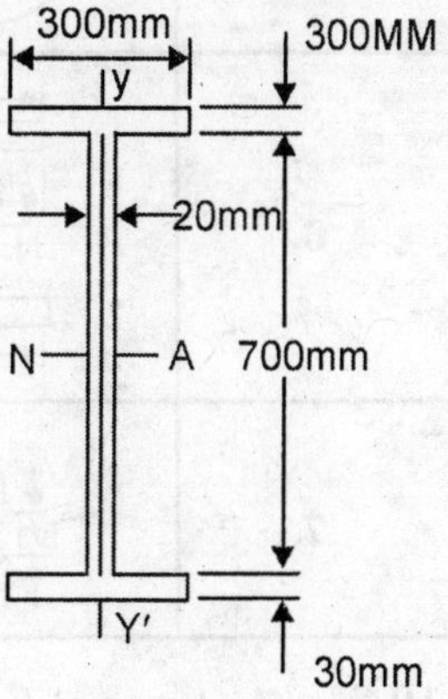

Fig. 10.2

For a simply supported beam with span l and a distributed load w, we have at the mid span

$$M_{max} = \frac{1}{8} wl^2$$

$$= \frac{5000 \times l^2}{8} = 3.91 \times 10^5 \text{ N-m}$$

i.e.
$$l^2 = \frac{3.91 \times 10^5 \times 8}{5000} = 625.6 \text{ N-m}$$

or
$$l = 25.01\text{m}.$$

TABLE 10.1

S.no	Section	Area	I_{NA}
1.	b, d, N, A	bd	$\frac{bd^3}{12}$
2.	a, a, N, A	a^2	$\frac{a^4}{12}$
3.	d, N, A	$\frac{\pi d^4}{4}$	$\frac{\pi d^4}{64}$
4.	d, D, N, A	$\frac{\pi}{4}(D^2 - a^2)$	$\frac{\pi}{64}(D^4 - d^4)$
5.	2a, 2b, N, A	πab	$\frac{\pi ab^3}{4}$
6.	B, b/2, b/2, d, D, N, A, B	$(BD - bd)$	$\frac{(BD^3 - bd^3)}{12}$
7.	N, A, h, h/3, b	$\frac{1}{2}bh$	$\frac{bh^3}{36}$

10.2 : *A strongest beam of rectangular section is to be cut out of a round wooden log of diameter D. Find the dimensions of the rectangular beam.*

Solution : Let the dimensions of such a beam be $b \times d$ (Fig. 10.3)

So that $$b^2 + d^2 = D^2 \quad \text{...(10.10)}$$

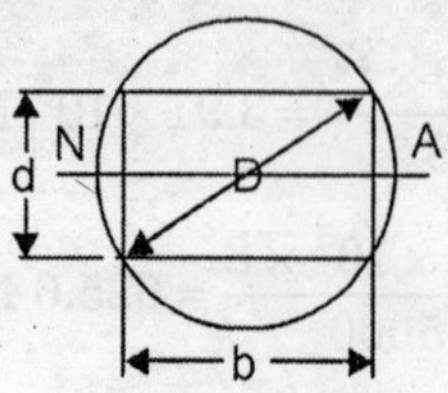

Fig. 10.3

For a strongest beam the resisting moment of the section must be greatest. From Equation (10.9),

$$M = \sigma Z$$

That is, for a given allowable stress in the beam, M will be greatest when Z is maximum,

Now $$Z = \frac{I}{y_{\max}} = \frac{bd^3 \times 2}{12d} = \frac{bd^2}{6}$$

Substituting equation (10.10) we obtain,

$$Z = \frac{b}{6}(D^2 - b^2)$$

For Z to be maximum,

$$\frac{dZ}{db} = 0$$

i.e. $$\frac{1}{6}(D^2 - 3b^2) = 0$$

or $$b = D/\sqrt{3}$$

and from equation (10.10),

$$d = \sqrt{\frac{2}{3}}D.$$

10.3 : *A beam of built up section shown in Fig. 10.4 is simply supported at its ends with a span of 5 m. It carries a concentrated load of 10 kN at a distance of 2 m from the left hand support. Calculate the maximum bending stress induced in the beam.*

Solution : $I_{NA} = 2$ [I_{NA} of top plate + I_{NA} of one channel]

Applying principle of paralel axes I_{NA} of top plate.

$$I_{NA} = 2\left[\frac{200 \times 5^3}{12} + (200 \times 5)(52.5)^2\right] + 2\left[\frac{40 \times 100^3}{12} - \frac{35 \times 90^3}{12}\right]$$

$$= 7.904 \times 10^6 \text{ mm}^4$$

From § 3.5.1 we have,

$$M_{\max} = \frac{10 \times 2(5-2)}{5}$$

$$= 12 \text{ kN-m}$$

$$= 12 \times 10^6 \text{ N-mm}$$

$$= 12 \text{ kN–m}$$

$$= 12 \times 10^6 \text{ N–mm}$$

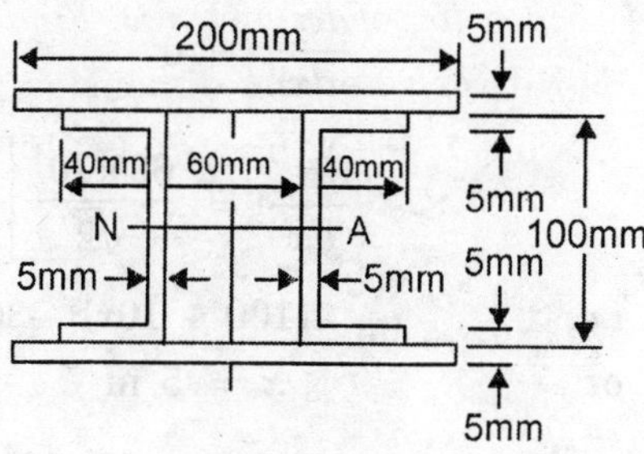

Fig. 10.4

EXAMPLE

E X A M P L E

$$\sigma_{max} = y_{max} \frac{M_{max}}{I_{NA}}$$

$$= \frac{55 \times 12 \times 10^6}{7.904 \times 10^6}$$

$$= 83.7 \text{ N/mm}^2$$

$$= 83.7 \text{MN/m}^2 .$$

10.4 : *An electric pole, standing 10 m above the ground level, is square in section throughout. The dimensions being 200 mm × 200 mm at the base and tapering uniformly to 100 mm × 100 mm at the top. The resultant force due to the tension in the transmission lines is 50 N in the direction along the diagonal of the section at the top. Determine the maximum stress induced in the poles.*

Solution : Consider a section XX at a distance x metre from the top. The side 'a' of this section will be given by,

$$a = \left(100 + \frac{x}{10} \times 100\right) = 100 + 10x \text{ mm}$$

If 'd' is diagonal, then,

$$d = \sqrt{2}a = \sqrt{2}\,(100 + 10x) \text{ mm}$$

B.M. at XX $\quad M_x = 50x$ N–m $= 50 \times 10^3 x$ N–mm

I_{XX} = Second moment of area about diagonal of the section $= a^4/12$

$$= \frac{(100 + 10x)}{12}$$

$$y_{max} = \frac{d}{2} = (100 + 10x)/\sqrt{2}$$

Thus, $\quad \sigma_x = (M_x)(y_{max})/I_{XX}$

Thus, $$\sigma = \frac{100 + 10x}{\sqrt{2}} \cdot \frac{50 \times 10^3 \times 12}{(100 + 10x)^4} = \frac{60 \times 10^4 x}{\sqrt{2}\,(100 + 10x)^3}$$

The location of the section having maximum bending stress will be found by setting,

$$\frac{d\sigma_x}{dx} = 0$$

$$\frac{d\sigma_x}{dx} = \frac{60 \times 10^4}{\sqrt{2}} \left[\frac{(100 + 10x)^3 - x \times 3\,(100 + 10x)^2 \times 10}{(100 + 10x)^6}\right] = 0$$

i.e. $\quad (100 + 10x)^3 - 30x\,(100 + 10x)^2 = 0$

or $\quad x = 5$ m

Therefore, $$\sigma_{max} = \frac{60 \times 10^4 \times 5}{\sqrt{2}\,(100 + 50)^3} = 141.42 \text{N/mm}^2 (\text{MN/m}^2)$$

10.3 COMPOSITE OR FLITCHED BEAMS

A composite or flitched beam is one whose section is built up of two different materials rigidly fixed to each other throughout its length, such as a timber beam reinforced by steel plates called the flitches. Since the two materials are connected rigidly, the strains at the common surfaces are equal. In addition both the materials will bend to the same radius.

Following are the two cases which arise due to the placement of the flitches with respect to the neutral axis of the main beam:

Case I

The flitches are placed symmetrically about the neutral axis of the beam such that the flitches alone will have the same neutral axis. Examples of such cases are shown in Fig. 10.5.

If M_1 and M_2 are the resisting moments of the two materials respectively, the resisting moment of the composite section will be given by,

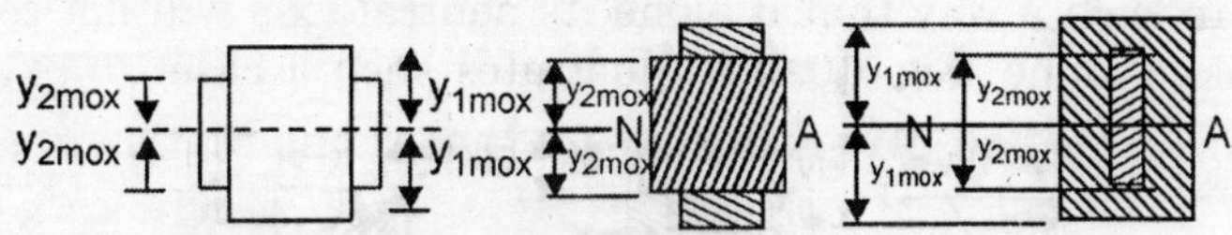

Fig. 10.5

$$M = M_1 + M_2 \qquad ...(10.11)$$

Also, the radii of curvature of the two parts are same,

i.e. $$R_1 = R_2$$

or $$\frac{E_1 I_1}{M_1} = \frac{E_2 I_2}{M_2}$$

or $$\frac{M_1}{M_2} = \frac{E_1 I_1}{E_2 I_2}$$

or $$M_2 = k\frac{I_2}{I_1} M_1 \qquad ...(10.12)$$

Where k is called the Modular ratio and given by,

$$k = E_2/E_1 \qquad ...(10.13)$$

Also, if σ_{1max} and $\sigma_{2\,max}$ are the maximum induced stresses in the respective materials

$$\sigma_{1\,max} = \frac{M_1}{I_1} y_{1\,max} \qquad ...(10.14)$$

and $$\sigma_{2\,max} = \frac{M_2}{I_2} y_{2\,max} \qquad ...(10.15)$$

Dividing equation (10.14) with (10.15),

$$\frac{\sigma_{1\,max}}{\sigma_{2\,max}} = \frac{M_1}{M_2} \cdot \frac{I_2\, y_{1\,max}}{I_1 y_{2\,max}}$$

$$= \frac{1}{k}\left(\frac{y_{1\,max}}{y_{2\,max}}\right) \qquad ...(10.16)$$

Thus, if the total moment acting on the composite section is known, the moments taken up, by the respective materials, can be evaluated with the help of equations (10.11) and (10.12). Further the induced stresses can then be evaluated with the help of equations (10.14) and (10.15).

Alternately if the permissible stresses for the two materials are given as σ_1 and σ_2 respectively and the resisting moment of the composite section is to be evaluated then first of all with the help of equation (10.16) the limiting stress out of σ_1 and σ_2 is ascertained such that the induced stresses in the two materials do not cross the respective permissible values. After having evaluated the maximum induced stresses the moments M_1 and M_2 are calculated with the help of equations (10.14) and (10.15) to get M.

Case II

The plate is placed in such a way that if alone its neutral axis will not coincide with that of the timber beam taken alone. Fig. 10.6(*a*) illustrates such a case.

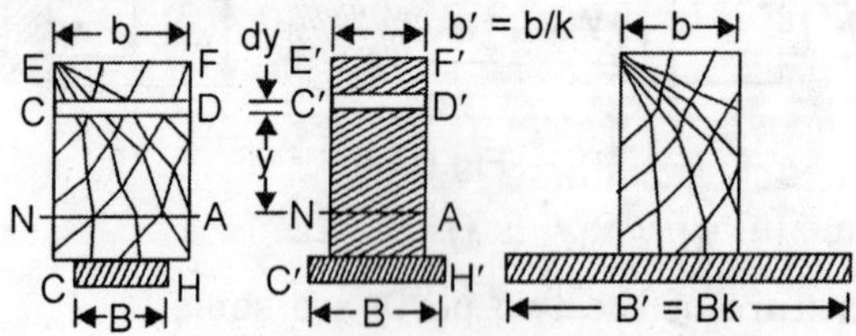

Fig. 10.6

The neutral axis of the composite beam will evidently not pass through the centroid of the wooden portion. This type of problem is tackled by replacing the given section with an equivalent section of one material (either of steel or wood) as shown in Fig. 10.6(*b*) and 10.6(*c*). The equivalent section is such that it will bend to the same radius and extend to the same height as the original beam. Under these conditions, the strains in any fibre *CD* and the corresponding fibre *C′D′* in the equivalent steel section will be same.

Thus if the height of the fibres *CD* or *C′D′* is *y* from the neutral axis, location of which is yet to be found, then assuming plane sections remaining plane after bending.

ε_{CD} = Longitudinal Strain at a distance *y* from N.A., in timber portion.

$$= \sigma_{CD}/E_t \qquad ...(10.17)$$

$\varepsilon_{C'D'}$ = Longitudinal strain at a distance *y* from N.A., in the equivalent steel portion.

$$= \sigma_{C'D'}/E_S \qquad ...(10.18)$$

Where E_t and E_S are the Young's modulii for timber and steel respectively. As these strains are equal,

$$\sigma_{C'D'}/E_s = \sigma_{CD}/E_t$$

$$\sigma_{C'D'} = \left(\frac{E_s}{E_t}\right)\sigma_{CD}$$

$$= k\sigma_{CD} \qquad ...(10.19)$$

where k is the Modular ratio E_S/E_t.

Thus, the stress at any point in the equivalent steel section is k times the stress at the corresponding point in the original section.

Also, since the resisting moment of the equivalent steel portion should be same as that of the original timber portion the normal force on the elements CD and $C'D'$ must be same

That is $\sigma_{CD}\,(bdy) = \sigma_{C'D'}(b'dy)$

$= (K\,\sigma_{CD})\,(b'dy)$ (From equation 10.15)

i.e., $b = kb'$ ...(10.20)

Thus, the width of the equivalent steel portion will be $1/k \times$ the width of the original timber portion.

Since $k(=E_S/E_t)$ is greater than unity, the width b' will be smaller than b.

After having found the equivalent steel section the position of the neutral axis and the value of second moment of area about neutral axis can be found. This can then be used to evaluate the stress $\sigma_{C'D'}$ with the help of equation $\sigma_{C'D'}/y = \dfrac{My}{I}$ if the total resisting moment M of the section is known. Alternatively if the total resisting moment is to be evaluated with allowable stresses being given for timber and steel, the stress $\sigma_{E'F'}$ is calculated with the help of equation (10.19) and is used to evaluate M as indicated above. The value of M is also checked for the allowable stress in steel at $G'H'(\sigma_{GH} = \sigma_{G'H'})$. It must be noted that for the corresponding points in the steel portion of the two sections the stress is equal or in other words in equation (10.19) modular ratio k is reduced to unity. The smaller of the two values of M is then the resisting moment of the section.

As an alternative an equivalent timber section could be found as shown in Fig. 10.6(*c*) and the required quantities evaluated.

EXAMPLE 10.5 : *Fig. 10.7(a) shows the cross–section of a compound bar which has been formed by brazing a brass strip 25 mm × 12 mm to a steel strip 25 mm × 6 mm. The compound bar is bent to a circular shape with neutral axis parallel to BC. Find the position of the neutral axis and the ratio of the maximum stress in steel to that in the brass. Also, find the radius of curvature of the neutral surface and the resisting moment of the section when the maximum stress in steel becomes 66 MN/m².*

Take $E_{steel} = 200 \times 10^3$ MN/m² and $E_{brass} = 80 \times 10^3$ MN/m²

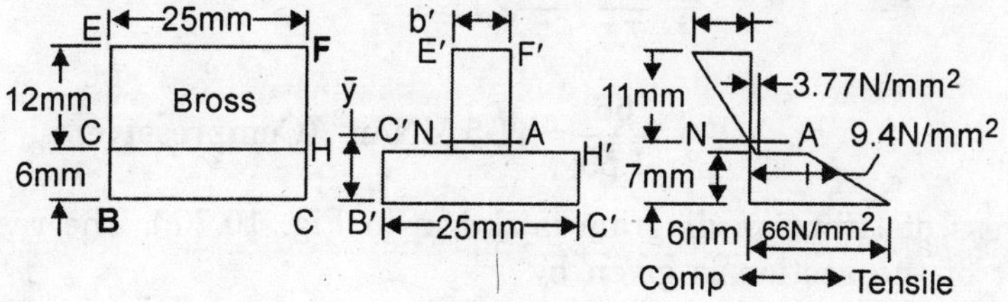

Fig. 10.7

Solution. $k = E_S/E_b = \dfrac{200\times10^3}{80\times10^3} = 2.5$

$b' = b/k = 25/2.5 = 10$ mm

Taking moment about $B'C'$ [Fig. 10.7(*b*)]

$$(25 \times 6 + 12 \times 10)\,\bar{y} = (25 \times 6) \times 3 + (12 \times 10)\,(6 + 6)$$

i.e. $$\bar{y} = 7.0\text{mm}$$

Now from bending equation $\sigma/y = E/R$, we have,

$$\sigma_{B'C'}/\sigma_{E'F'} = y_{B'C'}/y_{E'F'}$$
$$= 7/(18 - 7) = 7/11$$

But $$\sigma_{BC} = \sigma_{B'C'} \text{ and } \sigma_{EF} = \sigma_{E'F'}/k$$

Therefore, $$\sigma_{BC}/\sigma_{EF} = \frac{7}{11} \times k = \frac{7}{11} \times 2.5 = 1.59$$

It must be noted here that if σ_{BC} were tensile in nature, then σ_{EF} would be compressive in nature.

(*b*) For evaluating of the radius of curvature, we have,

$$R = Ey/\sigma$$

and since the maximum stress in steel occurs at BC or $B'C'$

$$R = E_S y_{B'C'}/\sigma_{B'C'} = \frac{200 \times 10^3 \times 7}{66} \text{ mm}$$

The stress distribution on the section can also be found

In the steel $$\sigma_{GH} = \sigma_{G'H'} = \frac{\sigma_{B'C'}}{y_{B'C'}} \cdot y_{G'H'} = \frac{66 \times 1}{7}$$

$$= 9.428 \text{ MN/m}^2 \text{ (Tensile)}$$

In Brass $$\sigma_{GH}\sigma_{G'H'} = \sigma_{G'H'}/k = \frac{1}{k}\left[\frac{\sigma_{B'C'}}{y_{B'C'}} \cdot y_{G'H'}\right] = \frac{9.428}{2.5}$$

$$= 3.77 \text{ MN/m}^2 \text{ (Tensile)}$$

$$\sigma_{EF} = \frac{\sigma_{E'F'}}{k} = \left[\frac{\sigma_{B'C'}}{y_{B'C'}} \cdot y_{E'F'}\right] \times \frac{1}{k}$$

$$= \frac{1}{2.5} \times \frac{66}{7} \times 11$$

$$= \frac{66}{1.59} = 41.5 \text{ MN/m}^2 \text{ (Compressive)}$$

The stress distribution diagram is shown in Fig. 10.7(*c*). The resisting moment of the section is given by,

$$M/I_{NA} = E/R$$

i.e. $$M = EI_{NA}/R$$

Applying theorem of parallel axes,

Now $$I_{NA} = \left[\frac{25 \times 6^3}{12} + 250 \times (4)^2\right] + \left[\frac{10 \times 12^3}{12} + 120 \times (5)^2\right]$$

$$= 8890 \text{ (mm)}^4$$

Therefore, $M = \frac{200 \times 10^3 \times 8890}{2100} = 84700 \text{ N/mm}$

$= 84.7$ N-m.

10.6 : *If the brass and steel strips of Example 10.5 were not brazed together and just placed over each other as shown in Fig. 10.7(a) so that they could bend independently and the bending stresses in brass and steel were limited to 66 MN/m² and 100 MN/m² respectively, find the resisting moment of the section.*

Solution. In this case each strip will bend independently about its own neutral axis. Also, it can be assumed for all practical purposes, that the radii of curvatures of the two neutral surfaces are same. Then for the individual strips.

$$\sigma_S/y_S = E_S/R_S = M_S/I_S$$

and $$\sigma_b/y_b = E_b/R_b = M_b/I_b$$

Now $$I_S = \frac{25 \times 6^2}{12} = 450 \text{ mm}^4$$

$$I_b = \frac{25 \times 12^2}{12} = 3600 \text{ mm}^4$$

From the bending equations, the maximum induced stresses in the two materials are related as,

$$\frac{1}{R_S} = \frac{1}{R_b} = \frac{\sigma_s}{y_s E_s} = \frac{\sigma_b}{y_b E_b}$$

i.e., $$\sigma_s = \sigma_b \cdot \frac{y_s}{y_b} \cdot \frac{E_S}{E_b}$$

$$= \sigma_b \left(\frac{3}{6}\right)(2.5)$$

or $$\sigma_s = 1.25\, \sigma_b$$

Thus, if the stress in steel were the limiting case, then the induced stress in brass would have been,

$$\sigma_b = 100/1.25 = 80 \text{ N/mm}^2 \text{ or } 80 \text{ MN/m}^2.$$

This is not admissible as the stress in brass is limited to 66 MN/m². Therefore the stress in brass will be the limiting stress corresponding to which the stress induced in steel will be

$$= 66 \times 1.25 = 82.5 \text{ N/mm}^2 \text{ or } 82.5 \text{ MN/m}^2$$

For these stresses in steel and brass the respective resisting moments are

$$M_s = \frac{\sigma_s I_s}{y_s} = \frac{82.5 \times 450}{3} = 12375 \text{ M} - \text{mm}$$

E X A M P L E

and $M_b = \frac{\sigma_b I_b}{y_b} = \frac{66 \times 3600}{6} = 3960$ N/mm

Thus, total resisting moment

$= M_s + M_b = 51975$ N–m

$= 51.975$ N–m.

10.7 : *A composite beam is made of two timber joints each 100 mm wide by 300 mm deep, with a steel plate 25 mm thick and 300 mm deep firmly fixed between them. Determine the resisting moment of the composite section if the maximum stress in timber is limited to 7.5 MN/ m². Also, find the maximum centrally applied load which the beam can carry on a simple supported span of 4 m. Take* $E_{steel} = 200 \times 10^3 MN/m^2$ *and* $E_{timber} = 10 \times 10^3 MN/m^2$.

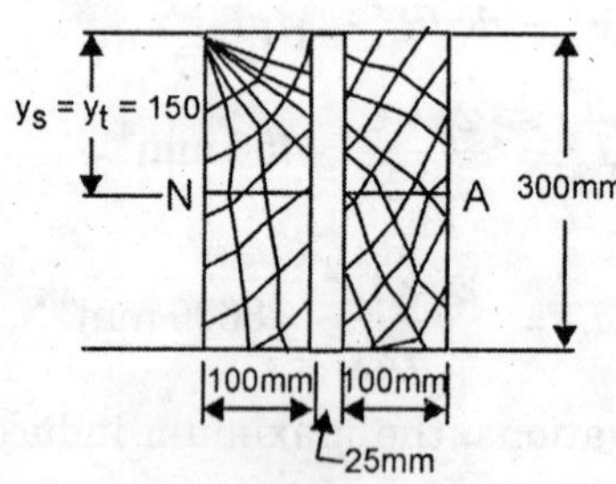

Fig. 10.8

Solution. Since both timber and steel are going to bend to the same radius of curvature, we have,

$$\frac{\sigma_s}{y_s} = \frac{M_s}{I_s} = \frac{E_s}{R}$$

and

$$\frac{\sigma_t}{y_t} = \frac{M_t}{I_t} = \frac{E_t}{R}$$

where suffix t stands for timber and s for steel

That is

$$\frac{\sigma_s}{y_s E_s} = \frac{\sigma_t}{y_t E_t}$$

or

$$(\sigma_s)_{\max} = (\sigma_t)_{\max} \frac{(y_s)_{\max}}{(y_t)_{\max}} \cdot \frac{E_s}{E_t}$$

$$= 7.5 \times \frac{150}{150}\left[\frac{200 \times 10^3}{10 \times 10^3}\right]$$

$$= 150 \text{ N/mm}^2 \text{ or MN/m}^2$$

Now

$$I_s = \frac{25 \times (300)^3}{12} = 56.25 \times 10^6 \text{ mm}^4$$

$$I_t = \frac{2 \times 100 \times (300)^3}{12} = 450 \times 10^6 \text{ N/mm}^4$$

Therefore, $$M_s = \frac{I_s \sigma_s}{y_s} = \frac{56.25 \times 10^6 \times 150}{150}$$

$$= 56.25 \times 10^6 \text{ N–mm}$$

and $$M_t = \frac{I_t \sigma_t}{y_t} = \frac{450 \times 10^6 \times 7.5}{150}$$

$$= 22.50 \times 10^6 \text{ N–mm}$$

∴ Resisting moment of section

$$= 56.25 \times 10^6 + 22.50 \times 10^6$$

$$= 78.75 \times 10^6 \text{ N–mm} = 78.75 \times 10^3 \text{ N-m}$$

(*b*) For a centrally applied load *W* on a simply supported beam of span 4 m, the maximum bending moment will be $(W \times 4)/4 = W$ Nm. (See Fig. 3.4)

This must be equal to the resisting moment of the section,

That is $$W = 78.75 \times 10^3 \text{ N.}$$

10.8 : *A wooden beam 100 mm wide by 300 mm deep is reinforced by two similar steel plates each 12 mm thick, one being secured to the upper face of the beam and the other to the lower face. Determine the necessary width of plates so that the composite section thus formed should be able to resist a bending moment of 8300 N–m in the vertical plane without the stress in wood exceeding 5.0 MN/m². Take* $E_{steel} = 200 \times 10^3$ *MN/m² and* $E_{wood} = 20 \times 10^3$ *MN/m².*

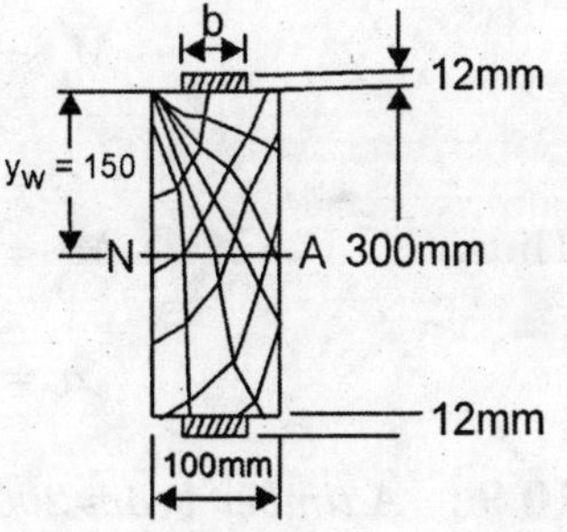

Fig. 10.9

Solution. Since both wood and steel will have the same radius of curvature, we have,

$$\frac{\sigma_s}{y_s} = \frac{M_s}{I_s} = \frac{E_s}{R}$$

and $$\frac{\sigma_w}{y_w} = \frac{M_w}{I_w} = \frac{E_w}{R}$$

where the suffix *s* stands for steel and *w* for wood.

Therefore, $$\frac{\sigma_s}{y_s E_s} = \frac{\sigma_w}{y_w E_w}$$

E X A M P L E

E X A M P L E

or $$(\sigma_s)_{max} = \sigma_w \frac{(y_s)_{max}}{(y_w)_{max}} \frac{E_s}{E_w}$$

$$= \frac{5.0 \times 162}{150} \times \frac{200 \times 10^3}{20 \times 10^3}$$

$$= 54 \text{ N/mm}^2 \text{ or } 54 \text{ MN/m}^2$$

Now if 'b' mm is the width of the steel plates (in mm) then by invoking principle of parallel axis,

$$I_s = 2\left[\frac{b \times 12^3}{12} + (b \times 12)(156)^2\right] \text{mm}^4$$

$$I_w = \frac{100 \times 300^3}{12} = 225 \times 10^6 \text{ mm}^4$$

Again from bending equation,

$$M_w = \frac{\sigma_w I_w}{y_w} = \frac{5 \times 225 \times 10^6}{150} = 7.5 \times 10^6 \text{ N} - \text{mm}$$

$$= 7.5 \times 10^3 \text{ N–m}$$

$$M_s = \frac{54 \times 27.89 \times 10^3\ b}{162} = 9.297 \times 10^3 b \text{ N} - \text{mm}$$

$$= 9.297b \text{ N–m}$$

Thus, $$M_w + M_s = 7.5 \times 10^3 + 9.297b = 8300$$

i.e. $$b = \frac{800}{9.297} = 86 \text{ mm}.$$

10.9 : *A timber beam 200 mm wide by 300 mm deep is reinforced with two rectangular steel plates each 250 mm deep by 10 mm thick. The steel plates are fixed rigidly on each side of the beam as shown in the Fig. 10.10(a). Find the position of the neutral axis and the resisting moment of the composite section, when the stresses in steel and timber are limited to 85 MN/m² and 5.5 MN/m² respectively. Take modular ratio as 20.*

Solution. Since the modular ratio is 20, the equivalent steel width for 200 mm wide timber section will be 200/20 = 10 mm. Accordingly the equivalent steel section is shown in Fig. 10.10(b). Taking moments of the area about the line $G'H'$.

$$\bar{y}(10 \times 300 + 2 \times 10 \times 350) = 10 \times 300 \times 150 + 2 \times 10 \times 250 \times 125$$

i.e. $$\bar{y} = 134.4 \text{ mm}$$

Now $$\frac{\sigma}{y} = \frac{M}{I} = \frac{E}{R}$$

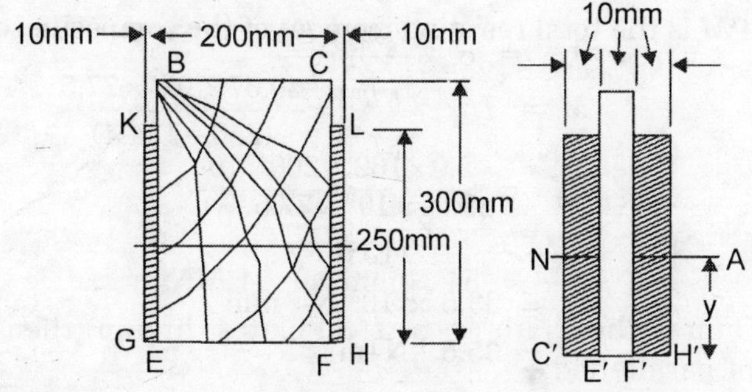

Fig. 10.10

Therefore, $\frac{\sigma_{B'C'}}{y_{B'C'}} = \frac{\sigma_{K'L'}}{y_{K'L'}}$

i.e. $\sigma_{B'C'} = \sigma_{K'L'} \frac{y_{B'C'}}{y_{K'L'}} = \frac{(300 - 134.4)}{(250 - 134.4)} \sigma_{K'L'}$

or $\sigma_{B'C'} = \frac{165.6}{115.6} \sigma_{K'L'}$

Now $\sigma_{KL} = \sigma_{K'L'}$

and $\sigma_{B'C'} = k\sigma_{BC} = 20\sigma_{BC}$

Therefore, $\sigma_{BC} = \frac{1}{20}\left(\frac{165.6}{115.6}\right)\sigma_{KL} = 0.071\,\sigma_{KL}$

Assuming the stress in steel to be the limiting stress

i.e. $\sigma_{KL} = 85 \text{ MN/m}^2$

Then corresponding induced stress in timber will be

$= 0.071 \times 85 = 6.04 \text{ MN/m}^2$

This is more than the allowable stress in timber. Therefore, the stress in timber will be the limiting case *i.e.* the stress

$\sigma_{BC} = 5.5 \text{ MN/m}^2$

and the corresponding induced stress in steel will be,

$\sigma_{KL} = \sigma_{K'L'} = 5.5/0.071 = 77.5 \text{ MN/m}^2$

From Fig. 10.12(*b*), using theorem of parallel axis we get,

$$I_{NA} = 2\left[\frac{10 \times 250^3}{12} + 10 \times 250\,(134.4 - 125)^2\right] + \left[\frac{10 \times 300^2}{12} + 10 \times 300\,(150 - 134.4)^2\right]$$

$= 26.44 \times 10^6 + 23.23 \times 10^6$

$= 49.67 \times 10^6 \text{ mm}^4$

EXAMPLE

Therefore, if M is the total resisting moment of the composite section, then

$$M = I_{Na}\left(\frac{\sigma_{KL'}}{y_{KL'}}\right) = \frac{49.67 \times 10^6 \times 77.5}{(250 - 134.4)}$$

$$= \frac{49.67 \times 10^6 \times 77.5}{115.6}$$

$$= 33.5 \times 10^6 \text{ N– mm}$$

$$= 35.5 \text{ kN–m.}$$

EXAMPLE **10.10 :** *A metallic tube of internal diameter 25 mm and 5 mm wall thickness is simply supported on a span of 1 m and the load at the mid span just sufficient to bring the stress to elastic limit level, is found to be 875 N.*

Four such tubes are firmly clamped to each other to form a single beam, the centres of the tubes forming a square of 35 mm side with two sides of this square horizontal. Calculate the maximum central load which this composite beam can carry if the stress is not to exceed the elastic limit. *(Panjab University)*

Solution. For the single tube with load at mid span,

$$M_{max} = \frac{W \times 1}{4} \text{ Nm}$$

$$\sigma_{max} = \frac{M_{max}}{I_{XX}} \cdot \frac{D}{2}$$

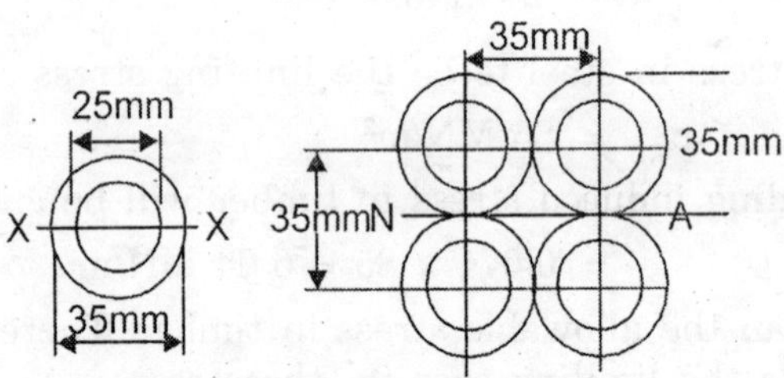

Fig. 10.11

Now $$I_{xx} = \frac{\pi}{64}(35^4 - 25^4)$$

and $$Z = I_{xx} \cdot \frac{2}{D} = \frac{\pi}{64}(35^4 - 25^4) \times \frac{2}{35}$$

$$= 3120 \text{ mm}^3$$

Therefore, $$\sigma_{max} = \frac{875 \times 1000}{4 \times 3120} \text{ N/mm}^2 \text{ or MN/m}^2$$

For the composite section,

$$I_{NA} = 4\left[\frac{\pi}{64}(35^4 - 25^4) + \frac{\pi}{4}(35^2 - 25^2)\left(\frac{35}{2}\right)^2\right]$$

$$\therefore \qquad Z' = \frac{I_{NA}}{35} = 2.245 \times 104 \text{ mm}^3$$

Thus, if W' is the load which can be carried by the composite tube

$$\frac{(W' \times 1000)}{4} = Z' \sigma_{max}$$

$$= 2.245 \times 10^4 \times \left(\frac{875 \times 1000}{4 \times 3120}\right)$$

i.e. $$W' = \frac{2.45 \times 10^4}{3120} \times 875 = 6.3 \times 10^3 \text{ N}$$

$$= 6.3 \text{ kN.}$$

10.11 : *A 25 mm diameter bronze rod is enclosed tightly in a steel tube of outside diameter of 50 mm. Find the resisting moment of the composite section assuming that the allowable stresses is steel and bronze do not* exceed 150 MN/m² *and 120 MN/m² respectively.* E_{steel} *= 200 × 103 MN/m² and* E_{bronze} *= 100 × 10³ MN/m².*

Solution. The two metals bend to the same radius of curvature. Therefore, from the bending equation,

$$\frac{1}{R} = \frac{\sigma_s}{E_s y_s} = \frac{\sigma_b}{E_b y_b}$$

i.e. $$\sigma_s = \sigma_b \left(\frac{E_s}{E_b}\right)\left(\frac{y_s}{y_b}\right) = \sigma_b \,(2)\,(2) = 4\sigma_b$$

Thus, if the stress in steel is the limiting case,

i.e. $$\sigma_s = 150 \text{ MN/m}^2$$

Then maximum induced stress in bronze will be

$$\sigma_b = 150/4 = 37.5 \text{ MN/m}^2$$

which being less than the allowable stress in bronze of 120 MN/m² is admissible.

Thus, $$M_s = I_s \,\sigma_s / y_s$$

$$= \frac{\pi}{64}(50^4 - 25)^4 \,\frac{150 \times 2}{50}$$

$$= 1.715 \times 10^6 \text{ N–mm}$$

and $$M_b = I_b \sigma_b / y_b = 0.0575 \times 10^6 \text{ N–mm}$$

Therefore, total resisting moment of the section

$$= M_s + M_b = (1.715 + .0575)\ 10^6 \text{ N–mm}$$

$$= 1.7725 \text{ kN–m.}$$

EXAMPLE

10.4 COMBINED DIRECT AND BENDING STRESSES

When a member of a structure is under the action of both a bending moment and axial force, the bending and direct stresses induced due to their action may be calculated separately and resultant stress at any point found by adding them algebraically. Thus, if σ is the resultant stress at any point of the section, then

$$\sigma = \pm \sigma_0 \pm \sigma_b \quad ...(10.21)$$

Where σ_0 is the axial force divided by the area of cross–section and σ_b is the bending stress as calculated from the bending equation (10.8). The signs of σ_0 and σ_b being taken positive or negative depending upon whether they are tensile or compressive.

10.5 ECCENTRIC LONGITUDINAL LOADS

Let the line of action of a force P be parallel to the axis of a bar and intersect the axis of symmetry XX of the cross–section $ABCD$ at a distance x_1 from the centroid O Fig. 10.12.

The equilibrium of the column will not be disturbed by introducing two equal and opposite forces P_1 and P_2 (both equal to P) at the centroid O as in the Fig. 10.12. The forces P and P_2 will constitute a bending moment Px_1 with the axis of bending as YY. Thus, the original force P applied at the point Q is equivalent to a bending moment Px_1 about YY and a direct load P_1 (= P) applied at the centroid of the section. The direct load P_1 will induce a direct compressive stress which will be uniform over the cross–section.

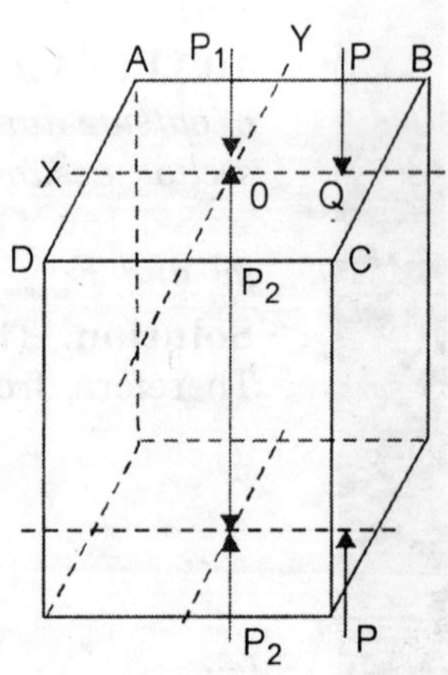

Fig. 10.12

$$\sigma_0 = -\frac{P}{bd} \quad ...(10.22)$$

where b and d are the breadth and the depth of the cross–section.

The effect of the bending moment Px_1 will be to bend the column in the plane passing through the point O and containing the force P, such that the centre of curvature will lie on the right side of the centroidal bending axis YY. Thus as discussed in article 10.2, the bending stress at any point of the section lying to the right of YY will be compressive and all the points of the section to the left of YY will experience tensile bending stresses.

The magnitude of the bending stress at any point at a distance x from the YY line will be

$$\sigma_b = x \times [(Px_1)/I_{YY}] = \frac{12\,Pxx_1}{db^3} \quad ...(10.23)$$

Thus σ_b in magnitude increases linearly with x. The maximum value of the bending stress will occur when x is greatest that is along the lines AD or CB where $x = b/2$

i.e. $$(\sigma_b)_{BC} = \frac{12Px_1}{db^3}\cdot\frac{b}{2} = \frac{6Px_1}{db^2} \quad \text{(Compressive)}$$

also, $$(\sigma_b)_{AD} = \frac{6Px_1}{db^2} \quad \text{(Tensile)}$$

Therefore, the magnitude of the resultant maximum and minimum compressive stresses over the section are given by,

$$\sigma = \frac{-P}{db} \pm \frac{6Px_1}{db^2} \quad ...(10.24)$$

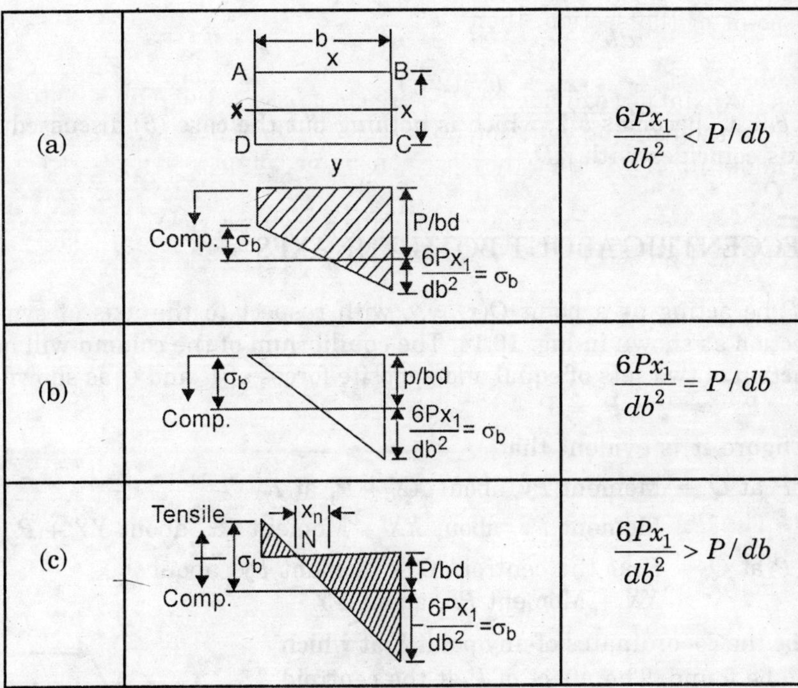

Fig. 10.13

Fig. 10.13(*a*), (*b*) and (*c*) show the possible forms of the resultant stress distribution curves. It is evident that in the case (*c*) some portion of the section will experience a tensile stress even though the nature of the applied load at *Q* is compressive. This type of situation is called "*the reversal of stress*". In the cases (*a*) and (*b*), the entire section is having the stress of type of the applied load. As a matter of fact for the non–reversibility of the stress (as in the case of cast iron or a masonary structure), the maximum eccentricity of the load is given by the condition,

$$\frac{6Px_1}{db^2} = \frac{P}{bd}$$

or

$$x_1 = b/6$$

Since this eccentricity can be on either side of *YY*, for the non–reversibility of stress the load point *Q* must lie within the middle third of the width of the column.

Similarly if the point of application of the load were lying on the other axis of symmetry *i.e.* on *YY* at a distance y_1 from *O*, the bending will take places about *XX*. Here, also to avoid reversal of stress the load point must be within the middle third of the depth *d*.

The position of the neutral axis of the section in the case of combined direct and bending stresses can be found graphically by projecting a line parallel to *YY* on the section from the point *N* (Fig. 10.13*c*) because the stress at all the points of the section on this line will be zero. Alternatively if the neutral axis is lying at a distance x_N on the left of YY (side opposite to that of load), then from the definition of the neutral axis,

$$\sigma_{NA} = 0$$

or $$(\sigma_b)_{NA} + (\sigma_0)_{NA} = 0$$

i.e., $$\frac{12Px_1\, x_N}{db^3} = \frac{P}{bd}$$

or $$x_N = b^2/(12\, x_1) \qquad ...(10.25)$$

When $x = b/6$, x_N becomes $b/2$, which is nothing but the case (*b*) discussed above, where the neutral axis coincides with *AD*.

10.6 LOAD ECCENTRIC ABOUT BOTH THE AXES

Let the load *P* be acting at a point $Q(x_1, y_1)$, with respect to the axis of symmetry of the rectangular section as shown in Fig. 10.14. The equilibrium of the column will not be affected by the introduction of two sets of equal and opposite forces at *R* and *O* as shown in Fig. 10.14 such that $P_1 = P_2 = P_3 = P_4 = P$.

From the figure it is evident that,

P at $Q \equiv$ Moment Py_1 about XX + P_1 at R

$\equiv$ Moment Py_1 about XX + Moment Px_1 about YY + P_3 at O

Thus, P at $Q \equiv P$ at the centroid O + Moment Py_1 about XX + Moment Px_1 about YY ...(10.26)

Let (h, k) be the co–ordinates of any point *S* at which the stress is to be found. The effect of *P* at the centroid will be to induce a compressive stress of uniform intensity $\sigma_0 = (P/bd)$ all over the cross–section including the point *S* (article 1.4).

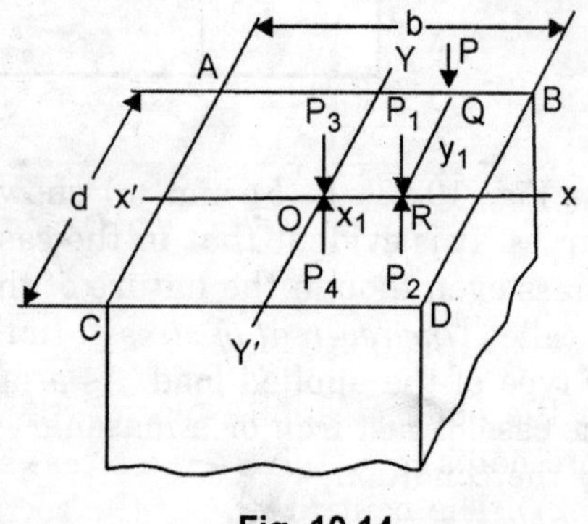

Fig. 10.14

Effect of bending moment Py_1 about XX will be to give rise to a compressive bending stress at the point *S*, lying in first quadrant.

$$\sigma_{S_1} = -k\frac{Py_1}{I_{xx}} = -\frac{12Py_1k}{d^3b}$$

Similarly the effect of the other moment Px_1 about YY will be to induce a compressive bending stress at *S*.

$$\sigma_{S_2} = -h\frac{Px_1}{I_{yy}} = -12\frac{Px_1h}{b^3d}$$

Thus, the resultant stress at *S* is given by,

$$\sigma_S = \sigma_0 + \sigma_{S_1} + \sigma_{S_2}$$

$$= -\frac{P}{bd} - \frac{12\, Py_1k}{d^3b} - \frac{12\, Px_1h}{b^3d} \qquad ...(10.27)$$

For the positive values of h, k, x_1 and y_1 the value of σ_S will be negative signifying that the resultant stress at S will be compressive. If the point S is in the third quadrant both h and k will be negative which will render the last two terms in equation (10.27) positive. Therefore, depending upon the values of other quantities appearing in equation (10.27) the sign of σ_S may turn out to be positive. Under these circumstances there will be the reversal of stress over the section. Since for the point D of the section the respective magnitudes of h and k are greatest, though negative, the reversal of stress will occur first at this point. Thus, the limiting condition for the non–reversibility of the stress will be to equate the resultant stress at the point D to zero.

That is
$$\sigma_D = \frac{P}{db} - \frac{12Py_1}{d^3b}\left(\frac{-d}{2}\right) + \left[\frac{12Px_1}{b^3d}\right]\left(\frac{-b}{2}\right) = 0$$

or
$$\frac{x_1}{(b/6)} + \frac{y_1}{(d/6)} = 1 \qquad \text{...(10.28)}$$

Thus, generalising relation (10.28), the locus of the point (x_1, y_1), lying in the first quadrant only, for the non–reversibility of stress will be the straight line of equation,

$$\frac{x}{(b/6)} + \frac{y}{(d/6)} = 1 \qquad \text{...(10.29)}$$

This, straight line is shown in Fig. 10.15 by EF.

Thus, if the load application point lies anywhere within the area EOF there will be no reversal of stress on the section. Similarly it can be shown that if the point Q is not necessarily within the first quadrant but goes in the other three quadrants as well, the condition for the non–reversibility of the stress will be that the point Q lies within the area $EFGH$. This area $EFGH$ is known as the Core or Kernel of the rectangular section $ABCD$.

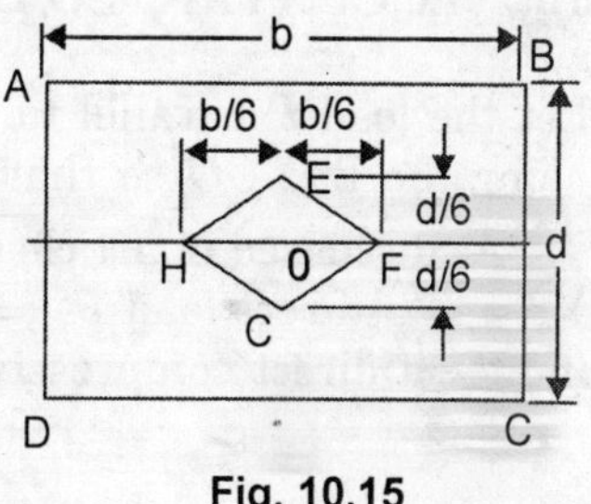

Fig. 10.15

Stress Distribution Curve

Let the point of application of the load, Q, be outside the core of the section and has the coordinates (x_1, y_1). Then the stress at any point S with coordinates (h, k) is given by equation (10.27). The neutral axis of the section can be found by finding the locus of S such that the stress $\sigma_s = 0$.

Therefore, from equation (10.27) by setting σ_s equal to zero we get the relation,

$$\frac{-12y_1k}{d^2} - \frac{12x_1h}{b^2} = 1 \qquad \text{...(10.30)}$$

Thus, generalising the relation (10.30), the equation of the neutral axis of the given rectangular section is,

$$\frac{x}{(-b^2/12x_1)} + \frac{y}{(-d^2/12y_1)} = 1 \qquad \text{...(10.31)}$$

This again is an equa'ion of a straight line having intercepts on x and y axes as $(-b^2/12x_1)$ and $(-d^2/12y_1)$ respectively. This is shown by NN' in Fig. 10.16. Evidently the stress on any point on NN' will be zero. The resultant stress on any point on the portion $ANN'CB$ will be compressive and on the portion $NN'D$ tensile. Moreover, the iso–stress lines will be parallel to NN'. In order to draw the stress distribution curve for the section produce NN' to a convenient point M_N and draw $M_DM_NM_B$ the base line of the stress distribution curve, at right angles to NN' M_N, OM_O, DM_D and BM_B are perpendiculars on the line M_DM_B. From equation (10.27), it is evident that at the centroid O, of the section, the resultant stress will only be the direct stress (P/bd). To some convenent scale measure M_0T_0 to represent $\sigma_0(= P/db)$. Join M_NT_0 to meet the line BM_B and DM_D at T_B and T_D respectively. $T_DM_NT_B$ is the required resultant stress distribution curve as shown in Fig. 10.16.

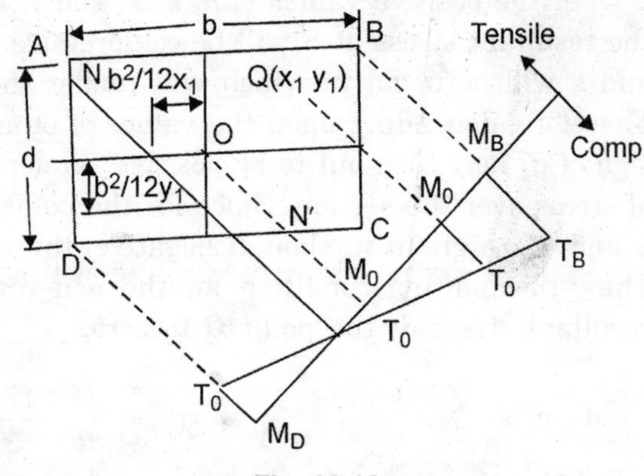

Fig. 10.16

The stress at any point Q is then represented by M_QT_Q which can be measured.

10.7 ECCENTRIC LOADING OF SOLID CIRCULAR SHORT COLUMN

Let the load P parallel to the axis of the bar be acting at any point Q. Marking the x–axis along the line OQ so that the eccentricity of loading is x_1 about the YY axis.

As discussed in article (10.5) the force P at Q is equivalent to a bending moment Px_1 about YY and a direct load P_1 (= P) applied at the centroid of the section. The direct load P_1 will induce a direct compressive stress, uniform over the cross–section

i.e., $$\sigma_0 = -\frac{P}{A} = -\frac{4P}{\pi D^2} \quad \text{...(10.32)}$$

where A is the area of cross–section of the circular section of diameter D.

The effect of the bending moment Px_1 will be to bend the column in the plane containing the force P and passing through the point O, such that the centre of curvature will lie on the right side of the centroidal bending axis YY. Thus, as discussed in article (10.1), the bending stress at any point of the section lying to the right of YY will be compressive and all the points of the section lying to the left of YY will experiences tensile bending stresses.

As per the discussion of the article (10.5), evidently the condition for non–reversibility of stress will be obtained by setting the resultant stress at A equal to zero.

i.e. $$\sigma_A = \sigma_0 + \frac{D}{2}\left(\frac{Px_1}{I_{yy}}\right) = 0$$

or $$\frac{-4P}{\pi D^2} + \frac{D}{2}\frac{P.x_1 \times 64}{\pi D_4} = 0$$

or $$x_1 = \frac{D}{8} \quad \text{...(10.33)}$$

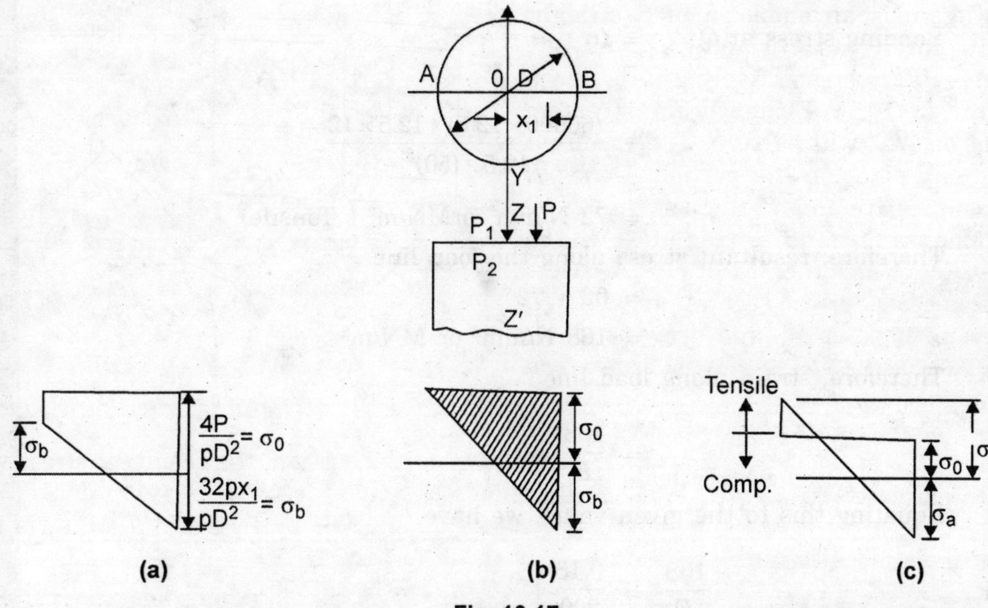

Fig. 10.17

Thus, for the non–reversibility of the stress, the load point must lie within the circle of radius *D*/8, *i.e.* within the middle quarter of the given column section.

For any given load position, the stress distribution curves can be drawn on the lines of article 10.5. These diagrams are shown in Fig. 10.17.

EXAMPLE

10.12 : ***A 50 mm × 12.5 mm flat steel bar was placed in a testing machine and subjected to a 60 kN load acting as shown in Fig. 10.18. An extensometer placed in line with the load recorded an extension of 0.16 mm on a gauge length of 200 mm. Calculate the maximum and minimum stresses set up, and the value of Young's Modulus.***

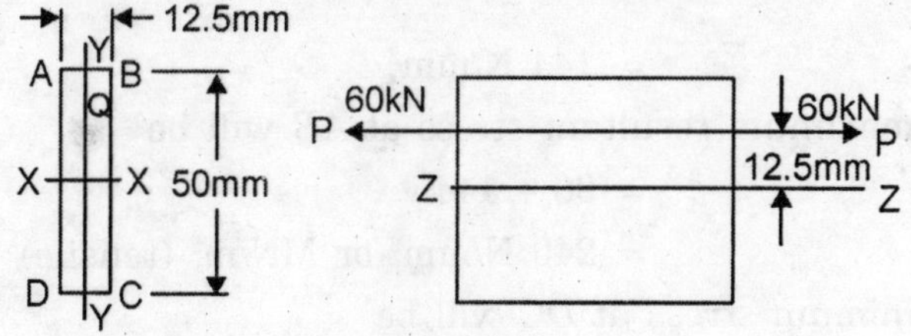

Fig. 10.18

Solution. Eccentricity about

$$XX = 12.5 \text{ mm}$$

$$P = 60 \text{ kN}$$

$$M = 60 \times 10^3 \times 12.5 \text{ N–mm}$$

Direct stress $\sigma_0 = \dfrac{60 \times 1000}{50 \times 12.5} = 96 \text{ N/mm}^2$ (Tensile)

Bending stress at Q $= (\sigma_b)_Q = \dfrac{M}{I_{xx}} \cdot y_Q$

$$= \frac{(60000 \times 12.5) \times 12.5 \times 12}{12.5 \times (50)^3}$$

$= 72$ N/mm² orMN/m² (Tensile)

Therefore, resultant stress along the load line

$= 96 + 72$

$= 168$ N/mm² or MN/m²

Therefore, strain along load line

$$= \frac{168}{E}$$

Equating this to the given value, we have,

$$\frac{168}{E} = \frac{0.16}{200}$$

or $$E = \frac{200 \times 168}{200}$$

$= 210000$ N/mm²

or $= 210 \times 10^3$ MN/m²

The bending stresses at AB and DC will be equal in magnitude and are given by,

$$(\sigma b)_{AB} = (\sigma b)_{CD}$$

$$= \frac{(60000 \times 12.5)}{12.5 \times (50)^3} \times 12.25$$

$= 144$ N/mm²

Thus, the maximum resultant stress at AB will be

$= 96 + 144$

$= 240$ N/mm² or MN/m² (tensile)

and the minimum stress at DC will be

$= 144 - 96$

$= 48$ N/mm² or MN/m² (Compressive)

10.13 : *Fig. 10.19 shows the boom of a cantilever type hydraulic crane. A load of 200 kN is supported by a single rope which passes over two pulleys, mounted on the boom as shown.*

After passing through the hollow rectangular pillar of the crane, it is then connected to the hydraulic apparatus. The axis of the pulley D is vertically above the centroid of the section AB. The outside dimensions of the hollow rectangular section AB are 760 mm × 380 mm and the material is 30 mm thick all round, and the longer dimension is in the direction AB. Find the maximum tensile and compressive stresses induced in the section AB. Also, find the location of the neutral axis.

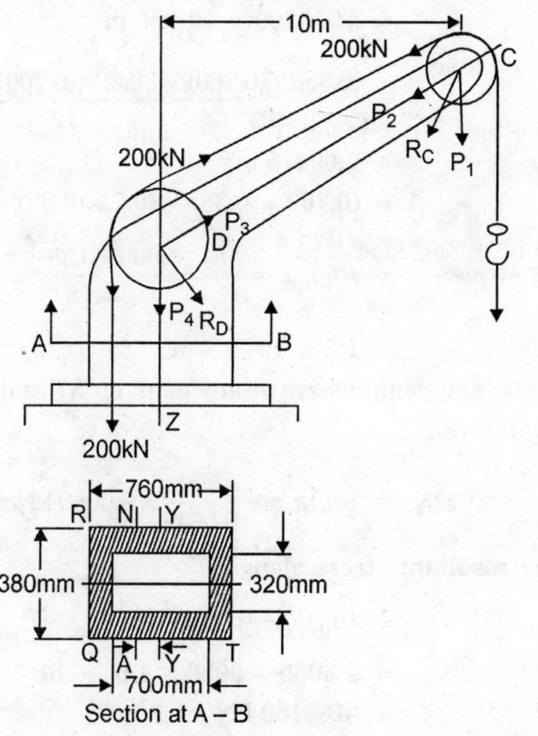

Fig. 10.19

Solution : Neglecting friction, the tension in the rope will be 200 kN everywhere. By drawing free body diagrams of the two pulleys, it can be seen that due to the tension in the rope, the reactions on the bearings of the pulleys at C and D will be R_C and R_D respectively. These reactions can be split up into loads, P_1 and P_2 at C, and P_3 and P_4 at D as shown in Fig. 10.19 such that $P_1 = P_2 = P_3 = P_4 = 200$ kN. Out of these four effective loads on the boom, P_2 and P_3 being equal and opposite will cancel out each other's effect on the section AB. Thus, the effect of the load supported by the hook, on the section AB is equivalent to the combined effect of the loads P_1 and P_4 acting at the points C and D of the boom respectively.

Effect of P_4. Its effect will be to induce a uniform direct compressive stress on the section AB of magnitude,

$$(\sigma_0)_{P_4} = -\frac{200}{A} \quad ...(1)$$

Effect of P_1. Its effect will be to induce direct as well as bending stresses. Direct compressive stress,

EXAMPLE

EXAMPLE

$$(\sigma_0)_{P_1} = -\frac{200}{A} \quad ...(2)$$

Bending moment about YY

$$= M = 200 \times 10 \text{ kN–m}$$

$$I_{yy} = \frac{0.380 \times (0.760)^3}{12} - \frac{0.320\,(0.700)^3}{12}$$

$$= 0.00475 \text{ m}^4$$

$$A = (0.760 \times 0.38) - (0.7 \times 0.32) = 0.0648 \text{ m}^2$$

Max. bending stress $= (\sigma_b)_{max} = \frac{My}{I_{yy}} = \frac{2000 \times 0.38}{0.00475}$

$$= 1.6 \times 10^5 \text{ kN/m}^2$$

This bending stress is compressive at any point on ST and tensile on QR.

Now from (1) and (2),

$$(\sigma_0)_{P_4} = (\sigma_0)_{P_1} = -\frac{200}{0.0648} = -3080 \text{ kN/m}^2$$

Therefore, the resultant stress along ST

$$= (\sigma_0)_{P_4} + (\sigma_0)_{P_1} - (\sigma_b)_{max}$$

$$= -3080 - 3080 - 1.6 \times 10^5$$

$$= -166160 \text{ kN/m}^2$$

Or the maximum compressive stress = 166.16 MN/m²

The resultant stress along

$$RQ = (\sigma_b)_{max} + (\sigma_0)_{P_4} + (\sigma_0)_{P_1}$$

$$= 1.6 \times 10^5 - 3080 - 3080$$

$$= 153840 \text{ kN/m}^2$$

Or the max. tensile stress along

$$RQ = 153.84 \text{ MN/m}^2.$$

Location of Neutral Axis. Let it be at a distance x to the left of YY. Then as the resultant stress along neutral axis is to be zero the magnitudes of the direct and bending stresses will be equal.

i.e. $$\frac{Mx}{I_{yy}} = (\sigma_0)_{P_1} + (\sigma_0)_{P_4}$$

or $$\frac{2000 \times x}{0.00475} = 3080 + 3080$$

i.e. $$x = \frac{6160 \times 0.00475}{2000} = 0.01465 \text{ m}$$

Thus, the neutral axis is 14.65 mm to the left of *YY* and parallel to it.

10.14 : *Fig. 10.20 shows a forged crank with a section along the line AB. The thrust P of the connecting rod acts on the crank at an angle θ as shown. Find the maximum allowable value of the thrust P, if the stress at the section AB were not to exceed 45 MN/m². Assume the axis ZZ to pass through the centroid of the section AB.*

Solution : Area of the section *AB*

$$= 150 \times 75 - 30 \times 100$$

$$= 8250 \text{ mm}^2$$

$$I_{NA} = \frac{75 \times (150)^2}{12} - \frac{30 \times (100)^3}{12}$$

$$= 1.86 \times 10^7 \text{ (mm)}^4$$

The thrust *P* can be resolved into two components, *viz.* $P_Z = P \cos \theta$ giving direct compressive stress on the section and $P_y = P \sin \theta$ giving rise to a bending moment of magnitude ($P_y \times 100$) over the section *AB*, due to which bending stresses will be induced.

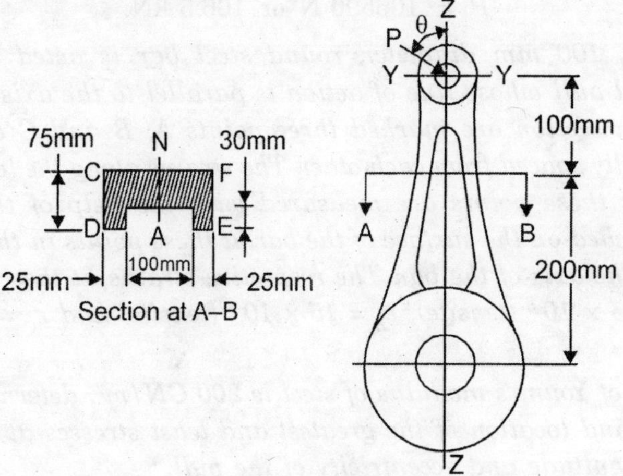

Fig. 10.20

For any value of θ the resultant compressive stresses along the line *FE* will be greatest and its magnitude is given by,

$$|\sigma_{EF}| = \frac{P_Z}{A} + \frac{100\,P_y}{I_{NA}} \cdot \left(\frac{150}{2}\right)$$

$$= \frac{P}{A}\cos\theta + \frac{7500}{I_{NA}} P \sin\theta = 0$$

For getting the absolute maximum value of σ_{EF}, ($d\sigma_{EF}/d\theta$) should be zero

i.e., $$-\frac{P}{A}\sin\theta + \frac{7500}{I_{NA}} P\cos\theta = 0$$

or $\tan\theta = \dfrac{7500\,A}{I_{NA}}$

$$= \frac{7500 \times 8250}{1.86 \times 10^7}$$

$$= 3.33$$

or $\theta = 73°\ 18'$

Substituting this value of θ, we have,

$$\sigma_{max} = \frac{P}{A}\cos 73°\,18' + \frac{7500}{I_{NA}} P \sin 73°\,18'$$

$$= \frac{P}{8250} \times 0.2874 + \frac{7500}{1.86 \times 10^7} \times P \times 0.9578$$

or $45 = 4.21 \times 10^{-4} P$

i.e. $P = 106500$ N or 106.5 kN.

EXAMPLE

10.15 : *A, 100 mm diameter, round steel bar is acted upon by a longitudinal pull whose line of action is parallel to the axis of the bar. At a certain section are marked three points A, B and C around the section equally spaced from each other. The strains along the longitudinal direction at these points are measured with the help of three strain gauges attached on the surface of the bar at these points in the direction parallel to the axis of the bar. The respective strains at these points are italic* $\varepsilon_A = 45 \times 10^{-5}$ *(tensile),* $\varepsilon_B = 15 \times 10^{-5}$ *(tensile) and* $\varepsilon_C = 37.5 \times 10^{-5}$ *(tensile).*

If the value of Young's modulus of steel is 200 GN/m², determine (a) the magnitude and location of the greatest and least stresses at the section ; (b) the magnitude and eccentricity of the pull.

Solution. The cross–section of the bar is shown in Fig. 10.21. The longitudinal stresses at the points *A*, *B* and *C* are,

$$\sigma_A = E\varepsilon_A$$
$$= (200 \times 10^3)(45 \times 10^{-5})$$
$$= 90 \text{ MN/m}^2$$
$$\sigma_B = E\varepsilon_B$$
$$= (200 \times 10^3)(15 \times 10^{-5})$$
$$= 30 \text{ MN/m}^2$$
$$\sigma_C = E\varepsilon_C$$
$$= (200 \times 10^3)(37.5 \times 10^{-5})$$
$$= 75 \text{ MN/m}^2$$

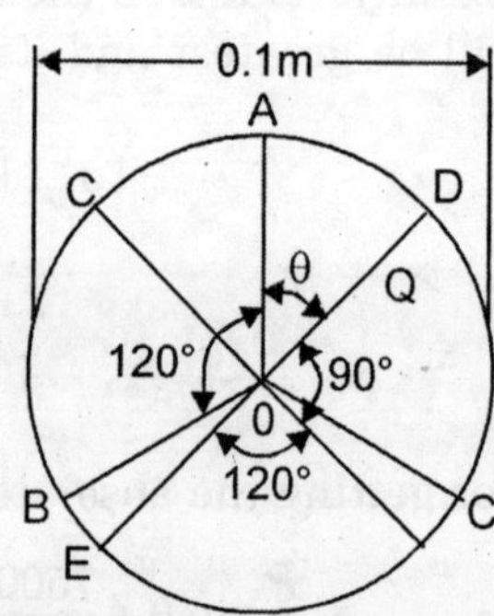

Fig. 10.21

Since the stresses at A and C are higher than that at B, it is expected that the point of application of the load, Q, will be on the radial line OD lying between OA and OC.

Let the load be P with the eccentricity OQ equal to x such that the axis of bending is GH. Assuming the inclination between OA and OD as θ, we have,

For the point A

$$y_A = OA \cos\theta \text{ and } I_{GH} = \frac{\pi d^4}{64}$$

$$\sigma_A = \frac{4P}{\pi d^2} + \frac{P \times x \times 64}{\pi d^4} \cdot \left(\frac{d}{2}\cos\theta\right) = 90$$

or $$\frac{4P}{\pi(0.1)^2} + \frac{32\, Px \cos\theta}{\pi(0.1)^3} = 90$$

or $$\frac{400\,P}{\pi}(1 + 80x\cos\theta) = 90$$

or $$1 + 80\,x\cos\theta = \frac{9\pi}{40P} \quad \text{...(A)}$$

Similarly for the point B

$$\sigma_B = \frac{400P}{\pi}[1 + 80x\cos(120° + \theta)] = 30$$

or $$1 + 80x\left[-\frac{1}{2}\cos\theta - \frac{\sqrt{3}}{2}\sin\theta\right] = \frac{3\pi}{40P}$$

or $$1 + 40x[-\cos\theta - \sqrt{3}\sin\theta] = \frac{3\pi}{40P} \quad \text{...(B)}$$

For the point C

$$\sigma_C = \frac{400P}{\pi}[1 + 80x\cos(120° - \theta)] = 75$$

or $$1 + 80x\left[-\frac{1}{2}\cos\theta + \frac{\sqrt{3}}{2}\sin\theta\right] = \frac{75\pi}{400P} \quad \text{...(C)}$$

Adding equations (B) and (C),

$$2 - 80x\cos\theta = \frac{10.5\pi}{40P} \quad \text{...(D)}$$

E X A M P L E

E X A M P L E

Again adding equations (*A*) and (*D*),

$$3 = \frac{19.5\pi}{120} = 0.51 \text{ MN}$$

or

$$P = \frac{19.5\pi}{120} = 0.51 \text{ MN}$$

Also, subtracting equation (*B*) from (*C*),

$$80\sqrt{3}\, x \sin\theta = \frac{4.5\pi}{40 \times 0.51} = 0.693 \quad ...(E)$$

and from equation (*A*),

$$80x \cos\theta = \frac{9\pi}{40 \times 0.51} - 1 = 0.386 \quad ...(F)$$

Therefore, dividing equations (*E*) and (*F*),

$$\tan\theta = \frac{0.693}{80\sqrt{3} \times 0.7193} = 0.0096$$

i.e.

$$\theta = 46°$$

From equation (*E*),

$$x = \frac{0.693}{80\sqrt{3} \times 0.7193} = 0.00695 \text{ m}$$

$$\sigma_{max} = \sigma_D = \frac{4P}{\pi(0.1)^2} + \frac{64P \,.\, x}{\pi(0.1)^4}\left(\frac{0.1}{2}\right)$$

$$= \frac{400\,P}{\pi}(1 + 80x)$$

$$= \frac{400 \times 0.51}{\pi}(1 + 80 \times 0.00695)$$

$$= 101 \text{MN/m}^2 \text{ (tensile)}$$

$$\sigma_{min} = \sigma_E = \frac{400\,P}{\pi}(1 - 80x)$$

$$= \frac{400 \times 0.51}{\pi}(1 - 80 \times 0.00695)$$

$$= 210.9 \text{ MN/m}^2 \text{ (tensile).}$$

10.16 : *A 30 m high brick chimney is 2.3 m outside diameter at the top and tapers uniformly to 3.3 m outside diameter at the bottom. The chimney weighs 2200 kN and is 0.675 m thick at the base. If a uniform* *horizontal wind pressure of 1 kN/m² acts on the projected area of the chimney, determine the maximum and minimum normal stresses on the base.*

Solution. Total wind force F

$$= 1 \times \text{Projected area.}$$

$$= 1 \times \frac{1}{2}(3.3 + 2.3) \times 30)$$

$$= 84 \text{ kN}$$

The force F acts at a height $\bar{y}$ from the base of the chimney and passes through the centroid of the projected area which is similar to the elevation as drawn in Fig. 10.22. To evaluate $\bar{y}$, take moments of the projected area about the base AB

$$\left[\frac{1}{2}(3.3 + 2.3) \times 30\right]\bar{y} = (2.3 \times 30)\,15 + 2\left[\frac{1}{2} \times .5 \times 30 \times \frac{30}{3}\right]$$

i.e. $\bar{y} = 110.1 \text{ m}$

Now $$I_{YY} = \frac{\pi}{64}[(3.3)^4 - (1.95)^4]$$

$$= 5.1 \text{ m}^4$$

$$A = \frac{\pi}{4}(3.3^2 - 1.95^2) = 5.56 \text{ m}^2$$

$$M = F\bar{y} = 84 \times 14.1 \text{ kN} - \text{m}$$

$$(\sigma_b)_B = \frac{(84 \times 14.1)}{5.1} \times \left(\frac{3.3}{2}\right)$$

$$= 383 \text{ kN/m}^2 \text{ (compressive)}$$

and $(\sigma_b)_A = 383 \text{ kN/m}^2$ (tensile)

Direct stress $$\sigma_0 = \frac{2200}{A}$$

$$= \frac{2200}{5.56}\ 395 \text{ kN/m}^2 \text{ (compressive)}$$

Fig. 10.22

Therefore, max. stress at B

$$\sigma_B = 383 + 395$$

$$= 778 \text{ kN/m}^2 \text{ (compressive)}$$

and minimum stress at

$$\sigma_A = 395 - 383$$

$$= 12 \text{ kN/m}^2 \text{ (compressive)}$$

REVIEW QUESTIONS

10.1. State assumptions made while deriving the equation for bending stress and explain their importance.

10.2. Define the neutral surface and neutral axis.

10.3. Explain the procedure as to how the neutral axis is fixed in position and direction. Has it any bearing on the axis of moment ? If yes, how ?

10.4. While determining the bending stresses at any point on the section, how it is decided whether it is tensile or compressive ?

10.5. Write down the expression for bending stress and radius of evaluature. Explain the symbols used.

10.6. The cross-section of a rectangular beam is 200 mm wide by 250 mm deep Fig (10.22*a*). This beam is simply supported at its ends and carries a central vertical load of 100 kN. If the span is 3m, find the maximum tensile and max compressive stresses induced in the beam and give the location of the max stress points.

Solution. Here, $b = 200$ mm

$d = 250$ mm

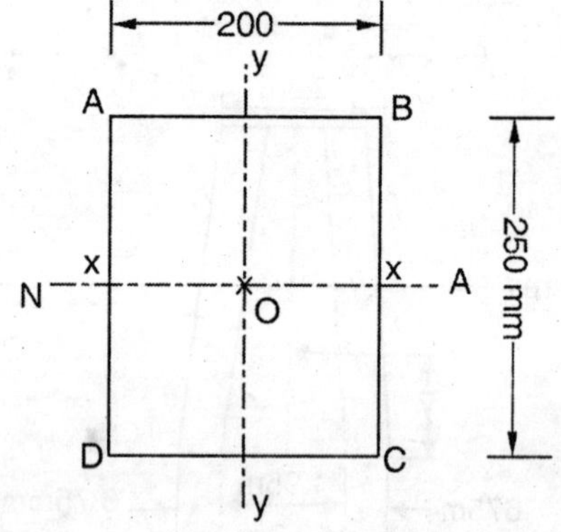

Fig. 10.22 (a)

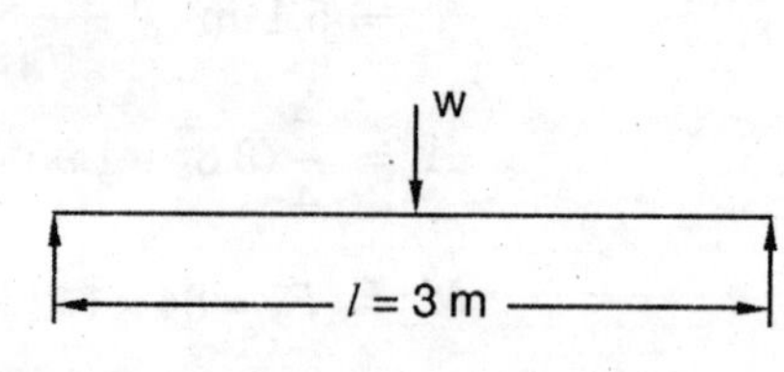

Fig. 10.22 (b)

$$I_{NA} = \frac{bd^3}{12} = \frac{200 \times (250)^3}{12} \text{ (mm}^4\text{)}$$

$$= 260.4166 \times 10^6 \text{ (mm}^4\text{)}$$

σ_{max} will occur along *AB* where it will be (comp) while along *CD* it will be maximum tensile

$$\sigma_{AB} = \frac{yM}{I}$$

Here, $$M = \frac{Wl}{4}$$

$$= \frac{100 \text{ (kN)} \times 3m}{4}$$

$$= 75 \text{ kN-m} = 75 \times 10^6 \text{ N-mm}$$

Now $$\sigma_{max} = \sigma_{AB} = \frac{yM}{I}$$

Here $\quad y = \frac{1}{2}$ depth $= 125$ mm

$$\therefore \quad \sigma_{AB} = 125 \times \frac{75\ 10^6}{260.4166 \times x^6} \text{ N/mm}^2$$

$$= 36.00 \text{ MPa}$$

$$= 36 \text{ MPa}$$

It will be compressive on points along the line *AB*. On the contrary it will be tensile on points along *DC* because it is crying sagging moment.

10.7. Find the value of I_{NA} for the section shown in Fig. (10.22*c*) and location of *NA*. If *B.M.* on the section is 1150 kN-m find the stress along *AD*.

Solution. Let $\bar{y}$ be the height of centroid from $BC = h$

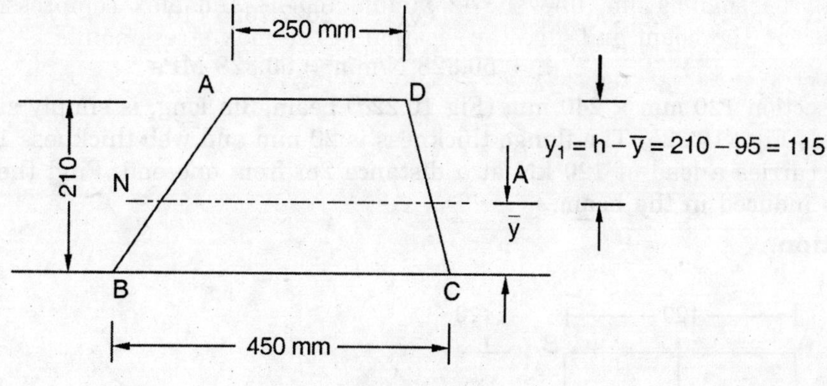

Fig. 10.22 (c)

Here, $\quad a = 250$ mm

$b = 450$ mm

$h = 210$ mm

Then from equation (*Q*) of article 10.2.7),

We have,
$$\bar{y} = \frac{h}{3}\left(\frac{b+2a}{b+a}\right)$$

$$= \frac{210}{3}\left[\frac{450+500}{450+250}\right]$$

$$= 70 \times \frac{950}{700} = 95.0 \text{ mm}$$

From eqn. (*R*) of § 10.2.7,

$$I_{NA} = \frac{h^3}{36}\left[(a+b) + \frac{2ab}{(a+b)}\right]$$

$$= \frac{(210)^3}{36}\left[(250+450) + \frac{2 \times 250 \times 450}{700}\right] \text{mm}^4$$

$$= \frac{(210)^3}{36}\left[700 + \frac{500 \times 450}{700}\right]$$

$$= \frac{(210)^3}{36}(700 + 321.4286) = \frac{1021.4286 \times (210)^3}{36}$$

$$= 2627.7625 \times 10^6 \text{ mm}^4$$

(*b*) Let the stress along the line *AD* due to bending moment *M* be σ_{AB} and then

$$\sigma_{AD} = \frac{M}{I_{NA}} \cdot y_{AD}$$

Now $\quad y_{AD} = 115$ mm and M = 1150 kN-m = 1150 × 10⁶ kN

Then $\quad \sigma_{AD} = y_{AD}\dfrac{M}{I_{NA}} = \dfrac{115 \times (1150 \times 10^6)}{2627.7625 \times 10^6}$

$$= 50.328 \text{ N/mm}^2 = 50.328 \text{ MPa}$$

10.8. An I-section 120 mm × 240 mm (Fig 10.22*d*) beam, 3m long, is simply supported at its ends Fig. (10.22*b*) The flange thickness is 20 mm and web thickness 10 mm. The beam carries a load of 120 kN at a distance 2m from one end. Find the maximum stress induced in the beam.

Solution.

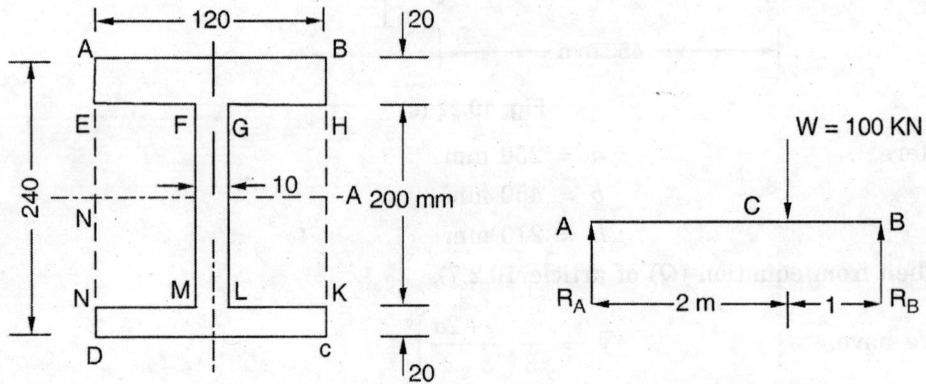

Fig. 10.22 (d) Fig. 10.22 (e)

$$I_{NA} = I_{NA} \text{ of } ABCD - 2I_{NA} \text{ of } EFMN$$

$$= \frac{120 \times 240^3}{12} - 2\left[\frac{55 \times 200^3}{12}\right]$$

$$= 13824 \times 10^4 - 7.333333 \times 10^4$$

$$= 6491 \times 10^4 \text{ mm}^4$$

$$M_{max} = \frac{Wab}{l}$$

when $W = 120$ kN

$a = 2$m

$b = 1$m

$l = 3$m

i.e., $$M_{max} = \frac{120\ kN \times 2m \times 1m}{6\ m}$$

$$= 40 \text{ kN-m} = 40 \times 10^6 \text{ N-mm}$$

and $$y_{max} = \frac{240}{2} = 120 \text{ mm}$$

Therefore, $$(\sigma_n)_{max} = y_{max}\frac{M_{max}}{I_{NA}} = \frac{120 \times 40 \times 10^6}{6491 \times 10^4} \text{ N/mm}^2$$

$= 73.949$ MPa (comp. along *AB* and tensile along *DC*)

10.9. The cross-section of a cantilever beam is shown in figure (10.22*f*). It carries a load of 300 kN at the free end. The length of the cantilever is 1.2 m. Find the maximum tensile and compressive stress induced in the section.

Solution. For finding the location of the centroid, take moments of areas about the line *AB*. Let NA be at a distance $\bar{y}$ from *AB*

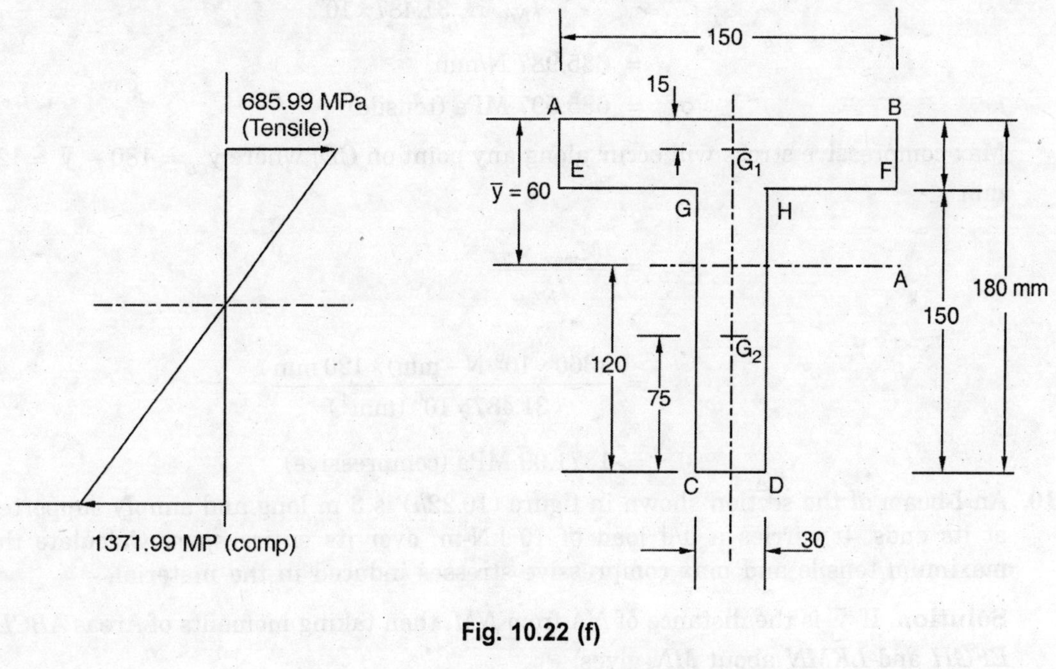

Fig. 10.22 (f)

Fig. 10.22 (g)

$$\bar{y}\,[(150 \times 30) + (150 \times 30)] = (150 \times 30 + 15) + 150 \times 30\,(30 + 75)$$

or $\bar{y} = 60$ mm

Then $I_{NA} = I_{NA}$ of $ABFE + I_{NA}$ of $GHCD$

$$= \left[\frac{150 \times 30^3}{12} + (150 \times 30)\left(\bar{y} - 15)^2\right)\right]$$

$$+ \left(\frac{30 \times 150^3}{12} + (150 \times 30)\,(180 - \bar{y} - 75)^2\right)$$

$$= 31.487 \times 10^6 \text{ mm}^4$$

$$M_{max} = (300 \times 10^3 \text{ N})\,(1.2 \times 10^3 \text{ mm})$$

$$= 360.0 \times 10^6 \text{ N-mm (Hogging)}$$

The max tensile stress will occur along AB where

$$y_{AB} = \bar{y} = 60 \text{ mm}$$

$$\therefore \quad \sigma_{AB} = y_{AB}\frac{M_{max}}{I_{NA}} = \frac{60 \times 360 \times 10^6}{31.487 \times 10^6}$$

$$= 685.987 \text{ N/mm}^2$$

i.e., $\sigma_{AB} = 685.997$ MPa (tensile)

Max compressive stress will occur along any point on CD, where $y_{CD} = 180 - \bar{y} = 120$ mm

$$\sigma_{CD} = \frac{M_{max}\,y_{CD}}{I_{NA}}$$

$$= \frac{(360 \times 10^6 \text{ N} - \text{mm}) \times 120 \text{ mm}}{31.487 \times 10^6 \text{ (mm}^4)}$$

$$= 1371.99 \text{ MPa (compressive)}$$

10.10. An I-beam of the section shown in figure (10.22*h*) is 3 m long and simply supported at its ends. It carries a u/d load of 10 kN-m, over its entire span. Calculate the maximum tensile and max compressive stresses induced in the material.

Solution. If $\bar{y}$ is the distance of NA from NM, then taking moments of Areas $ABCD$, $EFGH$ and $LKMN$ about MN gives,

$$\bar{y}\,[(100 \times 20) + (20 \times 100) + (200 \times 30)$$

$$= (100 \times 20) \times 140 + (100 \times 20)\,80 + (200 \times 30) \times 15$$

or $\bar{y} = 53$ mm

$$I_{NA} = I_{NA} \text{ of } ABCD + I_{NA} \text{ of } EFGH + I_{NA} \text{ of } KLNM$$

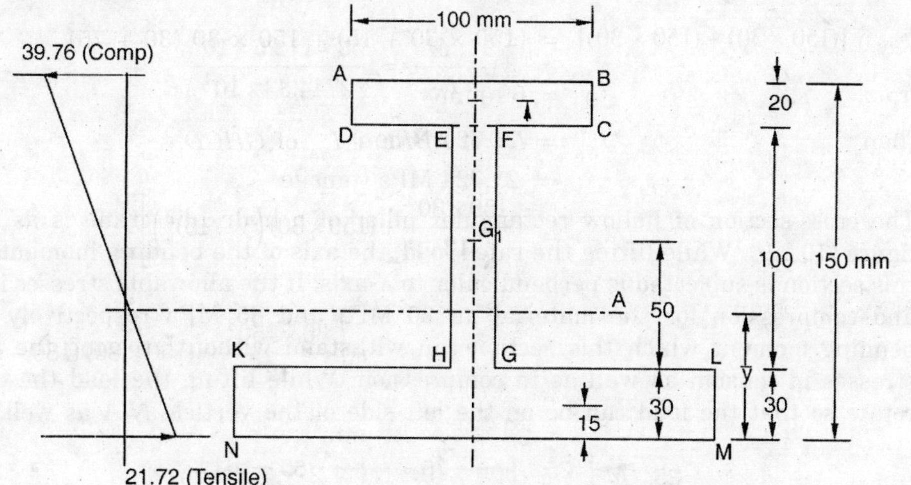

Fig. 10.22 (h)

or
$$I_{NA} = \left[\frac{100 \times 20^3}{12} + (100 \times 20)(140 - 53)^2\right] + \left[\frac{20 \times 100^3}{12} + (20 \times 100)(80 - 53)^2\right] + \left[\frac{200 \times 30^3}{12} + (200 \times 30)(53 - 15)^2\right]$$

$$I_{NA} = 6.66 \times 10^2 + 1513.8 \times 10^2 + 165.667 \times 10^2 + 145.8 \times 10^2 + 866.4 \times 10^2$$
$$= 2744.33 \times 10^4 \text{ mm}^4$$

$$\text{Max } BM = \frac{wl^2}{8}$$

where,
$$w = 10 \text{ kN-m}$$
$$l = 3\text{m}$$

$$= \frac{10\,(3)^2}{8} \text{ kN} - \text{m}$$

$$= 11.25 \text{ kN-m} = 11.25 \times 10^6 \text{ N-mm}$$

w = 10 kN\m

A B

3 m

Fig. 4.22 (i)

Now
$$\sigma = \frac{M}{I_{NA}}$$

At AB
$$\sigma_{AB} = y_{AB}\,\frac{M}{I_{NA}} = (150 - 53)\,\frac{11.25 \times 10^6}{2744.33 \times 10^4}$$

$$= \frac{97 \times 11.25 \times 10^6}{2744.33 \times 10^4} = 39.76 \text{ N/mm}^2$$

$$= 39.76 \text{ MPa} \qquad \text{(comp.)}$$

$$\sigma_{MN} = y_{MN}\frac{M}{I_{NA}} = \frac{53 \times 11.25 \times 10^6}{2744.33 \times 10^4}$$

$$= 21.724 \text{ N/mm}^2$$

$$= 21.724 \text{ MPa (tensile)}$$

10.11. The cross-section of hollow rectangular pillar of a hydraulic crane is as shown in figure (10.22*j*). While lifting the rated load, the axis of the bending moment to which this section is subjected is perpendicular to *X*-axis. If the allowable stresses in tension and compression for the material an 20 MPa and 85 MPa respectively, find the bending moment which this section can withstand without crossing the allowable stresses in tension as wall as in compression. While lifting the load the crane can rotate so that the load can be on the left side of the verticle *N-A* as well.

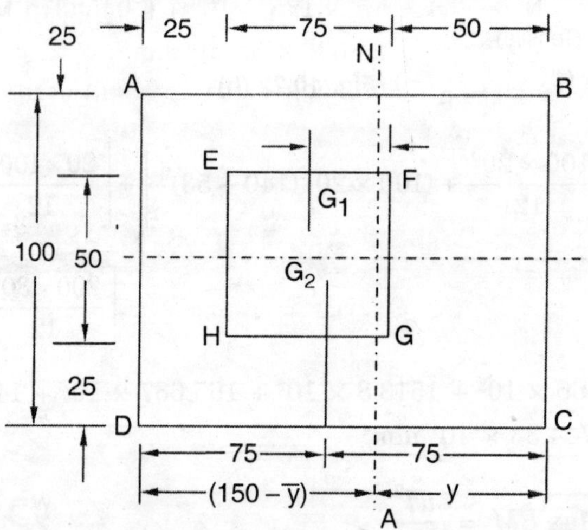

Fig. 10.22 (j)

Solution. From the problem, it is evident that the load to be lifted is such that the axis of *BM* is to be at right angles to *XX*. That is the centroidal bending axis or *N-A* of the section will be vertical.

In order to locate the centroid *G* of the hollow section take moments of areas about edge *BC*. Let its distance from *BC* be $\bar{y}$. Then taking the moment of areas *ABCD* and *EFGH* and subtracting the later from the former we have,

$$(150 \times 100 - 75 \times 50)\bar{y} = \left[(150 \times 100)\frac{150}{2}\right] - (75 \times 50)(50 + 37.5)$$

$$11250\,\bar{y} = 1.125 \times 10^6 - 0.328125 \times 10^6$$

or $$\bar{y} = 70.8 \text{ mm}$$

$$I_{NA} = I_{NA} \text{ of } ABCD - I_{NA} \text{ of } EFHK$$

Let G_1 be the centroid of *ABCD* and G_2 centroid of *EFHK*

Then $$GG_1 = \frac{150}{2} - \bar{y} = 75 - 70.8 = 10.2 \text{ mm}$$

and $GG_2 = (50 + 37.5) = 16.7$ mm

Then $$I_{NA} \text{ of } ABCD = \frac{(150)^3 \times 100}{12} + \text{Area } ABCD \times GG_1^2$$

$$= \frac{150^3 \times 100}{12} + (150 \times 100)(4.2)^2$$

$$= 210.125 \times 10^6 + 0.2646 \times 10^6$$

$$= 28.3896 \times 10^6 \text{ mm}^4$$

$$I_{NA} \text{ of } EFHK = \frac{50 \times 75^3}{12} + (50 \times 75)(16.7)^2$$

$$= 1.758125 \times 10^6 + 1.0458375 \times 10^6$$

$$= 2.803646 \times 10^6 \text{ mm}^4$$

There I_{NA} of given section $= 28.3896 \times 10^6 - 2.8036 \times 10^6$

$$= 25.586 \times 10^6 \text{ mm}^4$$

Case I : When load to be lifted is on the right of NA

In this case the stress on any point on line AD will be tensile and that on any point on BC will be compressive then for failure in tension because the material is weaker in tension

$$M = \frac{\sigma_{AD} I_{NA}}{y_{AD}}$$

Where, y_{AD} = distance of AD from NA = 79.2 mm

σ_{AD} = 20 MPa = 20 N/mm²

Therefore, $$M = \frac{20\,(25.586 \times 10^6)}{79.2}$$

$$= 6.46 \times 10^6 \text{ N-mm} = 6460 \text{ N-m}$$

Case II : When the load is on the left of NA, the stress on any point on BC will be tensile and on AD compressive taking failure in tension,

We have, $$M = \frac{\sigma_{BC} \times I_{NA}}{y_{BC}}$$ Here, $y_{BC} = \bar{y}$, $\sigma_{BC} = 20$ MPa

$$= \frac{20 \times 25.586 \times 10^6}{70.8}$$

$$= 7.2277 \times 10^6 \text{ N-mm} = 7227.7 \text{ N-m}$$

Thus, the safest moment which this section can take up is smaller of the two values *i.e.*, 6460 N-m.

10.12. A short hollow cylindrical column, 250 mm outside dia and 150 mm inside dia, carries an eccentric load of 500 kN whose line of action is parallel to cylinder axis but at a distance of 70 mm from the axis of the cylinder. Find the maximum values of the tensile and compressive stresses induced in the material.

Solution. Let the electricity = e = 70 mm of given compressive load

Given load $P = 500$ kN

A = Area of the cross-section

$= \frac{\pi}{4}(250^2 - 150^2)$

$\frac{\pi}{4}(62500 - 22500) = 40000 \text{ mm}^2$

$I_{NA} = \frac{\pi}{64}(D^4 - d^4)$

$= \frac{\pi}{64}(250^4 - 150^4)$

$= 166.897 \times 10^6 \text{ mm}^4$

σ_0 = direct stress = $\frac{P}{A}$

$= \frac{500 \text{ kN}}{40000}$

= 12.5 MPa (Compressive)

$M = P \times e$

= (500 kN) × 70

= 35000 kN-mm

Fig. 10.22 (k)

Resultant stress at B

$(\sigma_r)_B = \sigma_0 + \sigma_b$

= (12.5) MPa (Compressive)

But $\sigma_B = y_B \cdot \frac{M_B}{I_{NA}}$

$= \frac{250}{2} \times \frac{(35 \times 10^6)}{166.897 \times 10^6}$

= 26.2137 MPa (Compressive)

Thus, the resultant max comp stress at B

= 12.5 MPa + 26.2137 MPa

= 38.7137 MPa (Compressive)

At the point A the bending stress

= 26.2137 MPa (Tensile)

Thus, resultant tensile stress at A

= 26.7137 – 12.5 MPa

= 13.7137 MPa (Tensils)

10.13. The cross-section of a hollow pier is as shown in figure (10.22l)

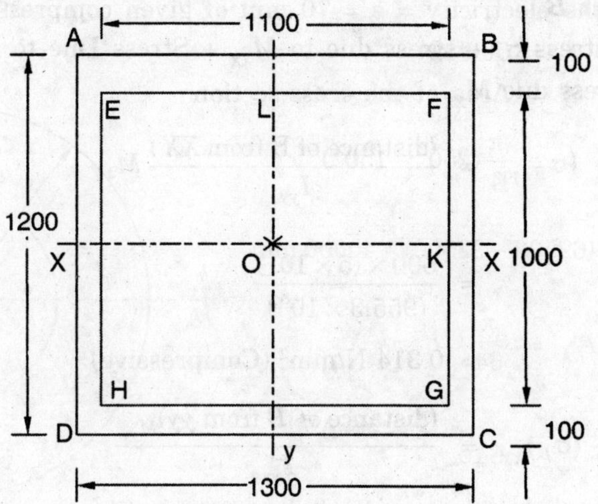

Fig. 10.22 (l)

A Compressive load of 100 kN normal to section acts at the point *F*.

Find the stresses at the four corners *A, B, C* and *D* of the section.

Let the load be = P = 100 kN = 10^5 N

The load line is eccentric with both the axes,

$$M_{XX} = BM \text{ about } XX = P \times FK$$

$$= (100 \times 10^3) \times 500 \text{ N-mm} = 5 \times 10^7$$

$$M_{yy} = \text{BM about yy} = \left(10^5 \times \frac{1100}{2}\right) = 5.5 \times 10^7 \text{ N-mm}$$

$$I_{XX} = \frac{1300 \times 1200^3}{12} - \frac{1100 \times 1000^3}{12}$$

$$= 1872 \times 10^8 - 916.7 \times 10^8$$

$$= 955.3 \times 10^8 \text{ mm}^4$$

and $$I_{YY} = \frac{1200 \times (1300)^3}{12} - \frac{1000\,(1100)^3}{12}$$

$$= 2197 \times 10^8 - 1109.2 \times 10^8$$

$$= 1087.8 \times 10^8 \text{ mm}^4$$

Direct stress over the entire section = $\dfrac{P}{\text{Area of the section}}$

$$\sigma_0 = \frac{P}{(1300 \times 1200) - (1100 \times 1000)}$$

$$= \frac{10^5\,N}{46 \times 10^4\,\text{mm}^2} = 0.2174 \text{ N/mm}^2 \text{ (Compressive)}$$

Stress at ponit B

(*i*) σ_B =Direct stress σ_0 + stress due to M_{XX} + Stress Due to M_{YY}

(*ii*) Bending stress due M_{XX}

$$(\sigma_B)_{YY} = \frac{(\text{distance of B from } XX)}{I_{yy}} M_{YY}$$

$$= \frac{600 \times (5 \times 10^7)}{(955.3 \times 10^8)}$$

$$= 0.314 \text{ N/mm}^2 \text{ (Compressive)}$$

$$(\sigma_B)_{XX} = \frac{(\text{distance } of\ B \text{ from } yy)_{M_{yy}}}{I_{yy}}$$

$$= \frac{(500)(5.500 \times 10^7)}{(1087.8 \times 10^8)}$$

$$= 0.2528 \text{ N/mm}^2 \text{ (Compressive)}$$

Therefore, $\sigma_B = (-2174) - (0.314) - (0.2528)$

$= 0.7842$ MPa (comp) N/mm² or MPa

(*ii*) Stress at the point *C*

σ_{BC} = Resultant of direct stress + Bending stress at *C* due M_{XX} + Bending stress at *C* due to M_{YY}

Now $\sigma_0 = 0.2174$ MPa (Compressive)

Bending stress at *C* due M_{XX} = (Bending stress at *B* are to M_{XX} because the distance of *C* from *XX* is same as that of *B*.]

i.e., $(\sigma_C)_{XX} = (\sigma_B)_{XX} = 0.314$ MPa (tensile)

Bending stress at *C* due M_{YY},

$$= (\sigma_C)_{YY} = (\sigma_B)_{YY}$$

$= 0.2528$ MPa (Compressive)

Therefore, resultant of these,

$$\sigma_C = -0.2174 + 0.314 - 0.2528$$

$= -0.1562$ MPa

or $= 0.1562$ MPa (Compressive)

(*iii*) Stress at the Point D

i.e., $\sigma_D = (\sigma_0) + (\sigma_D)_{XX} + (\sigma_D)_{YY}$

Now $\sigma_0 = -0.2174$ (Compressive)

And $(\sigma_D)_{XX} = (\sigma_C)_{XX} = 0.314$ MPa (tensile)

Also, $(\sigma_D)_{YY} = -(\sigma_C)_{YY}$

$= -(-0.2528)$

$= 0.2528$ MPa (tensile)

$\therefore \quad \sigma_D = -0.2174 + 0.314 + 0.2528$

$= 0.3494$ MPa = MPa (tensile)

(iv) Stress at the point A $= \sigma_0$ + stress due to M_{XX} + stress due to M_{YY}

$= -0.2174 - 0.314 + 0.2528$

$= -0.2786$ MPa

$= 0.2786$ MPa (Compressive)

10.14. A wooden beam of hollow section as shown in figure (10.22*m*) is simply supported at its ends with span of 3m. A load of 10kN acts at a distance 2m from left hand support. Calculate in maximum tensile and compressive stress on the mid-section which 1.5 m from the left support

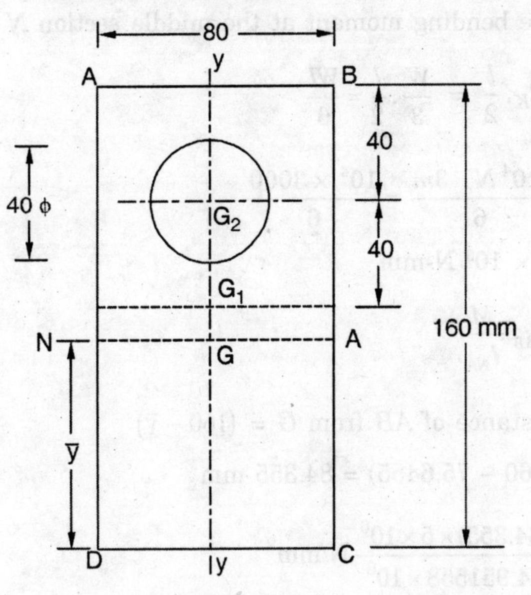

Fig. 10.22 (m)

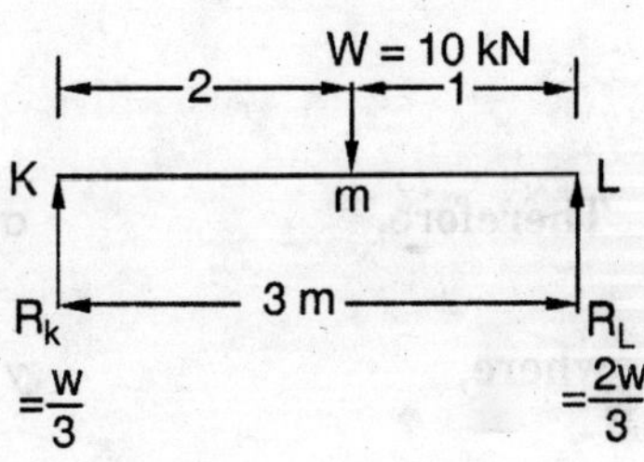

Fig. 10.22 (n)

Here, $R_K = \dfrac{W}{3} = \dfrac{10}{3} kN$

and $R_L = \dfrac{2W}{3} = \dfrac{20}{3} kN$

Solution. Let M be the BM at mid-section,

$$M = \frac{l}{2} R_K = R_k \times \frac{l}{2} = \frac{10}{3} \times \frac{3}{2} \text{ kN} - \text{m} = 5 \text{ kN-m}$$

Let $\bar{y}$ be the distance of the centroid of the given section from *DC*. Then taking moments of the individual areas about *DC* we get,

$\bar{y}$ [Area *ABCD* – Area of circular hole]

= Moment of area *ABCD* about *DC* – moment of circular area about *DC*

i.e., $\bar{y}\left[(80\times160)-\frac{\pi}{4}(40)^2\right] = \left[(80\times160)\times80-\frac{\pi}{4}(40)^2\,(80+40)\right]$

Therefore, $\bar{y} = 75.6455$ mm

Then $I_{NA} = \left[\frac{80\times160^3}{12}+(80\times160)\,(GG_1)^2\right]$

$-\left[\frac{\pi\times(40)^4}{64}+\frac{\pi}{4}(40)^2\times(G_2G)^2\right]$

$= 24.955 \times 10^6$ mm

For the simply supported beam the bending moment at the middle section *N* is

$$M_n = R_K.\frac{l}{2} = \frac{W}{3}.\frac{l}{2} = \frac{Wl}{6}$$

$$= \frac{(10^4\,N)\ 3m}{6} = \frac{10^4\times3000}{6}$$

$$= 5 \times 10^6 \text{ N-mm}$$

Therefore, $\sigma_{AB} = y_{AB}\,\frac{M}{I_{NA}}$

where, y_{AB} = distance of *AB* from $G = (160-\bar{y})$

$= (160 - 75.6455) = 84.355$ mm

or $\sigma_{AB} = \frac{(84.355)\times5\times10^6}{24.951558\times10^6}$ N/mm²

$= 16.904$ N/mm² or MPa (Compressive)

$\sigma_{CD} = y_{CD}\,\frac{M}{I_{NA}}$ Here $y_{CD} = \bar{y}$

$= 75.6455\times\frac{5\times10^5}{24.95\times10^6}$

$= 15.158$ N/mm² or MPa (tensile)

10.15. Cast iron channel 500 mm × 320 mm (Fig. 10.22) carries water. The channel is 10 m long and is simply supported at its ends. If the bending stresses in the channel are not to exceed 15.5 MPa in tension and 45.5 MPa in compression, estimate the depth of water in the channel. Density of *CI* is 7000 kg/m³ and that of water 1000 kg/m³.

Solution. Let the NA of channel section be at a distance of $\bar{y}$ from EF.

Then taking the moments of areas $ABCD$, $NCLM$ and $DEFK$ about FE

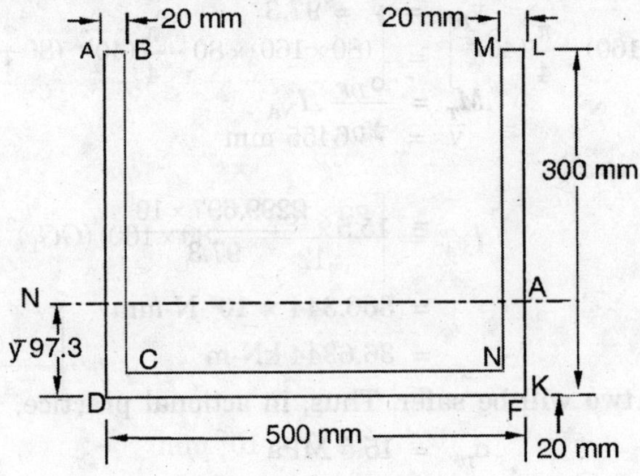

Fig. 10.22 (o)

We get $\bar{y}[(500 \times 20) + 2(300 \times 20)]$

$$= (500 \times 20) \times 10 + 2\,(300 \times 20)\,(20 + 150)$$

$$\bar{y} = 97.3 \text{ mm}$$

Then $I_{NA} = I_{NA}$ of area $DEFK + 2I_{NA}$ of $NKLM$

$$= \left[\frac{500 \times (20)^3}{12} + (500 \times 20)\,(97.3 - 10)^2\right]$$

$$+ 2\left[\frac{20 \times (300)^3}{12} + (20 \times 300)\,(170 - 97.3)^2\right]$$

or $I_{NA} = 3.3333 \times 10^5 + 762.2129 \times 10^5$

$$+ 2\,[450 \times 10^5 + 317.1174 \times 10^3]$$

$$= 2299.6971 \times 10^5 \text{ mm}^4$$

Resisting moment of channel section in compression is given by M_C as,

$$M_C = \frac{\sigma_{AB}\, I_{NA}}{y_{AB}}$$

But $\sigma_{AB} = 45.5$ MPa

and $y_{AB} = 320 - \bar{y}$

$$= (320 - 97.3) = 222.7 \text{ mm}$$

Therefore, $M_C = \dfrac{45.5 \times 2299.697 \times 10^5}{222.7}$ N-mm

$$= 469.853 \times 10^5 \text{ N-mm}$$

$$= 46.9853 \text{ kN-m}$$

Let Resisting moment of section in tension = M_t

Given $\sigma_{DF} = 15.5$ MPa

$$y_{DF} = \bar{y} = 97.3$$

$$M_T = \frac{\sigma_{DF}}{y_{DF}} \cdot I_{NA}$$

$$= 15.5 \times \frac{2299.697 \times 10^5}{97.3}$$

$$= 366.344 \times 10^5 \text{ N-mm}$$

$$= 36.6344 \text{ kN-m}$$

Smallar of the two will be safer. Thus, in actional practice,

$$\sigma_{DF} = 15.5 \text{ MPa}$$

$$\bar{y} = 97.3 \text{ mm} = y_{EF}$$

will prevail. Thus, if wkN-m is the weight per unit length of the channel (including that of water present at home) for the beam in question, then

$$M_{max} = \frac{wl^2}{8} \quad \text{where } l = 10 \text{ m}$$

$$= \frac{w \times 10^2}{8} = 12.5w \text{ kN-m}$$

Equating this moment to the resisting moment,

We get, $12.5w = 36.6344$

or $w = 2.9307$ kN-m

Now weight of CI channel per unit length (excluding water)

$$= [2(20 \times 300) + (20 \times 500)] \times 1000 \times \frac{7000}{(10)^9} \times 9.8$$

$$= 1510.74 \text{ N/m} = 1.51074 \text{ kN-m}$$

Therefore, weight of 1 m length of water column

$$= w - w_{CI} = 2.9307 - 1.51074$$

i.e., $w_{water} = 1.4133$ kN-m

But $$w_{water} = \left(\frac{h}{1000} \times \frac{460}{1000} \times \frac{1000}{1000}\right) \times (1000 \times 9.81) \text{ K/m}$$

$$= 4.5126h \times 10^{-3} \text{ kN/m}$$

Thus, $(10^{-3} \times 4.5126)h = 1.4133$

$$h = 0.31318 \times 10^3 \text{ mm}$$

$$= 313.18 \text{ mm}$$

PROBLEM FOR PRACTICE

1. A cast iron beam of section shown in Fig. 10.23, is simply supported at its ends and carries a load of 18 kN at mid span. Find the maximum allowable span, if the stress due to bending is not to exceed 30 MN/m² tension. Neglect the weight of the beam.

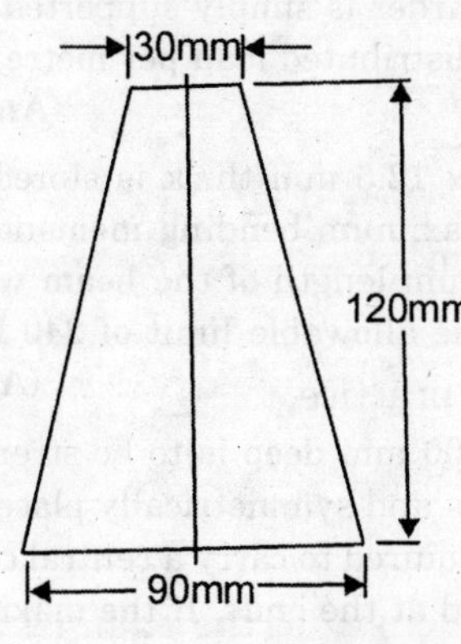

Fig. 10.23

Fig. 10.24

What will be the maximum compressive stress ? (**Ans.** 1.056 m ; 42 MN/m²)

2. Fig. 10.24 shows the section of a beam. What is the ratio of its moment of resistance to bending in the plane *YY* to that for bending in the plane *XX* if the maximum stress due to bending is the same in both cases ?

For a semi–circle of radius *r*, the centroid is at a distance of 4 *r*/(3π) from the centre.

(**Ans.** 2.85)

3. Find the ratio of the resisting moments of a square section beam, when in one case the two sides are vertical and in the other a diagonal is vertical. (**Ans.** $\sqrt{2}$:1)

4. Prove that the moment of resistance of a beam of square section with its diagonal in the plane of bending is increased by flattening the top and bottom corners, as shown in Fig. 10.25, and that the moment of resistance is a maximum when,

$$y = \frac{8}{9}Y$$

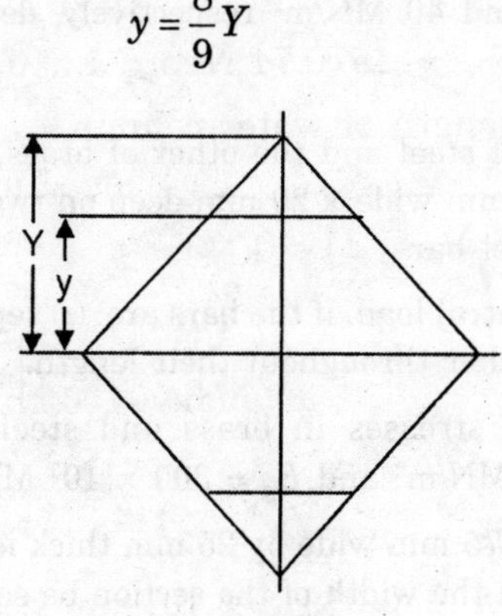

Fig. 10.25

5. A cantilever specimen for a fatigue testing machine is of circular cross–section throughout its length, but in a length of 75 mm the diameter decreases from 10 mm at the fixed end to 6.25 mm at the free end. Calculate the maximum stress due to bending when a static load of 300 N is applied at the free end in a direction perpendicular to the length of the specimen. (**Ans.** 231.4 MN/m^2)

6. An I–section girder 240 mm deep, has a web 20 mm thick. The top flange is 120 mm × 20 mm and the bottom flange is 160 mm × 20 mm. If the girder is simply supported at its ends, find the maximum span which can be used, if the total distributed load per metre run is 6000 N/m, and the maximum stress is limited to 70 MN/m^2. (**Ans.** 8.666 m)

7. A symmetrical I–section steel beam 100 mm × 100 m × 12.5 mm thick is stored by placing it on two simple supports in such a manner that the maximum bending moment to which it is subjected is the least possible. Determine the maximum length of the beam which can be supported in this manner without the stress crossing the allowable limit of 240 MN/m^2. The density of steel is 7800 kg/m^3. (**Ans.** 66.9 m)

8. A timber beam of rectangular section 150 mm wide × 300 mm deep is to be strengthened by two steel flitches 1 mm thick firmly attached to the sides and symmetrically placed about the horizontal centre line of the beam section. The beam is required to carry a central concentrated load of 52.5 kN on a span of 3 m, being simply supported at the ends. If the maximum stress in timber is not to exceed 10.5 MN/m^2, find the depth of the steel flitches required. $E_{steel} = 20\ E_{timber}$.

(**Ans.** 220 mm)

9. A timber beam 80 mm wide × 160 mm deep is to be strengthened with 5 mm thick steel plates. For the same value of maximum bending stress in the timber, compare the moments of resistance of the composite sections in the following two cases :

1. The plates are 160 mm deep and fixed to the vertical sides of the timber beam,
2. The plates are 80 mm wide and are fixed to the top and bottom surfaces of the timber beam. (**Ans.** 0.703)

10. A compound beam is formed by joining two bars rigidly together, one of steel and the other of brass, each bar being 40 mm wide. The bars are of thickness t_1 and t_2 respectively so that the total depth is $(t_1 + t_2)$. If E for the steel is twice that for the brass, find the ratio of t_1 to t_2 so that the neutral axis of the section is at the dividing line of the two bars.

If the total depth of the section is to be 20 mm and the stresses in the steel and brass are not to exceed 110 MN/m^2 and 40 MN/m^2 respectively, determine the maximum moment of resistance of the beam. (**Ans.** $1 : \sqrt{2}$; 125 Nm)

11. Two rectangular bars, one of steel and the other of brass, each 30 mm × 10 mm are placed together to form a beam 30 mm wide × 20 mm deep on two supports 0.75 m apart. The brass bar is on the top of the steel bar.

Determine the maximum central load, if the bars are: (*i*) separate and can bend independently, (ii) firmly secured to each other throughout their length.

The maximum permissible stresses in brass and steel are 70 MN/m^2 and 105 MN/m^2 respectively. $E_B = 85 \times 10^3$ MN/m^2 and $E_S = 300 \times 10^3$ MN/m^2. (**Ans.** 373.3 N, 825 N)

12. A bar of rectangular section 75 mm wide by 25 mm thick is subjected to an axial tensile force of 100 kN. By how much can the width of the section be safely reduced by removing material from one edge only, if the maximum stress over the reduced width may not exceed 100 MN/m^2. (**Ans.** 12.05 mm)

13. Fig. 10.26 shows a C–clamp which is tightened up to a load of 540 N. Calculate the stresses at the bottom and top edges of the section *XX*. Draw the stress distribution diagram.

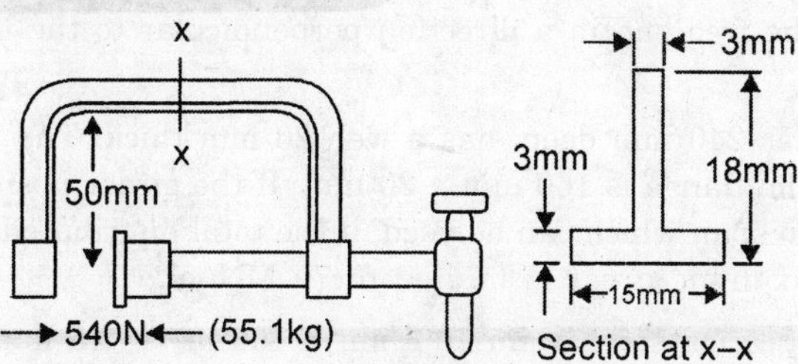

Fig. 10.26

(**Ans.** 63.6 MN/m² tensile, 109.2 MN/m² comp.)

14. A short, hollow, cylindrical column carries a compressive load of 490.5 kN. Find the maximum permissible eccentricity of this load, if the maximum permitted stresses are 75 MN/m² compressive and 20 MN/m² tensile. The diameters are 200 mm external and 124 mm internal.

(**Ans.** 63 mm from the axis)

15. A pillar 1.5 m high is of rectangular section 50 mm thick and tapers longitudinally from a width of 150 mm at the base to 50 mm at the top, as shown in Fig. 10.27. A compressive load of 100 kN acts through the centroid at the top end and is parallel to the vertical edge. Determine the magnitude of the maximum compressive stress and the cross section at which it will occur.

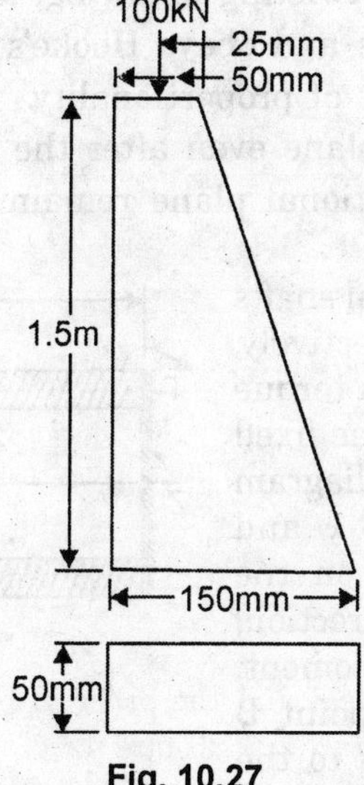

Fig. 10.27

(**Ans.** 55.33 MN/m² ; 0.375 m from top)

16. A short hollow pier, 1.2 m × 1.2 m outside and 0.75 m × 0.75 m inside, supports a vertical point load of 120 kN located on a diagonal and 0.69 m from the vertical axis of the pier. Neglecting the self–weight of the pier, calculate the normal stresses at the four outside corners on a horizontal section of the pier.

(**Ans.** 616.8 kN/m² comp. ; 136.8 kN/m² comp.; 136.8 kN/m² comp. ; 343.2 kN/m² Tensile)

❑❑❑

CHAPTER 11

TORSION

11.1 TORSION OF CIRCULAR SHAFTS

When a circular shaft is acted upon by an axial torque and is not carrying any bending moment or axial thrust, its each and every cross-section is known to be under pure shear. The shaft can be assumed to be composed of a large number of thin discs, face of each disc having a tendency to slip with respect to that of the adjacent disc. Total twist of the free end of the shaft is then equal to the sum total of the individual twists of all these elementary discs. The shearing stress induced in the material over any cross-sectional plane produces the resisting moment equal and opposite to the applied torque. The complementary shear stress on the longitudinal planes will cause a distortion of the elements which were originally lying along the longitudinal direction.

In evaluating the stress due to twisting following, assumptions are made :

1. The material is homogeneous and obeys Hooke's Law.
2. It is stressed within the limit of proportionality.
3. Plane cross-sections remain plane even after the application of torque.
4. All the radii in the cross-sectional plane remain straight and radial even after the application of torque.

Fig. 11.1 shows a hollow cylindrical shafts of inner and outer radii r and R respectively, fixed at one end and acted upon by a torque T on the other end, distance l from the fixed end. Figure 11.2 shows the free body diagram of an annular cylinder of radius x and thickness dx. AB is a line drawn on the surface of this cylinder in the axial direction. After the application of the twisting moment, the line AB becomes a helix with point B shifting to position C corresponding to the twist θ about ZZ of the section at B with respect to the fixed section at A. The inclination of the helix AC with the ZZ axis will give the shear strain (refer to article 1.5). It may be seen that the shear strain, and hence, the shear stress, on any of the cross-sectional planes of the annular cylinder along the length AB, will be same.

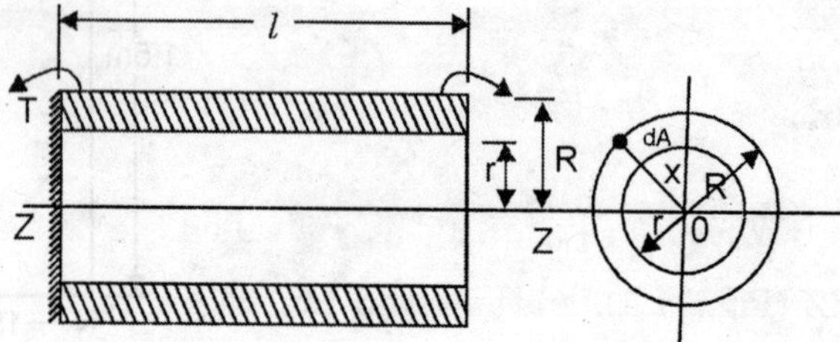

Fig. 11.1

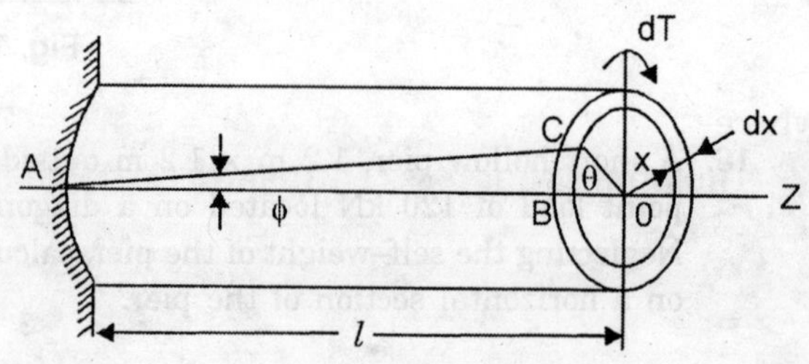

Fig. 11.2

If the shear stress is within the elastic limit, the shear strain angle ϕ and the angle of twist θ due to torque T, will be small, then,

$$BC = l\phi = x\theta$$

i.e.

$$\phi = \frac{x\theta}{l} \qquad ...(11.1)$$

Thus, if G is the modulus of rigidity of the shaft material, then the shear stress τ_x on the sectional plane of the annular cylinder with be given by,

$$\tau_x = \phi G$$

$$= Gx\frac{\theta}{l}$$

or

$$\frac{\tau_x}{x} = \frac{G\theta}{l} \qquad ...(11.2)$$

For any given shaft and given torque G, τ and l will be fixed *i.e.* the right hand side of equation (11.2) is constant. In other words, the ratio (τ_x/x) is constant. That is the shear stress goes on increasing linearly with radius 'x' and at the axis, where $x = 0$, it is zero. Therefore, if τ is the shear stress on the section at outer radius, R, it will be maximum and we will have,

$$\frac{\tau_x}{x} = \frac{\tau}{R} = \frac{G\theta}{l} = \text{constant} \qquad ...(11.3)$$

Again if dT is the torque transmitted by the annular cylinder, then the total resisting moment of the cylinder, which is equal to T is given, by referring to figure 11.1, as,

$$T = \int_{x=r}^{R} dT = \int_{x=r}^{R} (\tau_x \,.\, dA)\, x$$

$$= \int_{X}^{R} \frac{\tau}{R} x^2 \, dA$$

$$= \frac{\tau}{R}\int_{r}^{R} (2\pi x \, dx)\, x^2$$

or

$$T = \frac{\tau}{R} J \qquad ...(11.4)$$

Where $J = \int_{r}^{R} dA\, x^2$ and is called the polar second moment of area of cross-section about ZZ (Fig. 11.1). and is given by,

$$J = \int_{r}^{R} (2\pi x \, dx)\, x^2$$

$$= \frac{\pi}{2}(R^4 - r^4) = \frac{\pi}{32}(D^4 - d^4)$$

where

$$D = 2R \text{ and } d = 2r$$

Thus from equation (11.3) and (11.4), we have,

$$\frac{T}{J} = \frac{\tau}{R} = \frac{\tau_x}{x} = \frac{G\theta}{l} \qquad ...(11.5)$$

For a solid shaft

$$J = \int_{0}^{R} (2\pi \, x dx)\, x^2 = \frac{\pi}{2} R^4 = \frac{\pi D^4}{32}$$

Therefore, $$\tau = \frac{T}{J} \cdot \frac{D}{2} = \frac{16T}{\pi D^3} \qquad ...(11.6)$$

or $$T = \frac{\pi}{16} \tau D^3 \qquad ...(11.7)$$

11.2 COMPOSITE SHAFTS

A shaft which is composed of two or more materials rigidly fixed together so that they share the torque being transmitted, is called composite shaft (Fig. 11.3).

Let T_1 and T_2 be the torques shared by the two portions respectively, then

$$T_1 + T_2 = T \qquad ...(11.8)$$

Also, as the twist per unit length will be same for the two portions, we will have from equation (11.5),

$$\frac{\theta_1}{l_1} = \frac{\theta_2}{l_2}$$

That is $$\frac{T_1}{J_1 G_1} = \frac{T_2}{J_2 G_2}$$

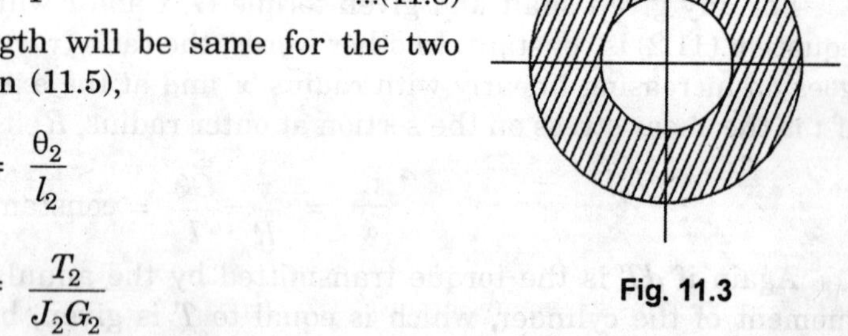

Fig. 11.3

Or $$\frac{T_1}{T_2} = \frac{J_1 G_1}{J_2 G_2} \qquad ...(11.9)$$

Knowing total torque T, equations (11.8) and (11.9) can be solved to evaluate T_1 and T_2. Using equation (11.5) the shear stress can then be calculated at any point in the cross-section.

11.1 : *A horizontal bar of 60 mm diameter is rigidly fixed at each end, the fixings being 1200 mm apart. A rigid bracket is fixed at 800 mm from one end of the bar, at right angles to its axis and in the same horizontal plane. Determine the maximum radius-arm of the bracket at which a vertical load of 1000 N can be suspended, if the deflection of the load is not to exceed 0.1 mm.*

Take $E = 200\ GN/m^2$ and $G = 80\ GN/m^2$.

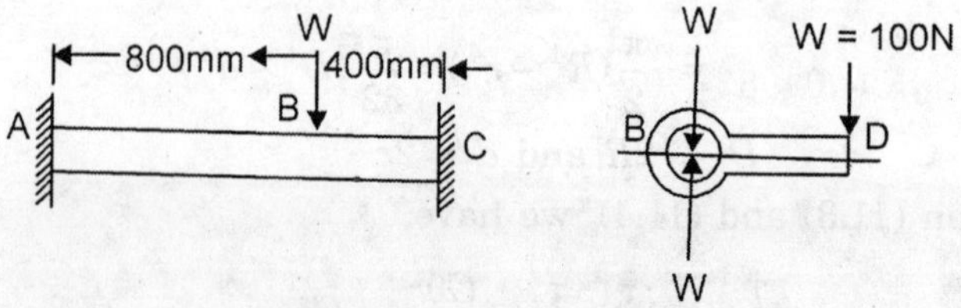

Fig. 11.4

Solution : Let a be the radius arm. The load at D will give rise to a load W in the vertical plane of the bar and a twisting moment equal to (W ×

(*a*). The downward movement of *D* will be equal to the downward deflection of the beam at *B* due to load *W* at *B* plus the movement of *D* due to rotation of the section *B*. ($\theta_{AB} = \theta_{BC}$)

Now $W = 1$ kN,

the deflection at *B* from Eq, 6.136 is

$$y_B = \frac{1000 \times (800)^3 \times (400)^3}{3 \times (200 \times 10^3)\left[\frac{\pi}{64} \times 60^4\right] \times (1200)^3} \text{ mm}$$

$$= 0.0496 \text{ mm}$$

Torques T_{AB} and T_{BC} sustained by the portions *AB* and *BC*

$$T_{AB} + T_{BC} = W \times a = 1 \times a \quad \text{...(1)}$$

$$\theta_{AB} = \theta_{BC}$$

$$\frac{T_{AB}\, l_{AB}}{JG} = \frac{T_{BC}\, l_{BC}}{JG}$$

or $$T_{AB} \times 800 = T_{BC} \times 400$$

i.e. $$T_{BC} = 2\, T_{AB} \quad \text{...(2)}$$

From equations (1) and (2),

$$T_{AB} = \frac{a}{3}$$

and $$T_{BC} = \frac{2a}{3} \quad \text{...(3)}$$

Therefore, $$\theta_{AB} = \frac{T_{AB}\, l_{AB}}{JG} = \frac{a}{3} \times 800 \times \frac{32}{\pi \times (60)^4} \times \frac{1}{(80 \times 10^3)}$$

$$= 2.62 \times 10^{-6} \text{ a radians}$$

Thus the downward movement of end *D*, due to twisting

$$= \theta_{AB} \times a$$

$$= 2.62 \times 10^{-6}\, a^2 \text{ mm}$$

Therefore,

$$2.62 \times 10^{-6}\, a^2 + .0496 = 0.1$$

or $$a^2 = \frac{0.0504}{2.62 \times 10^{-6}}$$

or $$a = 138.5 \text{ mm}$$

11.2 : *A hollow shaft having 50 mm outside diameter and 30 mm internal diameter. An applied torque of 1.6 kN-m is found to produce an angular twist of 0.4°, measured on a length of 0.2 m of the shaft. Calculate the value of the modulus of rigidity.*

E X A M P L E

EXAMPLE

Calculate Also, the maximum power which could be transmitted by the shaft at 2000 r.p.m. if the maximum allowable shearing stress is 65 MN/m².

Solution :

(*a*) $$\frac{T}{J} = \frac{G\theta}{l}$$

or $$G = \frac{Tl}{J\theta} = 1.6 \times .2 \times \frac{180}{0.4 \times \pi} \times \frac{32}{\pi\,(.05^4 - .03^4)}$$

$$= 86.4 \times 10^6 \text{ kN/m}^2$$

(*b*) $$\frac{T}{J} = \frac{\tau}{R}$$

or $$T = \frac{J\tau}{R}$$

$$= \frac{\pi}{32}(.05^4 - .03^4) \times \frac{65 \times 10^6 \times 2}{.05}$$

$$= 1.39 \times 10^3 \text{ Nm}$$

Therefore, $$\text{Power} = 1.39 \times 10^3 \times \frac{2\pi \times 2000}{60}\, W$$

$$= 292 \text{ kW}$$

11.3 : *The diameter ratio of a hollow shaft is 3 : 5. It is required to transmit 600 kW at 100 r.p.m., the maximum torque being 12% greater than the mean. If the twist in a length of 3 m is not to exceed 1° and the shearing stress is limited to 60 MN/m², calculate the minimum external diameter of the shaft satisfying these conditions. Take G = 80 × 10³ MN/m².*

Solution : $$\frac{D}{d} = \frac{5}{3}$$

or $$d = 0.6\,D$$

Now $$J = \frac{\pi}{32}(D^4 - d^4)$$

$$= \frac{\pi}{32} D^4\,(1^4 - 0.6^4)$$

$$= 0.0855\,D^4$$

Mean torque $$= \frac{\text{Power} \times 60}{2\pi N} = \frac{60 \times 1000 \times 60}{2\pi \times 100}$$

$$= 5.72 \times 10^4 \text{ Nm}$$

Therefore, $T = 1.12 \times 5.72 \times 10^4$

$= 6.41 \times 10^4$ Nm

Case 1. If the torsional rigidity were the limiting case, then

$$T = JG\,\theta/l$$

or $$6.41 \times 10^4 = 0.855\,D^4 \times (80 \times 10^9) \times \frac{\pi}{180} \times \frac{1}{3}$$

or $$D^4 = 16.11 \times 10^{-4}$$

i.e. $$D = 0.2001 \text{ m}$$

For this value of D, the maximum shear stress induced will be,

$$\tau_{max} = \frac{D}{2} \cdot \frac{T}{J} = \left(\frac{0.2001}{2}\right) \cdot \frac{0.41 \times 10^4}{0.0855 \times (0.2001^4)}$$

$$= 46.7 \times 10^6 \text{ N/m}^2$$

which is less than the allowable stress of 60 MN/m².

Case 2. If the shear stress is the limiting criterion, then

$$\frac{T}{J} = \frac{\tau}{R}$$

or $$\frac{6.41 \times 10^4}{(0.0855)\,D^4} = \frac{60 \times 10^6}{D} \times 2$$

or $$D^3 = \frac{6.41 \times 10^4}{0.0855 \times 120 \times 10^6} = 6.25 \times 10^{-3}$$

or $$D = 0.184 \text{ m}$$

For this value of D the twist, over 3 m length is

$$= \frac{6.41 \times 10^4 \times 3}{0.0855 \times (.184)^4 \times 80 \times 10^9} \text{ radians}$$

$$= 0.467 \text{ degrees.}$$

This also is less than the allowable limit of 1°. Thus, both the criterion are safe. Hence, the least external diameter is 0.2001 m which will satisfy both the restrictions.

11.4 : *A hollow shaft of diameter ratio 0.6 is required to transmit 1500 kW at a speed of 2500 rpm. It is to be fitted with a flanged coupling having 8 bolts on a circle of diameter twice that of the shaft. If the allowable shear stresses in shaft and bolts are 70 MN/m² and 55 MN/m² respectively, find the shaft and bolt diameters.*

Solution :

For shaft $$d = 0.6\,D$$

$$J = \frac{\pi}{32}(D^4 - d^4) = 0.0855\,D^4$$

EXAMPLE

$$\text{Torque} = T = \frac{\text{Power} \times 60}{2\pi N} = \frac{(1500 \times 1000) \times 60}{2\pi \times 2500}$$

$$= 5740 \text{ Nm}$$

$$\frac{T}{J} = \frac{2\tau}{D}$$

or $$\frac{5740}{0.0855\, D^4} = 2 \times \frac{2 \times 70 \times 10^6}{D}$$

or $$D = 0.0782 \text{ m } \textit{i.e. } 78.2 \text{ mm}$$

$$d = 0.6 \times 78.2 = 46.9 \text{ mm.}$$

Bolts

Shearing area of each bolt

$$= \frac{\pi}{4} D_b^2$$

Shearing force of each bolt

$$= \frac{\pi}{4} D_b^2 \times 55 \times 10^6 \text{ N}$$

Distance of bolt axis from shaft axis $= \frac{1}{2}(2D) = D$

Torque due to eight bolts

$$= \left(8 \times \frac{\pi}{4} D_b^2 \times 55 \times 10^6\right) D$$

$$= 110 \times 10^6\, \pi\, D_b^2 = 5740$$

Thus, $110 \times 10^6 \times \pi \times .0782 \times D_b^2 = 5740$

$$\therefore \quad D_b = 0.01454 \text{ m} = 14.54 \text{ mm.}$$

11.5 : *A solid alloy shaft of 50 mm diameter is to be coupled in series with a hollow steel shaft of the same external diameter. Find the internal diameter of the steel shaft, if the angle of twist per unit length is to be 75 per cent of that of the alloy shaft.*

Determine the speed at which the shafts are to be driven to transmit 200 kW, if the limits of shearing stresses are to be 55 MN/m² and 80 MN/m² in the alloy and steel respectively. G_{steel} *= 2.2 time* G_{alloy}.

Solution. As the shafts are in series, the torque transmitted by each will be same,

$$T_a = T_s \quad \text{...(1)}$$

$$\frac{J_a G_a \theta_a}{l_a} = \frac{J_s G_s \theta_s}{l_s}$$

Therefore, $$J_s = J_a\left(\frac{G_a}{G_s}\right)\left(\frac{\theta_a}{\theta_s}\right) \qquad (l_a = l_s = 1)$$

or $$\frac{\pi}{32}(50^4 - d^4) = \frac{\pi}{32}(50)^4\left(\frac{1}{2.2}\right)\left(\frac{1}{0.75}\right)$$

or $$d^4 = (50)^4\left[1 - \frac{4}{6.6}\right]$$

i.e. $$d = 50\left(\frac{2.6}{6.6}\right)^{1/4} = 39.6 \text{ mm}$$

(*b*) **Case 1**. If the stress in alloy is the limiting case, then,

$$\frac{T_a}{J_a} = \frac{2\tau_a}{D_a}$$

or $$T_a = \frac{2 \times 60 \times 10^6}{.050} = \frac{\pi}{32} \times (.050)^4$$

$$= \frac{200 \times 1000 \times 60}{2\pi N}$$

i.e. $$N = \frac{200 \times 1000 \times 60 \times 60}{2\pi^2 \times 160 \times 10^6 \times (.05)^3} = 976 \text{ rpm}.$$

At this speed stress in steel is found as follows :

$$T_s = \frac{200 \times 1000 \times 60}{2\pi \times 976}$$

$$= \frac{\pi}{3.2}(0.05^4 - 0.0396^4) \times 2 \times \frac{\tau_s}{0.05}$$

or $$\tau_s = \frac{200 \times 1000 \times 60 \times .05 \times 32}{2\pi^2 \times 9.76 \times 2(.05^4 \times .0396^4)}$$

$$= 132 \text{ MN/m}^2$$

which is more than allowable.

Thus, the assumption of case 1 is not valid.

Case 2. If the stress in steel is the limiting case, then,

$$\frac{T_s}{J_s} = \frac{2\tau_s}{D_s}$$

$$\tau_s = \frac{T_s D_s}{2I_s}$$

EXAMPLE

EXAMPLE

or $$80 \times 10^6 = \frac{200 \times 1000 \times 50}{2\pi N} \times \frac{.050 \times 32}{2 \times \pi\,(.05^4 \times .0396^4)}$$

or $$N = \frac{200 \times 1000 \times 60 \times .05 \times 32}{4\pi^2\,(.05^4 - .396^4) \times 80 \times 10^6}$$

$$= 1620\ rpm.$$

At this speed, the stress in alloy is given by,

$$\frac{2\tau_a}{D_a} J_a = \frac{200 \times 1000 \times 60}{4\pi\ N}$$

or $$\tau_a = \frac{200 \times 1000 \times 60}{4\pi\ NJ_a} \times D_a$$

Since τ_a is inversely proportional to N, the stress at 1620 r.p.m. will be smaller than that at 976 r.p.m. Thus, the assumption of case 2 holds good and the required working speed will be 1620 r.p.m.

11.3 COMBINED TORSION AND BENDING

In all, the practical cases of shafts transmitting power, they will be subjected to bending moments due to gravity or inertia loads. Stress will be set up in the shaft due to bending moment, torque and shearing forces. But the effect of the stresses due to shearing forces is negligible, particularly as its maximum value occurs at the neutral axis (Chapter 4) where there is no bending stress. Evidently both direct and shear stresses will be set up at any point of the cross-section of the shaft. For design purposes, it will be essential to find the principal stresses along with the maximum shear stress at any particular point in the cross-section.

Thus, if σ_b is the greatest bending stress and τ the greatest shear stress on the cross-sectional plane, then,

$$\sigma_b = \frac{32M}{\pi D^3} \qquad \text{...(11.10)}$$

and $$\tau = \frac{16T}{\pi D^3} \qquad \text{...(11.11)}$$

where, M = Bending moment at the section.

T = Torque being transmitted by the same section.

D = Diameter of the solid shaft.

Both σ_b and τ occur together at the ends of the vertical diameter for the bending moment to be in the vertical plane. Therefore, referring to the point diagram (Fig. 11.5 which is the top view) for the point near the end of the vertical diameter (line *BC* and *AD* are parallel to cross-sectional plane), we have the maximum principal stress as,

$$\sigma_{max} = \frac{1}{2}\left[\sigma_b + \sqrt{\sigma_b^2 + 4\tau^2}\right]$$

$$= \frac{1}{2}\left[\frac{32M}{\pi D^3} + \sqrt{\left(\frac{32M}{\pi D^3}\right)^2 + 4\left(\frac{16T}{\pi D^3}\right)^2}\right]$$

Fig. 11.5

or
$$\sigma_{max} = \frac{16}{\pi D^3}\left[M + \sqrt{M^2 + T^2}\right] \quad ...(11.12)$$

Suppose we had the pure bending moment M_e acting on the shaft such that the magnitude of the bending stress induced by it would have been σ_{max}, then,

$$\sigma_{max} = \frac{32M_e}{\pi D^3} \quad ...(11.13)$$

Equating (11.12) and (11.13), we get,

$$M_e = \frac{1}{2}\left[M + \sqrt{M^2 + T^2}\right] \quad ...(11.14)$$

This M_e is called the equivalent bending moment. It may be noted that the purpose of M_e is just to calculate the magnitude of the maximum principal stress. It does not give the location of the plane on which the maximum principal stress acts, which actually is inclined at $\left[\frac{1}{2}\tan^{-1}(-2\tau/\sigma_b)\right]$ with direction of σ_b.

Again the magnitude of maximum shear stress is given by,

$$\tau_{max} = \frac{1}{2}\sqrt{\sigma_b^2 + 4\tau^2} = \frac{1}{2}\sqrt{\left(\frac{32M}{\pi D^3}\right)^2 + 4\left(\frac{16T}{\pi D^3}\right)^2}$$

$$= \frac{16}{\pi D^3}\sqrt{M^2 + T^2} = \frac{16}{\pi D^3}T_e \text{ (from 11.10 \& 11.11)} \quad ...(11.15)$$

Thus, if τ_e, called the equivalent torque, acting alone on the shaft, induces the maximum shear stress of magnitude τ_{max}, then,

$$\tau_{max} = \frac{16}{\pi D^3}T_e \quad ...(11.16)$$

Therefore, from equations (11.15) and (11.16),

$$\tau_e = \sqrt{M^2 + T^2} \quad ...(11.17)$$

τ_e is also used to evaluate the magnitude of the maximum shear stress. It does not give the location of the planes of maximum shear stress which are at angles of 45° to the principal planes.

11.6 : *A solid steel shaft transmits 1000 kW at 100 r.p.m., the maximum torque being 50 percent greater than the mean. The shaft supported in ball bearings 1.5 m apart with a 7500 kg fly-wheel midway between bearings. Find the shaft diameter, if the maximum tensile stress is 60 MN/m².*

Solution.

$$\text{Mean torque} = \frac{(1000 \times 1000) \times 60}{2\pi \times 100} = 95600 \text{ Nm}$$

$$\therefore \quad T = 1.5 \times 9.56 \times 10^4 = 14.3 \times 10^4 \text{ Nm}$$

$$M = (7500 \times 9.81)\frac{1.5}{4} = 2.76 \times 10^6 \text{ Nm}$$

$$M_e = \frac{1}{2}[2.76 \times 10^4 + \{(14.3 \times 10^4)^2 + (2.76 \times 10^4)^2\}^{1/2}]$$

$$= 8.67 \times 10^4 \text{ Nm}$$

Now
$$\sigma_{max} = \frac{32\, M_e}{\pi D^3} \Rightarrow D^3 = \frac{32 M_e}{\pi \sigma_{max}}$$

or
$$D^3 = \frac{32 \times 8.67 \times 10^4}{\pi \times 60 \times 10^6} = .01475$$

i.e.
$$D = .246 \text{ m or } 246 \text{ mm.}$$

11.7 : *A solid phosphor-bronze shaft Of 50 mm diameter is rotating at 740 r.p.m. It is subjected to torsion only. An electrical resistance strain gauge mounted on the surface of the shaft with its axis at 45° to the shaft axis, gives the strain reading as 4.17 × 10⁴ . Find the power being transmitted, if the modulus of elasticity for bronze is 105000 MN/m². Take ν = 0.3.*

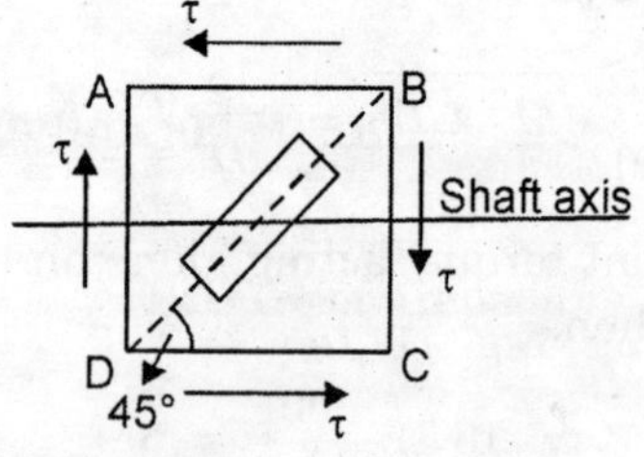

Fig. 11.6

Solution. If the maximum shear on the cross-sectional plane *BC* is τ then referring to Chapter II.

(*i*) Principal stress along *DB*

$$= \sigma_1 = -\frac{1}{2}\sqrt{4\tau^2}$$

$$= -\tau \text{ (i.e. comp.).}$$

(*ii*) Principal stress along *AC*

$$= \sigma_2 = \frac{1}{2}\sqrt{4\tau^2} = \tau \ (i.e.\text{Tensile})$$

Thus, magnitude of the compressive strain along *DB* is

$$\varepsilon_{DB} = \frac{\tau}{E}(1+\nu) + 4.17 \times 10^{-4}$$

i.e.
$$\tau = \frac{4.17 \times 10^{-4} \times 105000}{(1+0.3)} = 33.7 \text{ MN/m}^2$$

Torque being transmitted, then, is

$$T = \frac{\pi}{16}\tau D^3 = \frac{\pi}{16} \times 33.7 \times (50)^3 \text{ Nmm} = 826 \text{ Nm}$$

Therefore,
$$\text{power} = \frac{826 \times (2\pi)(740)}{60} W$$

$$= 64000 \text{ W} = 64 \text{ kW}$$

11.8 : *A propeller shaft of a ship having 0.45 m diameter and it supports a propeller of mass 15 tonnes. The propeller can be considered as a load concentrated at the end of a cantilever of length 2 m. When the speed of the ship is 32 km/hr, the propeller is revolving at 100 r.p.m. If the engine develops 15 MW, calculate the principal stress and maximum shear stress in the shaft. It may be assumed that the propulsive efficiency of the propeller is 85%.*

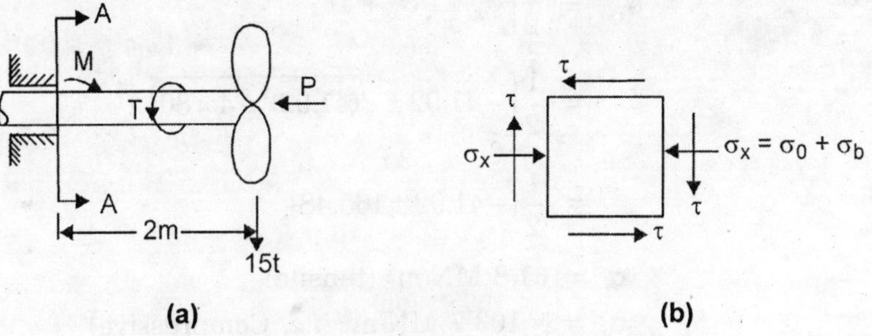

Fig. 11.7

Solution. At the bearing, Fig. 11.7(*a*),

Bending moment

$$M = (15000 \times 9.81) \times 2 \text{ Nm} = 0.2943 \text{ MN-m}$$

Torque
$$T = \frac{\text{Power} \times 60}{2\pi N} = \frac{15 \times 10^6 \times 60}{2\pi \times 100} \text{ Nm}$$

$$= 1.433 \text{ MN-m}$$

If *P* is the propulsive force and *v* is the linear speed of the ship, then,

$$\text{Engine Power} = \frac{Pv}{\text{efficiency}}$$

E X A M P L E

E X A M P L E

i.e. $$P = \frac{15 \times 10^6 \times 0.85 \times 3600}{32 \times 10^3} N = 1.435 \text{ MN}$$

Therefore, direct stress on the cross-section *AA*,

$$\sigma_0 = \frac{P}{\text{Area}} = \frac{1.435 \times 4}{\pi \times (.45)^2} = 9.02 \text{ MN/m}^2$$

(Compressive)

Max. Bending stress due to M, on the cross–section *AA*

$$\sigma_b = \frac{32M}{\pi D^3} = \frac{32 \times 0.2943}{\pi \times (.45)^3} = 32.9 \text{ MN/m}^2$$

(comp. at the lower most point)

Therefore, $\sigma_x = 32.9 + 9.02 = 41.92 \text{ MN/m}^2$ (Compressive)

Shear stress due to torque *T*,

$$\tau = \frac{16T}{\pi D^2}$$

$$= \frac{16 \times 1.433}{\pi \times (.45)^3} = 80 \text{ MN/m}^2$$

The point diagram for the bottom most point of the section *AA* is given in Figure 11.7(*b*). Thus, the principal stresses are given by (Chapter 2)

$$\sigma_{1,2} = \frac{1}{2}\left[\sigma_x \pm \sqrt{\sigma_x^2 + 4\tau^2}\right]$$

$$= \frac{1}{2}\left[-41.92 \pm \sqrt{(41.92)^2 + 4 \times 80^2}\right]$$

$$= \frac{1}{2}[-41.92 \pm 165.48]$$

i.e. $\sigma_1 = 61.8 \text{ MN/m}^2$ (tensile)

and $\sigma_2 = -103.7 \text{ MN/m}^2$ (*i.e.* Compressive)

$$\tau_{max} = \frac{\sigma_1 - \sigma_2}{2} = \sqrt{\sigma_x^2 + 4\tau^2}$$

$$= \frac{61.8 + 103.7}{2} = 82.75 \text{ MN/m}^2$$

11.9 : *A horizontal shaft, securely fixed at each end, has a free length of 10 m. Viewed from one end of the shaft, axial couples of 30 kN-m clockwise and 40 kN-m counter clockwise act on the shaft at distance of 4m and 7 m respectively from the viewed end. Determine the end fixing couples in magnitude and direction and find the diameter of the solid shaft for a maximum shearing stress of 60 MN/m². Draw the torque distribution diagram for the shaft.*

Also, draw a diagram to show how a line, originally parallel to the axis and on the outer surface of the shaft will appear after the application of the couples and find the position along the shaft where the shaft suffers no angular twist.

Solution. Let the reaction torque at the end P be T (clockwise) as seen from P towards S (Fig. 11.8).

Torque on $PQ = T$ kNm

Torque on $QR = T + 30$ kNm

Torque on $RS = T + 30 - 40 = T - 10$ kNm

For the above torques,

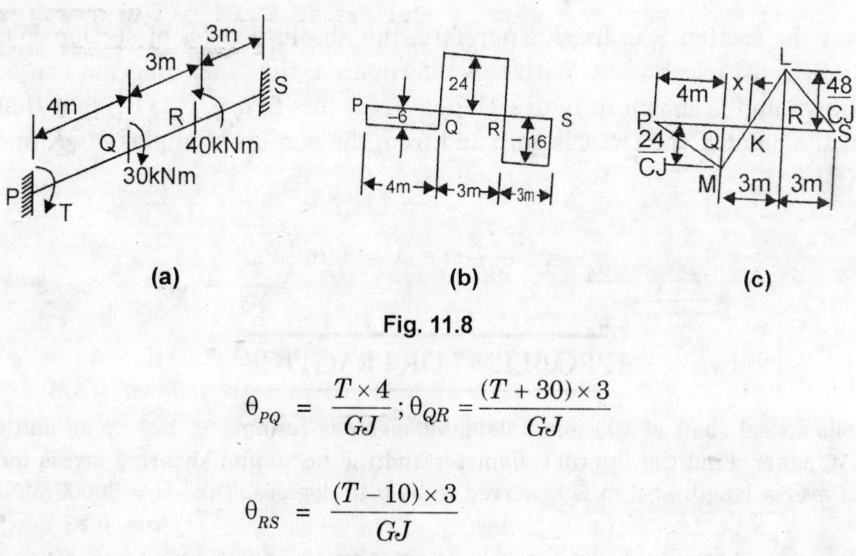

Fig. 11.8

$$\theta_{PQ} = \frac{T \times 4}{GJ}; \theta_{QR} = \frac{(T+30) \times 3}{GJ}$$

$$\theta_{RS} = \frac{(T-10) \times 3}{GJ}$$

But the ends being securely held, there is no relative twist between P and S. Therefore,

$$\theta_{PS} = \theta_{PQ} + \theta_{QR} + \theta_{RS} = 0$$

or $\quad 4T + 3\,(\mathbf{T} + 30) + 3\,(T - 10) = 0$

or $\quad T = -6$ kNm (*i.e.* anticlockwise as seen from P to S)

Substituting the value of T, we have,

$$T_{PQ} = -6, T_{QR} = 24 \text{ and } T_{RS} = -16 \text{ kNm}$$

These values enable us to draw the torque diagram of the shaft, shown in Fig. 11.8 *b*.

(*b*) From the torque diagram it is evident that the maximum torque occurs in the portion QR and is equal to 24 kNm. Corresponding value of the maximum shear stress tau> can then be found.

$$T = \frac{\pi}{16} \tau D^3$$

E X A M P L E

or $$D^3 = \frac{16 \times (24 \times 10^3)}{\pi \times (60 \times 10^6)} \quad i.e.\ D = 0.127 \text{ m}$$

(*c*) With respect to *P*, the twist at *Q* is,

$$\theta_Q = \frac{-6 \times 4}{GJ} = \frac{-24}{GJ} \quad \left(i.e. \frac{24}{GJ} \text{ anticlockwise} \right)$$

Similarly with respect to *R*, the twist of *S*,

$$\theta_S = \frac{-16 \times 3}{GJ} = \frac{-48}{GJ}$$

But the section S is fixed. Therefore, the absolute twist of section *R* is 48 over *GJ* (clockwise). With this information, the twist diagram can be drawn and is shown in figure 11.7(*c*). From this figure, it is evident that at the point *K* the twist is zero and from the similar triangles *MQK* and *KRL*.

$$\frac{x}{24} = \frac{3 - x}{48} \quad i.e.\ x = 1 \text{ m}$$

PROBLEM FOR PRACTICE

1. A hollow steel shaft of 400 mm outside diameter is running at 120 r.p.m. and transmitting 9 MW power. Find the internal diameter and the maximum shearing stress induced, if the twist over a length of 2 m is observed to be 0.45 degrees. Take $G = 80000$ MN/m^2.
   (**Ans.** 0.22 m ; 62.9 MN/m^2)
2. A torque of 1.6 kNm acting on a hollow shaft of 50 mm outside diameter and 30 mm internal diameter, is found to produce an angular twist of 0.4 degrees over a length of 0.2 m of the shaft. Find the value of the modulus of rigidity. Also, determine the maximum power which could be transmitted by this shaft at 2000 r.p.m., if the maximum allowable shearing stress is 65000 MN/m^2. (**Ans.** 86000 MN/m^2 ; 292 kW)
3. A hollow steel shaft transmits 1600 kW at 180 r.p.m. If the internal and external diameters of the shaft are 125 mm and 200 mm respectively, find the shear stress at the outer and inner radii. Take $G = 84000$ MN/m^2. (**Ans.** (64.5 MN/m^2 ; 40.5 MN/m^2)
4. A flange coupling is required for a shaft transmitted 220 kW at 240 rev./min. The bolts are 16 mm diameter and are to be arranged on a circle of 250 mm diameter, the working shear stress in the both must not exceed 60 MN/m^2. Determine the number of bolts required and the actual stress realised.

   Briefly describe and explain how the power transmitted by the shaft may be measured.
   (**Ans.** 6 ; 59.4 MN/m^2)
5. A shaft runs at 300 r.p.m. and transmits power from a pulley A at one end to two pulleys, *B* and *C* ; each of which drives a machine in a workshop. The distance between pulleys *A* and *B* is 3 m and that between *B* and *C* is 2.4 m (*i.e.* the total length of the shaft is 5.4 m). The shaft has a diameter of 50 mm between *A* and *B* and a diameter of 40 mm between *B* and *C*. If the maximum permissible shear stress in the shaft is 80 MN/m^2, calculate the maximum

power which may be supplied from each of the pulleys B and C, assuming that both machines would be in operation at the same time. Also, calculate the total angle of twist of one end of the shaft relative to the other when running on full load. G = 8000 MN/m².

(**Ans.** 30.1 kW ; 31.6 kW ; 13.75°)

6. A steel shaft *ABCD* having a total length of 1.5 m is made of three lengths *AB, BC* and *CD*, each of 0.5 m long. *AB* and *BC* are solid, having diameters of 60 mm and 70 mm respectively, and *CD* is hollow having outside and inside diameters of 70 mm and 45 mm respectively. When an axial torque of 2 kN-m is transmitted from one end of the shaft to the other, the total angle of twist from A to D is 1.2°. Determine : (*a*) the maximum shearing stress in the shaft and state where it occurs, (*b*) the angle of twist for each of the three lengths *AB, BC* and *CD*, (*c*) the modulus of rigidity of the material.

(**Ans.** 49.4 MN/m² in *AB* ; 0.548° ; 0.296° ; 0.356° ; 82.2 *GN*/m²)

7. A solid cylindrical shaft 50 mm diameter and 1.5 m long, is passed through the centre of a hollow cylindrical shaft of the same material 1.5 m long 55 mm and 75 inner and outer diameters. The ends of the two shafts are rigidly joined, with the shafts concentric. The composite shaft so formed is used to transmit 375 kW at a speed of 600 rev./min. Find the maximum and minimum stress in the two shafts. (**Ans.** 79.3 MN/m² ; 58.1 MN/m² ; 52.9 MN/m²)

8. A steel bar 19 mm diameter is encased in a closely fitting brass tube of 32 mm external diameter, securely fixed together at the ends. The compound bar is subjected to a torque of 520 Nm and the angle of twist measured on a gauge length of 250 mm is found to be 1.8°. If G for steel is assumed to 80 GN/m² calculate G for the brass.

Find also, the maximum shearing stresses in the two materials.

(**Ans.** 34.6 GN/m² ; 95.5 MN/m² ; 69.5 MN/m²)

9. A shaft is required for an engine which indicates 746 kW power at 80 r.p.m. The maximum twisting moment on the shaft is 1.8 × the mean twisting moment. The main bearings are 4500 mm apart, and the shaft carries a flywheel mid-way between bearings. The mass of the flywheel is 90000 kg. The bending moment due to the flywheel is additional to that due to steam pressure, the latter of which is 0.8 × the mean twisting moment. Find the diameter of the shaft to satisfy the condition that the maximum tensile stress in the material is not to exceed 62.78 MN/m². (**Ans.** 558 mm)

INDEX

IMPORTANT MATHEMATICAL FORMULAE WHICH OFTEN FIND THEIR APPLICATION

(1) TRIGNOMETRIC FORMULAE

$$\sin(A+B) = \sin A \cos B + \cos A \sin B$$

$$\sin(A-B) = \sin A \cos B - \cos A \sin B$$

$$\cos(A+B) = \cos A \cos B - \sin A \sin B$$

$$\cos(A-B) = \cos A \cos B + \sin A \sin B$$

$$\tan(A+B) = \frac{\tan A + \tan B}{1 - \tan A \tan B}$$

$$\tan(A-B) = \frac{\tan A - \tan B}{1 + \tan A \tan B}$$

$$\tan 2A = \frac{2\tan A}{1 - \tan^2 A}$$

$$\sin 2A = 2 \sin A \cos A = \frac{2\tan A}{1 + \tan^2 A}$$

$$\cos 2A = \cos^2 A - \sin^2 A = 1 - 2\sin^2 A$$

$$= 2\cos^2 A - 1 = \frac{1 - \tan^2 A}{1 + \tan^2 A}$$

$$2 \sin A \cos B = \sin(A+B) + \sin(A-B)$$

$$2 \cos A \sin B = \sin(A+B) - \sin(A-B)$$

$$2 \cos A \cos B = \cos(A+B) + \cos(A-B)$$

$$2 \sin A \sin B = \cos(A-B) - \cos(A+B)$$

$$\sin C + \sin D = 2\sin\frac{C+D}{2}\cos\frac{C-D}{2}$$

$$\sin C - \sin D = 2\cos\frac{C+D}{2}\sin\frac{C-D}{2}$$

$$\cos C + \cos D = 2\cos\frac{C+D}{2}\cos\frac{C-D}{2}$$

$$\cos C - \cos D = -2\sin\frac{C+D}{2}\sin\frac{C-D}{2}$$

(2) DIFFERENTIATION

$$\frac{du}{dx} = \frac{du}{dy} \cdot \frac{dy}{dx}$$

$$\frac{d}{dx}(uv) = u\frac{dv}{dx} + v\frac{du}{dx}$$

$$\frac{d}{dx}e^x = e^x$$

$$\frac{d}{dx}(\ln x) = \frac{1}{x}$$

$$\frac{d}{dx}\sin x = \cos\ x$$

$$\frac{d}{dx}\cos x = -\sin\ x$$

$$\frac{d}{dx}\tan x = \sec^2 x$$

$$\frac{d}{dx}(\sec x) = \sec x \tan x$$

$$\frac{d}{dx}(\sin^{-1} x) = \frac{1}{\sqrt{1-x^2}}$$

$$\frac{d}{dx}(\tan^{-1} x) = \frac{1}{1+x^2}$$

$$\frac{d}{dx}(\sec^{-1} x) = \frac{1}{x\sqrt{x^2-1}}$$

$$\frac{d}{dx}(\sinh x) = \cosh\ x$$

$$\frac{d}{dx}(\tanh x) = \text{sech}^2 x$$

(3) INTEGRATION

$$\int x^n dx = \frac{x^{n+1}}{n+1} \text{ (when } n \neq -1)$$

$$\int \frac{1}{x} dx = \ln\ (x)$$

$$\int e^x dx = e^x$$

$$\int \sin x dx = -\cos\ x$$

$$\int \cos x dx = \sin\ x$$

$$\int \tan\ x\ dx = -\ln\ (\cos\ x)$$

$$\int \sec\ x\ dx = \ln\ (\sec x + \tan x)$$

$$\int \sec^2 x dx = \tan\ x$$

$$\int \frac{dx}{a^2+x^2} = \frac{1}{a}\tan^{-1}\left(\frac{x}{a}\right)$$

$$\int \frac{dx}{a^2-x^2} = \frac{1}{2a}\ln\left(\frac{a+x}{a-x}\right)$$

$$\int \frac{dx}{x^2=a^2} = \frac{1}{2a}\ln\left(\frac{x-a}{x+a}\right)$$

$$\int \sqrt{a^2 - x^2}\,dx = \frac{1}{2}x\sqrt{a^2 - x^2} + \frac{a^2}{2}\sin^{-1}\left(\frac{x}{a}\right)$$

$$\int \sqrt{a^2 + x^2}\,dx = \frac{x}{2}\sqrt{a^2 + x^2} + \frac{a^2}{2}\sinh^{-1}\left(\frac{x}{a}\right)$$

$$\int \sqrt{x^2 - a^2}\,dx = \frac{x}{2}\sqrt{x^2 - a^2} - \frac{a^2}{2}\cosh^{-1}\left(\frac{x}{a}\right)$$

$$\int e^{ax}\sin bx\,dx = \frac{e^{ax}}{a^2 + b^2}\,(\text{a} \sin bx - b \cos bx)$$

$$\int e^{ax}\cos bx\,dx = \frac{e^{ax}}{a^2 + b^2}\,(\text{a} \cos bx + b \sin bx)$$

$$\int \sinh x\,dx = \cosh x$$

$$\int \cosh x\,dx = \sinh x$$

$$\int \tanh x\,dx = \ln(\cosh x)$$

$$\int \coth x\,dx = \ln(\sinh x)$$

$$\int \text{sech}^2 x\,dx = \tanh x$$

$$\int F\left(\frac{dF}{ax}\right)dx = \frac{F^{n+1}}{n+1} \quad (\text{when } n \neq -1)$$

and F is a function of x

$$\int_{-\pi/2}^{\pi/2} \frac{d\theta}{a + b\sin\theta} = \left[\frac{2}{\sqrt{a^2 - b^2}}\tan^{-1}\left\{\frac{a\tan\frac{\theta}{2} + be}{\sqrt{a^2 - b^2}}\right\}\right]_{-\pi/2}^{\pi/2} \quad (\text{where } a > b)$$

Definite integral in turns of Gamma function

where,

$$\sqrt{n+1} = n\sqrt{n}$$

$$\sqrt{1} = 1$$

$$\sqrt{\frac{1}{2}} = \sqrt{\pi}$$

Integration By Parts

$$\int F_1 F_2\,dx = F_1\int F_2\,dx - \int\left[\frac{dF_1}{dx}\left\{\int F_2\,dx\right\}\right]dx$$

$$\int_0^{\pi/2} \sin x^n \cos x^m\,dx = \frac{\overline{\left|\frac{n+1}{2}\right.}\;\overline{\left|\frac{m+1}{2}\right.}}{2\,\overline{\left|\frac{m+n+2}{2}\right.}} \quad (\text{when } n \text{ and } m \text{ are +ve integers})$$

where,

$$\overline{|n+1} = n\overline{|n}$$

$$\overline{|1} = 1$$

$$\overline{\left|\frac{1}{2}\right.} = \sqrt{\pi}$$